WASTE TREATMENT AND UTILIZATION

Theory and Practice of Waste Management

2

Also in This Series

M MOO-YOUNG & G J FARQUHAR: Waste Treatment and Utilization (Theory and Practice of Waste Management) 1

Other Related Pergamon Titles of Interest

Books

CURI Treatment and Disposal of Liquid and Solid Industrial Wastes

HENSTOCK Recycling and Disposal of Solid Waste

HENSTOCK & BIDDULPH Solid Waste as a Resource

JENKINS Treatment of Domestic and Industrial Wastewaters in Large Plants

JENKINS Water Pollution Research and Development (4 vol. set)

MOO-YOUNG *et al.* Advances in Biotechnology (4 vol. set)

MOO-YOUNG Comprehensive Biotechnology (Principles, Methods and Applications) (3 vol. set)

PAWLOWSKI Physicochemical Methods for Water and Wastewater Treatment

SUESS Examination of Water for Pollution Control (3 vol. set)

WHO Waste Discharge into the Marine Environment (Principles and Guidelines)

Journals

Chemical Engineering Science

Conservation and Recycling

Materials and Society

Water Research

Water Supply and Technology

WASTE TREATMENT AND UTILIZATION

Theory and Practice of Waste Management

2

Proceedings of the Second International Symposium held at the University of Waterloo, Waterloo, Ontario, Canada, June 18-20, 1980

Edited by

MURRAY MOO-YOUNG

Department of Chemical Engineering

CAMPBELL W. ROBINSON

Department of Chemical Engineering

and

GRAHAME J. FARQUHAR

Department of Civil Engineering,
University of Waterloo, Ontario, Canada

PERGAMON PRESS

OXFORD · NEW YORK · TORONTO · SYDNEY · PARIS · FRANKFURT

U.K.	Pergamon Press Ltd., Headington Hill Hall, Oxford OX3 0BW, England
U.S.A.	Pergamon Press Inc., Maxwell House, Fairview Park, Elmsford, New York 10523, U.S.A.
CANADA	Pergamon Press Canada Ltd., Suite 104, 150 Consumers Rd., Willowdale, Ontario M2J 1P9, Canada
AUSTRALIA	Pergamon Press (Aust.) Pty. Ltd., P.O. Box 544, Potts Point, N.S.W. 2011, Australia
FRANCE	Pergamon Press SARL, 24 rue des Ecoles, 75240 Paris, Cedex 05, France
FEDERAL REPUBLIC OF GERMANY	Pergamon Press GmbH, 6242 Kronberg-Taunus, Hammerweg 6, Federal Republic of Germany

First edition 1982

British Library Cataloguing in Publication Data

Waste treatment and utilization 2.
1. Refuse and refuse disposal — Congresses
2. Factory and trade waste — Congresses
3. Sewage disposal — Congresses
4. Waste products — Congresses
I. Moo-Young, Murray II. Robinson, Campbell W.
III. Farquhar, Grahame J.
628.4'4 TD785

ISBN 0 08 024012 7

In order to make this volume available as economically and as rapidly as possible the authors' typescripts have been reproduced in their original forms. This method unfortunately has its typographical limitations but it is hoped that they in no way distract the reader.

Printed in Great Britain by A. Wheaton & Co. Ltd., Exeter

CONTENTS

PREFACE

This second volume of the Waste Treatment and Utilization series is based on papers presented at the Second International Symposium on Waste Treatment and Utilization (IWTU-II) held at the University of Waterloo, Ontario, Canada, June 18 to 20, 1980. Like the first of these bi-annual Symposia, the proceedings of which were published as a monograph entitled "Waste Treatment and Utilization" by Pergamon Press in 1979, the IWTU-II technical programme and roundtable discussions were designed to provide a forum for critical review and information exchange regarding both current technology and new developments in a variety of waste treatment and utilization areas. In addition, we have hoped to facilitate inter-disciplinary contact, which sometimes is lacking at more specialized meetings, and thereby encourage fresh approaches to the solution of outstanding waste management problems common to industrial and agricultural operations throughout the world. The international and multi-disciplinary backgrounds of the contributors to this volume are indicative of the usefulness of this approach.

The papers in this volume deal with both theoretical and applied aspects of the management of gaseous, liquid and solid wastes using physicochemical or biological methods. Waste sources include the chemical, petrochemical, petroleum and mining industries, agricultural and food processing operations and municipal wastewaters. The papers were refereed by an international group of reviewers. 41 of the total of 58 papers presented at the Second Symposium are included in this volume. The remaining papers did not meet the revision deadline, were withdrawn by the authors or were deemed unsuitable for publication. The material should be of interest to students, teachers, researchers and practioners of waste management. It also is suitable as supplementary literature for post-secondary courses at the senior undergraduate or graduate levels in the relevant Science or Engineering disciplines.

We are grateful for the assistance of the many people who helped to make the Second Symposium and the publication of its proceedings a success. Special thanks are due to other members of the Organizing Committee: Edward Rhodes (Chairman), Robert Hudgins, Donald Spink and Donald Scott; to Judy Brown for her efficient organization and operation of the Symposium Secretariat; to the University of Waterloo for the provision of financial and logistical support; to the reviewers listed elsewhere in the volume; to Alex Gonzales-Valdes for his help with publication management and proof-reading; to Ursula Thoene and her assistants Penny Preis, Marianne Lamers and Janet Vaughan, the cheerful dedication of whom was of immeasurable assistance in bringing about the preparation of this monograph.

Waterloo, Canada
July, 1981

C.W. Robinson
M. Moo-Young
G.J. Farquhar

FOREWORD
THE RIGHT TO KNOW
PETROSAR'S COMMUNITY AWARENESS PROGRAM

Stanley R. Stephenson

President and Chief Executive Officer, Petrosar Limited, Sarnia, Ontario, Canada

I am pleased and honoured to be with you today ... to have been invited to participate in this symposium on industrial waste treatment and utilization. It addresses issues which are very important to Canadians -- and indeed, to all citizens of 20th century industrial societies. The University of Waterloo is to be commended for its efforts to contribute to the knowledge bank on these and other environmental concerns ... concerns that have become so critical in the increasingly uneasy balance between contemporary lifestyles and economic interests.

INTRODUCTION

With reference to this morning's session and the specific programs that are to follow, I'd like to present an overview that reflects an "up front" approach, one which anticipates the needs of a community, and responds. It accepts as a given, the right to know. In fact, if Canadian industry doesn't implement anticipatory measures, this approach will be forced upon us -- and at increased cost and decreased effectiveness.

There is an increasing need for government/business/community interface when it comes to the environment, especially in the achievement of balanced social/industrial interests. Perhaps by outlining Petrosar's Community Awareness Program, I can illustrate a measure of success that can be achieved, by giving consideration to the human side of the equation which is full of unknowns.

Petrosar, as many of you know, is a unique industrial complex in Canada and, as such, has some unique concerns relative to our community. I'll provide a little background, by way of preface to my remarks. Petrosar is a petrochemical refinery. This two-fold, integrated operation is Canada's first world-scale petrochemical plant designed at the outset to combine petrochemicals with refinery products such as gasoline, as well as domestic and industrial fuels. It produces some 2.7 billion pounds of primary petrochemicals annually, along with fuel co-products. Our operating shareholders are our principal customers. We also have additional outside markets, in Canada, and to a limited extent in the U.S. We're located in Sarnia, Ontario close to the St. Clair River. We're young. The company has just (during '79) completed its first full fiscal year of operation and we've had our share of start-up headaches and "breaking-in" challenges. We have also had advice from across Canada as to how we might improve our performance.

That said, I'd like to emphasize one other feature of Petrosar which, I believe, is significant to my subject this morning. At Petrosar, we have a participative style of management, based on a team-work approach -- the organization and employment of operative or task teams at all levels of the management process. This approach emphasizes the training and development of people in order to achieve the best operating results, while contributing directly to the vocational and personal fulfillment of the individual, as a person. Attitudes in the workplace, as in modern society generally, have changed. The exercise of individuality and the opportunities to realize full potential are serious occupational goals today. And really, these aspirations are the keystones to creating a work environment where human resources complement, and are not subordinate to, technological hardware.

From our beginnings in 1974, having committed ourselves to a team management philosophy that's highly people-oriented, it was logical and consistent that we should exercise a similar approach and behaviour with our most important "publics". The community about us -- where we are and where our people live --- is such an "important public". Social concerns, especially those that care for the environment, had to be treated with commensurate sincerity by Petrosar. They became, in fact, a prerequisite of management responsibility. Discharge of this responsibilit called for a conscious, long-term orchestration of attitude and action; hence our "Community Awareness Program".

Time will permit me to only highlight its features and content. I'll define "waste" in the man on the street sense of "anything that didn't need to happen".

PROGRAM PLANNING

Planning for the Petrosar operation began under a development study organization called Sarnia Olefins and Aromatics Project (SOAP). Its origins were in the late 60's. This organization settled on a specific location for the project. That location, in itself, reflected concern for public attitudes. The site selected in the Sarnia suburban community of Corunna covers some 1100 acres, of which less than half is now being used. This acreage was examined extensively insofar as topograph and geology were concerned. The soil was inferior for agriculture, although it had been used for that purpose.

A new highway -- Highway 40A -- had already been planned by the Department of Highways for this area. It had been located away from residential areas, through the region of the proposed plant site. This offsetting of the main access artery from residential subdivisions meant that tank truck and rail traffic -- which for the Petrosar operation is largely a daytime activity anyway -- would not be a daily irritant to local residents. Site planning called for location of the physical plant well back from the perimeter roads and the property boundaries. Most importa were the designed berms inside greenbelts, along two of the "public" property sides These berms, now all sodded and attractively landscaped with trees, shrubbery and flowers, were part of the total site layout and they assist in reducing fence line noise levels.

PROGRAM IMPLEMENTATION

It should be noted here that Petrosar's "Community Awareness Program" per se began early in our construction period, when we organized teams of people to personally visit and to personally respond to queries, complaints or whatever, from the residents around the plant ... the people who were to be our long term neighbours! Our purpose was not just to discuss possible concerns about noise and other pollution, but to deal with all other aspects of public interest -- environmental, social, political, and even medical.

Our public relations people, two in number, fully supported and aided by all depart ments, organized a plan and schedule of visits to the immediate neighbours (within two miles). Arrangements were also made to contact and visit -- if it was their wish -- the neighbours in a two-three mile radius beyond our plant site. Those living in a third area, of three miles or more from the site, were sent letters explaining our poposed operation. These residents were encouraged to contact our people to express any concerns or questions. Full disclosure was the key word. These visits and contacts were carried out over a period of many months in the construction period that spanned 1976-78. Our teams reported cordial receptions in all but a very few cases. We found that telephone calls to us were limited and generally more out of curiosity than concern.

All in all, the program was judged a success. It enabled us to present very early a genuine human "face" to our neighbours and to the nearby community, using an "up front" approach that was honest and sincere. This communication emphasis still exists and is today a real and important part of Petrosar's community public relations activity. Execution of the program enabled us to communicate and comment on myriad features about Petrosar -- the full range of features that come under "concern for the environment".

Air Emissions

Original design of the two major sulphur-bearing stacks called for components and heights to result in ground-level concentrations which would be half the Ministry of the Environment SO_2 standard. We added fifty feet to the height of these stacks and we have a sulphur dioxide level that is well below the current MOE standard. To be more specific, our average mean is 0.07 ppm -- measured at ground level -- compared to the Chemical Valley standard of 0.15 ppm and the overall Ontario MOE requirement of 0.3 ppm.

In addition, Petrosar joined the Sarnia area Lambton Industrial Society and provided $40,000 for a new air quality station in that organization's grid system. We have played a major role in the establishment this year of Bluewater Clean, the oil spill clean-up organization that will make available manpower, boats, booms and other equipment to protect the St. Clair River shoreline in the event of an emergency.

We have had no complaints about affected air quality, noticeable to the senses or otherwise.

Flare Stack

Burn-off of process gases is, of course, necessary at Petrosar and our flare stack, at 350 feet, is highly visible. Major flaring is sporadic, however, and this was explained in our program of public awareness. In these instances, while the most impressive part is the sight, the brief noise activity that's part of venting is also there -- even though the stack itself is equipped with a special $92,000 flare tip. We paid a $25,000 premium to obtain higher levels of smokeless burning and lower noise levels.

When we know in advance that flaring will occur, we telephone our immediate neighbours and explain to them when, where and why there will be flaring -- and for how long. This has produced a very positive attitude in the community and has cut complaints to a rare occurrence. When a complaint does come in, however, we investigate the problem and provide an explanation that stresses the necessity and safety aspect of the flare stack.

Cooling Tower

This 514 ft. long, 85 ft. high structure has cooling capacity for 150,000 gallons per minute of process water. The loss runs up to 5,000 gallons per minute and the resultant water vapour plume has been assessed and found to have no effect on hydro lines, roads, the community generally. The tower can be modified to reduce the plume, should this every prove necessary.

Water Supply

Petrosar's relatively low water needs and high recycle rate conserves a great deal of energy and has no effect on nearby municipal systems. As noted, the process water loss from the closed system, via the cooling tower, is up to only 5,000 gallon per minute; this amount is replaced from the St. Clair River.

We have a 5.5 million gallon capacity storm water lagoon which retains a minimum 3 million gallons for firefighting or other emergency purposes.

Waste Water

In the design of our water treatment system, all available, practicable technology was used -- at a cost of $40 million. This system includes:

- segregated sewers
- oil/water separation facilities
- stabilization through the use of holding ponds
- oil and floatable solids in intermediate treatment
- secondary treatment by biological oxidation
- tertiary treatment through the use of activated carbon filters
- storage ponds before discharging back to the St. Clair River

Petrosar is, in fact, the only plant in Chemical Valley that uses tertiary treatment, and is the only operation with a diffuser on the return water outlet into the river.

Minimal water use and high recycling rates have saved many dollars and reduced total costs -- offsetting the $300,000 higher cost of operation per year than a once throu plant. We're more than meeting MOE water quality objectives with respect to phenols ammonia, total organic carbon, oil and total suspended solids.

Solid Wastes

Petrosar employs Tricil Ltd. on a contract basis for the disposal of solid wastes. No imposition is made on the municipal systems. Needless to say, all Ministry of Health and Ministry of Environment requirements are met in this area of waste disposal.

Working Environment

We have a well-staffed Medical Centre, which works closely with our Industrial Hygie group to study the long-term health effects of our workplace. We are constantly evaluating possible hazards such as benzene and carcinogenics in general, and, as well, are continually monitoring noise and heat stress.

Measurement of noise levels for instance, is taken at intervals, and under a variety of conditions; then these measurements are logged, summarized and studied. By this careful procedure, potential health hazards can be averted and a safe working enviro ment for the employee can be maintained.

CONCLUSIONS

In conclusion, I'd like to touch on a couple of points that I think relate effective ly, if not directly, to our "Community Awareness Program". Our own employees -- all 750 of them -- are also very much a part of "the community". Company-wide communica tions meetings were initiated prior to the plant start-up. In fact, early meetings were held in the president's office. Dates and times are set to allow all employees to attend. We've had average voluntary attendance running at about 50% of total employees. The meetings -- now held three times yearly -- allow two-way dialogue between the employees and the executive officers. They allow our people to have regular "state of the operation" summaries and to pose questions for our executive to answer.

As chief executive officer, I take a position myself of encouraging our employees to participate in the social institutions of the surrounding communities; we co-operate with our people who take office or leadership roles in community events and organizations. Community responsibility by our employees has, in fact, manifested itself quite pointedly in two recent events. In last fall's United Way campaign, Petrosar employees were cited as having contributed the highest donation per employee in the Sarnia industrial sector. In another instance, the company and its employees responded to a special appeal to finance a burn care unit at the Sarnia General Hospital, resulting in the highest per capita contribution to this project.

We actively support local health and welfare causes, especially a number of youth oriented projects in the Chemical Valley. We are also active members in all of the more classical community, county, provincial and national environmental safety organizations. To be quite honest, this community involvement is simply not a philanthropic gesture. We believe it is to our own advantage as well in the long run.

We want to know what the community thinks of us and, in turn, we want to tell the community what we are doing and what we are about to do. Only through this up-front dialogue, as well as through a sincere effort to more than meet environmental standards, can we achieve accountability and credibility. Accountability and credibility are the two major concerns of today's society, and we must start putting self-imposed standards in place that will be a cornerstone for efficient industrial operation throughout this decade and beyond.

I offer you this thought in closing. We, technical and professional people, are prone to think that when we have solved a "waste" problem, we have done our duty to mankind - NOT SO. Your technical solution may be perfect, totally effective, and highly cost efficient but, if the public do not perceive that to be the case, you have failed. If you are not prepared to meet the public face to face and respond to their needs, then your success is questionable. I have no sympathy for those who say that the subject matter is too complex for the public to understand. I believe that is a cop-out by those who don't understand their own technology well enough to be able to explain it.

Petrosar's experience is one piece of proof that those industries which make an honest and open effort to tell their story to the public will be well received, will be respected, and will be believed -- providing you do it in advance and not as an excuse after the fact, and providing you are prepared to accept the public definition of waste.

GUEST EDITORIAL
ACID RAIN

H. C. Martin

LRTAP Scientific Program Office, Atmospheric Environment Service, 4905 Dufferin Street, Downsview, Ontario, Canada

ABSTRACT

This paper discusses our current assessment of the acid rain problem in eastern Canada, the elements which lead to serious environmental damage, and some of the federal research projects which are underway to examine the impacts of acidic precipitation on sensitive ecosystems. The outlook for eastern Canada is assessed in the context of the experience in Scandinavia over the last twenty years.

KEYWORDS

Acid rain; air pollution; long-range transport.

Recently The Chemical Institute of Canada met in Ottawa and I was surprised to find, when I received the program prior to the meeting, that there was so much disussion of acid rain and long-range transport. Among the world authorities that we had, there was a well-known researcher from Norway, Arne Henrikson, who started his talk with a slide of his favourite lake. The following day we had another Scandinavian, a Swede, William Dickson, who started his talk with a picture of his favourite lake. I too have my favourite lake. Occasionally, a fisherman goes out there and spends an afternoon drowning worms. Why? In the case of Henrikson, Dickson and myself, we have lakes which have no fish. They have new chemistry, they have impoverished biological communities. They have gone acid. They are essentially dead. There is little comfort for us Canadians to know that we only have a few hundreds of these acidified lakes, little comfort when we realize that Scandinivia has tens of thousands. But there is something coming.

We did a study last summer, north of Sault Ste Marie. Over fifty lakes were sampled; all the head water lakes were acid. As we move eastward, we have known for years that the LaCloche mountain lakes are acid. As we move further east into the Haliburton region, we have a watershed where the Ministry of Environment for Ontario has been doing work for several years. They have lakes with pH's of about 5.0 - 5.5, and above. Quite comfortable! But they are going acid, the pH is dropping. A short time ago I received a report from the province of Quebec, where we are cooperating with some scientists. I looked at some of the data they had from the past year. Focussing on the lakes within a hundred miles of the St.Lawrence basin, once you are into the pre-Cambrian Shield, every once in a while you pick up one with a pH below 5.0. Moving further eastward, Nova Scotia has probably ten salmon rivers which no longer support fish. They have gone acid. The lakes in the Adirondack mountains with an altitude above 600 metres are dead.

Fig.1. Regions of North America containing lakes that are sensitive to acidification by acidic precipitation.

This is where we are going. These are the types of watersheds we are looking at. They are watersheds which have very thin soils, the granitic rock pokes out, it is bare, it has no calcareous material; hence, the landscape has no buffering capacity. Acidic precipitation and dry deposition fall upon these ecosystems, and are not neutralized.

Figure 1 is a map which very roughly gives us an idea, based on geology, of the regions in eastern Canada which are potentially going to be damaged by acidic precipitation. The dark areas are non-calcareous rock; there is no buffering material available to neutralize acid rain falling on these areas. We can identify Sault Ste Marie, LaCloche mountains, Haliburton, the Quebec area, and central Nova Scotia. Because it is a Canadian map there is no U.S. data, but New York State, the Adirondack mountains area, would also be dark, no buffering capacity. You might ask what is happening. To understand what is happening we must regress to some years ago, say forty years.

About forty or fifty years ago, we had a local pollution problem. There were low emissions of high concentrations. The materials that came out of the stacks were pollutants that caused difficulty at street level in a small local area. Over the past thirty or forty years we have pretty well cleaned up this problem with regulations. But as our cities have grown we now have large urban regions producing low concentrations which, in fact, are changing the composition of whole air masses.

Another rather recent phenomenon concerns significant point sources. For large point sources we have built tall stacks which pump the material higher up into the atmosphere allowing it to disperse and move downwind. These large polluted airmasses and these high point sources have ushered us into the era of long-range transport. The materials from tall point sources or large regional sources are moving over long distances, causing damage somewhere downwind, maybe five hundred to a thousand kilometres or more from their origin.

The various elements of the process are illustrated diagramatically in Fig. 2. We have our sources, whether they are cities or the high stacks. The pollutants enter the atmosphere. It is essential that they have a long transit time, that is, a long period of time in which oxidation can occur. Once SO_2 and NO_x change into sulphate and nitrate, there is a reaction with the water vapour in the atmosphere to produce acid. Finally, there is a deposition. The acidic deposition has to impact on the sensitive areas shown in Fig. 1.

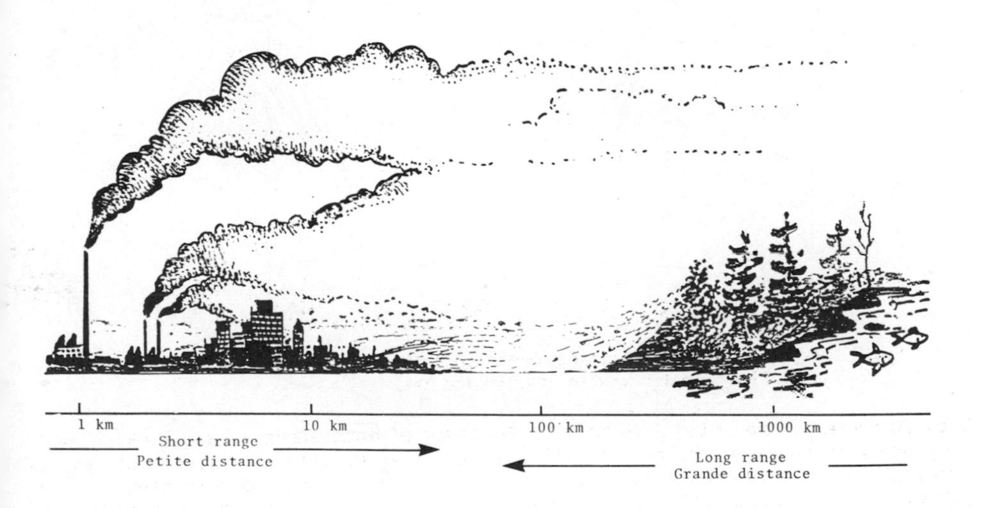

Fig. 2. The components of the acid rain problem: large sources, transformation during transport over long distances, deposition on sensitive ecosystems.

Well, we do have the emissions, we have 30 million metric tons of sulphur dioxide produced in eastern North America each year with highest emissions along the Ohio valley. Next we must ask: how does the pollution move? The mean summer time air flow carries the pollution from the main sources northward and eastward, to the worst deposition sites, to those regions which are most sensitive: central Ontario, Quebec, the Maritimes and the Adirondacks.

One of the first things we would like to do, knowing that the materials are coming from somewhere and falling in a sensitive area, is to go out there and actually measure the chemistry of rain. For several years, the CANSAP (Canadian Network for Sampling Precipitation) network has been providing monthly rain samples at selected sites. The samples are sent to a central lab and analyzed for pH and a number of other variables. There are 55 of these stations across Canada. They have been in operation for 4-5 years. When we get our information from the analysis, we can start drawing pH contours for eastern Canada and produce maps such as Fig. 3. Clean rain has a pH of 5.6, so most of the rain is acid. In the St. Lawrence valley, the pH is typically 4.2, about 40 times more acid than normal. If that water starts working through the ecosystems unchanged, we will eventually change the water of the collectors, of the lakes, to something below 5.0 with the consequences. This collecting network is just one of over a hundred projects that we have in the federal government and in the provinces. It is just the beginning. We also are measuring and cataloguing inventories of sulphur emissions, nitrogen emissions, and hydrocarbon emmisions, for the continent. This information is used in mathematical models, so that we can determine not only where materials are going, but also determine what would happen if we turned such and such a stack off, or if we doubled its emission. And we also, of course, have to get out into the environment. We have to go to the sites. We have selected four watersheds in eastern Canada where we are undertaking intensive studies - interdisciplinary studies involving the chemists, the biologists, the geologists, the meteorologists, the whole gambit, trying to get the people together at one

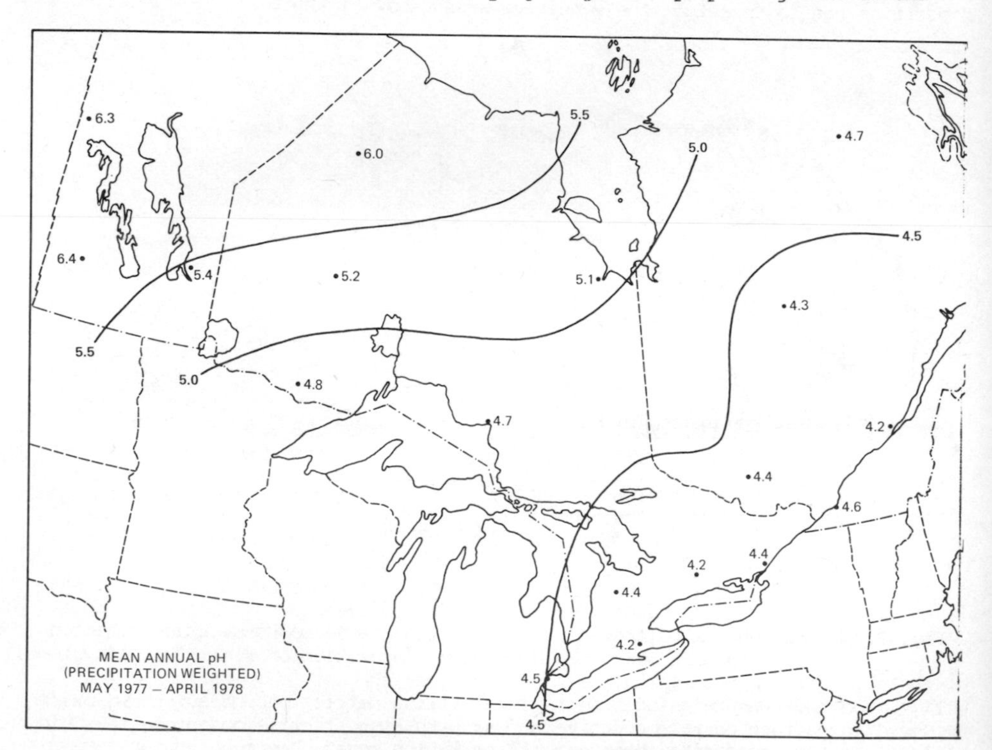

Fig. 3. Weighted mean annual pH values for precipitation in eastern Canada.

location, so that the information is collected simultaneously and as a group. One of these sites is north of Sault Ste Marie, the Turkey Lake watershed which has four lakes with a 1400 ha drainage basis. The pH of the headwater lake is 4.7. It has gone acid. The lakes down the chain, near the exit, are in better shape. There are many studies at such a watershed: we measure the daily precipitation; we try to estimate the dry deposition of material; we measure the chemistry, that is the chemistry of the soil and water; we take inventories of soils and forest cover; we look at the lichens and mosses. At the exit, which is a tiny stream, we will place a weir, and with that weir we will be able to determine the amount of water exiting. With our rain gauges we will know how much water is coming in. With the analysis of the rain we will know the chemistry of water coming in, and with samples at the weir we will know the chemistry of the water going out. We will start putting together an ion budget. At such a site we might have as many as forty projects going simultaneously, involving a number of departments. The Department of Fisheries & Oceans for example, is involved with the examination of fish populations and age classes, and the whole fish food chain. There is another calibrated watershed at Kejimkujik Park in central Nova Scotia. There we have another team essentially doing the same kind of inter-

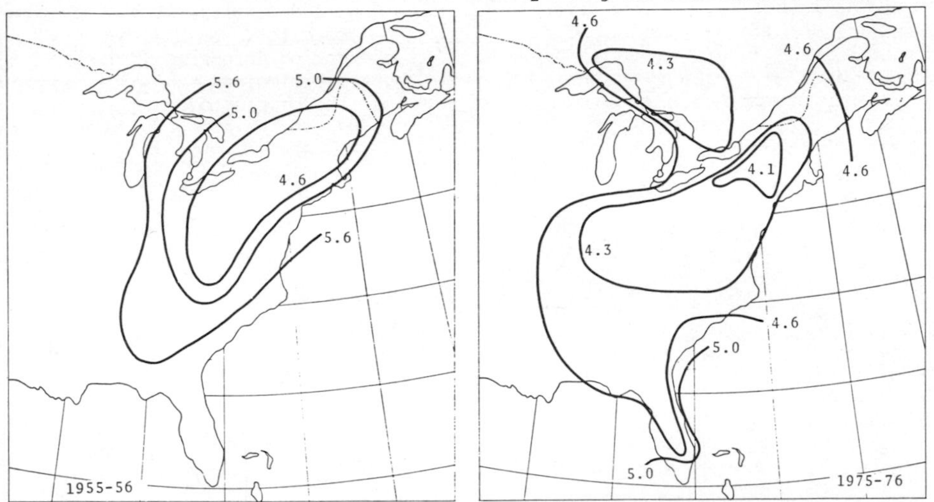

Fig. 4. The change in weighted mean pH of precipitation between 1955 and 1975

disciplinary work on a different ecosystem. There is a different atmospheric load falling in that region. We have a different climate, but we do have the same concern. You might ask: where are we going? Fig. 4 shows some old data, which has recently been updated. Within 10 years, the pH in the area with highest acidity has dropped to 4.1 while the overall acid rain area has greatly expanded. There is no longer clean rain to the south, and the clean rain isopleth is now somewhere across central Ontario/Quebec.

Given the course of events that we have seen over the last few years, and the energy scenarios we are anticipating, I think it is fairly certain where we are going. This situation will continue to degrade. Our friend from Norway, Arne Henrikson, made a sobering observation last week. He said "There is a new generation in Norway, those under 25, who are not so upset that the lakes in southern Norway have no fish - after all, to them, it has always been that way." Perhaps this will be the legacy we leave for our next generation.

SECTION I

POLICIES FOR ENVIRONMENTAL CONTROL

1
FEDERAL POLICIES FOR TOXIC CONTAMINANTS CONTROL

G. M. Cornwall

Policy Planning and Assessment Directorate, Environmental Protection Service, Environment Canada, Ottawa, Ontario K1A 1C8, Canada

ABSTRACT

Environmental protection issues are changing as we enter a new decade. As a result of increased public awareness of the adverse effects of toxic contaminants such as PCBs, chlorofluoromethanes, mercury, and others, environmental health -- including fear of involuntary exposure to invisible toxicants -- is a growing concern among many Canadians. These problems of a blobal nature require national and international cooperation.

Environment Canada is developing a new federal invironmental protection policy which, among other things, will put forward the need for new federal legislation to protect the environment and human health against toxic chemicals and their hazards. The policy and accompanying legislation will look at the environment as a whole rather than as a collection of disparate parts. Environment Canada is establishing a Toxic Chemicals Management Program to coordinate government activities, and make the delivery of government services more efficient.

INTRODUCTION

I have been invited to speak on concerns of the Canadian government with regard to toxic contaminants. Environment Canada has a number of activities and responsibilities related to control of toxic chemicals. The Environmental Contaminants Act, the Clean Air Act, the pollution provisions of the Fisheries Act, and the Ocean Dumping Control Act all address these toxic components. In addition, a number of other federal deprtments and agencies have legislation and programs that address toxic chemicals and their control. For example,

- National Health and Welfare administers the Food and Drug Act and shares administration of the Environmental Contaminants Act;
- Agriculture Canada administers the Pest Control Products Act;
- Indian and Northern Affairs has general legislative authority for environmental protection in the Canadian North, e.g., the Arctic Waters Pollution Prevention Act;

- Transport Canada is Proposing the Transport of Dangerous Goods Bill, which is currently before Parliament[1];

- The Atomic Energy Control Board administers the Atomic Energy Control Act, and regulates nuclear-related activities. It thereby protects health, safety, and the environment relative to radioactive elements.

The list could go on but I believe I have given adequate illustration of the fact that the vollection of federal government policies for toxic contaminants control is broader than I am able to address today. With that qualifier, let me begin.

CURRENT PROBLEMS

In the public mind, the management of toxic chemicals is becoming one of the most important issues facing Canadian society. The example of Love Canal is fresh in all our minds. Although it would be difficult to exaggerate the seriousness of the questions it raises, there is a great temptation here to say that "chemicals" is a dirty word and we should avoid using them whenever possible. But surely the answer to our problems does not lie in such simple and extreme solutions.

Although we created our environmental organizations along media lines when we set them up nearly a decade ago -- air pollution, water pollution, land pollution, and so forth -- we have learned a lot in 10 years. Environmental problems do not respect boundaries, whether between land, water, and air, or between provinces or countries. What concerned us then seems almost simplistic to us now. Some would say the lines have blurred since then, that it is harder to tell the good guys from the bad guys and that the issues seem to get foggier every week. I prefer to think that the "good guys - bad guys" distinctions we used to make are not very useful any more. After 10 years, we are learning to distrust simple scenarios and to look for a balance. We feel the need to coordinate our resources, to bring areas of jurisdiction and knowledge together to solve problems on a global scale and with a global perspective. The challenge to government, not just to the Canadian federal government but to governments all over the world, is a big one. From one side comes pressure for restraint -- do more with less; from another side comes the need to cope with major issues facing societies in the eighties.

Two reports have recently brought the problem of toxic contaminants into sharper focus. The first of these offered a definition of the word <u>toxic</u> -- in itself a difficult task as the word can be taken to include everything or nothing.

Webster's gives us the one-word definition "poison" and leaves it at that. The U.S.[2] report defines it this way:

"Toxic is a relative term. The effects of any chemical or mixture depends not only on its composition and basic properties but also on dosage, route and conditions of exposure, the susceptibility of the organism exposed and other factors....
Toxic chemicals are thus here defined as those chemical substances which, when released to the environment, or thereafter if chemically transformed through combination or otherwise, could pose a significant threat to natural ecosystems or to human health and well-being."

---

[1]The bill was passed on July 17, 1980, and will come into force upon proclamation.

[2]Report to President Carter by the Toxic Substances Strategy Committee, August, 1979.

The Canadian report[3] eloquently sums up the enormous impact of such chemicals on our lives and I will quote it at some length:

"One of the most daunting environmental problems of our time arises from the flood of man-made chemicals pervading our lives. The products or by-products of our industries, they are in every home in a multitude of forms -- floor coverings, insulation, varnishes, synthetic fabrics, washing agents, medicines, hair sprays and antiperspirants, toys, vinyl clothing, pest killers -- in variety too numerous to list. The ingenuity of those who have contrived new chemical compounds and devised ways of inserting them into our economy has had much to do with the improvement of the human state.

"But we have too frequently ignored the other side of the coin. To our distress, we have slowly learned that some of these products are damaging to human health. For some of these, we have developed restrictive legislation which we hope will protect us.

"But the ultimate fate of every compound is to be discharged via the sewer of the incinerator stack or by accident into the air, the water or onto the land, where singly or in combination they alter the environment. Species are destroyed, lakes and rivers lose their ability to support their normal faunas, vegetation changes; the habitats upon which life forms depend become less suitable as places for plants, animals and man to survive."

The documents I quoted from are a reflection of increasing public concern over the treats to human health posed by uncontrolled use of these chemicals. There is a strong interdependency between human health concerns and environmental quality and I am convinced that those of us with enviornmental protection responsibilities must intensify our cooperative partnership with those responsible for health protection.

POSSIBLE SOLUTIONS

What can we do? Scientists have listed some 3,000 compounds that present various degrees of hazard, and technicians are trying to determine which of an overwhelming number of combinations may create a danger. Hundreds of incidents, less spectacular and less well publicized than Love Canal, have caused the public to ask whether anyone -- producers, users or government -- can ensure the safety of human beings and the environment. This concern has been reflected in a number of institutional responses and calls for action here and abroad. In Canada, public awareness of the dangers of toxic chemicals is growing, and Canadians expect government to ensure that the health and safety of individuals isprotected.

The time has come to stop flagellating ourselves for the ignorance and carelessness of the past, and to look for solutions is as realistic and practical a way as possible -- not with accusations, however justified they may seem, but with constructive solutions.

One of these solutions, or the path to it, lies in the area of international cooperation. There is a growing realizat-on among the industrialized nations that information on chemicals and their potential hazards should be exchanged as a normal part of the trading process. There is also a realization that industrialized nations have a responsibility at least to notify a foreign country of hazards, if not to actually control the export of a hazardous chemical to that country. This responsibility particularly applies to notifying those small countries which do not have sophisti-

---

[3]Donald Chant and Ross Hall, "Ecotoxicity: Responsibilities and Opportunities," Canadian Environmental Advisory Council.

cated mechanisms to assess or control hazardous chemicals. For example, considerable effort is being made in the Organization for Economic CO-operation and Develop ment to develop a basis for regulatory action related to the control of toxic chemicals. Although some may describe the effort as one which seeks consistency, I prefer to see it as an attempt to share scientific data (Health and environment related) which has a known base. In any event, the objective is to establish international norms for testing and for provision of information on new chemicals. The intent is to create what could be called "chemical passports" for these substances. The concept would be gradually expanded to existing chemicals (of which there is now $140 billion a year in trade).

The need for all this can be ascribed simply to the "chemical revolution." That expression may seem overworked, but let me relate a few facts and figures assembled by the U.S. Environmental Protection Agency to set the stage:

- Until 1945 or thereabouts, most chemicals in common use were derived from naturally occurring materials, principally minerals and plants. Each of them had been "screened" by our environment and by some three million years of trial and error by humans as to which were safe and which were dangerous.

- Since that time -- post World War II -- the chemicals industry has developed synthetic substances never encountered before. In today's terms it takes a team of pathologists, 300 mice, two or three years and probably in excess of $300 000 to determine whether a single suspect chemical causes cancer.

- Who in 1945 could have guessed that millions of human beings spraying deodorants in their bathrooms might erode the ozone shield?

- Who in 1950 could have guessed that sulphur and nitrogen oxides rising from one nation might fall in another as acid rain, slowing forest growth, impoverishing soil, and killing lakes?

All of this brings me to an important point: our lack of knowledge.

In only a few short years, we have expanded our ability to measure from parts per million to parts per trillion. Indeed, it was not so very long ago that we could not tell the difference in the environment between DDT and PCBs. Yet the paradox in all this is that our knowledge about specific substances in the environment is rapidly outstripping our capacity to understand the implications of such knowledge

The terms "parts per million" and "parts per trillion" by themselves are difficult to picture. What do they really mean? Dow Chemical recently asked some of its research chemists to translate these measurements into plain English. The list they devised showed imagination, to say the least. One part per million is equal to one bad apple in 2,000 barrels. One part per billion equals a pinch of salt in 10 tons of potato chips. And one part per trillion equals a drop of vermouth in 250,000 hogsheads of gin, or one flea among 360 million elephants!

The point is that while we may be getting better and better at measuring, this has not necessarily increased our ability to understand what we measure. Meanwhile th chemical and petrochemical industries are producing hundreds, if not thousands, of new compounds each year.

In the ideal, we dream of science someday being able to tell policy makers how much of which substances will effect human health or the environment. In the ideal, we hope we will someday be able to control environmental pollutants by operating within clearly defined environmental or human health limits, while taking full advantage of the resilience of nature and of the human body. But the reality is that science alone cannot give us the data nor the mechanisms we need to understand the environment in this ideal way. Although our research efforts may eventually bring us closer to the ideal, science alone cannot provide all the answers. We will always be faced with the management of ignorance, and with the need for value judgements. In the interim, we must provide the most effective protection possible, within our limitations.

INTER-GOVERNMENTAL ROLES FOR TOXIC CHEMICAL MANAGEMENT

How can we use our growing knowledge to manage our ignorance in the coming decade? The first step is for governments -- provincial and federal governments and international organizations -- to define common objectives that will provide guidance and coherence.

Coordinating the goals of 10 constitutionally separate, but nonetheless interdependent, provincial governments is a continuing problem in the Canadian federal system. The goals are diverse, sometimes conflicting, from province to province, region to region, and even between areas in a single region. In fact, sometimes it seems that this interdependence we value so highly, instead of promoting cooperation, is a profound source of tension and conflict.

What are the appropriate federal-provincial roles for managing toxic chemicals? We need to define these mandates much more clearly if government resources are to be used to the fullest in addressing the environmental concerns of the 1980s.

Some progress has been made in this defining of relative responsibilities. Given the international character of the chemical industry, the movement of chemicals across boundaries in various environmental media, and the complexity of assessing hazards, the federal government has assumed responsibility within Canada for providing leadership in managing toxic chemicals. In this context, leadership means using the federal government's scientific resources for such activities as environmental measurements, for cause-and-effect research and for interpreting foreign information. It also means using the federal government's regulatory authority which spans the country.

This leadership by the federal government is exercised in cooperation with the provinces, both in assessing problems and in bringing about solutions. Cooperation includes relying on provinces for some of the necessary scientific work, a full information exchange to encourage common assessment of problems, and implementation by the provinces of federal controls on toxic chemicals.

Environment Canada is in the process of developing a federal environmental protection policy. Among other things, this policy will call for new federal legislation, to allow governments to protect the environment and human health against involuntary exposure to toxic chemicals and their hazards.

While this approach could be considered a new tactic, it is new only in the sense that the policy, and the expected legislation, will look at the environment as a whole rather than as a collection of disparate parts. I can also assure you that Environment Canada has assigned high priority to concerns about toxic contaminants. We are taking steps to coordinate Environment Canada's activities in this area by establishing a Toxic Chemicals Management Program.

TOXIC CHEMICALS MANAGEMENT PROGRAM

At the beginning of this discussion, I named some of the numerous federal programs that are involved in control of toxic contaminants. It will be the aim of our Toxic Chemicals Management Program to coordinate these many and various programs, along with Environment Canada's own activities, so there is better consistency in terms of priorities, assessment activities, and so forth.

A preliminary set of objectives has been defined. Although they need further discussion and refinement, it may be useful to quote them here. The four preliminary objectives are:

(1) To prevent or control the entry of harmful quantities of toxic chemicals to the various components of the environment -- air, water, land, biota, and man. For example Environment Canada's position is that more effective testing is required and it is the responsibility of the manufacturer to assure the safety of his product to human health and the environment.

(2) To minimize or ameliorate the adverse effects of chemicals now in the environment.

(3) To encourage the development and promotion of alternate products, processes and technologies which are less harmful to the environment.

(4) To inform the public on the origin, pathways, accumulation, and effects of toxic chemicals released to the environment and on strategies for their control such as regulation, alternate products, and processes.

Obviously the whole question of federal-provincial coordination and cooperation will play a key role in assuring that scarce resources, both human and dollar, are used as effectively as possible.

Control of toxic products is only one way to safeguard the public and the environment. The development of alternative, more benign, technologies offers a number of promising possibilities. The scientific and technical endeavours of those who are directly involved in research on waste treatment and utilization applies here. Without the scientific advances to which their work contributes, regulatory controls could hardly progress.

A broad range of control options will be used in Environment Canada's Toxic Chemicals Management Program. These will range all the way from outright banning to regulation, and substitution of alternate products and processes. The socio-economic impact of any control program will be a key consideration. As a matter of policy, public opinion will be sought, and made part of the decision-making process. The balance that is struck will, as always, be the result of a tug-of-war among issues and interests, but it will reflect our values and priorities as a society.

Risk analysis, and the establishment of acceptable risk, will play a central role in setting priorities. A thing will be considered safe if its risk is judged to be acceptable. We must remember, however, that many toxic chemicals, particularly in combination with other chemicals in the environment, defy establishment of "safe" levels. In fact, establishing such levels when the actual effects are in doubt can create a false climate of confidence. Therefore the need to encourage minimal release into the environment, and to take proper account of the socio-economic effects of such encouragement, must be recognized.

There will always be problems connected with the disposal of some of these toxic materials. As one example, there have been controversies over the incineration at high temperatures of PCBs in cement kilns. Last October the Mississauga City Council rejected the idea of destroying PCBs by incinerating them in local cement kilns. Editorials and columns in several area newspapers criticized the town's "not-in-my-backyard you don't" attitude about waste disposal. Yet a real difficulty arose because various levels of government -- government in general, if you will -- did not involve the people of Mississauga early enough or completely enough to provide assurance that the technology was safe. The negative reaction set in, and solidified.

Whatever we conclude from this, one thing is certain: the technology of control and disposal has to keep pace with the development of new products. For governments, policies and legislation must keep pace with our growing knowledge about toxic chemicals. The public demands it, and the next generation requires it.

2
LONG-RANGE TRANSPORT IMPLICATIONS FOR CONTROL OF AMBIENT AIR QUALITY

F. R. Pearson

Science and Technology Research Office, Minnesota State Legislature, St. Paul, Minnesota 55155, USA

ABSTRACT

This paper discusses long-distance transport of pollutants and the implications of long-range transport on efforts to control pollution. After being emitted, pollutants may be transported over long distances to areas sensitive to their affects. This paper consists of an introduction to the problem and discusses control of pollution and government activities.

INTRODUCTION

Each year millions of tons of pollutants are discharged into the atmosphere. In the United States the major pollutants are those associated with fossil fuel consumption, either from stationary or mobile sources. Stationary sources, such as electrical power plants and industries, release many pollutants, but special attention is given to sulfur oxides (SO_x), nitrogen oxides (NO_x), carbon monoxide (CO), and tiny solid particles that become suspended in the air. Pollutants produced by mobile sources (primarily automobiles) are CO, NO_x, and a variety of incompletely burned hydrocarbons. In addition, some newer cars release sulfuric acid (H_2SO_4) as a fine mist, or aerosol. In the atmosphere, these pollutants undergo complex chemical reactions which are influenced by meteorologic situations. These reactions can lead to the production of sulfates, nitrates, and photochemical oxidants such as ozone (U.S. Environmental Protection Agency, 1979).

Because the pollutants are either gases or suspended particles, they may move easily in the atmosphere. Wind currents transport these pollutants hundreds or even thousands of miles from where they originated. Some sulfur compounds have travelled several hundred miles in one day by atmospheric transport. During transport, these pollutants can cross many state or even national boundaries (Fisher, 1975; Park and co-workers, 1978). Therefore, the air pollution standards of one state or country can have a direct impact on the environment of another state or country (Hirsch and Abramovitz, 1979). In North America these pollutants generally move northward. For example, the pollutants from the Ohio River basin can constitute a serious problem to areas of the northeast and northcentral United States and eastern and central Canada. Therefore, many localities far removed from the source have costs and environmental problems placed on them from pollutants that originate far away (Carter, 1979).

With the tightening of controls on power plants and industries in Europe and North America, it became a standard procedure to meet new regulations by heightening the stacks used for emitting pollutants. These higher stacks enabled the plants to pass the new regulations, because measurements for pollutants were made at the ground level adjacent to the source. The higher stacks give the pollution to someone else due to the long-range transport of pollutants.

THE PROBLEM

Research conducted in the past several years in Europe and North America indicates that a number of pollutants (commonly gases and totally suspended particulates--TSP) are capable of rising into the atmosphere and undergoing long-range transport (Park and co-workers, 1978). Due to these pollutants, the following two problems are of the most importance: (1) the pollutants can rise into the upper atmosphere an effect the ozone layer, or (2) they can return to the earth as acid precipitation (wet or dry deposition). These can cause environmental, health, materials, and economic problems. Acid precipitation is currently of the greatest concern, followed by pollution effects of ozone.

Acid Precipitation

The pollutants SO_x, NO_x, CO, etc., are emitted into the atmosphere where they may react with moisture and other chemicals. Many of these pollutants are chemically changed into acids and then brought back to the ground in the form of acid precipitation, both rain and snow (see Fig. 1). Certain other harmful compounds in lesser amounts are also found in the "acid rain", including ozone, nitric acid (HNO_3), heavy metals, hydrocarbons, and various other compounds (Babich, Davis and Stotzky, 1980).

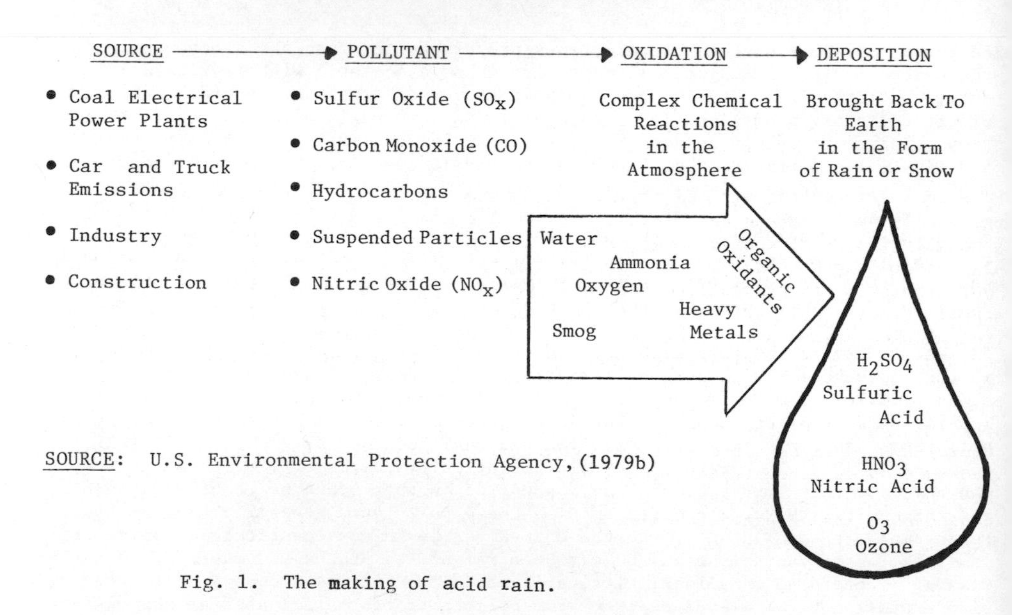

Fig. 1. The making of acid rain.

Rain normally has a slightly acidic pH of 5.6, due to the action of carbon dioxide in the air. (pH is a logarithmic measure, which means that for a change of one unit in pH the concentration of hydrogen ions changes tenfold. A pH of 4.6 is ten times more acidic than a pH of 5.6 and 100 times more acidic than a pH of 6.6) However, if air becomes polluted with sulfates and nitrates, the pH of rain may be substantially lower. Currently, the rainfall in the eastern half of the United States has an average pH between 4.0 and 4.5 (West, 1980). Some rainfall has been measured with a pH as low as 3.0 (see Fig. 2)(Likens and co-workers, 1979).

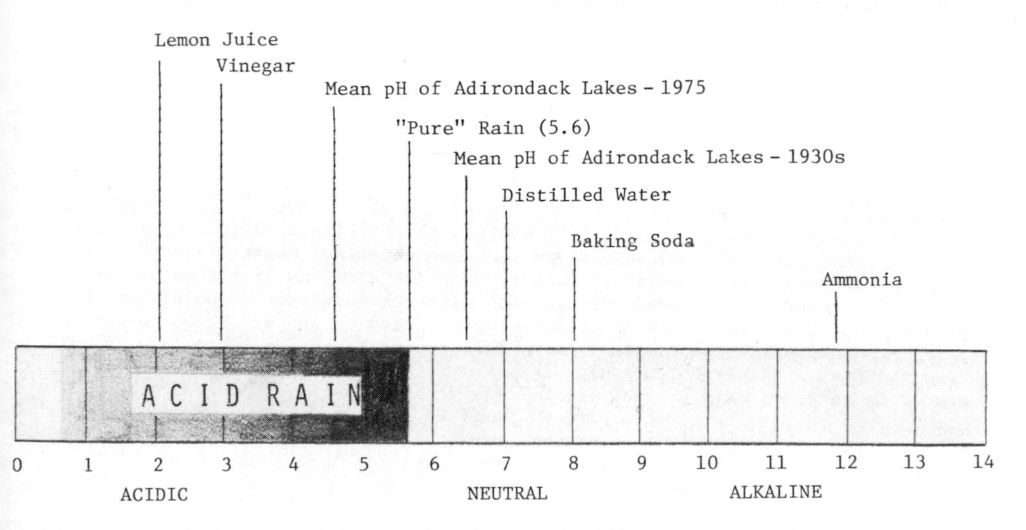

SOURCE: U.S. Environmental Protection Agency (1979b)

Fig. 2. pH scale comparison for understanding acid rain.

Due to mixing of the air and deposition, the concentration of an air pollutant generally decreases as the distance from the source increases. However, despite the dilution of air pollutants during mixing of the air, these pollutants accumulate over long periods of continuous deposition to the ground. The effect of these pollutants is often magnified due to a synergistic or additive effect caused by the interaction of various pollutants (Glass and colleagues, 1979). Therefore, the problem arises not because of local air pollutants alone, but because of the cumulative effect of air-transported pollutants onto the ecosystem, usually in the form of acid precipitation (Glass and colleagues, 1979).

The full extent of the acid precipitation problem is not known. So far, the evidence shows that acid precipitation can limit or prevent fish reproduction, can reduce crop and forest yields, can damage steel and stone structures (Varshney and Dochinger, 1979), and can cause several indirect human health problems through processes such as increased leaching of trace metals into the drinking water (DeRoos, 1980).

Ozone

Ozone is produced in the lower atmosphere when emissions of hydrocarbons and nitrogen oxides react through a photochemical oxidation reaction. Ozone consists of two types--that generated by vegetation and that generated by man-made industrial pollutants. Vegetative ozone contributes less to the problem than man-made ozone. Usually ozone remains close to the earth and can be a severe irritant to many human respiratory systems. Also, as a pollutant of the lower atmosphere, it comprises about 90 percent of the photochemical oxidants that are formed and is the main component of urban smog (Cox and co-workers, 1975).

Ozone can be returned to the ground as a part of acid precipitation or as a dry deposition. It also has been shown to depress photosynthesis, leading to decreased plant growth (Cox and co-workers, 1975).

Human Health Effects

Air pollutants have been shown to cause adverse health effects. In controlled laboratory experiments, using healthy adults, these pollutants (ozone, sulfated, nitrates, and particulates) were shown to cause coughing, chest discomfort, substantial soreness, and impaired pulmonary function after as little as one or two hours of exposure. Suspended sulfates and nitrates have been associated with increased occurrence of asthma in epidemiological studies and are known to have other effects on bronchial functioning. In a 1978 report by the American Lung Association (Health Effects of Air Pollution), air pollutants were shown to cause increased mortality rates on days when air stagnation greatly increased concentrations of atmospheric pollutants. Other effects found to increase because of air pollutants were respiratory diseases, lung cancer, cardiovascular disease, and some sensory, neurological, and behavioral disorders (American Lung Association, 1978; U.S. Department of Health, Education, and Welfare, 1976).

CONTROL OF POLLUTION

Polluted Air

Definitions of air pollution vary from community to community, but a representative pollution standard is given in Table 1 for the gaseous pollutants discussed in this paper.

TABLE 1 Comparison of Gaseous Pollutant Concentrations
(*ppm = parts per million)

	Clean Air	Polluted Air	Ratio Polluted to Clean
CO	0.1 ppm*	40-70 ppm	400-700
SO_x	0.0002 ppm	0.2 ppm	1,000
NO_x	0.001 ppm	0.2 ppm	200
O_3	0.02 ppm	0.5 ppm	25

SOURCE: Urone, "The Primary Air Pollutants-Gaseous: Their Occurrence, Sources, and Effects," in Air Pollutant V. 1, Authur C. Stern (Ed.), Academic Press, New York 1976-1977, Page 36.

The data in Table 1 is only suggestive. No precise definition of "clean air" or polluted air has been established. But even without a rigid definition of clean air, the differences between naturally-occurring clean air and "polluted air" is striking. The high levels of polluted air can be assumed to be man-made and if so, then in theory they are controllable.

While solid suspended particles and aerosols can be released in a wide variety of ways, gaseous pollutants are usually emitted through some type of exhaust system or smoke stack. Add-on mechanisms like "scrubbers" or post-combustion types of systems can be designed to either convert the gaseous pollutants to less harmful compounds or filter them. Some current and developing technologies are discussed later to exemplify methods of control.

Sulfur Oxide

Technologies for sulfur control are probably the most advanced of any of the gaseous pollutants. Sulfur appears to be the pollutant in most need of controls (U.S. Environmental Protection Agency, 1979a). Utilities and industry, in the past, have used several approaches to reduce sulfur emissions.

Low-sulfur coal. Utilities and industry have used low-sulfur coal either by buying low sulfur coal from the western U.S. and shipping it east or by physically cleaning the coal--a method that is often cost-effective. Up to 80 percent of the inorganic sulfur can be removed by physical removal techniques. However, the organic sulfur, which is chemically bound to the coal, cannot be removed by physical methods. It has been estimated that about 14 percent of the U.S. coal reserves are of the type that can be physically cleaned (U.S. Environmental Protection Agency, 1979a).

Flue gas desulfurization. Flue gas desulfurization technology, or "stack gas scrubbers", plays a central role in sulfur pollution control and is able to remove 95 percent of the SO_2. There are two general types of flue gas desulfurization. The first is the non-regenerable or throwaway system, in which the effluent goes through the scrubbing or absorption process and the waste product is discarded. The second is the regenerable or recovery system, in which the sulfur is combined with other chemicals and marketed as saleable products (i.e., elemental sulfur, liquid SO_2, and sulfuric acid)(U.S. Environmental Protection Agency, 1979a).

Fluidized bed combustion. This is a newer process and not yet in general use which involves the combustion of pulverized coal with granular limestone or dolomite (a carbonate of calcium and magnesium). The coal is burned in a closed container that contains a bed of pulverized coal and the limestone/dolomite solvent. The SO_2 is absorbed by the limestone as the coal burns. Research is currently in progress to improve this system, so that 85 to 89 percent of the sulfur will be removed.

Other technologies. Coal gasification and liquefaction are two emerging technologies that are expected to be ready for commercial use in the mid-1980s. These technologies could not only reduce sulfur pollution (95 percent removal is possible), but will also produce relatively clean burning fuels which can be used in place of natural gas. The sulfur removed by this method can also be marketed (U.S. Environmental Protection Agency, 1979a).

Nitrogen Oxides

Nitrogen oxide compounds result whenever air is used as the oxidant in fossil fuel combustion. Oxygen and nitrogen from the air combine at combustion to form nitric oxide.

The problem of reducing nitric oxides from automobiles is quite complex. Catalytic converters are now required on most newer U.S. cars (i.e., 1975 and later) to remove NO_x from automobile exhausts. How well they work remains controversial.

For stationary combustion sources, NO_x emissions can be reduced by (a) removing excess air not needed for combustion, (b) by two-step combustion, limiting the air in the first step, followed by heat removal (to lower the reactivity of nitrogen with oxygen), then burning the fuel completely in the second step, (c) utilizing flue-gas technology (as with SO_2), and (d) using low-temperature steam or water in the stack to provide a diluting effect on the NO_x. The two-stage combustion is probably best (showing a 90 percent reduction in NO_2) for electric utilities and industry. The water-injection technique may be one of the better ways to reduce NO_x emissions for stationary internal combustion engines (Van Nostrand's Scientific Encyclopedia, 1976).

Hydrocarbons

Pollution from hydrocarbons usually originates from the following two sources: (1) incomplete combustion of hydrocarbon fuels, usually in internal combustion engines, or (2) various high-carbon chemicals (i.e., chemical parts of paints and solvents released by industry). The first can be most effectively controlled by changing engine design and keeping cars in tune, and the second by either eliminating the use of high-hydrocarbon chemicals or by restricting the release of the chemicals by industry (which many states are already doing)(Van Nostrand's Scientific Encyclopedia, 1976).

Carbon Monoxide

Carbon monoxide is another pollutant that is associated with incomplete combustion. Control of this gaseous pollutant can best be achieved by assuring complete combustion of fuels in electrical power plants, industrial plants, and internal combustion engines.

Particulates

Numerous mineral and organic particles are released into the air by coal-fired power plants, industry (i.e., cement and steel manufacturing plants), and construction enterprises. Research shows that in the atmosphere, particles tend to form a stable aerosol (suspended 'mist') having a size ranging from 0.1 to 1.0 micrometers. However, aerosols of larger and smaller particles do occur. The larger particles tend to fall back to earth while the smaller particles tend to clump together to form larger particles that remain suspended. As mentioned before, these particulates are extremely hard to control. Present control systems seem to perform extremely well on the particles in the 2 to 3 micrometer size, but below this size the effectiveness of control decreases as the size decreases. Control becomes minimal between 1.0 and 0.1 micrometers and then increases again for particles smaller than this (Van Nostrand's Scientific Encyclopedia, 1976). This shows that particulates hardest to control are those that are the most common, the 0.1 to 1.0 micrometer size.

GOVERNMENTAL ACTIVITIES

The affects of long-range transportation of air pollutants are receiving more attention by legislative bodies and by international, national, and state agencies than ever before. This has been especially evident since the later 1960s. The first major act to control air pollution in the United States was the Clean Air Act of 1970. Under the Act, the U.S. Environmental Protection Agency (EPA) developed New Source Performance Standards (NSPS) for new facilities. The EPA was also asked to determine what changes were needed for existing facilities, but the EPA had no power to require those changes (Carter, 1979).

In 1977, amendments to the Clean Air Act were passed by the U.S. Congress. One of the most important aspects of the amended Clean Air Act was to give states responsibility for developing State Implementation Plans (SIPs), within a set time, to meet the National Ambient Air Quality Standards (Silverstein, 1979). Some states (e.g., California, Minnesota) have made great strides in developing and enforcing such plans, while other states have done little. In some states (i.e., Ohio), there has been great pressure from electric power utilities and industry to keep the air pollution standards as low as possible (McAvoy, 1980). Weak standards hurt not only the states involved but other states and sometimes even neighboring nations, because of long-range transport of pollutants.

The 1977 amendments attempted to resolve interstate pollution problems by adding a section which prohibits sources from emitting pollutants that would prevent another state from attaining or continuing to meet established pollution and visibility standards. The polluting states are required to revise their planned abatement measures to assure that regional compliance with National Ambient Air Quality Standards is met on a set schedule (Council on Environmental Quality, 1979).

In Europe, with the identification of low pH and fish kills in Scandinavian lakes, the Organization for Economic Cooperation and Development (OECD) started a cooperative technical program to study the long-range transport of air pollutants. Completed in 1977, the final report confirmed that long-range transport did occur and that the air quality of each European country is affected by the pollutants emitted from the other countries (Barnes, 1979; Organization for Economic Cooperation and Development, 1977).

With the problem becoming more evident in North America, Canada in 1976 identified long-range transport of air pollutants as a top priority environmental issue. A program was developed, and in 1978 U.S. researchers joined with Canadian researchers to study pollutant emission, transport and acid precipitation (United States-Canada Research Consultation Group, 1979).

The U.S. EPA has developed an integrated program to investigate acid precipitation. Research is being done to determine (1) sensitivity of numerous farm crops to sulfuric acid, (2) effects of acid precipitation on forest production, (3) impacts on fish populations in certain sensitive lakes and streams, and (4) impacts of coal-fired power plants and their pollutants on watersheds. To ensure the completion of these projects and others, the U.S. House and Senate Conference Committee agreed, on May 19, 1980, to create a comprehensive ten-year program on acid precipitation effects (U.S. Environmental Protection Agency, 1980).

Minnesota is currently studying the local impact of air pollution and especially acid precipitation with funds made available from the Minnesota Legislature. The University of Minnesota and the EPA in Duluth have developed models of Minnesota lakes and weather patterns to aid in predicting the effects of acid precipitation. Using precipitation pH data, they predict that if acid rain continues at current levels in northeastern Minnesota, fish in some lakes will become endangered in the next 10 to 20 years and in others within the next 100 to 200 years. This is based on assumptions that sulfate deposition would double by about 1990, and the contribution from nitrogen oxides is negligible (Ritchie, 1979).

Seven stations for monitoring sulfur and precipitation chemistry have existed in

Sherburne, Wright, and Stearns counties (central Minnesota) since 1976. Of 282 single samples of rainfall collected at seven different sampling locations during the period May 15 to October 15 of 1976, 1977, and 1978, 86 percent were shown to have sulfates and nitrates in the form of a neutralized salt. Acidity (pH) in these samples ranged from 3.76 to 7.90. Only about 30 samples had a pH value less than 4.5. Excess sulfate and nitrate were found to be correlated with hydrogen ion (pH) only in a few samples (Krupa, 1980).

Analysis indicates that in 90 percent of cases of acid rain, over 70 percent of the sulfates and nitrates were in the air for a long time--long enough to say that these air masses and the pollutants in them originated outside of Minnesota. Ambient sulfur dioxide and nitrogen dioxide concentrations in the study area are not sufficient to produce the sulfate and nitrate concentrations measured in the majority of the rain samples. Air mass trajectory tracing shows many air masses originate in the Ohio and Mississippi Valley--over one thousand miles away. A study on transport of photochemical oxidants (ozone), completed for the Minnesota Pollution Control Agency (MPCA) by Pacific Environmental Services in 1979, indicates that 61 to 85 percent of the ozone pollutant in Minnesota originates in eastern industrial centers (Pacific Environmental Services, Inc., 1980).

Recently, three state agencies (Department of Health, Pollution Control Agency, and Department of Natural Resources) have been commissioned to do a two-part study on the effect of acid precipitation on Minnesota. The first part is a literature study to determine what monitoring system would be suitable for Minnesota's problem and the second part would be to install monitoring systems through the state to determine the extent of the problem in Minnesota.

CONCLUSIONS

No one really knows what air pollution will cost the environment in the future. The ultimate question is "How much pollution are we willing to put up with?" This question must be addressed by legislative bodies. Legislatures set environmental policies which are rapidly becoming among their costliest expenditures. For example, according to the state Pollution Control Agency, Minnesota has spent over one million dollars to study acid precipitation alone (Minnesota Pollution Control Agency, 1980).

There are still many questions regarding the long-range transport of pollutants which must be answered in order to control pollutants and/or reduce their impacts to Minnesota and other states. Some of the basic questions being addressed in Minnesota are: (Ayres, 1980; Science and Technology Research Office, 1979)

1. Where do the gaseous and suspended particulates that pollute Minnesota originate?
2. Are air pollutant violations single episodes, or are they long-term, due to inadequate controls? Are pollutants accumulated in the air over long periods of time?
3. What are the best methods for evaluating the impacts of these pollutants?
4. What technologies are available for Minnesota and other states to control these pollutants or their effects?
5. What are the most effective actions that could be implemented on a state and regional basis to control the long-range transport of pollutants?
6. What are the meteorological conditions that are associated with air quality and pollutant change?
7. Does the absence of observed effects on agricultural and forestry production genuinely reflect "no damage", or does it reflect our inability to detect damage?

8. What effect does the leaching of heavy metals, due to acid deposition, have on water quality?
9. What are the options available for reducing the effects of acid precipitation?

The effects of acid precipitation and ozone may be reversible if action is taken soon. The identification of the source of the pollutant is necessary in order to reduce the emission at its source. Once the source of the pollutants has been identified, states like Minnesota could ask the EPA to require states and industries responsible to reduce emissions so as to meet federal standards. Seeing the need for these controls, EPA recently identified specific options available to reduce acid precipitation (U.S. Environmental Protection Agency, 1980).

If the EPA's regulations do not work, states like Minnesota could take legal action to require that sources of the pollutants help pay for the clean-up and the restoration of the damaged areas (Berle, 1980). If the battle against long-range transport of air pollutants cannot be won by cooperative actions, then it may have to be settled in the courts.

REFERENCES

American Lung Association (1978). Health Effects of Air Pollution, New York, N.Y.
Ayres, R. (1980). MCAQ study to assess causes, effects of long range particle transport. Air Waves, p. 1.
Babich, H., L. Davis and G. Stotzky (1980). Acid precipitation: causes and consequences. Environment, pp. 6-13.
Barnes, R.A. (1979). The long range transport of air pollution: a review of European experience. J. Air Pollution Association, 1219-1235.
Berle, P.A.A. (1980). U.S. Environmental Protection Agency, Federal control programs with Clean Air Act Authority. Inside EPA, Weekly Report, 1, 28.
Carter, L.J. (1979). Uncontrolled SO_2 emissions bring acid rain. Science, 204, 1179-1182.
Council on Environmental Quality (1979). Environmental Quality 1979. Tenth Annual Report, Washington, D.C.: U.S. Government Printing Office, pp. 17-19.
Cox, R.A. and co-workers (1975). Long-range transport of photochemical ozone in North Western Europe. Nature, 255, 118-121.
DeRoos, R. (1980). Statement to Minnesota Department of Health, Minneapolis, Minnesota.
Fisher, B. (1975). The long range transport of sulphur dioxide. Atmospheric Environment, 9, pp. 1063-1070.
Glass, N.R., G.E. Glass and P.J. Rennie (1979). Effects of acid precipitation. Environ. Sci. Technol., 13, 1350-1355.
Hirsch, K.L. and Abramovitz (1979). Cleaning the air: some legal aspects of interstate air pollution problems. Duquesne Law Review, 18, 53-102.
Krupa, S.V. (1980). Chemical Analysis of Rain in Central Minnesota. Paper prepared at the University of Minnesota, Department of Plant Pathology.
Likens, G.E. and co-workers (1979). Acid rain. Scientific American, pp. 42-51.
McAvoy (1980). Statement before the House of Representatives Subcommittee on Oversight and Investigation of the Committee on Interstate and Foreign Commerce, Ohio Environmental Protection Agency.
Minnesota Pollution Control Agency (1980). St. Paul, Minnesota.
Organization for Economic Cooperation and Development (1977). The OECD Program on Long Range Transport of Air Pollutants: Summary Report.
Pacific Environmental Services, Inc. (1980). Ozone.
Park, D.H. and co-workers (1978). Meteorology of long-range transport. Atmospheric Environment, 12, 425-444.
Ritchie, I. (1979). Statement to Department of Environmental Health, University of Minnesota.

Science and Technology Research Office (1979). Acid Precipitation Research. Inquiry Response No. 95, p. 6.
Silverstein, M.A. (1979). Interstate air pollution: unresolved issues. Harvard Environmental Law Review, 3, 291-297.
United States-Canada Research Consultation Group (1979). The LRTAP Problem in North America: A Preliminary Review.
U.S. Department of Health, Education and Welfare (1976). Human Health and the Environment, Some Research Needs. Washington, D.C.: U.S. Government Printing Office, pp. 10-14.
U.S. Environmental Protection Agency (1979a). Sulfur Emission Control Technology and Waste Management. Washington, D.C.: U.S. Government Printing Office, pp. 7-10.
U.S. Environmental Protection Agency (1979b). Research Summary: Acid Rain. Washington, D.C.: U.S. Government Printing Office, p. 2-4.
U.S. Environmental Protection Agency (1980). Federal control programs with Clean Air Act Authority. Inside EPA, Weekly Report, 1, 10-12.
Van Nostrand's Scientific Encyclopedia (1976). Van Nostrand Reinhold Co., New York, 5th ed.,1807, 1809-1910.
Varshney, C.K. and L.S. Dochinger (1979). Acid rain: an emerging environmental problem. Current Science, 48, 338-340.
West, S. (1980). Acid from heaven. Science, 117, 70-78.

3
AIR POLLUTION CONTROL AT A TAR SANDS PLANT

A. Kumar

Environmental Affairs, Syncrude Canada Ltd., 10030, 107 Street, Edmonton, Canada

ABSTRACT

The air pollution control at a tar sands processing plant is described. Using a generalized oil sands processing sequence, the requirements for air pollution control facilities are identified. A detailed discussion on the use of air pollution control facilities, ambient air quality monitoring, emission monitoring and air quality research is included with special reference to the world's largest tar sands plant, that of Syncrude Canada Ltd. Syncrude's air quality management program's designed to ensure compliance with the law and to fulfill its committment to minimize the effects of such a large development on air, land, water and people.

KEYWORDS

Air pollution control; ambient air quality monitoring; emission monitoring; air quality management; oil sands processing; tar sands.

INTRODUCTION

During the development of any new energy facility in Alberta, a number of stringent government regulations have to be complied with, for the disposing of the waste produced during various processes. Tar sands processing plants are no exception. So far these projects have been the subject of an intensive environmental impact assessment and reviewed by the Energy Resources and Conservation Board and by the Alberta Environment. (Syncrude, 1973, 1978; G.C.O.S., 1967, 1978; Alsands, 1978; E.R.C.B., 1974, 1975). During the planning phase, potential impacts were identified, and control features and mitigation measures were developed. This paper deals with the air pollution control at a tar sand processing plant with special reference to Syncrude Canada Ltd.'s $2.2 billion plant.

Using a generalized oil sands processing sequence, the requirements for air pollution control facilities are identified. The overall air quality management program at Syncrude may be grouped into three categories:

(i) use of air pollution control facilities for pollutants, such as SO_2, H_2S, nitrogen oxides, hydrocarbons and particulates;
(ii) ambient air quality monitoring and emission monitoring; and

(iii) air quality research.

A detailed discussion on Syncrude's efforts in each category is given in the following sections.

TAR SANDS EXTRACTION

Alberta tar sand is composed of sand, heavy oil (bitumen), mineral-rich clays and water. Sand makes up about 84% of the tar sand by weight and consists predominantl of quartz; heavy oil averages about 11% by weight, with clays and water contributin the remaining 5%. There are two basic approaches to recovering bitumen from the ta sand: (i) open-pit mining - in this technique the tar sand is mined and transporte to a processing plant where the bitumen is extracted and the sand, in the form of tailings, is discharged, and (ii) *in situ* recovery - the separation of bitumen from sand is carried out in place; techniques used are similar to those used in the seco ary recovery of conventional oil.

Open-pit mining techniques are currently being used by the two operating plants (Suncor Inc. and Syncrude Canada Ltd.) in the area, producing approximately 150,000 barrels of oil per day. A third plant (Alsands Project Group) is in the proposal/ approval stage and will use the same technique. Amoco Canada Ltd. is running a pilot plant based on *in situ* thermal recovery technology.

A generalized processing sequence of a tar sands plant is given in Fig. 1. As discussed above, bitumen may be recovered by using either an open-pit mining technique or an *in situ* recovery technique.

The hot water extraction process is the most commonly used process for extraction of bitumen from the tar sands, if an open-pit mining scheme is employed. The wet bitumen froth produced is further subjected to thermal or electric dehydration or centrifuging in order to obtain hydrocarbon bitumen. In order to obtain marketable synthetic crude oil, hot bitumen is upgraded. The upgrading schemes are generally based on thermal or catalytic cracking of the bitumen; this is followed by a partia conversion of the heavy residual fraction of the bitumen to lower-boiling hydrocarbons.

The off-gases from upgrading units passes through a sweetening system for the recovery of sulphur content and for production of a synthetic gas. Normally an amine cascade, followed by a Claus unit(s) is used as the sweetening system. Synthetic gas from the amine plant may be used as a fuel for utility boilers.

The Claus plant treats H_2S/CO_2 acid gas and produces elemental sulphur. Residual gases from the Claus plant are either oxidized to sulphur dioxide, or the sulphur is removed by using a tail gas clean up unit or by applying a flue gas desulphurization technique. Before the effluents are released to the atmosphere, electrostat precipitators may be used to remove ash material and fine coke, to meet air quality standards for particulate matter.

Selection of a particular air pollution control technology will depend on the technical aspects, economic considerations, public concerns and regulatory requirements for the plant. Addition of such a unit to an existing plant requires detailed engineering studies as well as studies of the economics/effectiveness for the reductior of a pollutant. For example, let us consider the case of the addition of a tail ga clean up unit, on a Syncrude-size tar sands plant, designed to improve the sulphur recovery efficiency of the Claus plant from 95% to an overall efficiency of 98%. The capital cost for such a modification can easily run in the range of $5 million and the operating cost per year to run such a unit may be around $2 million.

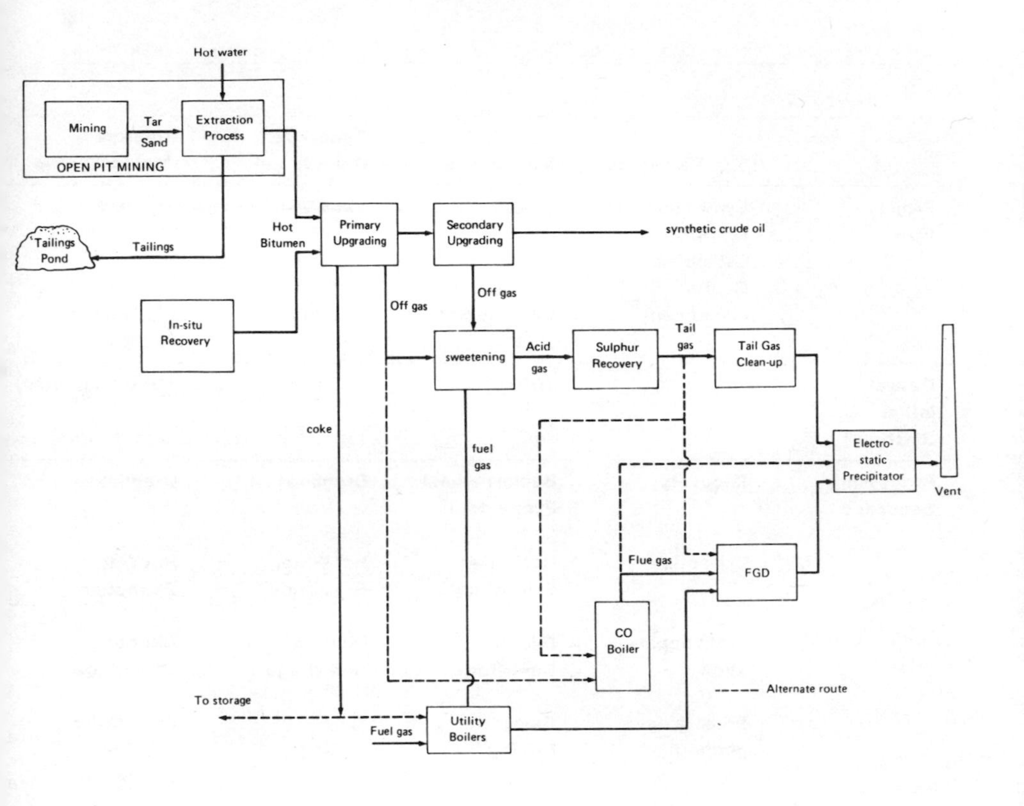

Fig. 1. Generalized processing sequence of a tar sands plant.

Note that this addition may reduce sulphur emissions by only 15%. Thus the question of using air pollution control devices is as "complex" as recoveing bitumen from the tar samds, particularly if retrofit is a consideration.

COMPARISON OF COMMERCIAL TAR SANDS PLANTS

A comparison of the tree commercial plants mentioned above is shown in Table 1. Basically, the same extraction process is used in these plants. However, the processing of the diluted bitumen at Suncor is carried out using delayed coking (the diluted bitumen is heated in furnaces and cracked in large vessles to separate the lighter oils from the coke residue) while at Syncrude, thermal cracking of diluted bitumen takes place in fluid cokers. (Alsands selected fluid coking with the possibility of phasing to coke gasification five to eight years after start-up, assuming satisfactory development of this technology). Note that the efficiency of sulphur recovery for each plant is gradually increasing; this is due to advances in best practicable technology for sulphur extraction. Further, Alsands proposes the use of two units of SCOT process in combination with the Claus plant to achieve an overall sulphur removal efficiency of 99.9%.

TABLE 1 Comparison of Commercial Tar Sands Plant in Athabasca Tar Sands Area.

	Variables	Suncor Ltd.	Syncrude Canada Ltd.	Alsands Project Group
Project Size	Synthetic Crude Oil (BPD)	65000	125000	140000
	Estimated Capital Investment	$230 million (1963)	$2 billion (1975)	$6.7 billion (1980)
Date of Initial Start-up		1967	1978	1986 (expected)
Processing Sequence	Recovery	Bucket-wheel Excavators	Draglines	Draglines
	Extraction	Hot Water Extraction	Hot Water Extraction	Hot Water Extraction
	Froth Separ-ation	Dilution Centrifuge	Dilution Centrifuge	Dilution Centrifuge
	Primary up-grading	Delayed Coking	Fluid Coking	Fluid Coking (+coke gasification?)
	Secondary upgrading	Hydrodesulphur-ization	Hydrodesulphur-ization	Partial Oxidation (POX)
	Sweetening	Amine Treatment	Amine Treatment	Amine Cascade
	Sulphur Recovery	2-train Claus 2-stages 93-95% eff. (3rd stage to be added in 1981 - 96% eff.)	2-train Claus 3-stages 95% eff. (min)	Claus 3-stages (2-train)
Air Pollution Control Equipment	Claus Tail-gas Cleanup	None	None	Two units of SCOT process (99.9% overall eff
	Boiler Flue-gas cleanup	Electrostatic Precipitators (added in 1979)	Electrostatic Precipitators	None

The other major visible difference, from an air pollution point of view, between the Suncor plant and the Syncrude plant is the number of stacks used for releasing pollutants. Suncor uses two main stacks (utility plant stack and incinerator stack) for venting SO_2 emissions while Syncrude combines the waste gases from the two CO boilers and the two tail gas incinerators for the two sulphur plants. Alsands is planning to use one stack.

Suncor uses coke (byproduct from delayed coking) for their utility plant while Syncrude uses fuel gas (by-product from the amine treating unit) supplemented by natural gas. The Alsands plant will use natural gas and fuel gas.

AIR QUALITY PROTECTION FACILITIES AT SYNCRUDE

Potential air contaminants generated during the processing of oil sands include sulphur, particulates, hydrocarbons, carbon monoxide and normal products of combustion. The air pollution control facilities are designed and installed at the plant to meet the regulatory requirements. Measures for emergency situations are also incorporated in the overall design.

Sulphur Emission Control

A simplified flow diagram of sulphur recovery facilities is shown in Fig. 2. Syncrude is allowed to emit up to 287 long tons of SO_2 per day as long as Alberta's Clean Air Regulation is not violated. The regulation stipulates that the average ground level concentration must not exceed 0.20 ppm. over any half hour period.

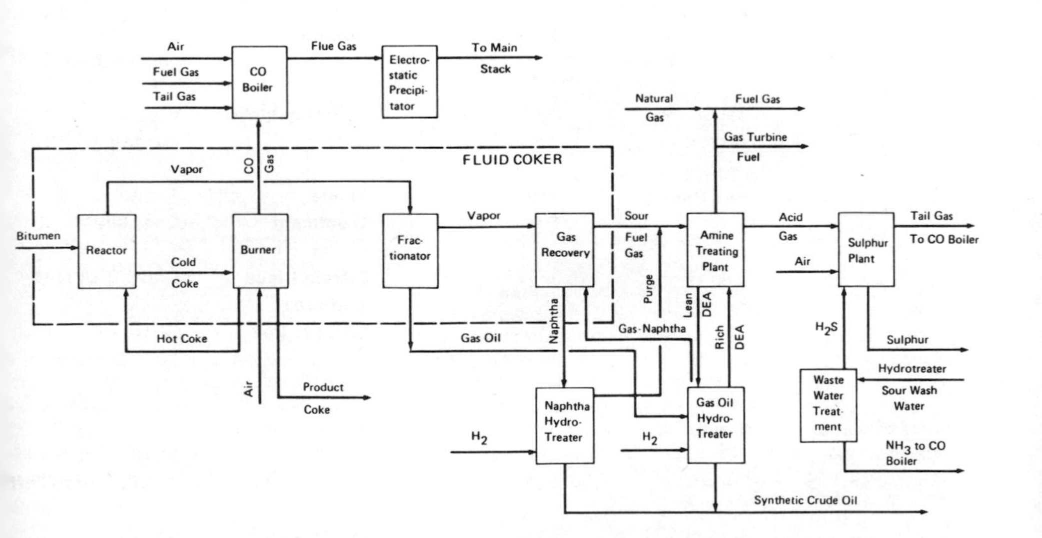

Fig. 2. Sulphur removal and recovery facilities at Syncrude.

Tar sand bitumen contains approximately 4.8% sulphur. During the upgrading process, sulphur is removed in elemental, fixed and gaseous form. Details on Syncrude's sulphur balance are given in Fig. 3. It is interesting to note that only 143 T/day of sulphur from a total input of 1321 T/day is emitted to the atmosphere under normal operating conditions.

Claus Sulphur Recovery Plants. The sulphur unit is a conventional Claus plant consisting of a thermal conversion and a series of three catalytic reactors. Sulphur condensation and gas reheating are provided to yield optimum sulphur recovery. Acid gases from both the amine plants and the wastewater treatment plant flow through knockout drums prior to entering the sulphur recovery plants. One-third of the H_2S in the acid gas feed is burned to form SO_2 in a reaction furnace. For optimum sulphur recovery, air supply to the furnaces is controlled to maintain a 2:1 ratio of hydrogen sulphide to SO_2. The reactions involved in the combustion of H_2S to SO_2 are highly exothermic. The hot gases produced are used to generate 600 psig steam in the waste heat boiler. The H_2S and SO_2 react to produce sulphur, converting abou two-thirds of the total sulphur initially entering the recovery unit. Effluent gas from the boiler is further cooled by a condenser. The remaining gas is reheated, catalytically reacted, and cooled to remove additional sulphur. This process is repeated three times, once in each of the three reactors operating in series.

The molten sulphur flows by gravity into hydrostatistically sealed seal-pots which then overflow into a sulphur pit. Liquid sulphur is then pumped from the pit to storage blocks.

The units are designed to effect 98% recovery of sulphur. Under present start-up operating conditions, sulphur recovery efficiencies of 97.5% have been achieved.

Amine Treating Plant. Sour fuel gas from the gas recovery unit and sour hydrogen purge gas from the naphtha hydrotreater is routed to amine treating for removal of hydrogen sulphide (see Fig.3.) The sour gases are water washed to remove traces of NH_3 and HCN, then contacted with lean DEA (diethanolamine) which absorbs the acid gas. The hydrogen sulphide contained in the rich DEA solution is steam stripped in an amine regenerator to restore the DEA solution to a "lean" condition. The sweetened gas is water washed for removal of entrained DEA and is used as fuel (along with natural gas) in the utility boiler and various process heaters. The amine treating unit also supplies lean diethanolamine to the amine contactors in the gas oil hydrotreater unit. The sweet hydrogen purge gas from these units is added to the fuel gas system.

Sour-Water Strippers. A single sour water treating plant is used to strip out H_2S and NH_3 from the sour water produced in the naphtha and gas oil hydrotreaters. The H_2S produced is fed to two sulphur plants for conversion of sulphur and the NH_3 is incinerated in the CO boilers. The wastewater treating unit is essentially a two-column distillation process. Note that H_2S and NH_3 are removed using reboiled stripping operation. The stripper overhead H_2S is used as sulphur plant feed and the NH_3 leaving the system for incineration in the CO boilers is relatively free of H_2S (see Fig. 2). Stripped water from the bottom of the NH_3 stripper is cooled and utilized as wash water in the hydrotreaters and the amine plant. In order to avoid a build up of impurities in the sour water/stripped water system, some of the stripped water circulation is continuously bled to the API separator and replaced by deaerated boiler feed water.

Dispersion of SO_2 emissions. A single 183 m high stack is designed to disperse the waste gases from the two CO boilers and the two tail gas incinerators. The purpose of combining all these emissions is to better disperse the plume by achieving higher plume rise.

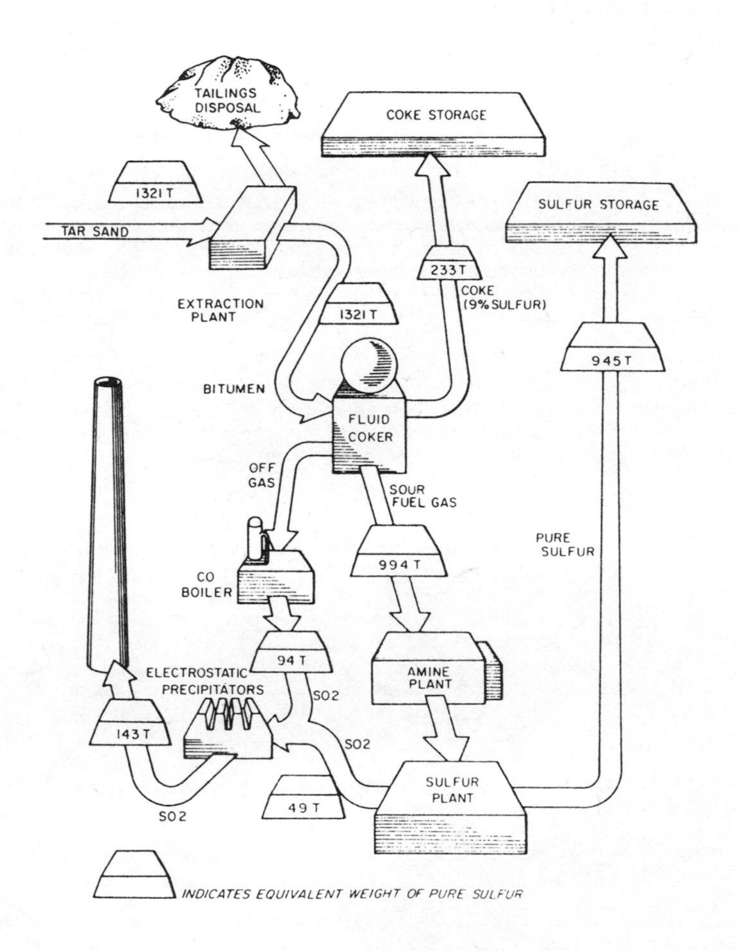

Fig. 3. Syncrude's sulphur balance (daily tonnages).

WTUZ - D

The stack has been designed by using the air quality guidelines issued by Alberta Environment. Syncrude used 0.04 ppm as the maximum ground level concentration for one-half hour for the basis of calculations (this value is less than the specified concentration of 0.06 ppm in Syncrude's permit to construct the plant). A later study (Murray and Kurtz, 1976 a) has confirmed that the stack height should be adequate for dispersing pollutants at the expected emission rates.

Control of Other Plant Emissions

Nitrogen oxides emissions. Nitrogen oxides are produced from elemental oxygen and nitrogen during high temperature combustion of hydrocarbon fuels in fired heate There are no facilities for direct control of these emissions because the predicted maximum ground level concentrations (0.004 ppm to 0.017 ppm) from such emissions ar well below the ambient air quality standards of Alberta (see Table 5.7-3 in Syncru 1978). Approximately 87% by volume of these emissions are diverted to the main sta while the remainder is emitted through various process heaters and the reformer furnace.

Hydrocarbons. The plant is designed such that there is no continuous venting of hydrocarbons. All hydrocarbon stocks having a true vapor pressure greater than 86 kPa are stored in closed vessels or pressure spheres. Floating roof tanks or th equivalant are normally provided for hydrocarbons in the range of 21-86 kPa. Occasionally, some abnormal process conditions may result in the release of hydrocarbon through a relief and blow down system. These hydrocarbons will be burned in a smokeless flare.

Traces of hydrocarbons may also be released to the atmosphere from sewers and ponds The API separator forebay has been provided with a floating cover and skimming facilities to minimize hydrocarbon emission. It is expected that the release of hydrocarbons or other polluting vapors from open water surfaces will be insignificant.

Particulate Emissions. There are two types of pollution control equipment installe

(a) Primary and secondary cyclones - the two stage cyclones remove most of the sand coke and fly ash from the flue gases. The fluid coker burner is equipped with 18 parallel primary cyclones paired in series with 18 secondary cyclones.

(b) Electrostatic precipitators - before gases are discharged to the main stack, ove 95% of the remaining particles are removed from the CO boiler flue gases. The precipitators are made up of four modules, with two modules servicing each of the two CO boilers. The designed particulate levels in the stack gases are less than 0.2 kg of particulates per 1000 kg of exit stack gas.

Dust. Adequate filters and dust handling equipment have been installed at various locations in the plant where chemical dust is considered a potential hazard.

Smoke. Smoke emissions are controlled by proper combustion control of boilers and by installing "smokeless" flare stack tips to ensure complete combustion.

Odours. Localized odours may arise from storage tanks, separators, safety valve discharges, valve packings, etc. If strong or offensive odours are detected, odour surveys are carried out as required by Alberta Environment and appropriate actions are taken to correct the problem.

Fog. Low temperature fog may result near the Syncrude plant from water vapours fro sources such as Mildred Lake, the tailings pond, the cooling towers, the tumbler screen, the extraction plant vents and the hot tailings stream. During the summer months, water vapour from various sources of the plant generally will evaporate quickly. On the other hand, during winter the atmosphere typically can absorb only about three percent of the moisture that it can absorb during summer months. Thus, the possibility of local fog episodes at ground level during cold weather cannot be ruled out. Such situations may result in restricted visibility (intensity will depend on weather conditions and amount of water vapour) on some roads surrounding the plant.

Regional ice fog (at low temperatures -40°C) has been recorded for 4 - 5 days per year near Fort McMurray Airport (Lamb and Croft, 1977). Ice fog layers are usually 10 to 30 meters thick. The plume from the main stack is not likely to mix with the ice fog (Murray and Kurtz, 1976 b), as the plume would be embedded in the stable air above the fog layer. Western Research and Development Ltd. (1976) conclude in a report prepared for Alberta Environment that the Syncrude plume "should have a negligible effect on ground level air quality" during any ice fog episodes.

Control of Emergency and Transitory Conditions

The following precautions have been taken to handle emergency and transitory conditions at the plant:

(i) the plant has been designed with two separate closed emergency relief and blow down systems - (a) a system for normal hydrocarbon relief and (b) a system for relief of streams with a high H_2S concentration.

(ii) a 83 m tall bypass stack has been designed to handle CO gas from coker burner, in the event of a failure of a CO boiler.

(iii) in the case of sulphur plant failure, immediate relief would be to the H_2S flare. Design provisions have been made to divert the acid gas to a CO boiler for incineration.

(iv) if a compressor or power source fails in the hydrogen plants, the product will vent to the atmosphere through a hydrogen vent muffler system.

(v) the waste water treatment plant is designed to operate in a single stage stripping mode if the H_2S stripper is out of service for an extended period. For short term shutdown, (6 days), a feed inventory tank has been incorporated in the plant design.

AIR QUALITY MONITORING

Syncrude's air qulity monitoring consists primarily of the following interrelated activities: monitoring of baseline meteorological conditions, ambient air quality monitoring (consists of continuous monitoring network, static air quality monitoring network, sulphur dustfall stations and biological air quality monitoring network), and source emission monitoring.

Syncrude has established five continuous ambient air monitoring stations (see Fig.4) within 15 km of the main stack in order to measure wind direction and speed, sulphur dioxide, hydrogen sulphide and oxides of nitrogen. All data, at one minute intervals, are transmitted to a central station computer for recording and/or display.

In addition to the continuous monitors, a network of 40 stations ("candles") was established between Fort McMurray and Fort MacKay, to measure total sulphation and H_2S sulphation, and four sulphur dust fall stations will be installed at locations around the sulphur storage area.

Also, a biological air quality monitoring network consisting of a system of 56 permanent plots laid out in a symmetrical radiating pattern from the Syncrude plant site has been established. This system utilizes two lichen species *Parmelia sulcata* and *Hypogymnia physodes*, as indicators of vegetative responses to changes in air quality.

Emissions from the plant are determined regularly, and the following instruments have been installed at the 92 m level of the main stack: for particulates, Lear Siegler Model RM41 Visible Emission Monitoring System; for emission rate, volume flow rate and exit temperature, Western Research and Development CSEM System for SO_2. A block diagram for the stack gas monitoring system is given in Fig. 5. SO_2 concentrations are measured continuously and all the parameters are displayed in the main control room and the operations laboratory.

AIR QUALITY RESEARCH

Early research efforts by Syncrude were directed towards satisfying regulatory requirements, collection of baseline information and answering public concerns (e.g. is there any potential for killer fogs due to Syncrude emissions?) Examples of programs are:

Dispersion Climatology Study.

The objectives of this study were to determine the frequency of occurence of various atmospheric despersions conditions from mini-sonde data and surface meteorological observations, and to predict the ground level concentration patterns which are associated with the observed dispersion conditions as a result of the emissions from the main stack. This study concluded that violations of the Alberta Clean Air Act (SO_2GLC's) would occur approximately 1.4% of the time at the maximum designed SO_2 emission rate (336.3 gram/sec) (Murray and Kurtz, 1976 a). These violations are mainly associated with limited mixing conditions of the atmosphere.

Fog Study

This study was designed to determine the potential for ice and low temperature fogs near the Syncrude plant. The incidence of winter fogs is expected to increase from the natural level due to the water vapour emitted by plant operations. There is no possibility of a "killer fog" occuring since, under the very stable atmospheric conditions associated with fog, the plume is unlikely to mix with the fog (Murray and Kurtz, 1976 b); the plume will remain above the fog layer (<100 meters).

Wind Profile Study

This study was carried out to establish wind climatology near the plant. The winds considered were restricted to those from 100 m to 600 m elevation. A power law was chosen to represent the winds. The value of the exponent of the power law varied most with the atmospheric stability and time of day, while variations due to season, wind direction and reference height were relatively unimportant. Typical values of the exponent were 0.11 for limited mixing, 0.18 for an unstable atmosphere, and 0.[illegible] for stable air using a reference height of 183 m. For dispersion calculations, the wind speed may be considered as constant in the case of limited mixing conditions. The surface winds had minimum values during mid-morning and maximum values at the time of maximum heating. Low level jets were shown to exist, most often from 200 m to 500 m. Such jets may have considerable effect on transportation and dispersion of pollutants. (M.E.P. Company, 1977)

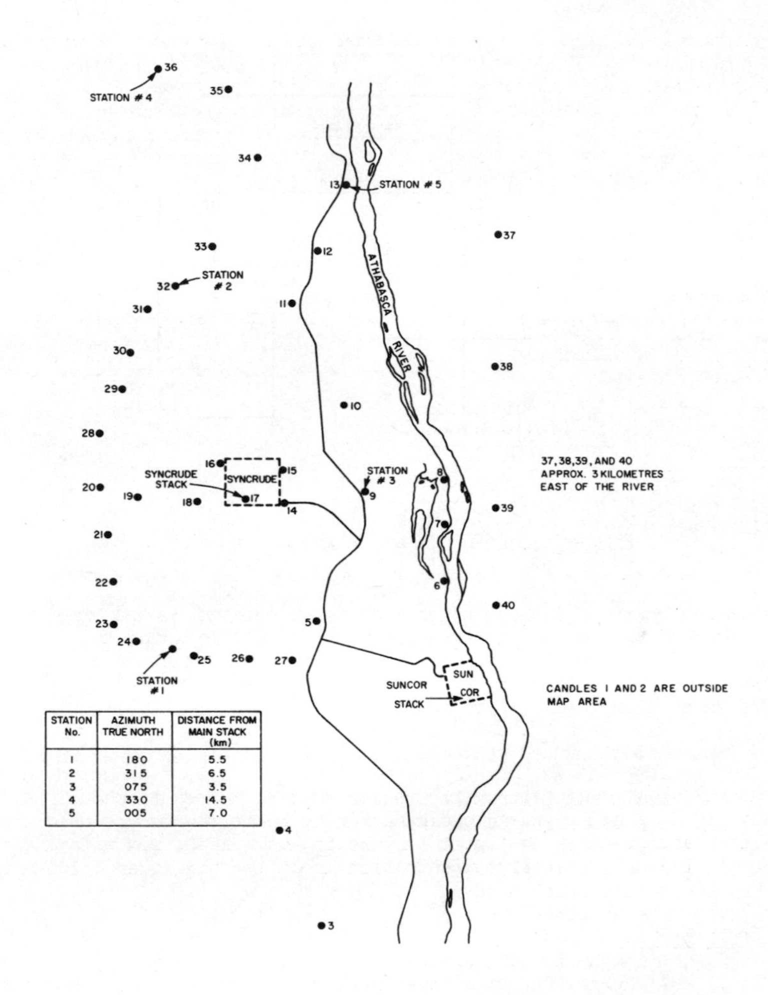

STATION No.	AZIMUTH TRUE NORTH	DISTANCE FROM MAIN STACK (km)
1	180	5.5
2	315	6.5
3	075	3.5
4	330	14.5
5	005	7.0

Fig. 4. Ambient air monitoring network and static air quality monitoring stations (Syncrude Canada Ltd.).

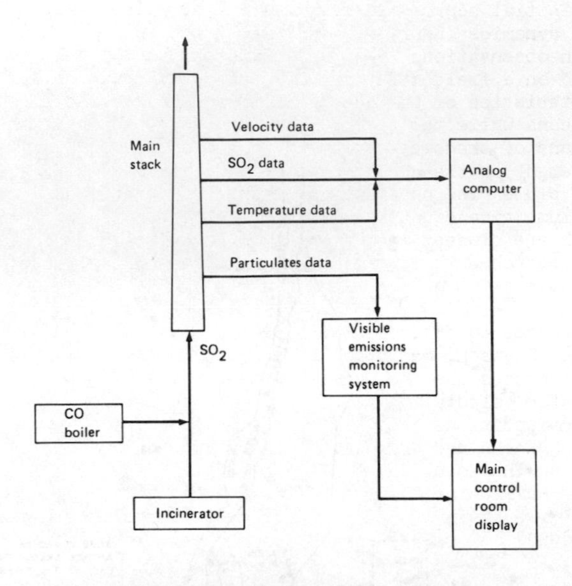

Fig. 5. Continuous stack gas monitoring system.

The research work carried out since 1977 is orientated towards improving the understanding of dispersion characteristics in the area, refining the models to compute ground level concentration, installation of new instruments such as an acoustic sounder for improving the input data for concentration modelling, and the preparation of an environmental impact assessment.

The various topics covered are:

. Pollutant Dispersion Studies
  - plume rise / plume dynamics
  - plume spread rates (σ_y and σ_z)
  - Ground level concentration
  - Turbulence study
  - Wind tunnel work

. Atmospheric Boundary Layer Studies
  - Thermal boundary layer development
  - Modelling of velocity and temperature fields
  - "Pollutant-prone" months

. Deposition of pollutants and long-range transport
. Advance air quality warning system
. Alternative emission control technologies for the Syncrude plant
. Effects of atmospheric pollutants on soil, water, vegetation, wildlife and humans

Developments in the area of plume rise modelling include: the extension of a no wind shear case to a wind shear case (Djurfors and Netterville, 1978), the exclusion of Boussinesq approximation to solve plume rise equations (Kumar, 1978 a) (to account for high initial source temperature of Syncrude emissions∿200°C), the study of vertical plume dynamics (Djurfors, 1977 a), and a comparison of various plume rise formulae with observations (Slawson, Davidson, and Muddukuri, 1980). Observations were based on a field study carried out by Syncrude in 1977 to study the dispersion characteristics of the Suncor plume; plume rise measurements were made by photographic means while the measurements of the geometry of the plume were carried out by means of transects using air-borne instruments. There is, as yet, no unanimous agreement on the use of a set of analytical/empirical formulae for computing maximum plume rise under all atmospheric conditions. However, a numerical solution of the plume rise equations with no Boussinesq approximation seems to provide results which are closest to the observed data (Slawson, Davidson, and Muddukuri, 1980). Current research work is to investigate the penetration of plumes through elevated inversions.

Attempts were also made to develop a turbulent diffusion typing scheme using aircraft measurements (Slawson, Davidson, and Muddukuri, 1980; Kumar, 1979 a). To date, very little success has been achieved on the use of a particular set of sigma curves applicable for calculating ground level concentrations under sub-arctic conditions. Direct computations of σ_y and σ_z (as a function of stability, source height, surface roughness and downwind distance) were also attempted using a numerical model based on the solution of convective-diffusion equation (Kumar, 1979 b). Efforts are now being made at Syncrude to use an acoustic sounder for computing sigmas (on a real time basis) based on atmospheric turbulence measurement (Balser and Netterville, 1980). Some new results are:

- Effects of cross-wind shear on plume dispersion are felt beyond 5 - 10 km for a typically stable plume;
- At large downwind distances (≃100 km) the stable plume disperses more than the unstable plume (this has also been observed recently by Millan, 1979, using dispersion data from INCO plumes)

A modified Gaussian model has been used to compute ground level concentrations of pollutants with time of day (Kumar and Djurfors, 1977; Kumar and Djurfors, 1978).

The current research trend is towards the development of air pollution frequency distribution models (Netterville, 1979) and extrapolation models based on observed concentration data. Syncrude commissioned a wind tunnel study of plume dispersion at its site (Wilson, 1979). The purpose was to assess the influence of tailings pond dike height on plume dispersion. The study concluded that the failings pond dike should have a negligible effect both on low level sources and on the main stack plume; because the wind remains attached to the downwind side of the tailings pond dike and thus avoiding a large flow separation and high turbulence levels.

Syncrude has developed two models (an analytical model, Kumar and Djurfors, 1977 and a numerical model, Kumar, 1979 c), for predicting diurnal variations of the thermal boundary layer height near our Fort McMurray plant. Both these models were further site-tuned using observed data collected during 1976 to 1979 (Kumar and Djurfors, 1977; Kumar, 1979 c, 1979 d). A predictive scheme for surface heat flux was also incorporated into the models (Kumar, 1978 c, 1979 c, 1980 a). Syncrude has also developed a numerical model for predicting velocity and temperature fields eight hours in advance of real time (Slawson and others, 1980). Future anticipated work in the area of atmospheric boundary layer physics includes the development of a statistical extrapolation model to predict wind and temperature profiles using the information from local, regional and national networks in the tar sand area. It is hoped that as a result of these studies a realistic composite forecast (numerical model plus statistical/extrapolation model) can be made for specifying velocity field to compute ground level concentrations of pollutants.

A study of "pollution-prone" months was also carried out using the data collected during the atmospheric boundary layer work (Kumar, 1980 b). Results indicate that (i) "Pollution-prone" months due to the emissions from ground based sources are November, December, January, and February, and (ii) "Pollution-prone" months due to emissions from the main stack are likely to be from March to September.

The question of deposition of pollutants in the tar sands area has also been studied since acid precipitation and heavy metals, which might be subject to long-range transport, may have potentially adverse effects on lakes in northeastern Alberta and northwestern Saskatchewan. A number of dispersion related projects are currently underway in co-operation with Alberta Oil Sands Environmental Research Program (AOSERP) and Oil Sands Environmental Study Group (OSESG). Research efforts are directed in three sectors: (i) a literature review of the deposition of gases and particulates by dry deposition precipitation scavenging (Denison, McMahon and Kramer, 1979), (ii) development of a computer model for the study of long range transport of pollutants and (iii) a study of the SO_2 assimilation capacity of the ecosystems in the tar sands area.

Syncrude has not announced any plans for the development of an advance air quality warning system. However, the research directions taken to date strongly indicate that the basic elements (monitoring network, meteorological data acquisition system and predictive models) for the development of such a system are "now available", and such a system may be a reality in the eighties. Figure 6 shows an inter-relationship of basic elements of an early warning system.

Efforts are made from time to time to study the possible technologies for air pollution control at the Syncrude plant (Djurfors, 1977 b; Goforth 1977). Options such as fuel switching, flue gas desulphurization, tail gas clean-up, production cut-back and intermittent control are considered. The studies are in progress and the results are not yet available.

A summary of effects of SO_2 concentration on soil, water, vegetation, wildlife and humans is given in a report by Syncrude (1978). It is interesting to note that:

(i) Syncrude emissions might reduce soil sulphur deficiencies in the vicinity of the plant. It is difficult to comment on the potential degree of soil acidification; however, the rate of soil acidification by SO_2 is generally very slow

(ii) It is highly unlikely that any significant alteration of the pH of lakes in the area around the Syncrude project will result from the emission of acid forming substances,

(iii) Low doses of SO_2 on vegetation are generally harmless and in some cases even beneficial since sulphur is a major component of amino acids and protein of plants. However, large doses of SO_2 may cause a disruption of photosynthesis, transpiration and other metabolic processes. The predicted levels of SO_2 from the Syncrude emissions are far below the level at which plants are damaged,

(iv) Any impact of low level SO_2 from Syncrude's plant on wildlife would be subtle, and would not be specific to the wildlife component of local ecosystems,

(v) The literature does not appear to contain reports of any significant health effects of SO_2 below the level of 0.6 parts per million in air.

Further, Syncrude is co-operating with OSESG and the Petroleum Association for the Conservation of the Environment (PACE) in developing research plans to study some of these areas; however, the plans have not yet been announced.

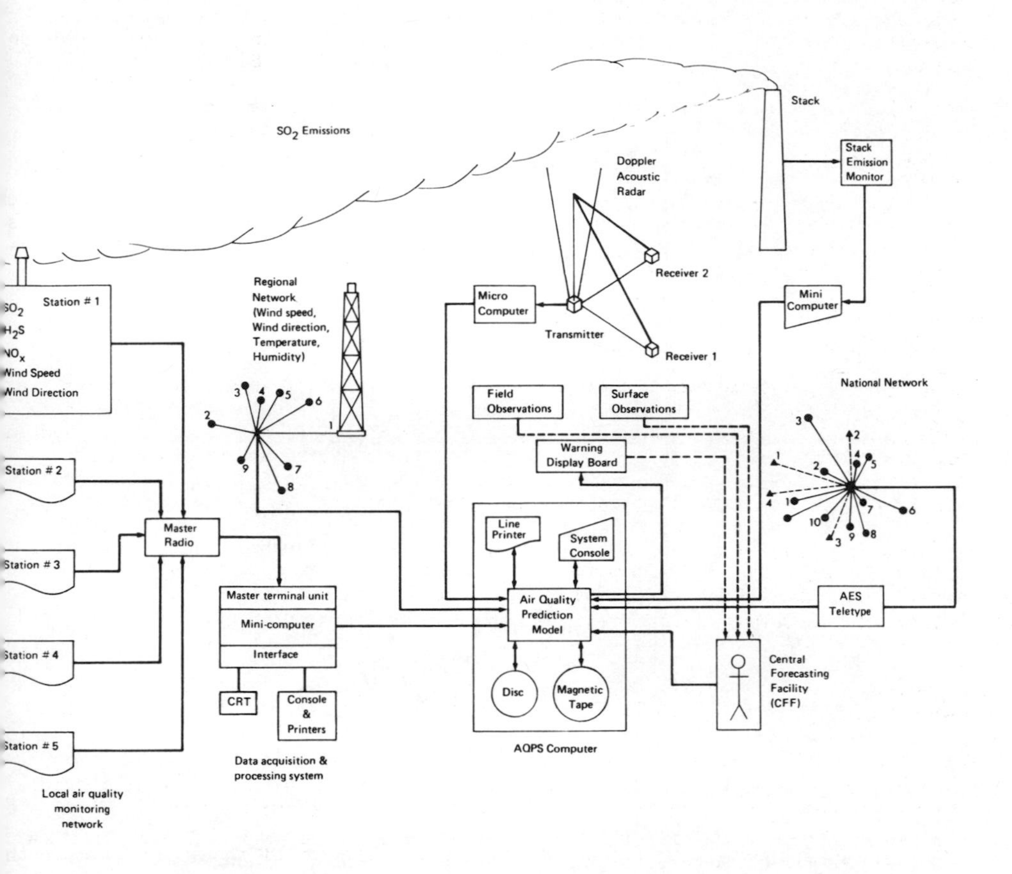

Fig. 6. Inter-relationship of basic elements of an early warning system.

CONCLUDING REMARKS

We have witnessed increasingly stringent envirionmental control being required of successive tar sands plants. For example, only 0.6 LT of SO_2 for every 1000 bbl of synthetic crude oil production may be permitted for the third tar sands plant as compared to the 5.3 LT of SO_2 for every 1000 bbl of synthetic crude oil production which was approved for the first tar sands plant (see Fig. 7). To cope with the regulations, current air quality management programs at tar sands processing plants consists of (i) incorporation of the air quality regulations during planning phase, (ii) installation of air pollution control equipment during operating phase, (iii) monitoring of environmental parameters during operating phase, and (iv) ongoing research on unsolved air quality problems.

Syncrude's air quality management program is designed to ensure compliance with the law and to fulfill its committment to minimize the effects of such a large development on air, land, water and people.

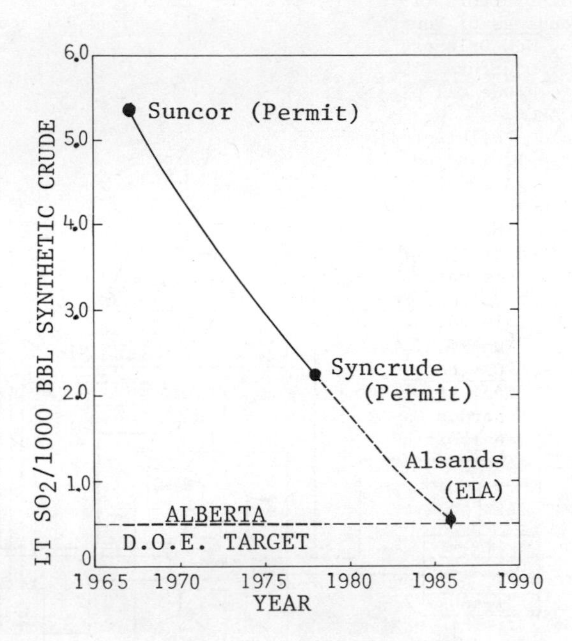

Fig. 7. SO_2 emissions per unit of synthetic crude oil from three tar sands plants

REFERENCES

Alsands Project Group (1978). Environmental Impact Assessment. Calgary, Alberta.

Balser, M. and D. Netterville (1980). Measuring wind turbulence with Doppler - Acoustic radar. Submitted to JAM.

Denison, P.J., T.A. McMahon and J.R. Kramer (1979). Literature review on pollutar deposition processes. Professional Paper 1979-5, Syncrude Canada Ltd. 1-261.

Djurfors, S. (1977 a). On the rise of vertical plumes. Professional Paper 1977-8 Syncrude Canada Ltd. 1-14.

Djurfors, S. (1977 b). Possible technologies for air pollution control at the Syncrude plant. Proceedings of Alberta Sulphur Gas Research Workshop III, Alberta Environment, 85-93.

Djurfors, S. and D. Netterville (1978). Buoyant plume rise in non-uniform wind conditions. APCA Journal, 28(8), 780-784.

Energy Resources Conservation Board (1974). In the matter of an application of Petrofina Canada Ltd., Pacific Petroleum Ltd., Hudson's Bay Oil and Gas Company Ltd., Murphy Oil Company Ltd. and Can Del Oil Ltd., under part 8 of the Oil and Gas Conservation Act. E.R.C.B. Report 74-x.

Energy Resources Conservation Board (1975). In the matter of an application of Hc Oil Company Ltd. and Alminex Ltd. under part 8 of the Oil and Gas Conservation Act. E.R.C.B. Report 1975-H.

G.C.O.S. (1967). Application to construct a bitumen extraction and upgrading plar

G.C.O.S. (1978). Clean Air Act license application.

Goforth, R.R. (1977). Application of best practicable technology in air pollution control for the Syncrude project. Proceedings of the Second Pacific Chemical Engineering Congress (PACHEC 1977), Denver, Col., 1287-1294.

Kumar, A. and S.G. Djurfors (1977). A model to predict violation of clean air regulations. Proceedings of Fourth Joint Conference on sensing of Environmental Pollutants, ACS, New Orleans, 281-288.

Kumar, A. and S.G. Djurfors (1978). Maximum and critical ground level concentrations from a synthetic crude Oil plant. Paper #78-48.5, Presented at the 71st Annual Meeting of the APCA.

Kumar, A. (1978 a). Pollutant dispersion in the planetary boundary layer. Professional Paper 1978-1, Syncrude Canada Ltd., 40-72.

Kumar, A. (1978 b). A simple radiation model for the tar sands area. Professional Paper 1978-3, Syncrude Canada Ltd.

Kumar, A. (1979 a). The effects of cross-wind shear on horizontal dispersion. Paper #79-29.3, Presented at the 72nd Annual Meeting of the APCA, Cincinnati.

Kumar, A. (1979 b). Estimation of atmospheric dispersion coefficients for elevated releases. Preprint Volume, Fourth Symposium on Turbulence, Diffusion and Air Pollution (A.M.S.), 19-26.

Kumar, A. (1979 c). Numerical modelling of the thermal boundary layer near a synthetic crude oil plant. APCA Journal, 29(8), 827-832.

Kumar, A. (1979 d). Evaluation of inter-facial mixing constant for a boundary layer evolution model. Syncrude Canada Ltd., 1-25.

Kumar, A. (1979 e). Analysis of 1978 net radiation data. Syncrude Canada Ltd.

Kumar, A. (1980 a). Analysis of 1977 and 1978 incoming radiation data. Syncrude Canada Ltd.

Kumar, A. (1980 b). Air quality studies in the tar sands development area. Presented at the 72nd Annual Meeting of the APCA, Montreal, Paper #80-59.1.

Lamb, A. and B. Croft (1977). The magnitude of fog occurence and associated problems in the oil sands area. Proceedings of the 2nd Pacific Chemical Engineering Conference (PACHEC 1977), Aug. 23-31, Denver, 1295-1298.

M.E.P. Company (1977). The vertical wind profile at Mildred Lake, Alberta. Syncrude Canada Ltd. Professional Paper 1977-1, 1-61.

Millan, M. (1979). Personal Communication. Barringer Research Ltd.

Murray, W. and J. Kurtz (1976 a). A predictive study of the dispersion of emissions from the Syncrude Mildred Lake plant. Environmental Research Monograph, 1976-1, Syncrude Canada Ltd., 1-128.

Murray, W. and J. Kurtz (1976 b). A study of potential ice fog and low temperature water fog occurrence at Mildred Lake, Alberta. Environmental Research Monograph 1976-4, Syncrude Canada Ltd., 1-63.

Netterville, D.D.J. (1979). Concentration fluctuations in plumes. Environmental Research Monograph 1979-4, Syncrude Canada Ltd.

Slawson, P.R., G.A. Davidson and C.S. Muddukuri (1980). Dispersion modelling of the G.C.O.S. plume. Syncrude Canada Ltd. (Draft copy).

Syncrude Canada Ltd. (1973). Environmental Impact Assessment. Vol. 1 to Vol. 4, Edmonton, Alberta.

Syncrude Canada Ltd. (1978). Environmental Impact Assessment Addendum to the 1973 Report. Vol. A to Vol. C, Edmonton, Alberta.

Western Research and Development Ltd. (1976). Evolution of Pollution Abatement Technology as Applied to the Alberta Oil Sands. Prepared by WRD for Alberta Environment, 1-162.

Wilson, D.J. (1979). Wind tunnel simulation of plume dispersion at Syncrude Mildred Lake site. Environmental Research Monograph 1979-1, Syncrude Canada Ltd.

SECTION II

REUSE OF WASTE MATERIAL

4
THE ROLE OF WASTE EXCHANGES IN RECOVERY OF VALUES

R. G. W. Laughlin

Ontario Research Foundation, Sheridan Park Research Community, Mississauga, Ontario, Canada

ABSTRACT

The role which industrial waste information exchanges can play in finding uses for waste or surplus materials is discussed. Much of the information is based on author's experience as manager of the Canadian Waste Materials Exchange since its inception in 1977. Comparisons are made between the approach taken in Canada to that of other countries, notably the United States. Some accomplishments of the CWME are presented. Present limitations to recovery opportunities are discussed, emphasizing the need for concurrent development of proper disposal facilities.

KEYWORDS

Industrial; waste; reuse; exchange; resource; recovery; disposal.

INTRODUCTION

This paper examines the role of waste materials exchanges in the recovery of values from waste streams. It then discusses the history and establishment of the Canadian Waste Materials Exchange, comparing it to other exchanges throughout the world. Finally, the performance of the Canadian Waste Materials Exchange since its inception in November, 1977, is reviewed in terms of its contribution to waste management in Canada.

CONCEPT

The concept of a waste exchange is predicated on the old adage that "one man's meat is another man's poison", or, as it might be restated today, "one man's garbage is another man's gold". Waste industrial materials may well prove to be a useful feedstock for another company.

In order that companies may consider using a waste material, they must first know of its existence. This is achieved by a waste materials exchange which may be defined as a vehicle by which the availability of waste materials or by-products is made known to potential users. Other less formal definitions might be "industrial flea market" and "industrial bargain hunters' press".

Large companies with many processes and skilled chemical engineers are likely to find numerous recycling opportunities within their own manufacturing facilities. However, even engineers in large national companies are not likely to recognize all waste transfer opportunities outside of their own industry. Thus the concept of spreading the word about the availability of particular waste streams is attractive in that it increases the number of people examining possible uses for the waste.

ROLE

The role of a waste materials exchange is to help return as much of what is now regarded as waste to an alternative industrial use. This may be achieved directly by one industry "buying" waste as a substitute raw material, or it may occur via some intermediary such as a reprocessor or scrap dealer.

The objectives for such an exchange are:

(1) To save valuable raw materials;

(2) To save energy by not having to process raw materials;

(3) To avoid environmental damage:

   (a) in the winning of raw materials and energy

   (b) in the avoidance of having to dispose of the waste.

It would be very naive to imagine that a waste exchange will eliminate all problems of waste disposal. There are many waste materials for which no use is ever likely to be found. A 1976 study (Anon, 1976; Arthur D. Little, Inc., 1976) of waste exchanges for the U.S. EPA concluded that, of all the industrial wastes generated by primary manufacturing industries ($\sim$ 206 million tons/year), 3% of the mass had potential value. While it is important to recognize the limitations on the waste reuse concept as a solution to all our waste disposal problems, it is also important to realize that the 3% figure of 1976 may grow substantially with changes in economic and energy and material supply situations. The economic situation in favour of waste reuse is improving rapidly, because there are three factors all working in favour of reuse. These are:

- Increases in raw materials costs
- Substantial increases in energy costs
- Substantial increases in waste disposal costs.

Thus, the importance of a good information exchange vehicle existing is more and more necessary to ensure that the rapidly increasing potential for waste reuse is actually realized.

Waste exchanges are most likely to be of most benefit to the transfer of wastes from one industry sector to another. For example, the petroleum industry probably already has examined the opportunities for reuse of its own waste materials within its own industry. It is, however, unlikely to be aware of all of the opportunities within, say, the pulp and paper industry for reuse of petroleum industry wastes.

Waste transfers will take place most readily from industries employing large complex continuous flow processes to smaller industries often operating more flexibly with batch processes. Transfers will occur from industries having very high purity requirements to those with slightly less stringent requirements, e.g. waste solvents from the pharmaceutical industry to paint formulators.

There are two distinctly different approaches to the operation of a waste materials exchange:

1. An active brokerage type approach, where the exchange actually buys and sells commodities, operating on a 'for profit' basis.
2. A passive information dissemination approach, where the exchange does not actually deal in anything except information, often operating on a 'not for profit' basis.

There is a need for both activities. The first tends to deal only in marketable wastes, while the second can afford to deal in all wastes trying to establish a marketplace for materials not presently reused. The Canadian Waste Exchange is of the second type, and it is this approach I will be discussing. It is interesting to note, however, that many of our most active participants are actually operating as waste brokers. So thus the information type exchange can, in fact, be a valuable tool for the more active exchange.

HISTORY OF THE CANADIAN WASTE MATERIALS EXCHANGE

The Canadian Waste Materials Exchange was established as the result of a study (Laughlin and Golomb, 1977) undertaken in 1976/77 by the Ontario Research Foundation for Environment Canada. In this study, we reviewed the operation of all the waste materials exchanges existing as of January, 1977. A list of these appears in Table 1, accompanied by a list of the waste exchanges that we have heard about being established since that time.

Many aspects of the operations of the existing exchanges were examined, and detailed descriptions are available in our study report (Laughlin and Golomb, 1977). The most pertinent factors are listed below with the range of values found for the entire 17 exchanges studied:

(a)	Circulation of information	600 - 700,000
(b)	Waste listings per year	5 - 5,000
(c)	Enquiries received per waste listed (average)	0.5 - 5.25
(d)	Proportion of wastes listed which are transferred (percentage of numbers)	10% - 45%

The most successful exchange was found to be that operated by Deutsche Industrie und Handelstag (DIHT). The DIHT has the advantage of a 700,000 circulation, since membership in DIHT is a legal requirement for registered companies in Germany. Since there was no equivalent organization in Canada with compulsory membership, the second most successful exchanges were looked at, as possible models for a Canadian Waste Materials Exchange. These were the U.K. Waste Exchange and the Nordic (Scandinavian) Waste Exchange. Both the U.K. and Nordic Exchanges are being operated by technologically based organizations who are able to offer advice and help on waste reuse and reprocessing, which was felt to be an important aspect of the operation of a waste materials exchange.

All of the exchanges had only been operating for a relatively short time, the oldest being the Dutch VNCI exchange which started in January, 1972. Most include two sections in their exchange literature - an Available and Wanted section - so that companies looking for substitute raw materials can also list with the exchange. Because some of the compositions of waste streams to be listed on the exchange could be either embarrassing to the lister or could indicate what process he is using to his competitors, all but two of the exchanges

TABLE 1 - Waste Materials Exchanges

1. Established Prior to January, 1977

Holland:	Vereniging van de Nederlandse Chemische Industrie (VNCI)
Belgium:	Ecochem Bourse
Switzerland:	Schweizerische Gesellschaft für Chemische Industrie (SGCI)
Italy:	Associazione Nationale dell Industria Chimica (ANIC)
Austria:	Handelskammer des Ober Österreich (HKOO)
West Germany:	Verband der Chemischen Industrie (VCI) Deuscher Industrie-und Handelstag (DIHT)
Scandinavia:	Nordic Exchange
United Kingdom:	National Industrial Materials Recovery Association (NIMRA) U.K. Waste Exchange
France:	Nuisances et Environnement (Technical magazine)
Canada:	Canadian Chemical Processing (Technical magazine) Ontario Ministry of Industry and Tourism
U.S.A.:	St. Louis Regional Commerce and Growth Association Iowa Industrial Waste Information Exchange Houston Chemical Recycle Information Tennessee Dept. of Public Health

2. Established since January, 1977

Australia:	Victoria Waste Exchange
New Zealand:	Dept. of Scientific and Industrial Research
Canada:	Canadian Waste Materials Exchange
U.S.A.:	California Waste Exchange Georgia Waste Exchange Illinois Liquid Waste Haulers Association Minnesota Waste Exchange New Jersey State Waste Exchange Industrial Material Bulletin Erkam Corp, New York Syracuse Waste Exchange (Allied Chemical) Ohio Industrial Waste Information Exchange Ohio Resources Exchange (ORE) Portland Recycling Team Western Environmental Trade Association (Oregon) World Association for Solid Waste Transfer and Exchange (WASTE)

list entries anonymously. Enquiries are sent to the exchange which passes them on to the lister who can decide whether the enquiry is bonafide or not before he makes contact with the enquirer.

The Nordic Exchange also includes a section on disposal and reprocessing or treatment capacity available for waste materials. It was felt that this was a valuable addition to the bulletin to let companies know what reprocessing facilities are available.

As a result of the study, the establishment of a Canadian Waste Materials Exchange was recommended, to serve the whole of Canada. This approach is different to that being taken in the United States, where a number of regional exchanges have been set up. Canada is a large country geographically, but quite small in terms of population. We recognized that wastes are unlikely to transfer from Vancouver in the west to St. Johns, Newfoundland, in the east. Information, however, can readily travel across the country. Taking the regional approach only three or possibly four or five centres across the country would have sufficient industry to warrant the establishment of an exchange. With a national exchange, all regions have access to the service, with very little additional cost. The costs of administration, publication and printing are minimized by this centralized approach. A national waste exchange also has the advantage of being able to establish a higher "profile" in terms of media and public interest.

While recommending the centralization of the mechanics of operating the Exchange, it was also recommended that the help of provincial and regional governments, industry associations, etc., be sought to promote local participation.

In October, 1977, Environment Canada contracted Ontario Research Foundation to operate the Canadian Waste Materials Exchange according to these recommendations for a two-year period, 1978 and 1979.

In late November, a brochure introducing the Exchange was mailed to 41,000 companies across Canada in the S.I.C. Manufacturing sector. This covered all manufacturers with five or more employees. We asked companies to indicate an interest in receiving our bi-monthly bulletin by returning a post-paid reply card. We also solicited wastes for inclusion in our first bulletin.

The first bulletin was mailed out in January, 1978. This was again mailed to the original 41,000 companies, plus to anyone who had indicated that they would like to receive it. It contained a total of 282 entries:

227 wastes available
27 wastes wanted
28 services available

The wastes are listed under 10 categories:

1. Organic Chemicals and Solvents
2. Oils, Fats and Waxes
3. Acids
4. Alkalis
5. Other Inorganic Chemicals
6. Metals and Metal Containing Sludges
7. Plastics
8. Textiles, Leather and Rubber
9. Wood and Paper Products
10. Miscellaneous

A typical page from a more recent bulletin is included as Table 2 to show the listing format. All titles and instructions are in both official languages, French and English. Waste are listed in whichever of the two languages they are submitted. This minimizes the printing and mailing costs for the bulletins. Each entry has: a code number (e.g. A 006 is Available waste No. 006); and a regional identifier (e.g. 0), one of 23 geographic regions shown on a map on the bulletin's front cover. It then shows the quantity available and a brief description of the waste. In some cases, a more detailed geographic location is also identified.

COMPARING THE CANADIAN AND U.S. APPROACHES TO WASTE EXCHANGE

In contrast to the single national exchange programme in Canada, the United States EPA has encouraged the growth of localized regional waste exchanges. There are advantages and disadvantages to both approaches.

The regional approach is in aggregate more expensive to operate because of the duplication of administrative centres. It does not result in as high a profile as would be enjoyed by a national exchange. It may also miss some transfer opportunities because of its limited geographic region. For example, some of our 1,000 mile plus transfers may not have occurred with a regional approach. The biggest advantage is the accessibility of the exchange when it is serving a limited area. Also, the exchange operator can have a much more comprehensive knowledge of the industry within its region than is possible with a nationwide exchange.

The World Association of Solid Waste Transfer and Exchange (WASTE) of California is attempting to set up a computerized link of all existing waste exchanges in the U.S. and elsewhere. They have not been successful in realizing this aim to date.

I do not believe that there is a "right" and a "wrong" approach to waste exchange operation. With the much larger industry base in the U.S., the regional approach is, perhaps, more logical. I think that the national approach and, perhaps more particularly, Environment Canada's funding of the exchange, allowed the CWME to become rapidly established and to make a fairly significant impact in Canada.

PERFORMANCE OF THE CANADIAN WASTE MATERIALS EXCHANGE TO JUNE 1, 1980

We have tried, on a continuing basis, to analyze the performance of the Canadian Waste Materials Exchange to determine the effectiveness of the programme. Table 3 summarizes the operation from January, 1978, to June 1, 1980. We have about 5,000 participants in the programme. While we have never sought foreign participants, we have a number of U.S. companies using the CWME and even a few in Japan, India and the Phillipines.

We have listed just over 500 wastes per year for a total of 1,238. 88% of those wastes were of sufficient interest to generate at least one enquiry. The average number of enquiries per listing is about 4.9. The number of waste transfers we have recorded is 179 or 14.5% of the listings on a numbers basis. Because we do not get involved in the negotiations between two companies considering an exchange, we have to rely on one of the companies letting us know that a transfer has occurred. This is not always done as rapidly as we would like. We continue to list wastes in successive bulletins until they transfer or the listers ask us to remove them from the lists. Thus, it is important that we know about transfers to keep our lists current.

The final two figures in Table 3 are, of necessity, approximations. We have

TABLE 2

WASTES AVAILABLE : DECHETS DISPONIBLES

Use Form 3 to enquire about these wastes if you think you could use them.

Veuillez utiliser la formule 3 pour obtenir des renseignements sur les déchets que vous pourriez utiliser.

1. ORGANIC CHEMICALS AND SOLVENTS : PRODUITS CHIMIQUES ORGANIQUES ET SOLVANTS

	Region/Région	Quantity/Quantité	
A 006	O	- - -	Crude Naphthenic Acid still bottoms (Toronto)
A 029	O	3 ton/week	Organic hatchery waste (Brantford)
A 058	O	60 gal/month	Toluol with silicone paint
A 170	O	5000 lbs/week	Slush from solvent processing
A 201	O	12,000 gal/year	90% Toluol; 10% phenolics (Scarborough)
A 221	O	2000 gal/week	Light solvents, sludge (Toronto)
A 298	R	80,000 gal/week	Molasses Stillage (Montreal)
A 375	U	3000 lbs/year	Activated charcoal and Oils and Phenols
A 415	G	- - -	Mixed rubber cure accelerators
A 416	N	12,000 gal/year	Methylene chloride & water & animal fat lubricant
A 528	O	1000 gals/month	Spent solvent ketones, alcohols, etc. (Hamilton)
A 547	O	100 lbs	U.C.L.-77 silicone (Toronto)
A 594	R	2000 gals/mois	Latex usé (Drummondville)
A 627	N	2 tons/week	1,2 Diamino cyclohexane/Hexamethylene diamine mixture (Maitland)
A 674	O	7000 gal/year	Blue quick dry enamel (Windsor)
A 689	O	11,000 lbs	Calcium acetate powder
A 690	O	500 lbs	Calcium formate powder
A 708	O	20-25,000 U.S.gal/week	Mixed pentanes/pentenes/isobutylenes
A 711	O	4000 lbs	Dow Corning - Antifoam "B" liquid
A 713	O	300 lbs/day	Latex compound and water sludge (Markham)
A 736	O	- - -	Used acetone and oil mixture
A 783	G	2000 gals/month	Fusel Oil (Calgary)
A 788	O	15 x 45 gal/month	Recovered and fresh solvents (not mixed)
A 816	S	3 x 45 gal	Varsol avec graisses
A 820	N	700 gal/month	Contaminated M.E.K.
A 851	N	100 gal/week	Glycol & Mineral oil mixture (trace H_2O, dirt)
A 852	N	4 drums/year	Thermally degraded polyalkylene oxide H.T. fluid
A 856	O	6 ton/month	Wetted paint sludge (alkyd resin & pigment) (Windsor)
A 878	O	1800 gal/year	Insulating varnish
A 918	R	- - -	Obsolete solvent based paints suitable for baking
A 929	O	5000 gal/month	Resin plant wash solvent
A 930	O	1000 gal/month	Water - resin plant solvent mixture
A 931	O	1000 gal/month	Resin and paint plant sludge
A 932	O	1000 gal/month	Defective resins and paints
A 938	N	6250 Kg/month	Filter press cake containing 50% plasticizers & 50% Carbon and filteraid
A 939	N	17,500 Kg/3 months	Distillation residue, phthalic anhydrides and higher acids
A 945	O	3000 lbs.	Trichloroacetic acid contaminated with Fe + H_2O (Sarnia)
A 962	O	2000 lbs.	High heat silicone resin (DC803-50X)
A 966	E	9000 gal/day	Aqueous organic acids (5% acetic 1% formic + trace proprionic)
A 974	O	6500 lbs	Trichlorobenzene dielectric grade, new, pure (London)
A 984	O	/97 gal	Baked on enamel paint
A 989	O	1500 gal/month	Waste solvents, paint, plant sludge (Burlington)
A 991	O	1100 lbs	Oil of sassafras (Fort Erie)
A 992	O	10,000 lbs	Alkali strippable polyester floor finish resin (Fort Erie)

Indicates the start of new listings in each category → **Indique le début des nouvelles inscriptions pour chaque catégorie**

attempted to estimate the annual tonnage of material transferred in those 179 recorded exchanges. Our best estimate is about 152,000 tons. The "replacement" value of that 152,000 tons has been estimated at $4.12 million/year. This figure is computed by estimating the value of the raw material which the waste has replaced. Conservative estimates have been used throughout; for example, for food waste, the value of its energy equivalent at $1/million Btu has been assumed. This replacement value estimate can be readily criticized, and it should be used with extreme caution. It does, however, we believe, give us a reasonable estimate of an order of magnitude estimate of the economic impact of the Canadian Waste Materials Exchange.

TABLE 3

Summary of the Operation of the

Canadian Waste Materials Exchange to June 1, 1980

Number of Participating Companies	5000
Number of Wastes Listed	1238
Number of Wastes Enquired About	1091 (88.1% of listings)
Number of Enquiries	5972 (4.8 per listing)
Number of Wastes Transferred	179 (14.5% of listings)
Annual Tonnage of Wastes Transferred	151,744 tons
Value of Wastes Transferred	$4.12 million/year

In Table 4, data are presented by category. As can be readily seen from this table, the performance of the Exchange varies with different categories. We have been quite successful in encouraging transfers of organic chemicals, plastics and oils. We have been much less successful with acids and inorganic wastes.

The geographic breakdown of listings, enquiries and transfers shown in Table 5 pretty well reflects the distribution of industry across Canada. The enquiries and transfers are skewed in favour of Ontario, which may be explained by one of two reasons: either (a) the Exchange is located in Ontario, therefore it is easier for Ontario industry to participate; or (b) the concentration of industry in Ontario increases the opportunities for transfers.

In Table 6, the distances over which wastes transferred are summarized. While a large proportion travelled quite short distances, some travelled over 1,000 miles. This range of distances recorded for waste transfers further strengthens the original recommendation for a national Canada-wide exchange.

TABLE 4 - Analysis of Enquiries and Transfers of Available and Wanted Wastes in Bulletins 1 - 14 by Categories

Category	No. of Wastes (A)	No. of Wastes Enquired About (B)	No. of Enquiries (C)	Enquiries Per Listing (C)/(A)	Transfers (D)	Transfers Per Listing (D)/(A)
1. Organic Chemicals and Solvents	111	92	528	4.8	35	0.32
2. Oils, Fats and Waxes	68	62	375	5.5	16	0.24
3. Acids	40	37	167	4.2	3	0.07
4. Alkalis	52	47	233	4.5	6	0.12
5. Other Inorganic Chemicals	108	91	386	3.6	5	0.05
6. Metals & Metal Containing Sludges	153	139	866	5.7	19	0.12
7. Plastics	95	87	691	7.3	19	0.20
8. Textiles, Leather and Rubber	191	169	853	4.5	20	0.10
9. Wood & Paper Products	288	257	1378	4.8	36	0.12
10. Miscellaneous	132	110	495	3.8	20	0.15
	1238	1091	5972	4.9	179	0.145

TABLE 5 - Geographical Breakdown of Listings, Enquiries and Transfers: Bulletins 1 - 14

	Listings		Enquiries		Transfers	
		%		%		%
West	221	17.9	804	13.5	27	15.1
Ontario	632	51.0	3550	59.4	103	57.5
Quebec	340	27.5	1458	24.4	42	23.5
East	45	3.6	160	2.7	7	3.9
	1238	100.0	5972	100.0	179	100.0

TABLE 6 - Analysis of Distance Travelled by Wastes Transferred

		% of Those Stating Distance
< 50 miles	64	48.1
50 - 100 miles	17	12.8
100 - 200 miles	20	15.0
200 - 500 miles	13	9.8
500 - 1000 miles	7	5.3
> 1000 miles	12	9.0
Did not state	46	-
	179	100

SUCCESSES, FAILURES AND FUTURE PLANS

Vhen the Canadian Waste Materials Exchange programme was launched, we had hoped :o attract 5,000 participants listing about 500 wastes a year and achieve a .5-20% transfer rate for those waste listings. As discussed previously, we ave achieved the first two goals and came close with a transfer rate of 14.5%. ntario Research has established a national exchange which enjoys industrial ecognition across the country. Environment Canada have received praise for heir continued sponsorship of the programme from both industry and environmental-st groups.

n the negative side, we had intended to try to make the project self-financing rom 1980 onwards. This has not been achieved to date. We have presently enerated about 40% of our required operating budget for 1980, by participants ontributing $20 per year, and listers in the Services Available section paying 80 per year for their listings. The remaining 60% will be provided by nvironment Canada, the Province of Ontario, and possibly by some of the other rovinces. This is an excellent example of Federal/Provincial/Industry co-peration to tackle a problem affecting all of us.

e have been less successful than we had hoped at finding uses for some of the roblem wastes like contaminated acids, or dilute alkalis. The way in which e operate the Exchange with a bi-monthly bulletin makes it difficult for us to elp industries with an immediate pressing disposal problem. Negotiation of a uccessful transfer often takes many months before both parties are satisfied ith terms and safeguards as to quantity and quality of the waste.

uture plans call for attempts to expand operations by making appropriate links ith U.S. exchanges. In May, 1980, all bulletins going to participants in ritish Columbia included a copy of the Oregon/Washington State bulletin.

eographically, north-south transfers make more sense than east-west in the Pacific-orthwest. We hope that perhaps this experiment might be repeated with other border-ng U.S. states, e.g. New York and Pennsylvania with Ontario. We intend also to xpand the editorial section of the bulletin to include short articles on novel ses for waste materials, and on research and development programmes being conducted n this field.

THE CONTINUING NEED FOR WASTE DISPOSAL FACILITIES

hile we might think that, theoretically, through programmes such as the Waste xchange, we could re-use all of our industrial wastes, this is unlikely ever to be ractical. Our present transfer rate of about 15% of the wastes listed on the xchange can be restated as 85% of the wastes listed not having found a suitable re-se opportunity. This 85% of our listings plus the other Canadian industrial waste ot listed on the Exchange must, therefore, be discarded. The need for proper safe reatment and final disposal facilities is perhaps one of the most urgent and complex roblems facing industry today.

he general public's mistrust of both industry and government has led to the situa-ion where it is virtually impossible to establish proper treatment facilities in anada. The alternative to treatment is storage or illegal dumping, which is what as led to the recent highly publicized horror stories from the United States. I eel that it is imperative that we deal with the unusable wastes as they are gener-ted, to destroy them or neutralize and detoxify them prior to any final disposal. his will be expensive, but most of the companies I have dealt with through the aste Exchange seem willing to bear the expense providing that the facilities are vailable.

There is a generally accepted order of desirability for approaches to dealing with industrial waste:

1. Reduce the amount of waste produced
2. Re-use as much as possible
3. Treat and dispose in an environmentally sound manner
4. Store or dump

Reduction of the amount of waste generated must be achieved by companies in-house. Programmes such as the Canadian Waste Materials Exchange try to maximize re-use opportunities. The establishment of proper waste treatment facilities in Canada will avoid the problems caused by the least desirable method for industrial waste disposal.

REFERENCES

Anon (1976). Chem. Eng. Prog. 72 (12), 58-62.

Arthur D. Little, Inc. (1976). Waste Clearinghouses and Exchanges, New Ways for Identifying and Transferring Reusable Industrial Process Wastes. Prepared for the U.S. Environmental Protection Agency under Contract Number 6B-01-3241, October.

Laughlin, R.G.W. and A. Golomb (1977). The Methodology for the Operation of a Waste Materials Exchange in Canada. Prepared for the Solid Waste Management Branch, Environment Canada, published as EPS Report, EPS-EC-77-8, March.

5

EXTRACTION OF METALS FROM SEWAGE SLUDGE

D. S. Scott, H. Horlings and A. Soupilas

Department of Chemical Engineering, University of Waterloo, Waterloo, Ontario, Canada

ABSTRACT

Results are described for a process whereby metals are removed from thickened municipal sludge or sludge filter cake by treatment with dilute sulfuric acid for 10 minutes at room temperature at pH 0.8 to 1.5. Both anaerobic and aerobic sludge samples were used. A high degree of extraction of phosphorous, and of metals with soluble sulfates, was achieved for all sludges tested. Brief heating of anaerobic acidified sludges improved filterability. Less than 10% of the organic matter was extracted, so that the residual sludge had a significantly lower ash content and a higher calorific value.

KEYWORDS

Sewage sludge; metal extraction; phosphorous recovery from sludges.

INTRODUCTION

Recent years have seen the upgrading of many municipal wastewater treatment facilities. This upgrading is characterized by the generation of increasing quantities of sludge residue. With the advent of phosphorous removal requirements, especially in the Great Lakes areas of both Canada and the United States, so-called chemical sludges are, in particular, forming a larger proportion of the total. A chemical sludge results when a specific phosphate precipitant, normally iron, alum or lime, is added in the treatment plant. This chemical addition markedly increases the amount of sludge to be handled and changes its nature. Most of the heavy metals, as well as the phosphate, will appear in the sludge and a higher ash content and a lower calorific value result.

Various alternatives exist for the treatment and disposal of municipal wastewater sludge. In the past, the least expensive was usually chosen with little concern for long range environmental effects. Today, the most common methods used are agricultural land application, land reclamation, sanitary land filling and incineration. As land areas for drying beds, lagoons and landfills are becoming more remote from urban centres and more costly to acquire, incineration is growing in favor for large urban systems. Agricultural use remains a major disposal method in more rural areas. If incineration is employed to dispose of the organic part of the sludge, the ash which may be 30%-50% of the solids weight still remains as a disposal problem. It cannot be assumed to be an innocuous material, since the heavy metal and phosphate contents can be quite high and are subject to leaching into ground waters. Similarly, the heavy metals content when applied to agricultural land can enter the food chain with undesirable effects on crops or animals. The full dimensions of possible problems arising in a given disposal method from the heavy metals or phosphate content of final sludges are not fully understood. However, it is certain that these problems are likely to be greatest with chemical sludges, particularly those from industrialized urban areas.

An earlier paper by the authors (Scott and Horlings, 1975) introduced the concept of acid treating of thickened municipal sewage sludge, or of sludge filter cake, t remove metals and phosphates prior to sludge disposal. The process is particularl applicable to anaerobically digested sludges obtained from treatment plants which include a phosphate precipitation step. In these sludges, high concentrations of both metals and phosphates are likely to be found. The authors showed in this earlier paper that high degrees of extraction of phosphate and of cations having soluble sulfates could be achieved by treatment of "as-is" thickened sludge from a digester with sulfuric acid to a pH of 1.0 to 1.5 for about 20 minutes at room tem erature. Preliminary results were given to show that metals and phosphates could be separated from the extraction solution into specific metals fractions by controlled pH precipitation.

Since the original work was done, some additional studies have been done by others on the subject of the metal contents of sludge solids. Oliver and Carey (1976) ex tended the original work of Scott and Horlings (1975) by testing the metal and pho phate extraction from 7 sludges (all but one anaerobically digested) using differe acids and a titration procedure. They concluded that reduction of pH to 1.5 was adequate as the sludge then began to exert a buffering action and little additiona extraction occurred. There was little to choose between the extraction performanc of nitric, hydrochloric or sulfuric acid, in agreement with the results of Scott and Horlings (1975). They reported that 50-80% of the metals were extracted by their procedures, and about 10-15% of the organic carbon was leached by the acid. Oliver and Carey (1976) also attempted basic extractions using hot NH_4OH or NH_4OH-$(NH_4)_2SO_4$ solutions, with and without an oxygen atmosphere. Extraction was poor even for those metals expected to form soluble amines.

Some studies (Oliver and Carey, 1976; Diosady, 1975; Fowlie and Stepko, 1978; Schroeder and Cohen, 1978; Gabler and Neyland, 1978) have also considered the recovery of metals or phosphates from the ash resulting from the incineration of municipal sludges, since incineration is now the major disposal method for large treatment plants. Significant percentages of metals such as cadmium, nickel and zinc can be leached by neutral water from this ash, the ease of leaching apparentl being very dependent on incinerator operating conditions (Oliver and Carey, 1976) In general, if reasonably high recovery of metals from ash is desired, ash must b leached with a large excess of hot concentrated (50%-60%) sulfuric acid. Excess acid can apparently only be recovered by distillation (Oliver and Carey, 1976), a the process has been judged to be uneconomic unless disposal benefits justify add costs (Fowlie and Stepko, 1978).

A considerable amount of work has also been done in recent years on the problem o the affects on agriculture of heavy metals when sewage sludge is used for agricultural purposes (for example, Environmental Protection Service, 1977; Bates and Beauchamp, 1978; Bryant and Cohen, 1978). Amounts of cadmium, nickel, lead and copper have now risen to levels in some Ontario soils such that the Ministry of t Environment proposes to limit the quantity of sewage sludge that may be applied a a fertilizer.

Finally, some concern has been expressed that large amounts of some metals, parti larly iron or calcium, may detrimentally affect the softening point of ash formed in sludge incinerators and thus lead to the severe operational problems with clinker formation which have been reported for some installations. In addition, high metal contents in incinerator feed means more auxiliary fuel must be used, more ash must be disposed of, and gas cleaning problems are more serious.

In summary, evidence accumulated in recent years reinforces the belief that numerous environmental and disposal problems can arise because of the high content of metals and phosphate found in digested, chemically-treated sewage sludge. It now seems to be clear that for some treatment plants with above normal metals intake, some treatment may have to be provided to sequester these metals. Either the source of metal contamination must be eliminated, or metals must be removed from sludge before disposal. The acid extraction process may be one way of accomplishing this latter objective.

This present report describes additional and more comprehensive results from the acid extraction process for removing metals from sewage sludge. Three sludges prepared by different waste treatment processes were used, two from large municipal plants practising incineration. One of these was a thickened, anaerobically digested sludge to which ferric chloride had been added as a phosphate precipitant, and the other a thickened, aerobic activated sludge to which alum had been added to precipitate phosphate. The third sludge was from a small plant in which alum was used as a phosphate precipitant and the sludge was anaerobically digested.

EXPERIMENTAL MATERIALS AND METHODS

All three sludges reported in the present work were thickened, and contained from 3-6% dry solids. One of the anaerobically digested sludges ($FeCl_3$ used as a phosphate precipitant) came from the Ashbridge's Bay plant of Metropolitan District of Toronto, Canada (180 mgpd). This sludge was thickened after digestion, and the sludge sample (A) was taken from the digester thickener underflow. This sludge was then treated with polymeric coagulants and de-watered on rotary vacuum filters prior to incineration in multiple hearth incinerators. The filter cake feed to the incinerator was also sampled and used in some tests (Cake B). Sludge samples were also obtained from the municipal treatment plant of the city of Warren, Michigan (a suburb of Detroit). This sludge was a waste activated sludge to which alum had been added as a phosphate precipitant, and the sludge subsequently thickened in a flotation-type unit. This thickened sludge was then mixed with an approximately equal amount of settled primary sludge, and the resulting sludge was dewatered on rotary vacuum filters with the aid of polymeric flocculants. The resulting filter cake (Cake D) was then incinerated in multiple hearth units. The third sludge sample was obtained from the anaerobic digester underflow of the treatment plant of the city of Guelph, Ontario (Sludge E). This sludge had been treated with alum for phosphate precipitation and had undergone thorough anaerobic digestion. It was then disposed of as agricultural sludge, or de-watered and the cake sent to land disposal.

Experimental methods and analyses were the same as described by Scott and Horlings (1975) with two exceptions. A recent report by Van Loon (1976) recommends a digestion procedure using a nitric acid - hydrochloric acid mixture for the analysis of metals content of sewage sludges. This procedure was adopted for the work.

Extraction tests were all carried out by the addition of concentrated sulfuric acid to an "as-is" sludge sample of 500 grams at room temperature until a desired final pH value (usually either 1.5 or 0.8) had been reached. Reaction was allowed to proceed for 20 minutes with stirring, and the extract solution was then brought to a boil and boiled for five minutes. After cooling to 65°C, the mixture was filtered, the residue washed several times, and filtrate and washings diluted to 1000 mL. It was this resulting extract solution which was used for all elemental analyses. The extraction was essentially complete after 20 minutes at room temperature, and the boiling added little additional soluble material. This step was carried out primarily to improve filterability and washing of the residual cake. It was quite effective with anaerobically digested sludges, but gave little or no benefit with an aerobic sludge. Nevertheless, a uniform extraction procedure was used for all sludges. When filter cake was used, it was reslurried with water to give a consistency approximately the same as the original sludge, or was extracted to a maximum feasible solids concentration. Sewage sludge slurries of up to 11% solids content are sufficiently fluid for extraction and will yield stronger extract solutions, of course.

Ash contents of the sludges were determined by the standard method of ashing in oxidizing conditions at 800°C in a muffle furnace. As Van Loon (1976) has pointed temperatures above 450°C will result in some loss of arsenic, cadmium, lead, mercury and zinc from the ash. However, since all these elements are present in quantities much less than 1%, little error in ash content results from the higher temperature ashing.

RESULTS

Analysis of Sludge Samples

The characteristics of the sludge samples used are summarized in Table 1 for the materials studied.

TABLE 1 Characteristics of Sludge Samples

Description	% dry Solids	% Ash	pH	v.s.
A. Thickened underflow from anaerobic digester, Ashbridge's Bay, Toronto. $FeCl_3$ added.	3.16	47.6	7.9	53%*
B. Sludge filter cake (rotary vacuum) from above thickened sludge, polymeric flocculants added	14.5	46.4	7.6	--
C. Flotation thickened aerobic waste sludge, Warren, Mich. Alum added (71 ppm)	4.09	34.9	6.7	57.2%
D. Sludge filter cake (rotary vacuum) from above sludge plus approximately equal amounts of settled primary sludge. Polymer flocculants added.	13.4	32.5	6.7	61.3%
E. Thickened underflow from anaerobic digester, Guelph, Ont. Alum added (45 ppm)	5.67	--	7.2	40.9%

* Plant value

Elemental analysis was carried out on all five samples. However, Cake B has essentially the same chemical composition as Sludge A, since only a small amount of polymeric flocculant is added to the sludge. Hence, only the analysis for Sludge A is given (Table 2).

TABLE 2 Elemental Analysis of Sludge
g/kg dry sludge solids

	Sludge A (and Cake B)	Sludge C	Cake D	Sludge E
Ca	40.1	4.42	5.77	65.6
P	32.8	39.8	32.5	32.9
Fe	52.4	31.0	27.7	27.4
Cu	1.10	0.83	0.81	2.9
Pb	2.04	0.42	0.43	1.28
Ni	0.082	1.28	1.28	0.01
Zn	3.33	5.9	5.9	6.6
Al	9.27	35.9	29.0	34.10
Cr	3.30	3.76	3.49	2.4
Mn	0.53	0.95	0.89	0.23
Mg	5.20	4.24	3.49	9.80

These metal analyses show fairly normal levels of phosphorus and metals content. Iron is high in Sample A because of the ferric chloride added to precipitate phosphorus, while for the same reason aluminum is high in Samples C, D and E. In the anaerobic sample (A), some lime has obviously been added for pH control. Comparison of samples C and D indicates that the primary sludge is much poorer in phosphorus than the aerobic sludge, but metal contents are comparable for the two. The higher lead in A and high nickel in C are reflections of specific industrial activities.

Extraction Tests

Many extraction tests were performed using the procedures described and the results of some of these tests are given in Table 3. All extractions are given as percentages, based on the analyses in Table 2. The pH given is the final pH value in the extraction.

TABLE 3 - Extraction Tests - Percent Extracted

Element	Sludge A pH 1.5	Sludge A pH 0.8	Cake B pH 1.5	Sludge C pH 1.5	Sludge C pH 0.8	Cake D pH 0.8	Sludge E pH 1.5
P	100	97.5	78.0	78	92	83	96.5
Fe	98.6	100	96.9	81	98	99.6	92.0
Al	73.5	80.2	58.9	96.8	100	93.9	100
Zn	89.2	97.0	70.0	100	100	97.9	57.6
Mg	100	100	---	100	100	96.3	96.5
Ca	---	---	---	39	74	66	17.5
Ni	---	---	---	100	100	100	---
Cu	0	2.7	0	94.5	100	73.1	1
Mn	100	100	100	100	100	100	100
Pb	32.3	33.3	9.3	0	0	0	8.6
Cr	76.6	97.9	55.5	100	98.8	91.9	91.7

These results illustrate several points. The copper and chromium tend to be much more readily extracted from aerobic systems in which these metals are in higher valence states than from anaerobic systems where presumably lower valence states giving less soluble sulfates exist. It makes little difference, as would be expected, whether iron is in the ferrous or ferric state. The results for calcium and lead reflect the solubility phenomena of the slightly soluble lead sulfates and calcium sulfates formed. The lower extractions for some components for cake B (anaerobic filter cake) are due largely to a concentrated solution and insufficient acid addition.

The residues from the extraction step were examined to determine their character, as were the extract solutions. The results of two typical tests are shown in Table 4, one for sludge A extracted at a final pH of 1.5 and one for sludge C extracted also to a final pH of 1.5.

TABLE 4 Analysis of Extracts and Residues
Final pH - 1.5

	Solids % of original*	Ash %	Volatiles %	Calorific Value** kJ/kg	Ash Removal
Extract - Sludge A	77.5	30.9	51.8		
Residue - Sludge A	59.5	30.5	67.7	15876	61.9%
Extract - Sludge C	56.0	44.8	42.9		
Residue - Sludge C	73.1	21.0	62.2 *	23505	56%

* Because of the added sulfate ion, the solids in extract plus residue total more than 100% of the original sludge solids.

** Calorific value of original sludge A = 12318 kJ/kg
Calorific value of original sludge C = 18028 kJ/kg

In Table 4, the ash percentage, volatiles percentage and calorific value are all expressed on the basis of dry extract or residue solids, as appropriate.

Total organic carbon and COD analyses were also carried out on several samples and results for the extract and residues obtained after extraction at pH 1.5 are listed in Table 5.

TABLE 5 TOC and COD Analyses
Final pH 1.5

	TOC	COD
	g/kg original sludge solids	
Original Sludge A	214.3	729.9
Extract Sludge A	22.5	62.8
Residue Sludge A	172.7	631.9
Original Sludge C	204	970
Extract Sludge C	25	157
Original Sludge E	240.4	703
Extract Sludge E	16.0	48.8

From the foregoing results, some interesting conclusions can be drawn. It is clear that, although a substantial amount of inorganic material is dissolved from the sludge, relatively little organic matter is extracted. One result of the extractio therefore, is to increase the calorific value of the residual solids by some 25% over the original sludge. The mass of ash which would result from incineration of extraction residue would be only about 50% of the amount formed from the original digester sludge, and it would be largely freed of metal content. Similarly, the extract solution would be low in organic matter, most of it of lower molecular weight, and correspondingly low in COD. Chemical treatment of this extract to recover phosphate or metals should not, therefore, pose any great problems in practice. Results shown in Tables 3, 4 and 5 were typical of extraction tests with all sludges or cakes tested.

Filterability Tests

The filterability of sludges is always a problem in sludge treating processes. Filterability tests were carried out using both the Buchner funnel method and a Dorr-Oliver test filter leaf (Water Pollution Control Federation, 1969). Anaerobic sludges filtered more readily than the aerobic sludge. Further, anaerobic sludge filterability could be improved by the acid treatment and brief heating (to 80°C) by a factor of two to twenty times. On the other hand, neither the acid treatment nor brief heating altered the filtering characteristics of the aerobic sludge, which remained poor. As a consequence, these latter sludges present a real problem in washing residual solids after extraction, whereas washing is accomplished fairly readily with anaerobic sludges. Therefore, residual cakes after extraction not only filter faster for the anaerobic sludges but have a lower moisture content than those from aerobic sludge.

Results obtained from Buchner funnel tests for three thickened sludges are shown in Table 6. The Buchner test results are given because this has become a nearly standard evaluation technique, and so direct comparisons to literature values can be made. A useful (even if not very exact) degree of correlation has been found between laboratory Buchner funnel results and plant filtration rates.

The standard Buchner funnel test was carried out using a 7 cm. diameter funnel and a fast quantitative filter paper as the medium. Extract solutions were filtered at 60°C under a constant pressure differential. Filtrate volume vs. time readings were taken, and from these data the specific cake resistance was calculated as described in Manual No. 20 Sludge Dewatering of the Water Pollution Control Federation (1969).

The specific resistance of the cake is given by the equation $r = \frac{2PA^2m}{\mu C}$

where r = specific cake resistance m/kg
- P = pressure differential, N/m^2
- A = filter area, m^2
- m = slope of the straight line obtained from a plot of t/V vs. V, where t is the time in sec. and V is the cumulative volume of filtrate, m^3
- μ = filtrate viscosity, kg/m·s
- C = slurry concentration, dry solids per unit volume, kg/m^3

None of the sludges in Table 6 had been conditioned for de-watering, either by chemical or flocculant additions. Similarly, the acidified sludge had had no treatment other than the acid addition and brief heating before filtration.

Tests on anaerobic sludges such as types A and E showed that filterability improved in the acidified sludge unless it became strongly acidic e.g., pH < 1.0. A pH value of 1.5 is nearly optimal for filterability in these sludges. On the other hand, the aerobic sludge appeared to behave somewhat more poorly after acidification to any degree.

TABLE 6 Specific Cake Resistances

	Specific Resistance m/kg
Sludge A "as-is"	6.62×10^{13}
Sludge A acidified pH 1.5	2.10×10^{13}
Sludge C "as-is"	3.46×10^{14}
Sludge C acidified pH 1.5	10.2×10^{14}
Sludge E "as-is"	3.90×10^{14}
Sludge E acidified pH 1.5	4.43×10^{13}

DISCUSSION

The results given in Table 4 show that some 60% of the ash forming constituents are removed from the sludge by the acid extraction step. Hence, the mass of ash resulting from incineration would be only 40% of that produced from the untreated sludge. Table 3 shows that high percentage solubilization of metal ions and phosphorous also occurs, with extraction at a pH of 1.5 generally adequate. For Sludge A, this corresponds to an acid usage of 0.48-0.50 kg/kg of dry sludge solids, and for Sludge C about 0.35 kg acid/kg dry sludge solids. Other anaerobic sludges tested in which iron content was high gave values for extractions of over 80% of the iron and phosphorus when acid usage was from 0.45-0.65 kg/kg dry solids. If the sludge is high in calcium and iron, then amounts of acid in the range 0.55-0.65 kg/kg dry sludge solids will be needed. Sludge A is average with respect to its content of these two elements, whereas Sludge C is low in these values. In fact, extraction of a variety of sludges shows that the optimum pH of 1.5 is directly proportional to the equivalents of the major cations in the sludge (Fe, Ca, Al, Zn, Mg) and represents about a 20%-30% excess over the stoichiometric requirement for these elements.

Sulfuric acid does not solubilize significant percentages of calcium, copper or lead in anaerobically digested sludges, due to the formation of insoluble sulfates. Good extraction of copper or lead can be obtained by using hydrochloric acid for leaching. Copper can be solubilized by adding a small amount of ferric salt to the sludge, which acts to oxidize the copper to the divalent state and render it soluble in sulfuric acid. If chromium is not extracted readily, the ferric chloride will also give improved chromium extraction in sulfuric acid solutions.

The removal of about 60% of the inorganic matter from the sludge solids while removing only 7-12% of the organic matter increases the calorific value of the extracted residual solids by 29% for sludge A and 30% for sludge C. Similar results were obtained for other sludges or filter cakes. This gain in calorific value must result in some reduction of auxiliary fuel usage in incineration.

The filterability of anaerobically digested sludges, whether treated with iron salts or alum for phosphorus removal, appears to be somewhat better after acid extraction and brief heating than is that of the thickened underflow from the digesters. The values in Table 6 for the "as-is" sludges are well within the typical ranges for specific cake resistances reported by Metcalf and Eddy(1979).

In summary, the acid treated sludge residue will yield only 40% of the ash and will have a 30% higher calorific value than an untreated sludge. The ash will also be more innocuous, being largely free of metallic elements. Ash fusion tests show that removal of the metals and phosphates has only a minor effect on softening temperatures.

Anaerobically digested sludges would appear to be the most suitable for acid extraction of metals and phosphates. The metal and phosphate concentrations are highest in these sludges, they extract readily and filterability and washing of the extracted residue are feasible. While aerobic sludge can be readily extracted, it is difficult to filter or wash residues.

Recovery of Chemicals

The selective precipitation of metals and phosphates at controlled pH was described by Scott and Horlings (1975) for extracts obtained from an iron-rich anaerobically digested sludge (sludge A). In Figure 1 similar results are shown for an extract made from an aluminum-rich anaerobically digested sludge (Sludge E). The principal difference with this extract from the iron-rich case, is that there is sufficient aluminum and magnesium to almost completely precipitate all the phosphate at a pH of 3.5. At this pH value, 70% of the iron and zinc are still in solution. Hence, a ready separation could be made into a solid fraction containing over 90% of the aluminum and phosphate, as well as most of the lead, chromium and magnesium, and a solution containing about 70% of the iron and zinc. These fractions could then be treated for recovery of aluminum for recycle in the sewage treatment plant and for the separate recovery of a phosphate compound.

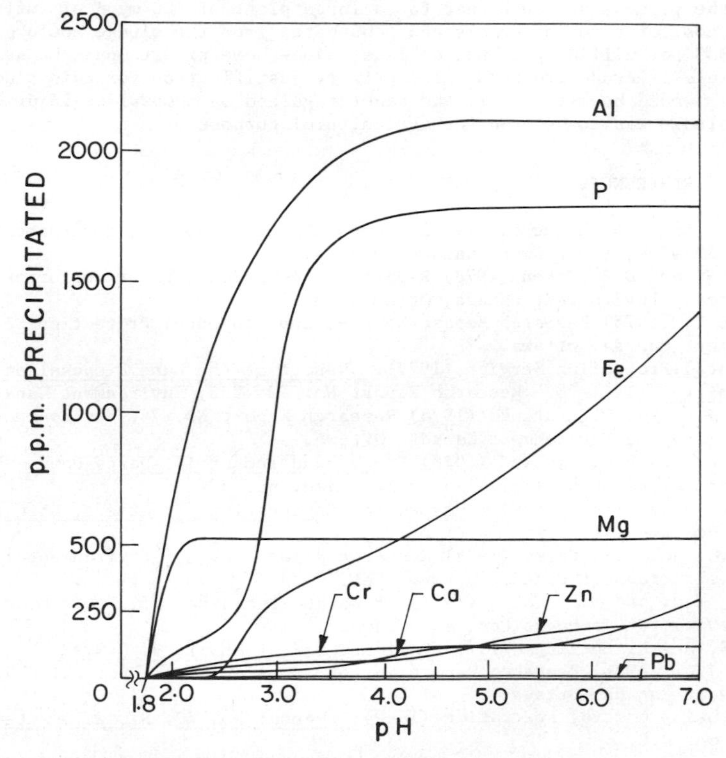

Fig. 1 Selective precipitation of metals from extract solution prepared from an alum treated anaerobic sludge

Cost of Acid Extraction Process

A preliminary cost estimate was made for two plant sizes, 30 mgpd and 180 mgpd, assuming an anaerobically digested sludge cake was produced and that iron salts had been added for phosphate removal. The process was assumed to consist of acid extraction with sulfuric acid, heating of the slurry followed by filtration and washing of the cake which was then sent to ultimate disposal. The extract solution was assumed to be neutralized by lime in two stages, producing one cake containing aluminum, magnesium and chromium and about 40% of the phosphate, and a second cake containing most of the iron and zinc and the balance of the phosphate. The cost of this degree of processing was about $30 per million Imperial gallons of influent for the large plant and about $46 per million gallons for the small plant. The amounts of iron, aluminum and phosphate in the two cakes are worth recovering, but the value of the products is only slightly more than the probable total cost of the refining steps. Credit could also be taken for the reduced amount of residual sludge or incinerator ash and, in the case of sludge incineration,for the increased calorific value of the extracted residual solids.

While heavy metals are also recovered, their amounts are such that it is not likely to be worthwhile to recover them, although they might be sold as crude materials. The amount of zinc and chromium recovered from even the large municipal plant (180 mgpd) assumed here was only 91 and 78 metric tonnes per year, respectively. Other heavy metals are present in even smaller amounts.

Costs of the process are such that for a large plant of 100 mgpd of influent or more, the cost of removing metals and phosphates from the sludge would be of the order of $30 per million gallons, or less, since some credit would be available from the sale of crude products. The primary justification for this sludge treating process would have to lie in the benefit gained by removal of injurious metals when the sludge was to be used for agricultural purposes.

REFERENCES

Bates, T.E. and E.G. Beauchamp (1978) Research Report No. 73, Environmental Protection Service, Environment Canada, Ottawa.

Bryant, D.N. and D.B. Cohen (1978) Report EPS 4-WP-78-3, Water Pollution Control Directorate, Environment Canada, Ottawa, April.

Diosady, L.L. (1975) Research Report No. 19, Environmental Protection Service, Environment Canada, Ottawa.

Environmental Protection Service (1977). Report of the Land Disposal of Sludge Sub-Committee, 1971-78. Research Report No. 70, EPS, Environment Canada, Ottawa

Fowlie, P.J.A. and W.E. Stepko (1978) Research Report No. 74, Environmental Protection Service, Environment Canada, Ottawa.

Gabler, R.C. and D.L. Neyland (1978) Proc. 32nd Industrial Waste Conf., Purdue Uni Ann Arbor Science Pub. Inc., Ann Arbor, Mich. pp. 39-49.

Metcalf and Eddy, Inc. (1979). Wastewater Engineering: Treatment, Disposal, Reuse 2nd ed. McGraw-Hill, N.Y.

Oliver, B.G. and J.H. Carey (1976) Research Report No. 33, Environmental Protectio Service, Environment Canada, Ottawa, Feb.

Schroeder, W.H. and D.B. Cohen (1978) Research Report No. 75, Environmental Protec tion Service, Environment Canada, Ottawa.

Scott, D.S. and H. Horlings (1975) Environ. Sci. Technol. 9, 849.

Van Loon, J.C. (1976) Research Report No. 51, Environmental Protection Service, Environment Canada, Ottawa.

Water Pollution Control Federation (1969). Manual No. 20, Sludge Dewatering, Washington, D.C.

6
RECOVERY OF SULFURIC ACID AND SALT CAKE FROM SMOOTH ROCK FALLS CHLORINE DIOXIDE GENERATOR EFFLUENT

D. Lobley* and K. L. Pinder**

**Multifibre Process Ltd., 852 Derwent Way, Annacis Island, New Westminster, B.C. V3M 5R1, Canada*
***Department of Chem. Eng., University of B.C., Vancouver, B.C. V6T 1W5, Canada*

ABSTRACT

This paper describes the basic principles of the Acid Recovery Process installed at the Smooth Rock Falls mill of Abitibi Paper Co. Ltd. and discusses some of the problems encountered. The process is designed to separate neutral sodium sulfate and sulfuric acid from waste liquor emanating from a chlorine dioxide generator operating with the Solvay process. The original plant was designed to be a Mathieson process but the change in reducing chemical did not alter the acid recovery process which will add on to any chlorine dioxide generator.

The sulfuric acid is recycled back to the generator rather than being incinerated, and the saltcake is used for chemical makeup within the Kraft recovery cycle. Theories and practices of saltcake crystallization are examined, as well as the future of this technology in other chemical processes.

KEYWORDS

Acid Recovery Process; chlorine dioxide effluent; saltcake recovery; chemical recycle; energy saving.

INTRODUCTION

The Smooth Rock Falls Mill of the Abitibi Paper Co. has installed the first commercial Acid Recovery Process (ARP) plant for the recycle of sulfuric acid to the chlorine dioxide generators and the recovery of saltcake. This process was developed by Multifibre Process Ltd., in New Westminster, B.C. and is designed to 'tie-on' to any chlorine dioxide generator system. By recycling the main chemicals in the effluent the pollution load is reduced and there is a net saving in energy.

In Kraft mills where chlorine dioxide bleach is almost universally used, the ClO_2-generator effluent is neutralized with black liquor and burned in the recovery furnace. However, because of improvements in the brown stock washer systems and the increased use of chlorine dioxide per ton of pulp, a sulfidity imbalance can occur in these mills. This can result in excess sulfidity in the cooking liquors and in a greater loss of sulfur compounds up the stack.

WASTE ACID CONTRIBUTION TO SULFIDITY

The waste acid from a typical Mathieson chlorine dioxide generator (no NaCl used) contains 28% H_2SO_4, 24% Na_2SO_4 and 48% water. Or, in other terms, for each ton of chlorine dioxide produced there are 1.25 tons of sulfuric acid and 1.1 tons of sodium sulfate. If all this is neutralized with black liquor or caustic soda, 2.91 tons of sodium sulfate will be produced per ton of chlorine dioxide.

Figure 1 shows schematically the feed streams and product streams for the Mathieson and the modified Mathieson process (NaCl added), with and without an ARP process. As can be seen, the sodium in the salt also contributes to the saltcake produced from the effluent of the modified process

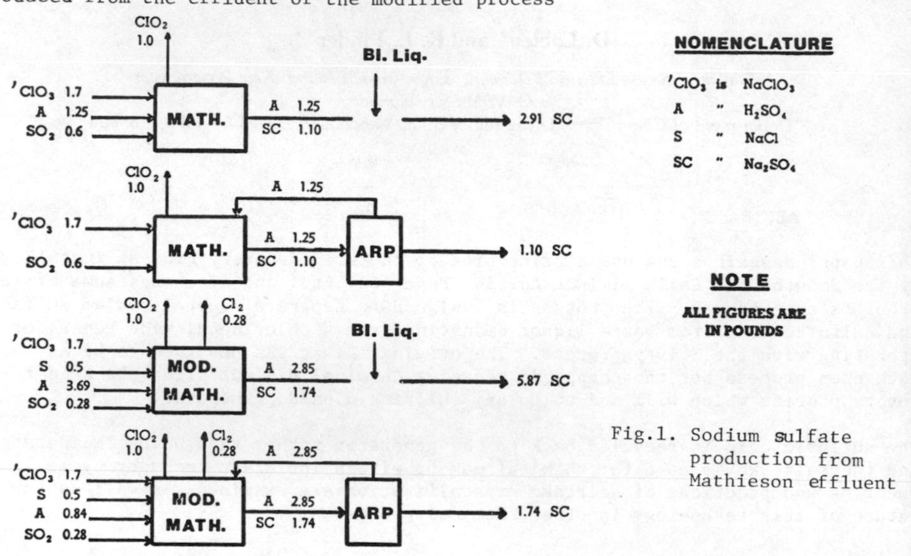

Fig.1. Sodium sulfate production from Mathieson effluent

Figure 2 is a composite diagram which shows the saltcake production, sulfuric acid and SO_2 consumption, and chlorine production per pound of chlorine dioxide production, as a function of the amount of sodium chloride used. The left ordinate represents the Mathieson process, the right the R2, R3 and SVP processes. Between these two extremes is the so called modified Mathieson process.

ACID RECOVERY PROCESS

Other routes have been sought for the removal of the sodium sulfate from the generator effluents- as Glauber salts (Electric Reduction 1964, 1967a; Schribner and Rapson, 1964), as anhydrous sodium sulfate (Read, 1967; Electric Reduction 1967b, Electric Reduction and Rapson, 1967) or as the acid sulfate (Electric Reduction, 196 These techniques are specific to the R2 or ER2 process in which chloride is the reducing agent.

The acid recovery process, patented by Multifibre Process Ltd. (Howard and Lobley, 1978) is not specific for any chlorine dioxide generation process and recovers the sodium as the neutral sulfate. The basic principles of the ARP process will be

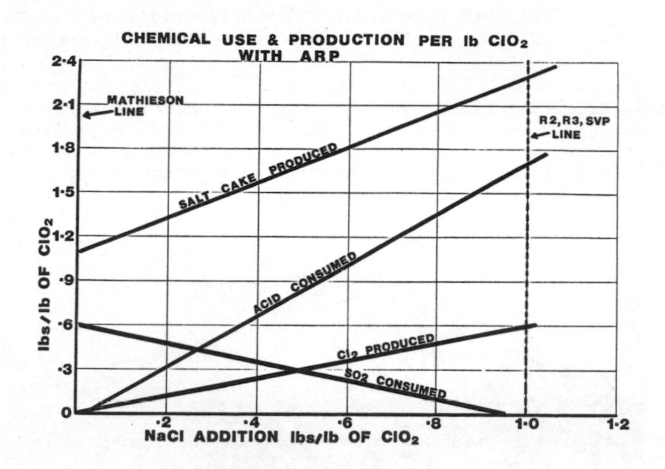

Fig. 2. Chlorine dioxide generator material balances

described with reference to Fig. 3. The variations in the Abitibi plant will then be considered later.

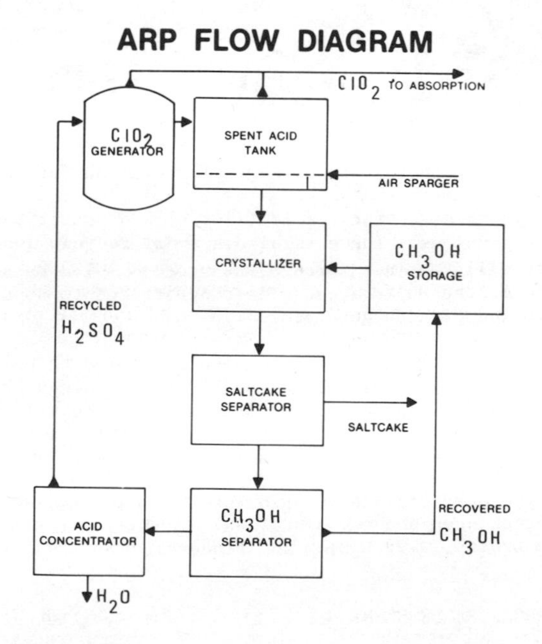

Fig. 3. Schematic flow sheet of acid recovery process

The spent acid is first warmed and degassed with sparged air to remove dissolved chlorine dioxide. A small amount of methanol is added in the stripping tank to reduce any remaining traces of chlorates. This step increases the ClO_2 production per unit of chlorate and also results in a chlorate-free effluent as feed to the recovery plant.

In the crystallizer, the effluent is mixed with methanol and water to precipitate neutral sodium sulfate which flows as a slurry to a separator; either a filter or a centrifuge. The rhombic crystals shown in Fig. 4 are easily dewatered and flow continuously to a dissolving tank. This saltcake is make-up for the Kraft furnace and so it is dissolved in black liquor and sent to the recovery furnace.

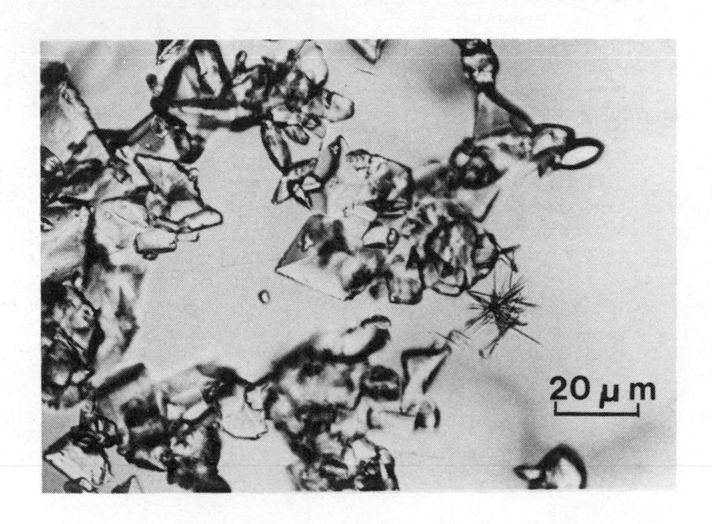

Fig. 4. Microphotograph of neutral sodium sulfate crystals

The filtrate is a mixture of sulfuric acid (11wt %) and water and methanol in roughly equimolar proportions. There is also a small recycle load of unprecipitated sodium sulfate which will continue through the process. The methanol is distilled from this mixture; then, the sulfuric acid is concentrated to about 67 percent and returned to the chlorine dioxide generator.

Thus this section of the mill has become completely closed to the environment without burdening the cooking cycle with extra sulfur. The energy previously used to completely evaporate and decompose the sulfuric acid in the recovery furnace is eliminated. Of course, some heat is used to recover the methanol and to concentrate the sulfuric acid, but there is an appreciable net saving in energy.

This has been a very superficial description of the basic process. More details about the precipitation step and the distillation column design are given in previous reports (Howard and Lobley, 1978; Lobley and Pinder, 1979).

SMOOTH ROCK FALLS PLANT

The chlorine dioxide plant at the Abitibi mill in Smooth Rock Falls, Ontario, was designed to produce 6.6 tons per day by the Mathieson process. Selection of the ARP process to close the cycle was based on a number of factors, not the least of which was the smaller amount of saltcake produced per ton of chlorine dioxide, compared to the established R3 or SVP systems. Probably the most important consideration as far as mill operation is concerned is the purity of the gas emanating from the generator. The Smooth Rock Falls bleachery demands a chlorine dioxide gram/atom purity of 90% for good control in their bleaching operations. This can be achieved by either the Solvay or Mathieson processes.

Since the ARP plant would be a prototype at Smooth Rock Falls, another important consideration entered into the decision making. This was the fact that the acid recovery process operates independently of the chlorine dioxide system, and failure of the recovery section of the plant will not hinder the mill operation or the generation of chlorine dioxide gas.

PROCESS CHANGE

The chlorine dioxide purity and previous experience proved to be the overriding considerations of the engineers, since, after all the vessels for a Mathieson plant were completed and shipped, operational staff decided to use methanol as the main reducing agent (Solvay process) because of their success with this process in their old plant. They experienced fewer 'dioxide decompositions' when using methanol, partially because of product purity. In the case of sulfur dioxide, the mill seemed to be plagued with dirty or impure shipments and this, in addition the possibility of iron contamination from rather old vaporizers, was another operational detraction. Methanol precludes the troublesome chore of vaporization, is easy to store under all weather conditions and is also less expensive per unit of reducing power. Figure 5 shows the material balance for the chlorine dioxide generators and the ARP process for the Solvay process used in the Smooth Rock Falls plant.

Figure 5 is based on the following reactions:

$$4NaClO_3 + CH_3OH + 2H_2SO_4 \rightarrow 4ClO_2 + HCOOH + 2Na_2SO_4 + 3H_2O$$

$$6NaClO_3 + 5CH_3OH + 3H_2SO_4 \rightarrow 3Cl_2 + 5CO_2 + 3Na_2SO_4 + 13H_2O$$

$$2NaClO_3 + 2CH_3OH + H_2SO_4 \rightarrow HCl + Na_2SO_4 + 2CO_2 + 4H_2O$$

$$NaClO_3 + 2HCl \rightarrow ClO_2 + \frac{1}{2} Cl_2 + NaCl + H_2O$$

The first is the basic production equation used in these calculations. Another basic equation which, for the sake of simplicity, was not used is

$$2NaClO_3 + H_2SO_4 + CH_3OH \rightarrow 2ClO_2 + HCHO + Na_2SO_4 + 2H_2O$$

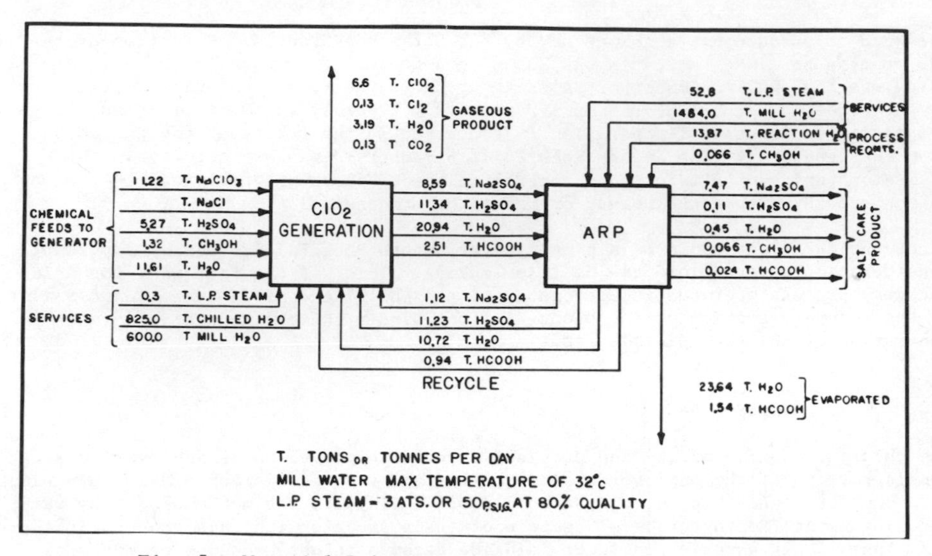

Fig. 5. Material Balance

As can be seen from Figure 5, not all the saltcake is precipitated in the ARP crystallizer but is carried through the chlorine dioxide circuit by the sulfuric acid. For normal operation, 87 percent of the saltcake is precipitated per cycle, the sulfuric acid recovery is 99 percent and the methanol loss is about 0.066 tons per day.

The acid recovery process demands an input with low chlorate residual. When the stripper is working well (see Fig. 6) the residual chlorate is between 0 and 1 gram per litre. This is obtained by direct steam injection into the waste acid along with a small additional dose of methanol to reduce the chloric acid. One of the problems encountered is with the material of construction for the stripping vessel. Titanium normally is suitable metallurgically when residual chlorate is present; however, when all the chlorate is reduced, the reducing/oxidizing conditions change making titanium unacceptable. Lab tests showed that as little as 2 grams per litre of chlorate was sufficient to prevent destruction of the rutite film coating the titanium. At 85°C, corrosion rates as high as 900 mils per year were observed with the lower chlorate concentrations. Fiberglass with a suitable resin is now used in this service, while titanium is still being used for the primary and secondary generators.

The original Solvay process recommendations for reduction with methanol indicate a primary and secondary generator of equal volume, with provisions for cooling in the primary and heating in the secondary. Since the Smooth Rock Falls generators were designed for sulfur dioxide reduction, in which approximately 85% of the reaction occurs in the primary and 15% in the secondary, the two reactors supplied were sized accordingly. It quickly became apparent that different technology was required to run this equipment, since normal operation yielded a fairly high chlorate residual from the secondary reactor because of this size factor. It still runs fairly high (about 9 to 12 grams per litre of residual chlorate) compared to a

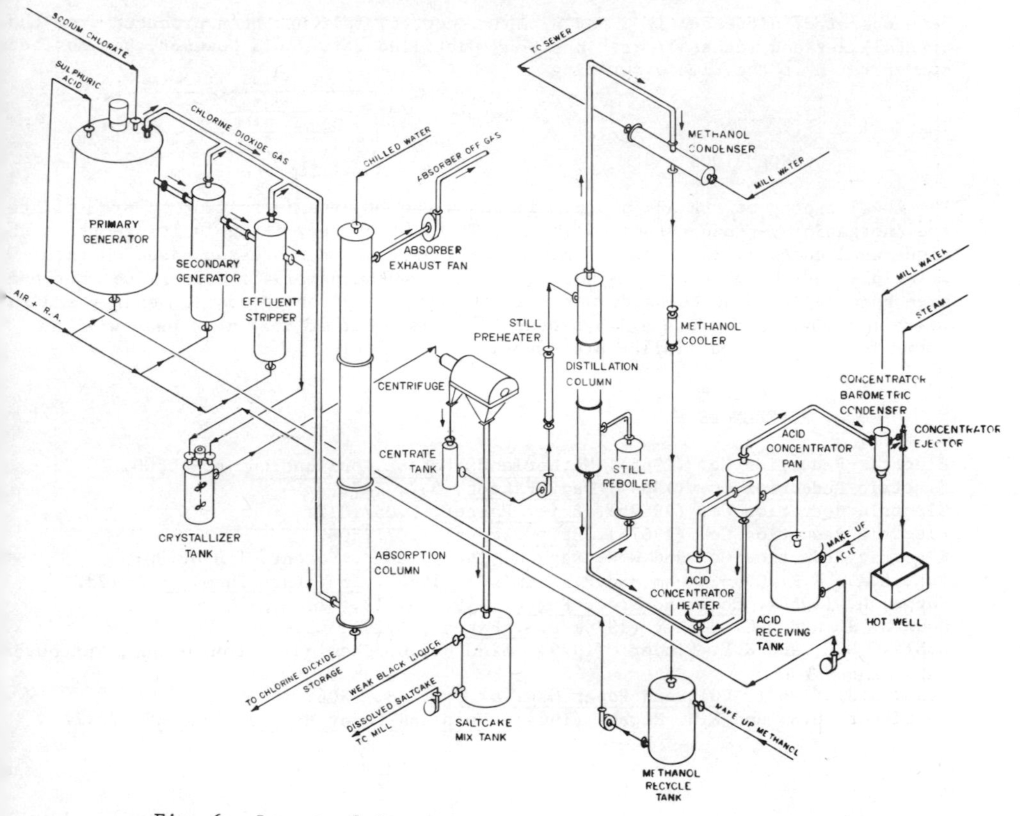

Fig. 6. Layout of Smooth Rock Falls Chlorine Dioxide Plant

normal Mathieson system, but the stripper is so sized that the generator waste acid has the same total retention time as previously in the secondary generator. The additional residence time plus the live steam and methanol produce a water-clear, gas-free and chlorate-free feed for the ARP process.

The plant was laid out in a conventional "Mathieson manner" with the acid recovery equipment fitted and arranged as well as the 50-year old reinforced concrete building, with its wide thick beams and slabbed floors, permitted. Pipe runs presented something of a problem for the same reasons, and pumps had to be carefully sized to account for pressure drops through multitudinous bends and elbows.

Selection of materials of construction to handle hot acids with strengths between 2 and 9 normal was a difficult chore because of the possibility of trace elements such as chloride and chromium (trivalent) being present in spent generator liquors. Liberal use was made of fibre impregnated graphite and furane resins for heat exchange and distillation column manufacture, both materials having been proven over the years in similar or more severe service.

In the case of the piping handling hot acid under high vacuum, chemical lead homogenously bonded to steel pipe was used exclusively. We can not yet report in detail about plant operation on a continuous basis, but elements of the system (ARP) have been operated independently. For example, neutral salt has been produced from the crystallizer and methanol/water has been distilled off. Acid however, has not been concentrated at the time of writing.

CONCLUSIONS

The novel aspect of the ARP process is the use of an organic solvent to precipitate the inorganic by-product from solution. This, of course, is the 'salting-out' technique used commonly in organic lab preparations. Other processes based on this principle have been proposed (Gee, Cunningham and Heindl, 1948; Hoppe, 1968) but none have been reported as being commercial. With the increasing cost of energy and the need to recover and reuse all byproducts, it is expected that more uses will be found for the ARP and similar processes.

REFERENCES

Electric Reduction Co. (1965). Netherlands Patent Application, 65-10600.
Electric Reduction Co. (1964). Can. Patent, 547, 227.
Electric Reduction Co. (1967a). Brit. Patent, 1,057,017.
Electric Reduction Co. (1967b). Brit.Patent, 1,077,306.
Electric Reduction Co. and W.H. Rapson (1967). Brit. Patent, 1,056,790.
Gee, E.A., W.K. Cunningham and R.A. Heindl (1948). Ind. Eng. Chem. 39, 1178.
Hoppe, H. (1968). Chem and Proc. Engrg., 49, No. 12, 61.
Howard, J. and D.G. Lobley (1978) U.S. Patent, 4,104,365.
Lobley, D.G. and K.L. Pinder (1979). 62nd C.I.C. Chemical Conference, Vancouver B.C. June 3-6.
Read, R.J. (1967). Pulp and Paper Mag. of Can. 68, T506.
Schribner, H.S. and W.H. Rapson (1964). Pulp and Paper Mag. of Can. 65, T297.

7

CONVERSION OF REJECT BANANA INTO ANIMAL FEED USING SOLID SUBSTRATE FERMENTATION

R. P. Sethi* and J. M. Grainger**

**Department of Microbiology, P.A.U., Ludhiana, India*
***Department of Microbiology, University of Reading, UK*

ABSTRACT

An attempt has been made to convert waste banana into animal feed by 'solid substrate fermentation' with *Aspergillus niger* M1. In small-scale experiments using 500-mL conical flasks, the optimum temperature, time of incubation and moisture content of the substrate for the maximum yield of mould protein were 35°C, 72 hours and 60%, respectively, under non-sterile conditions. The substrate was supplemented with nitrogen as ammonium sulphate and phosphorus as dipotassium hydrogen phosphate. Maximum fungal protein was produced with 250 mg nitrogen/50 g reject banana. The addition of phosphorus produced a marginal increase in protein production. Large-scale experiments were done in metal trays. The final product (i.e., mould + residual substrate) was fed to 3-day-old chicks. The results showed that mould protein could be satisfactorily substituted for fishmeal, an expensive source of protein in poultry feed, up to 50 per cent.

KEYWORDS

Conversion; reject banana; animal feed; solid substrate fermentation; *Aspergillus niger*.

INTRODUCTION

India is the third largest producer of banana in the world (FAO, 1976). The fruit being highly perishable, its waste occurs at many stages, particularly at the time of storing the fruits. This waste can be used as substrate for the growth of non-toxic mould, such as *A. niger*, for upgrading its protein value. The other competitiv advantages which the organism possesses are rapid growth, thermotolerance (40°C), high pepsin digestibility (74) and a good potential for use in animal feeding. A complete process has been developed to produce single-cell protein from agricultural and food industries' wastes using *A. niger* M1 (Imrie and Vlitos, 1974; Morris and co-workers, 1973 and Righelato and colleagues, 1976). With this in view, solid substrat fermentation of over-ripe reject banana (i.e., flesh and skin) under non-sterile conditions has been developed. The method is simpler and cheaper than the conventional submerged culture fermentation (Hesseltine, 1977a, 1977b).

MATERIALS AND METHODS

Organism and Inoculum

Aspergillus niger M1 from the Tate and Lyle, U.K. culture collection was used throughout the study. Stock culture was maintained on potato dextrose agar medium, subcultured every fortnight and stored at 4°C. For inoculating the substrate, spore suspension of mould in 0.01% Tween 80 was obtained after incubation of organism at 30°C for 4 days on the following medium:

Rice, 30g; asparagine, 0.1%; glycerol, 3%; water, 20 mL.

The medium was dispensed in 500mL conical flasks and sterilized by autoclaving at 121°C for 20 minutes. The volume of the spore suspension used for inoculating the substrate was 10% of the weight of the substrate. The number of spores in the inoculum was counted using a haemocytometer; it ranged from 4×10^6/mL to 5×10^6/mL. The results of the analysis of most of the experiments were checked on the bas s of carbon and nitrogen balance equation (Righelato and collegues, 1976).

Substrate

Reject over-ripe banana from Cameroon was collected from the warehouse of Rowe and Co., Groveland Road, Reading. It was considered to be unfit for human consumption in the U.K.

Preparation of substrate for fermentation. Fifty grams of banana (over-ripe) were cut into slices (5-7 mm thick) using a knife. The moisture content of the substrat was adjusted before inoculation to different values (40-70%) by drying in a hot-air oven at 100°C. The substrate was placed in a single layer in 500mL conical flasks. It was supplemented with nitrogen and phosphorus and inoculated with spores of *A. niger*. The change in moisture content of the substrate due to water added with the inoculum was taken into account when calculating the results for all experiments (Table 5). The contents were stirred with a glass rod.

Large-Scale Production of Microbial Protein

Five kg of reject, over-ripe bananas was cut into slices 7-9mm thickness, adjusted to optimum moisture content (60-62%) in an oven before inoculation and spread into a 65 x 50 x 5 cm aluminium tray (as far as possible in a single layer) which was lined and covered with tin foil. The nutrient solution containing the appropriate concentrations of nitrogen and the spore inoculum ($4\text{-}5 \times 10^6$ spores/mL) were spraye over the substrate. The mixing was done manually. Incubation was done in a hot room at 35°C ± 0.5 for 48 hours. All the experiments were conducted under non-sterile conditions, except the inoculum preparation.

Analytical Procedure

The entire contents of flasks or trays were dried overnight in an oven at 90°C, the mixed and a representative sample taken for analysis. Cellulose, hemicellulose, as and lignin were measured according to methods described in the Agriculture Handbook (Goering and Van Soest, 1970) and lipid by the ether extract procedure (AOAC, 1975) Initially, total and reducing sugars were analysed by the method of Fuller (1966). Subsequently, total carbohydrates were estimated by the phenol sulphuric acid metho (Norris and Ribbons, 1971).

Determination of starch was done by the technique of Clegg (1956). Crude protein was estimated by multiplying Kjeldahl nitrogen by 6.25. For true protein determina tion, the method of Potty (1969) was used successfully. The pH of the substrate wa determined with a Beckman pH meter. Amino acids were analysed by an automatic amin acid analyzer (Japan Electron Optic Laboratory Co. Ltd.).

Feeding Trials

Diets I and II which were prepared according to the formula used in Punjab (India) for poultry rations are given in Table 1 (Sethi and Virk, 1977). In diet II, half of the fish-meal was replaced by test product, i.e., mould plus residual substrate. The crude protein level in diets I and II was adjusted to approximately 22%. The diets were mixed thoroughly by a Hobart 6 kg batch mixer and pelleted in a pelleting machine. The size of pellets was 10-15 mm; this was reduced to 4-5mm in later stages of the trial.

Feeding trails were carried out in the Department of Physiology and Biochemistry, University of Reading, U.K. with 108 three-day-old chicks. The birds were selected by rejecting weak birds and those at the extremes of body weight. They were individually weighed to the nearest gram. Weights were recorded in increasing order by computer, commencing with the lightest. The selected birds were distributed into 12 cages of 9 birds each. Four cages were allocated to each diet. For the first three days, the chicks received only commerical diet. Then, each of the two experimental diets were given *ad libitum* to quadruplicate groups of 9 birds. The increase in weight of each bird was recorded separately every week. The growth of chicks was expressed as Protein Efficiency Ratio (PER).

$$PER = \frac{\text{Weight gain (g)}}{\text{Protein eaten (g)}}$$

Table 1. Ration for Chick Experiments

Ingredients	Crude protein (% wt.)	Amount % of weight	
		Standard diet (I)	Test diet (II)
Maize	8.30	50.00	50.00
Groundnut meal	46.00	25.67	25.67
Fish meal	60.43	10.00	5.00
Microbial product	23.94	-	12.60
Dicalcium phosphate	-	3.00	3.00
Salts + Vitamins	-	1.20	1.20
Starch	-	10.13	2.53
Analytical (crude protein %)		22.10	21.69

RESULTS

Chemical analysis of substrate. The average composition of five samples of overripe fruit shows that it is a rich source of sugars (44.50%) and poor in protein (6.88%). The detailed chemical analysis of green, mature and over-ripe fruit is given in Table 2. The pH of the reject fruit was 6.5 to 7.0. The results reported below are average of four separate determinations.

Table 2. Composition of Banana (g/100 g dry weight)

Constituent	Green fruit	Mature fruit	Over-ripe reject fruit
Dry matter	21.25	18.70	17.70
Total sugars	3.55	33.60	44.50
Reducing sugars	2.95	24.86	28.99
Starch	32.78	6.75	5.81
True Protein	3.08	5.13	5.44
Crude protein	4.75	6.13	6.88
Ash	5.00	4.90	6.00
Cellulose	6.18	22.18	14.40
Hemicellulose	25.08	2.95	4.91
Lignin	5.41	6.10	7.17

Effect of incubation period. *A. niger* (4-5 x 10^6 spores/mL and volume of 10% of the weight of the substrate) was inoculated on sliced banana having 50% moisture content supplemented with 200 mg N/flask as ammonium sulphate, 30 mg P/flask as dipotassium hydrogen phosphate and incubated at 25°C. The fungus grew rapidly on the substrate and white mycelial growth was evident in the flasks. The substrate appeared to be thoroughly ramified by the fungus as judged by visual examination. Therefore, a 3-day period was chosen for incubation in further experiments. The mould yield of true protein was 15.50% and crude protein 21% at the end of 4 days growth.

The decrease in dry weight (36%), and the utilization of total carbohydrates (73.8%) reflected the growth of the mould on banana. The fall in pH was from 6.9 to 2.1 at the end of fermentation. Although, as is clear from Table 3, growth had not ceased after 4 days of incubation, it was not continued for a longer period in subsequent experiments because of extensive sporulation which could be a health hazard and undesirable in animal feed.

Table 3. Effect of Incubation Period on Changes in pH, Dry Weight, Substrate Total Carbohydrate and Protein Yield (Crude and True) of *Aspergillus niger* M1 when grown on Slices of Over-ripe Reject Banana having 50% Moisture at 25°C.

Incubation period (days)	Decrease in dry weight (%)	Utilization of total carbohydrates (%)	Crude protein (%)	True protein (%)	Final pH
1	20	59.5	7.6±0.90	5.8	5.5
2	28	69.5	13.9±0.35	10.5	4.2
3	32	72.6	17.3±0.88	13.2	3.2
4	36	73.8	21.0±0.70	15.5	2.1

Effect of temperature. The flasks were incubated at 25, 30, 35 and 40°C for 3 days. The maximum amount of protein was produced at 30 and 35°C. The yields of crude protein (18.9%) and true protein (14.0% and 14.1%) were almost the same at both the temperatures (Table 4). The amount of total carbohydrates utilized and the increase in dry weight corresponded to protein yield at the end of 72 hours fermentation. The fall in pH was more or less the same at both the temperatures. An incubation temperature of 35°C was selected for further study.

Table 4. Effect of Temperature of Incubation on Changes in pH, Dry Weight, Substrate Total Carbohydrate and Protein Yield (Crude and True) of *A. niger* M1 when Grown for Three Days on Slices of Over-ripe Reject Banana Having 50% Moisture.

Temperature of incubation (°C)	Decrease in dry weight(%)	Utilization of total carbohydrates (%)	Crude protein (%)	True protein (%)	Final pH
25	27.3	62.8	14.7±0.73	10.5	3.1
30	35.6	72.0	18.9±0.73	14.0	2.4
35	36.0	72.2	18.9±0.89	14.1	2.3
40	16.4	52.4	10.3±0.60	8.2	4.5

Effect of moisture content. The effect of various values of moisture content of banana on fermentation is shown in Table 5. The maximum crude protein (19.8%) and true protein (14.6%) were produced by the fungus at 60% moisture level. The differences between protein content of the mycelium growth at 50, 60 and 70% moisture were not much, but were significantly greater than with 40% moisture content of substrate.

Table 5. Effect of Moisture Content of Substrate on Changes in pH, Dry Weight, Substrate Total Carbohydrate and Protein Yield (Crude and True) of *A. niger* M1 when Grown for 3 Days on Slices of Over-ripe Reject Banana at 35°C.

Moisture content of banana (% w/w)	Decrease in dry weight(%)	Utilization of total carbohydrates (%)	Crude protein (%)	True protein (%)	Final pH
40	35.0	70.0	12.5±0.57	10.8	2.9
50	36.9	70.0	18.3±0.57	14.3	2.8
60	37.2	74.0	19.8±0.33	14.6	2.8
70	37.7	72.8	18.5±0.62	14.5	2.7

The corresponding utilization of total carbohydrates and decrease in dry weight and fall in pH were in agreement with protein production by the organism. For the experiment on further optimization of protein production, 60% moisture content of substrate was used.

Effect of concentration of nitrogen. Various concentrations of ammonium sulphate (150, 200, 250, 300 mg N/50g of reject banana) were used to see the effect on sugar consumption and protein production by the fungus (Table 6). The increase in nitrogen concentration from 150 mg N to 250 mg N/50g banana increased the true protein (10.50% to 15.80%) and crude protein (14.6% to 21.2%) content of *A. niger*. A further increase up to 300 mg/50g banana did not increase the protein production. It was, therefore, decided to use 250 mg N/50g banana in future experiments.

Table 6. Effect of Concentration on Nitrogen as Ammonium Sulphate on Changes in pH, Dry Weight, Substrate Total Carbohydrates and Protein Yield (Crude and True) of *A. niger* M1 when Grown for Three Days on Slices of Over-ripe Reject Banana having 60% Moisture at 35°C.

Nitrogen (mg) added/ 50g banana	Decrease in dry weight(%)	Utilization of total carbohydrates (%)	Crude protein (%)	True protein (%)	Final pH
150	30.0	70.0	14.6±0.98	10.5	3.0
200	37.2	72.0	19.8±0.49	14.8	2.2
250	38.0	75.0	21.2±0.34	15.8	2.1
300	38.2	76.0	21.2±0.08	15.5	2.2

Effect of phosphorus concentration. The effect of different concentrations of phosphorus (30, 50, 75 mg/50g banana) on the growth of fungus is presented in Table 7. True protein contents of 14.8% and 14.5% were observed at concentrations of 50 mg an 75 mg phosphorus, respectively. Thus, the differences in protein synthesis due to the addition of different amounts of phosphorus were marginal. On the other hand, if the control (12.6%) is compared with the best protein yield(14.8%) with 50 mg phosphorus, the difference did not correspond with the amount of K_2HPO4 added. There was a corresponding decrease in dry weight, fall in pH and utilization of carbohydrates.

Table 7. Effect of Concentration of Phosphorus as Dipotassium Hydrogen Phosphate on Changes in pH, Dry Weight, Substrate Total Carbohydrates and Protein Yield (Crude and True) of *A. niger* M1 when Grown for Three Days on Slices of Over-ripe Reject Banana having 60% Moisture at 35°C.

Phosphorus added (mg/ 50gm banana)	Decrease in dry weight(%)	Utilization of total carbohydrates(%)	Crude protein (%)	True protein (%)	Final pH
0	32.5	72.8	16.8±0.89	12.6	2.5
30	35.7	75.0	18.9±0.94	14.1	2.5
50	36.2	76.5	19.3±0.29	14.8	3.0
75	36.0	76.7	19.3±0.54	14.5	3.0

Large-scale production of animal feed in aluminum trays. It was observed that 5 kg of reject banana per tray was the optimum quantity to obtain good mycelial growth and to avoid sporulation up to 48 hours of incubation. It was also recorded that thin spreading of the substrate helped achieve good growth of the organism which then penetrated thoroughly into the substrate. In order to reduce the cost at the expense of a marginal loss of protein yield, phosphorus was not added in these experiments. The final product had crude protein content of 20 to 25% and true protein of 14.5 to 17.8%. The average crude and true protein contents of the final product were 23.94% and 16.88%, respectively.

Table 8. Amino Acid Composition of *A. niger* M1 + Residual Over-ripe Reject Banana Compared to Fish Meal.

Amino acid	Microbial product g amino acid/16 g nitrogen	Fishmeal g amino acid/16 g nitrogen
Arginine	4.83	5.00
Cystine	1.48	1.00
Histidine	2.37	2.30
Leucine	7.03	7.30
Isoleucine	4.48	4.80
Lysine	4.79	7.00
Methionine	1.52	2.60
Phenylalanine	5.21	4.00
Threonine	9.08	4.20
Tryptophan	N.D.	2.90
Tyrosine	2.58	2.90
Valine	2.26	5.20
Glycine	3.60	3.48

Fermentation conditions: Temperature of incubation, 35°C ± 0.5; incubation period, 2 days; mosture content of substrate, 60-62%.

*Tannenbaum and Wang (1975)

Table 9. Protein Efficiency Ratio of Diets

Replicate	Diet I	Diet II
1	2.53	2.25
2	2.48	2.21
3	2.25	2.31
4	2.31	2.40
Mean	2.39±0.066	2.29±0.041

Amino acid analysis of the final product. The amino acid content of the mould plus the residual substrate is shown in Table 8. The product is particularly low in lysine (4.79g/100g protein), methionine (1.52g/100g protein) and tyrosine (2.58g/100g protein), when compared with an expensive and quality protein, fishmeal. The other amino acids, such as leucine, isoleucine and arginine, were present at slightly lower levels and the rest were higher than fishmeal.

Feeding trials. The average PER in chicks fed diets I and II were 2.39 and 2.29 respectively at the end of three weeks. The difference was found statistically to be non-significant (Table 9).

DISCUSSION

The results of the present experiments show that over-ripe banana is suitable for the growth of *A. niger* M1 and for conversion into animal feed by the solid substrate fermentation process. The optimum temperature for the growth of *A. niger* M1 on banana was 35°C. This mould has been used for the production of SCP from agricultural and food processing waste by submerged cultivation because of its rapid growth (Righelato and co-workers, 1976). However, the solid substrate fermentation has been employed by Tate and Lyle (1977) who have developed a low technology process to convert citrus peel and other fruit wastes into animal protein at a pilot plant in Belize using the same fungus at the above temperature. The process consists of composting peel and supplementing it with nutrients under non-sterile condition.

Although the difference in protein yield by *A. niger* was not significant at different moisture values (50%, 60% and 70%), the moisture content is one of the important factors for solid substrate fermentation. It plays an important role in the growth of moulds by affecting the availability of nutrients and oxygen (Hesseltine, 1977a,) Solid substrate fermentation of feed lot waste combined with feed grains to provide a basis for microbial protein production for animal feed was carried out at 35-42% moisture content of the final mixture (Rhodes and Orton, 1975), whereas pilot scale semi-solid fermentation was carried out at 70% substrate moisture for 20-40 hours a 20 to 40°C (Grant and colleagues, 1978).

When the M1 strain was grown on reject fruit, an increase in the nitrogen supply up to 250 g N/50 g banana enhanced protein production by the organism. Trevelyn (1974 using urea as a nitrogen source, reported increased protein yield up to a urea concentration of 1% of the weight of cassava flour. With 2% urea, a substantial proportion of the nitrogen added was not assimilated, but accumulated in the form of ammonia produced from the complete hydrolysis of the urea.

The most important criterion for judging an animal feed is its nutritive value as demonstrated by feeding trials. When the present product was included in the diets there was no statistical difference in the PER measured in the two groups of chicks These results suggest that deficiencies created by the test product in diet II might be counter-balanced by high amount of essential amino acids present in the groundnut meal, maize and fishmeal of the ration.

Imrie and Vlitos (1973) carried out feeding trials with carob-grown M1 fungus and observed no toxic, carcinogenic or teratogenic effects on rats. Feeding tests on broiler chicks showed no adverse effects of the use of M1 protein at the end of 8 weeks. The feed efficiency ratio of the test diet was better than that of the control. They concluded that M1 had a good potential for use in animal feeding.

The solid substrate fermentation under non-sterile conditions was quite successful in the present study and this may be attributed to the fall in pH and the rapid growth of the M1 strain as compared with any contaminant prevailing in the environment during fermentation. A short doubling time (7-8 hours) of *A. niger* when it was grown on carbon extract was observed by Morris and co-workers (1973). Righelato and colleagues (1976) produced SCP from agricultural wastes using the above fungus under non-sterile condition.

The present experiments indicate that the solid substrate fermentation is suitable for conversion of reject over-ripe banana into animal feed. The results also show that 50% fishmeal, an expensive source of protein in poultry feed, can be satisfactorily replaced in part by the final product. The process holds promise for developing countries, particularly for India, where large quantities of reject fruit and cheap labour are available.

REFERENCES

Association of Official Agricultural Chemists (1975). *Official Methods of Analysis*, Washington, D.C.

Clegg, K.M. (1956). The application of the anthrone reagent to the estimation of starch in cereals, *J. Sci. Food Agr.*, *7*, 40-44.

Food and Agriculture Organisation (1976). *Monthly Bulletin of Agric. Economics and Statistics*, *25*, 54.

Fuller, K.W. (1966). Automated determination of sugars. *Automation in Analytical Chemistry*, *11*, 57-60.

Goering, H.K. and P.J. van Soest (1970). *Forage and Fibre Analysis*, U.S. Department Agric. Handbook No. 379.

Grant, C.A., Y.W. Han and A.W. Anderson (1978). Pilot scale semi-solid fermentation of straw. *App. Environ. Microbiol.* *35(3)*, 549-53.

Hesseltine, C.W. (1977a). Solid state fermentation Part-I, *Process Biochem.*, *12(6)*, 24-27.

Hesseltine, C.W. (1977b). Solid State Fermentation Part-II, *Process Biochem.*, *12(9)*, 29-32.

Imrie, F.K.E. and A.J. Vlitos (1973). Production of fungal protein from Carob (*Ceratonia siliqua*, L.) In *Single Cell Protein II*, S.R. Tannenbaum and D.I.C. Wang (Ed.), MIT Press, Cambridge, U.S.A., pp. 223-235.

Morris, G.G., F.K.S. Imrie and K.C. Phillips (1973). The production of animal feed stuffs by the submerged culture of fungi on agricultural wastes. II Int. Conf. of Global Impacts of Applied Microbiology, Saõ Paulo, Brazil, 23rd July, 1973.

Norris, J.R. and D.W. Ribbons, Eds. (1971). *Methods in Microbiology*, *5B*, Academic Press, London.

Potty, V.H. (1969). Determination of proteins in the presence of phenols and pectins, *Analytica Biochem.*, *29(3)*, 535-539.

Rhodes, R.A. and W.L. Orton (1975). Solid substrate fermentation of feed lot waste combined with feed grains. *Trans. ASAE.* *18*, 728-33.

Righelato, R.C., F.K.E. Imrie and A.J. Vlitos (1976). Production of single cell protein from agricultural and food processing wastes. *Resource Recov. Conserv.*, *1*, 257-269.

Sethi, R.P. and R.S. Virk (1977). Evaluation of *Penicillium crustosum* as poultry feed. *Indian J. Nutr. Dietet.* *14*, 345-348.

Tate and Lyle (1977). Group research and development, *Annual Report*, Tate and Lyle Ltd., U.K.

Tannenbaum, S.R. and D.I.C. Wang (1975). *Single Cell Protein II*, The MIT Press, Cambridge, U.S.A.

Trevelyn, W.E. (1974). The enrichment of cassava with protein moist solid fermentation. *Tropical Science*, *16(4)*, 179-194.

ACKNOWLEDGEMENT

The authors are grateful to the Association of Commonwealth Universities for the award of a Commonwealth Scholarship to RPS and to Dr. A. J. Forage, Tate and Lyle Ltd., U.K. for useful discussions. The work was done in the Department of Microbiology, University of Reading, U.K.

8
EXTRACTION OF VANADIUM FROM OIL SANDS FLY ASH

P. J. Griffin* and T. H. Etsell

Department of Mineral Engineering, University of Alberta, Edmonton, Canada

ABSTRACT

The Suncor (Great Canadian Oil Sands) oil sands processing operation in northern Alberta produces coke which is subsequently burned to produce fly ash that contains over 2% vanadium. The vanadium is at least partially present as an iron vanadium oxide and concentrated on the surface of the fly ash spheres. The optimum extraction process involves roasting of the ash with sodium chloride and subsequent leaching with hot water to extract 85-90% of the vanadium which can then be precipitated by the addition of ammonium hydroxide.

KEYWORDS

Vanadium, fly ash, oil sands, Athabasca oil sands, salt roasting, leaching, ammonium metavanadate.

INTRODUCTION

The oil sands of northern Alberta contain approximately 143 x 10^9 tonnes of raw bitumen from which it is expected (Berkowitz and Speight, 1975) that at least 250 x 10^9 bbl of synthetic crude oil will be produced. The two plants currently operating, Syncrude and Suncor, can produce 19 x 10^4 bbl annually.

The raw bitumen contains 150-200 ppm vanadium; processing of the bitumen produces coke with approximately 0.1% vanadium. Subsequent boiler firing of the Suncor delayed coke produces fly ash at a rate of 77 tonnes/day containing 2.3% vanadium. This amounts to over 550 tonnes of vanadium annually, twice the current Canadian usage.

Industrially,most vanadium is recovered from uranium-vanadium ores or vanadiferous slags occasionally supplemented by enriched ashes. In Canada the only commercial recovery of vanadium was in the early 1970's by acid leaching of fly ash derived from Venezuelan crude oil at the Petrofina refinery in Quebec.

The extraction of vanadium from combustion residues has been examined many times for various ashes but never comprehensively or with any attempt to relate extraction to ash chemistry. The only study (Stemerowicz and others, 1976) of the possible

* Present address: Esso Resources Canada Ltd., Calgary, Alta.

recovery from the Athabasca oil sands ash involves direct smelting of the ash to form FeV/FeNi/FeVNi alloys.

Other reports (Walker, Luhning and Rashid, 1976; Jack, Sullivan and Zajic, 1979) have suggested that the oil sands ash is not amendable to direct leaching but no conclusive information is available. It also appears that ash produced from a flexicoker (Matula, Molstedt and Ryan, 1975) or by low temperature ashing (Jack, Sullivan and Zajic, 1979) is readily amendable to direct acid leaching of vanadium and nickel.

EXPERIMENTAL

The initial analysis of the ash and the analysis of the leaching residues was carried out by atomic absorption spectroscopy after digestion of the solid.

The primary technique used was that suggested by Perkin Elmer (1973). Approximately 0.1 g of ash was heated with 10 mL of hydrofluoric acid until near dryness. After cooling, 10 mL of perchloric acid was added and the solution steamed for 2 min. Then, after cooling again, a few drops of concentrated HCl was added and the solution heated until near dryness. If no undissolved ash remained, the residue was dissolved with 15-25 mL of 10% HCl and made up to 100 mL with distilled water and 5 mL of 4% lanthanum chloride solution to reduce ionization interferences To dissolve any remaining ash, the perchloric acid addition was repeated.

The determination of vanadium by atomic absorption spectroscopy can be affected by enhancing interferences from dissolved titanium and aluminium. The titanium and aluminium dissolved in the optimization studies was low enough (<1 ppm) to have negligible effect on the vanadium determination. The preliminary extraction soluti and their associated vanadium atomic absorption standards were made up to contain 500 ppm titanium and aluminium. At these levels the enhancement has been maximized

The vanadium, nickel and iron content of the ash was confirmed by the digestion technique of Nadkarni (1980); the vanadium content was additionally confirmed by instrumental neutron activation analysis. The silica content was determined by the gravimetric technique of Rigg and Wagenbauer (1964).

In the extraction studies, 20 g ash samples were reacted with 60 mL of leachant wit mixing under reflux conditions at a temperature of 97°C for 4 hours. Additional mixing in a high temperature shaker bath or for longer periods of time had a neglig ible effect on the vanadium extraction.

The roasting was carried out in a tube furnace for approximately 6 hours. The apparatus designed by Skeaff (1979) was used for the chlorination experiments; roasting under neutral and reducing conditions was carried out in a quartz tube with a gas flow of argon or hydrogen.

ASH STRUCTURE

The major elemental components of the Suncor fly ash on which all the extraction studies were carried out are shown in Table 1.

TABLE 1 Major Elemental Components of Suncor Fly Ash

Element	Percentage
Al	6.19
C	47.8
Ca	0.38
Fe	3.66
K	1.05
Mg	0.54
Na	0.57
Ni	0.68
S	2.65
Si	10.3
Ti	2.90
V	2.39

Most of the fly ash exists as aluminosilicate spheres and carbon granules as shown in Fig. 1. Cenospheres and pleurospheres can also be found in the ash. X-ray diffraction studies have identified an iron vanadium oxide, $Fe^{+3}_{2.6}V^{+3}_{5.5}V^{+5}_{1.6}O_{16}$ and a series of silicates of the general form x $NaAlSi_3O_8$, (1-x) $CaAl_2Si_2O_8$.

A previous study (Bueno, Rempel and Spink, 1980) has indicated that an excess of 7% of the ash is made up of microcrystals rich in iron, vanadium and titanium attached to the surface of the spheres. In the ash samples used for this study, these microcrystals were only very rarely located.

The ash has a density of 2.24 g/cm^3 determined by an air phychnometer and a mean particle diameter, using a Coulter Counter Particle Size Analyzer, of 15.7 microns. Further particle sizing was carried out on a Warman Cyclosizer; each fraction was then chemically analyzed and sized on the Coulter Counter. Figure 2 shows the variation in vanadium content of the sized ash samples and their mean spherical diameter. However, on ashing at 500°C, the carbon distribution in the fractions is such that the vanadium is 4-5% in every sized sample.

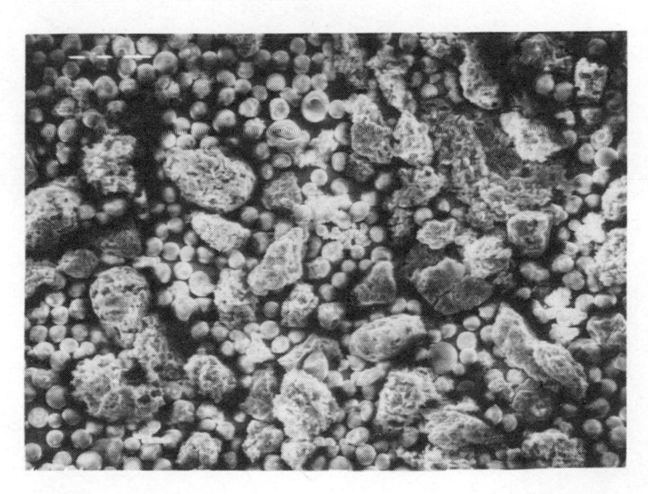

Fig. 1 Electronmicrograph of Suncor fly ash (Mag; 380).

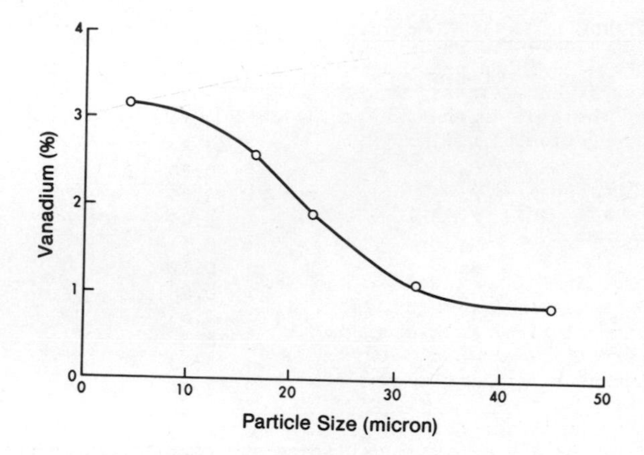

Fig. 2 Relationship between particle size and vanadium concentration.

VANADIUM EXTRACTION

The initial series of vanadium extraction tests involved leaching, with a wide variety of reagents, of the following ashes:-

1. original ash
2. ash roasted in air at 600°, 800° and 1200°C
3. ash roasted with sodium chloride or sodium carbonate

A review of these results is presented in Table 2.

TABLE 2 Vanadium Extraction(%) By Various Roast/Leach Systems

Leachant	Roasting Temp. (°C) & Additive (25%)				
	None None	800° None	1200° None	800° NaCl	800° Na_2CO_3
H_2O	0.2	5.7	0.4	68.8	63.0
10% Na_2CO_3	6.7	38.4	3.7	75.2	59.1
10% NH_3	4.5	20.2	0.7	58.6	40.1
10% HNO_3	13.5	57.3	5.3	*	*
10% HCl	13.9	45.7	9.0	*	*
10% H_2SO_4	17.3	54.0	5.7	*	*
10% NaOH	27.2	29.7	40.0	79.5	46.4

*gelatinous product

These results and an examination of the leaching of iron and nickel can be summar as:

- direct leaching does not extract a significant amount of any metal
- air roasting and acid leaching extracted over 50% of the vanadium and less than 5% of the iron and nickel

- the low extraction with ash roasted at 1200°C is due to vanadium becoming entrapped during melting of the silicates
- the gelatinous product formed from acid leaching of salt-roasted ash appears (Tarasevich and others, 1973) to be due to activation of the clay in the ash by the dissolution of MgO and the eventual production of silica gel.

Further extraction studies using ash roasted under neutral or reducing conditions, and attempting direct chlorination of the original ash all produced less than 5% vanadium extraction.

The highest extraction of vanadium was achieved by initial roasting of the ash with sodium chloride or sodium carbonate. Roasting with other salts such as potassium chloride, calcium chloride, calcium carbonate and sodium sulphate produced less than 50% vanadium extraction. Also a sodium chloride/sodium carbonate mixture did not significantly increase leaching recovery.

Optimization of the roasting temperature was carried out for the air roast-sulphuric acid leach, sodium chloride roast-sodium hydroxide leach and sodium chloride roast-water leach systems. The optimum vanadium extractions for these systems were found to be 60%, 79% and 82% at temperatures of 770°, 875° and 905°C, respectively. Figure 3 shows the dependence of vanadium extraction efficiency on roasting temperature for the sodium chloride roast/water leach system. The decrease in vanadium extraction at high roasting temperatures appears to be due to the entrapment of vanadium in partially molten silicates; the increase between 800° and 900°C due to the formation of molten sodium chloride with decreasing viscosity and better reactivity.

Further evaluation of this process led to the optimum usage of a 14 wt % sodium chloride addition to the ash and the results are shown in Fig. 4. The decrease in extraction at high salt contents may be due to the formation of a low melting point sodium silicate.

The effect of leaching temperature also was studied, as shown in Fig. 5. Additionally, the vanadium extraction was found to decrease at a phase ratio below 1 mL/g of roasted ash. Thus the optimum extraction conditions were determined to be:

roasting temperature: 905°C
roasting atmosphere: air
roasting time: approx. 6 hours
roasting additive: 14 wt.% sodium chloride
leaching solution: water
leaching temperature: 97°C
leaching time: approx. 3 hours
phase ratio: 1 mL/g of roasted ash

This procedure produces a solution containing 10 g/L vanadium and:

0.00 - 0.01 g/L: titanium, nickel
0.01 - 0.10 g/L: iron, aluminium
0.10 - 1.00 g/L: magnesium
1.00 - 3.00 g/L: calcium, potassium, sodium

The vanadium can then be recovered by precipitation of ammonium vanadate under the conditions suggested by Whindgren, Bauerle and Rosenbaum (1962) and used industrially.

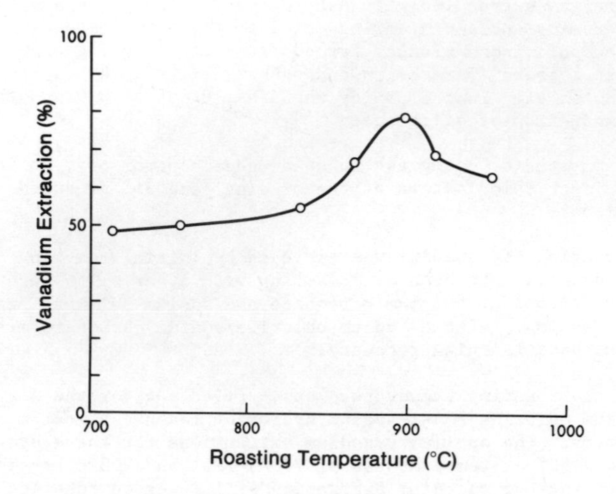

Fig. 3 Dependence of vanadium extraction on roasting temperature. (10% NaCl)

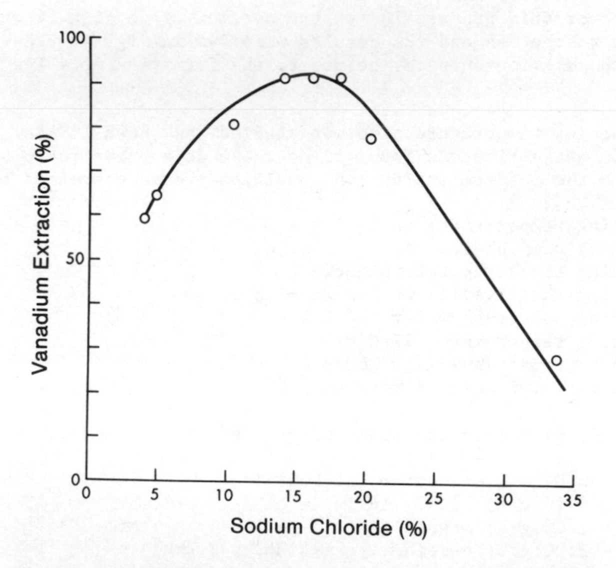

Fig. 4 Dependence of vanadium extraction on NaCl additions.

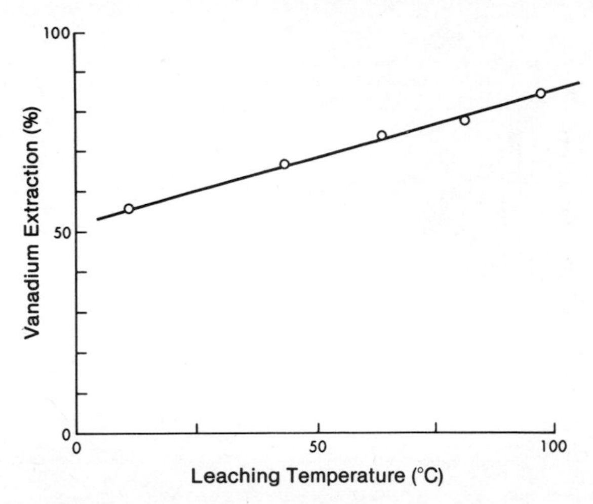

Fig. 5 Effect of leaching temperature on vanadium extraction.

REFERENCES

Berkowitz, N. and J. G. Speight (1975). The oil sands of Alberta. Fuel, 54, 138-149.

Bueno, C. O. G., G. L. Rempel and D. R. Spink (1980). Physical and chemical characterization of Athabasca tar sands fly ash. CIM Bulletin, 820, 147-151.

Jack, T. R., E. A. Sullivan and J. E. Zajic (1979). Leaching of vanadium and other metals from Athabasca oil sands coke and coke ash. Fuel, 58, 589-594.

Matula, J. P., B. V. Molstedt and D. F. Ryan (1975). Flexicoker proto-type demonstrates successful operation. Amer. Petrol. Inst. Div. Refining Proc., 54, 225-229.

Nadkarni, R. A. (1980). Multitechnique multielement analysis of coal and fly ash. Anal. Chem., 52, 929-935.

Perkin Elmer Canada Limited (1980). Analytical methods for atomic absorption spectrophotometry.

Rigg, T. and H. A. Wagenbauer (1964). Analysis of silicate rocks. Report 64-1, Research Council of Alberta.

Skeaff, J. M. (1979). Chlorination of uranium ore for extraction of uranium, thorium and radium and pyrite removal. CIM Bulletin, 808, 120-125.

Stemerowicz, A., R. W. Bruce, G. V. Siranni and G. E. Viens (1976). Recovery of vanadium and nickel from Athabasca tar sands fly ash. CIM Bulletin, 768, 102-108.

Tarasevich, Y. I., F. D. Ovcharenko, F. A. Belik, I. I. Martsin and V. P. Vasilenko (1973). Influence of acid activation on the structure and adsorptive properties of clay minerals. Kolloidnyi Zhurnal, 35, 467-475.

Walker, L. A., R. W. Luhning and K. Rashid (1976). Potential for recovering vanadium from Athabasca tar sands. 26th Can. Chem. Eng. Conf., Toronto.

Whindgren, C. J., L. C. Bauerle and J. B. Rosenbaum (1962). Preparing metal-grade vanadium oxide from red cake and mill solutions. USBM, R of I 5937.

SECTION III

ENERGY RECOVERY

9

THE USE OF STILLAGE IN THE FERTI-IRRIGATION OF SUGARCANE CROP LAND DEDICATED TO FUEL ALCOHOL PRODUCTION

C. Costa Ribeiro and J. R. Castello Branco

Centro de Tecnologia Promon, 22210 Rio de Janeiro, RJ., Brasil

ABSTRACT

The production of fermentation ethanol from biomass (sugarcane and mandioca) is being fostered by the Brasilian Government as an alternative for imported fossil fuels. Stillage or bottom slops - the main liquid effluent derived from fermentation ethanol distillation - is produced in Brasilian conventional sugarcane and mandioca distilleries at a rate 12 to 13 times the alcohol volumetric output. Because of this high volume output and low solids content (2 to 8%), mainly in dissolved form, stillage represents a potential threat as a water pollutant.

Conventional alternatives for stillage treatment as a residue to comply with pollution control legislation may represent a drawback in the economics of ethanol production as a replacement for gasoline. On the other hand, the water, organics and minerals present in stillage may be used to advantage in the ferti-irrigation of agricultural feedstock crops such as sugarcane and mandioca.

This paper discusses the replacement of conventional NPK fertilizer by in natura stillage in sugarcane and/or mandioca cultivation. Technical and economic problems and opportunities based on the Brasilian experience of this stillage recovery alternative are assessed.

KEYWORDS

Alcohol stillage slops; ferti-irrigation; sugarcane cultivation; mandioca cultivation; pollution abatement; fuel ethanol

INTRODUCTION

Brasil's National Alcohol Program (PROALCOHOL) was set up in 1975 as a result of two main factors:

- the acute, sudden decline in international sugar prices which discouraged Brasilian sugar exports;
- the concurrent hike in world oil prices, which worsened Brasil's already dependent position as importer of over 80% of its fossil fuel needs.

WTUZ - H

Together with other social and political factors, those two events were the main stimulants of the program. They took on greater importance than the foreign exchang savings from the utilization of ethyl alcohol to replace gasoline, which is proportionately small when considered in terms of Brasil's overall balance of payments (Centro de Tecnologia Promon, 1979a).

Optimistic estimates of PROALCOHOL's evolution indicate fermentation ethyl alcohol output in 1985 at no less than 10.7 million m^3. If achieved, that volume of production will make Brasil the world's largest alcohol-producing nation. Of the 16 million m^3 estimated for world production in 1985, Brasil will provide more than 60% (Centro de Tecnologia Promon, 1979a).

As a consequence, the annual production of stillage (vinasse, slop, bottom slop or still residue), the bottom stream of the first ethanol distillation column, could reach at least 130 million m^3 by 1985. It is unlikely that simple and direct solutions for recovery of stillage products in Brasil will be implemented in the short run. However, it can be stated that a nationwide total solution will certainly require large sums of subsidized financing. It is therefore of utmost importance that the various feasible solutions optimize economic and social cost/benefit ratio If the economic recovery of stillage products is not possible, suitable processing only for pollution abatement purposes could substantially boost the selling price of alcohol, thus jeopardizing its competitiveness with respect to gasoline.

Absolute alcohol in Brasil is produced primarily from fermentation of sugarcane molasses (also known as blackstrap molasses) in distilleries adjacent to sugar mill (byproduct or molasses distilleries).

Sugarcane is usually washed prior to milling. For production of white crystal suga the juice must be clarified with SO_2 (clarification step). When raw sugar (demerar is the end-product, the juice is not treated with SO_2. Liming is the next step, whereby the sulfited juice is treated with calcium hydroxide to correct pH and to precipitate proteins, fibers, waxes and other interfering substances.

The treated juice is then concentrated in multiple-effect evaporation systems where crystallization of the sugar occurs. Next, centrifugation separates the crystalliz sugar from molasses, which contains about 60% total reducing sugars.

In the batch fermentation step, molasses is diluted with process water or cane juic diverted directly from the mill. After inoculation with yeast, under adequate temp erature and pH conditions the sugars are converted into alcohol. The fermented mas results twelve hours later. This mash is then centrifuged and the yeast milk can b recycled for another fermentation cycle (Melle Boinot process).

The first distillation step of the centrifuged mash then yields ethanol (100 proof or 50°GL) and stillage. Depending on whether water or cane juice is used to prepar the mash, the stillage is called "molasses stillage" or "mixed stillage", respectiv ly. In addition to producing ethanol and stillage, the process yields fusel oil an flegmass. The ethanol (50°GL) is further distilled for production of anhydrous ethanol. Part of the cane juice in byproduct distilleries has traditionally been diverted to molasses dilution whenever sugar prices are down or when the mill reach its sugar production quota before the end of the harvest season.

Absolute alcohol can also be produced directly from sugarcane juice, in so called sugarcane independent distilleries. In such distilleries, before being sent to the fermentation section of the plant, the juice is sterilized to avoid infection by wild yeasts which can reduce the process alcohol yield. After fermentation, ethano is separated from the fermented wort by distillation, as are fusel oil, flegmass an stillage, which in this case is called "cane-juice stillage".

At both independent and molasses distilleries, bagasse is the fuel for raising steam used to drive the milling equipment and for consumption in other process steps.

PROALCOHOL considers mandioca as Brasil's second most important potential feedstock for fuel alcohol production. At independent mandioca distilleries, the mandioca roots are first washed and peeled for mash preparation. Cooking, liquifaction and saccharification steps are needed for transformation of starch into fermentable sugars. The sugars are then fermented and the broth sent to the distillation section for production of absolute alcohol where mandioca stillage is co-generated (Centro de Tecnologia Promon, 1977)

An analysis of the currente profile of daily ethanol production capacities at PROALCOHOL - authorized distilleries shows that the majority of the units lie in the range of 30-60 m^3 alcohol per day. However, this profile will change as Proalcohol - approved projects are gradually implemented through expansion of existing units and erection of new distilleries. Figure 1 shows this profile based on 260 PROALCOHOL projects of which 144 are already in operation. According to the PROALCOHOL schedule, the 260 distilleries would be operational for the 1984/85 harvest (Centro de Tecnologia Promon, 1979b).

CHARACTERISTICS OF BRASILIAN STILLAGES

Brasilian production of fermentation alcohol has always been subsidiary to the production of sugar. Throughout the last 15 years, 90% of the alcohol production in Brasil came from sugarcane molasses, as a byproduct of the manufacture of sugar.

Production of mixed stillage has occurred on a random basis and no information is available whereby to establish the precise volume of alcohol produced from the mixed mash (molasses diluted with cane juice).

The composition of in natura stillage from various origins is shown in Table 1. (Centro de Tecnologia Promon, 1977.) In all types considered, the organic fraction represents about 80% of the total solids content. They are mainly composed of proteins, dead yeast cells, residual sugars and ethyl alcohol, as well as substances such as waxes, fats and fibers originally present in sugarcane. Thus, several substances present in stillage are in the form of complex molecules (although in Table 1 they are expressed as oxides, corresponding to their most stable chemical form).

The characteristics of mandioca stillage, as it is known, differ substantially from those of molasses, cane juice and mixed stillages. Its low sulfur and calcium content (Centro de Tecnologia Promon, 1979b) are worth noting, since liming and sulfitation are not required in this alcohol-producing process.

As compared to other types of stillage, in most sugarcane derived stillages a large percentage of the solids present are in the dissolved form as shown in Table 2 (Centro de Tecnologia Promon, 1979b).

REVIEW OF TECHNICAL ASPECTS RELATED TO THE USE OF STILLAGE IN THE FERTI-IRRIGATION OF SUGARCANE FIELDS

Like any conventional fertilizer, stillage alters the soil's original properties as well as the productivity and characteristics of the crops grown on it. The industrial recovery of the end-products of such crops is also affected.

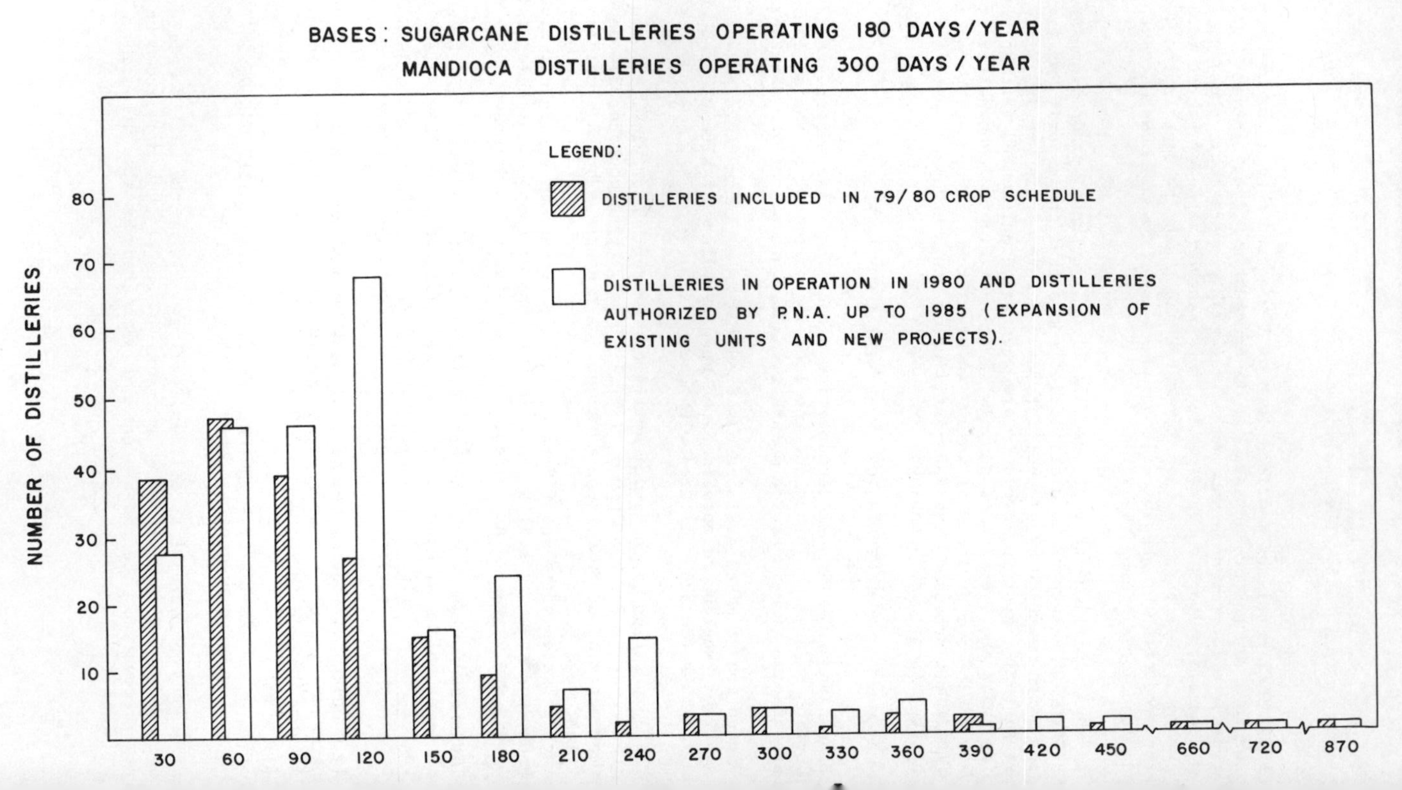
FIGURE 1
DAILY CAPACITY SIZE DISTRIBUTION OF ETHANOL DISTILLERIES
- BRASIL -
BASES: SUGARCANE DISTILLERIES OPERATING 180 DAYS/YEAR
MANDIOCA DISTILLERIES OPERATING 300 DAYS/YEAR
LEGEND:
DISTILLERIES INCLUDED IN 79/80 CROP SCHEDULE
DISTILLERIES IN OPERATION IN 1980 AND DISTILLERIES AUTHORIZED BY P.N.A. UP TO 1985 (EXPANSION OF EXISTING UNITS AND NEW PROJECTS).
NUMBER OF DISTILLERIES
80
70
60
50
40
30
20
10
30
60
90
120
150
180
210
240
270
300
330
360
390
420
450
660
720
870

TABLE 1

MEAN COMPOSITION OF *IN NATURA* STILLAGES PRODUCED IN BRASILIAN ETHANOL DISTILLERIES

(g/L)

Parameter	Type of Stillage			
	Molasses	Cane-Juice	Mixed	Mandioca
Total Solids	81.5	23.7	52.7	22.5
Volatile Solids	60.0	20.0	40.0	20.0
Fixed Solids	21.5	3.7	12.7	2.5
Carbon (as C)[a]	18.2	6.1	12.1	6.1
Reducing Substances	9.5	7.9	8.3	6.8
Crude Protein[b]	7.5	1.9	4.4	2.5
Potassium (as K_2O)	7.8	1.2	4.6	1.1
Sulphur (as $SO_4^=$)	6.4	0.6	3.7	0.1
Calcium (as CaO)	3.6	0.7	1.7	0.1
Chlorine (as NaCl)	3.0	1.0	2.0	0.1
Nitrogen (as N)	1.2	0.3	0.7	0.4
Magnesium (as MgO)	1.0	0.2	0.7	0.1
Phosphorus (as P_2O_5)	0.2	0.01	0.1	0.2
BOD	25.0	16.4	19.8	18.9
COD	65.0	33.0	45.0	23.4
Acidity[c]	4.5	4.5	4.5	4.5

(a) Carbon Content = Organic Solids Content ÷ 3.3

(b) Crude Protein Content = Nitrogen Content x 6.25

(c) Expressed as pH

The characteristics and yields of industrialized products of sugarcane, mandioca an
other crops fertilized with stillage undergo modifications that require individual
study and assessment. In sugar production, for example, it is important to know ho
stillage fertilization can affect the agricultural yield, i.e., sugarcane productio
per hectare (t/ha), cane characteristics and, consequently, the properties of the
juice from which sugar will be extracted.

Table 3 lists the main parameters influencing the yield of sugar and alcohol from
sugarcane, all of which are affected by stillage applied as fertilizer (Centro de
Tecnologia Promon, 1979b; Spencer, 1963).

Organic matter is the main fraction of the total solids present in the different
types of stillage (Table 1). It is therefore to be expected that its application
to the soil corresponds basically to an organic fertilization with the following
major effects generally observed when organic matter is added to the soil:

Reduction of soil acidity. The shift in the pH of the soil develops in two stages.
First, there is a fungi proliferation over the organic matter which acts as a substrate. The pH value drops sharply during the first ten days following addition of
the organic matter. During the second phase, the bacteria present in the soil grow
fast, accompanied by a rapid rise in pH which remains high for weeks.

Increased availability of some nutrients. Some partially immobilized soil nutrient
like phophorus, copper, iron and zinc become assimilable by plants through the dire
action of complementary organic substances or the indirect action of the lower soil
acidity (higher pH). Iron and aluminum, for example, are generally present in soil
particles as $AlPO_4$ and $Fe_3(PO_4)_2$, which plants cannot absorb. The presence of orga
ic acids and amino acids form complexes with iron and aluminum and render them
assimilable through plant roots. This effect takes place, for example, when natura
organic fertilizer is applied to the soil as manure or compost.

Increased retention of cations. Organic matter in colloidal form (as it is present
in stillage) promotes rapid increase of the soil's cation exchange capability. Thu
the efficiency of the fertilizing process will be increased by reducing the lixivia
tion of the exchangeable cations present in the soil (Gloria, 1976).

Increased water retention capacity. The presence of organic matter in the soil improves the water retention capacity of the surface layers through capillarity, thus
extending the beneficial effects of irrigation.

Improved physical structure. Organic substances act directly as a binding agent of
soil particles. This effect increases soil porosity, improving soil aeration. The
same effect could be caused indirectly through proliferation of the soil microflora
and microfauna.

Decrease in available nitrogen content. Addition of organic matter to soil enhance
proliferation of denitrifying bacteria. Thus, there is a consumption of the nitrat
present in the soil before it can be taken out by plant roots.

Increase in microbial population. Organic matter added to the soil causes rapid
proliferation of fungi. This effect is intense during the thirty days subsequent t
application, whereupon it tends to decrease.

All these effects generally are common to any type of organic matter used as fertil
izer, regardless of its origin, and have been observed to a lesser or greater degre
when stillage is applied to the soil (Gloria, 1976).

TABLE 2

REPORTED VALUES OF SOLIDS CONTENT, BOD AND COD IN RAW STILLAGES

Type of Stillage	Origin	Total (g/L)	Solids		BOD (g/L)	COD (g/L)
			Suspended (%)	Dissolved (%)		
Cane Molasses	USA	84	7	93	NR	100
Cane Molasses	Brasil	66	30	70	20	50
Cane Molasses	Cuba	60	7	93	NR	55
Mixed	Brasil	26	17	83	13	15
Cane Juice	Brasil	17	23	77	10	20
Pomace (a)	USA	32	59	41	NR	NR
Pomace (a)	USA	31	58	42	18	NR
Pomace (a)	USA	22	74	26	10	NR
Lees (b)	NR	68	87	13	20	NR
Lees (b)	NR	61	89	11	20	NR

(a) Stillage obtained in distillation of brandy (apples);
(b) Stillage obtained in distillation of brandy (grapes).

NR = Not Reported

TABLE 3

PARAMETERS INFLUENCING OPERATIONS AND YIELDS IN SUGAR & ALCOHOL PRODUCTION

Parameter	Unit	Implications
Cane Productivity	t/ha	Agricultural yield
Apparent Sugar Productivity	pol/ha	Sugar and alcohol yields
Apparent Level of Sugar in Cane	pol % cane	Sugar and alcohol yields
Solids Content in Juice	°Brix	Process consumption at mill/distillery; Stillage composition
Apparent Purity	pol/Brix	Process consumption at mill/distillery; Stillage composition
Starch Content in Juice	% (w/v)	Juice filterability
Reducing Sugars in Juice (RS)	% (w/v)	Sugars yield; crystallization efficiency
Ash content in Juice	% (w/v)	Molassigenic effect
Fiber Content in Cane	% (w/v)	Bagasse produced; steam availability for mill/distillery and stillage recovery

Stillage does not contain viruses or any kind of pathogenic bacteria. Some heavy metals are present, although in concentrations not higher than those found in conventional process water. Organic polychlorates are absent in stillage (see Table 1). Thus, contamination of underground water arising out of the ferti-irrigation of sugarcane soils with this liquid waste cannot occur. Furthermore, the low organic solids concentration found in stillage (20-60 g/L; Table 1) and its low rate of application (50m^3/ha) for ferti-irrigation of sugarcane crop land correspond to the addition of no more than 100 to 300 grams of organic matter per square meter of land. Such a relatively small mass of organic substances when compared to the amount used when natural organic fertilizers are applied to soil in the form of manure or compost cannot be considered as a potential cause for underground water contamination.

Stillage also contains potassium, sulfate and calcium in relatively high concentrations. Thus, its use as a fertilizer will increase the concentration of these ions in the soil. All are important sugar cane nutrients, especially potassium. Stillage application therefore corresponds also to fertilization with potassium, calcium and sulfate.

Thus, the ideal quantity of stillage to be applied to the soil as fertilizer is a function of its composition and the type of soil. In identifying the types of soil which respond well to stillage as a fertilizer, the following parameters should be taken into account:

Clay content. Soils with high clay content generally have a greater ion immobilizing capability that render them less susceptible to leaching. The addition of stillage will increase that retention power and could add considerable quantities of salts to the soil.

Organic matter content. Organic matter present in the soil would tend to immobilize the ions present in the stillage more than would occur in clayey soils alone, thus incurring a risk of salinization within a short period of time.

Original concentration of salts. Stillage should generally not be applied to soils with saline characteristics.

The experience accumulated in Brasil has demonstrated that applciation on some types of soil of up to 50 m^3 of cane molasses stillage per hectare does not cause soil salinization. This rate of application corresponds in fact to conventional fertilization with potassium, calcium and sulfur as practiced in other countries at levels often higher.

The different types of soils dedicated to sugarcane growing in Brasil and their stillage receptivity is presented in Table 4 (Centro de Tecnologia Promon, 1979b).

Among several others, research conducted by the Agronomics Division of Copersucar (1978) concluded that in natura stillage fertilization of sugarcane grown on dark red lateritic soil apparently increases cane/ha yield and, consequently, sugar/hectare yield. However, stillage fertilization also raises ash % Brix, while reducing apparent purity and delaying cane maturation. It was also shown that supplemented mineral fertilization can be excluded without impairing sugarcane yield and quality. Table 5 and Figure 2 indicate the conditions and results of some of the experiments conducted by Copersucar (1978).

The availability of cane stillage, whether produced from molasses, cane juice or a mixture of both, coincides with the cane harvest. Molasses is not usually stocked for ethanol production between harvests, owing to the high costs of storing required amounts and to the lack of fuel (cane bagasse) to run the distilleries. Stillage is, therefore, not available as fertilizer during the soil-preparation and cane-planting periods.

TABLE 4

RECEPTIVITY TO STILLAGE BY TYPE OF SOIL

- Sugarcane areas in Brasil -

Receptivity to Stillage	Type of Soil	Main Characteristics
Good	Lateritic	Low to average fertility requiring mineral fertilizer
	Podzolic	
	Tertiary (Tableland)	
Limited	Alluvial; Structured purple	High fertility
Limited-Low	Hydromorphic	Saline soils, with high content of Ca, Mg, Na and K

Since 75% of the cane grown in Brasil is of the 1st. ratoon type, almost all the stillage produced can be used to fertilize its plots. Furthermore, the water content of stillage will play an important role in cane cultivation since it will correspond to a "point in time" irrigation during a period when the freshly cut plant needs it most.

The cane growing periods in Brasil vary from region to region. Figure 3 illustrates the cane harvest calendar and simultaneous production of ethanol in the North-Northeastern and Central-Southern regions of Brasil (Instituto do Açúcar e do Álcool, 1977). It can be seen that alcohol production coincides with the dry season in both these regions of the southern hemisphere.

Thus the presence of water in stillage is as important as the fertilizer effect of the organic matter, calcium, sulphur and potassium also present in stillage. That is why the term ferti-irrigation is commonly used to express the beneficial effects of stillage application on sugarcane fields.

TABLE 5

STILLAGE APPLICATION CONDITIONS

Experiment	Stillage[(a)] Application Rate (m^3/ha)	Conventional Fertilizer Applied (kg/ha)		
		N	P_2O_5	K_2O
A	-	-	-	-
B	-	60	40	120
C	15	30	40	-
D	30	-	-	-
E	30	30	-	-
F	30	-	40	-
G	30	30	40	-
H	45	30	40	-
I	60	30	40	-

(a) Composition of stillage (grams/liter):
0.79N; 0.14 P_2O_5; 2.41 K_2O; 1.26 CaO; 0.67 MgO;
3.10 $SO_4^=$; 1.61 C; 21.4 organic matter

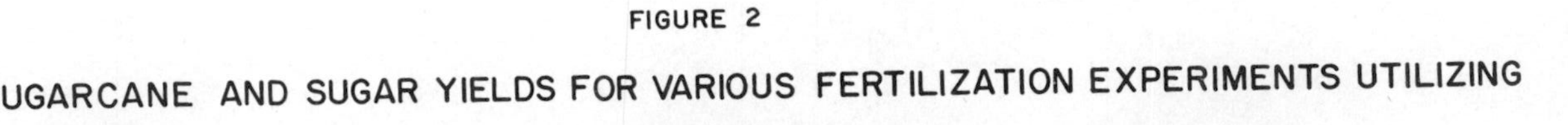

FIGURE 2

SUGARCANE AND SUGAR YIELDS FOR VARIOUS FERTILIZATION EXPERIMENTS UTILIZING IN-NATURA STILLAGE

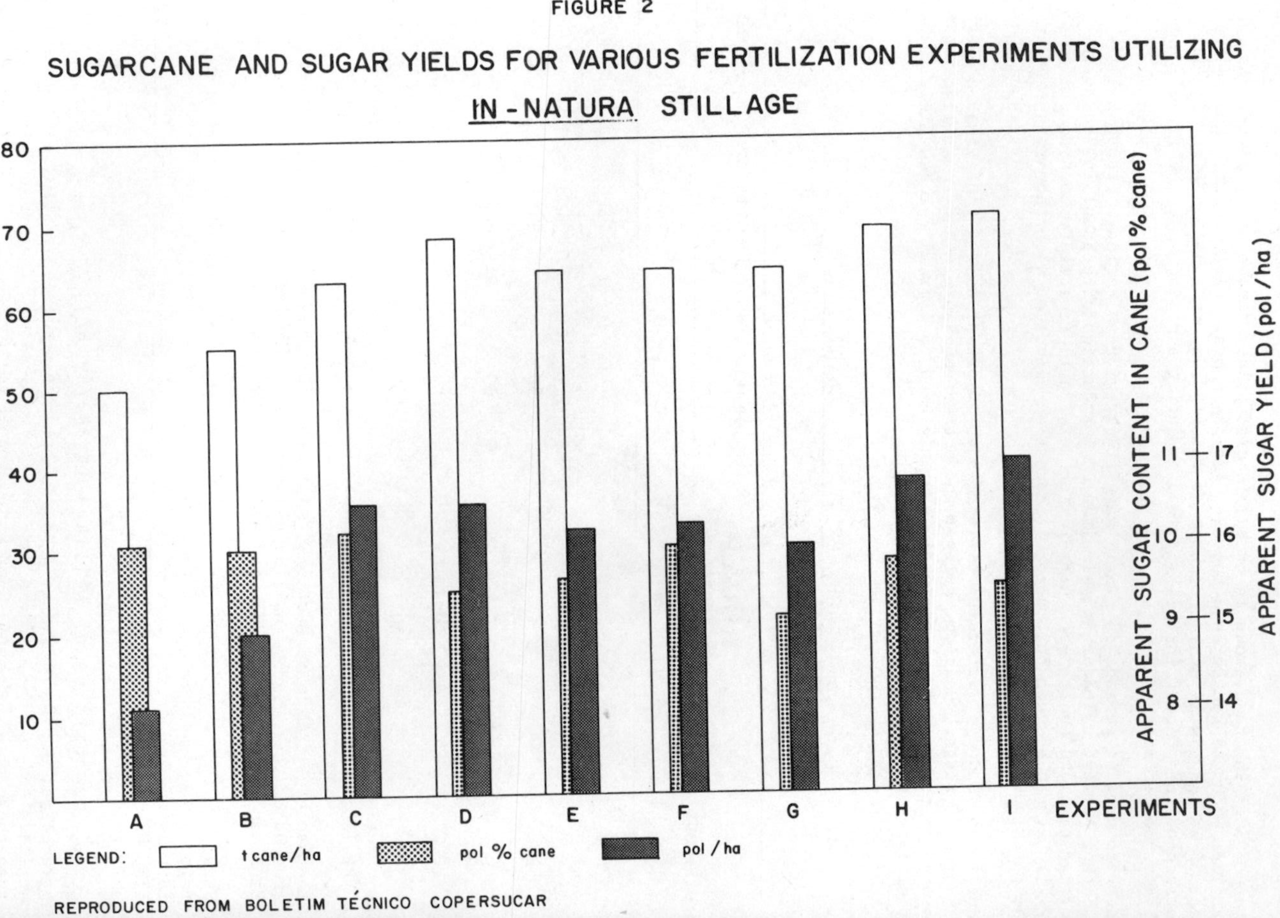

REPRODUCED FROM BOLETIM TÉCNICO COPERSUCAR

FIGURE 3

PERIODS OF SUGARCANE AGRICULTURAL OPERATIONS AND ALCOHOL/STILLAGE PRODUCTION IN BRASIL

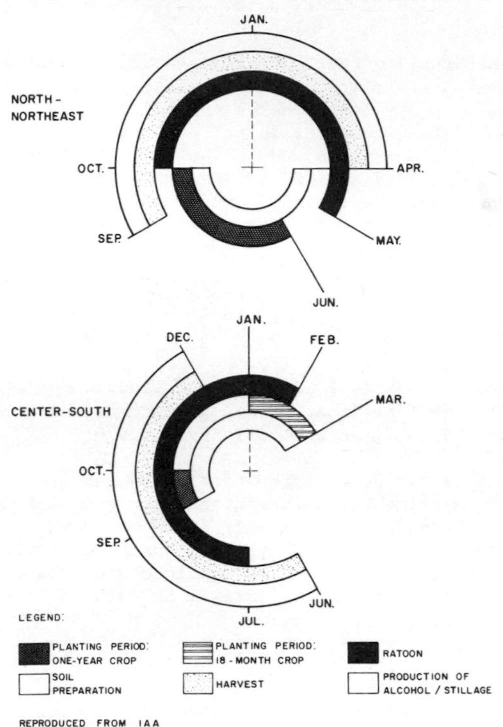

REPRODUCED FROM IAA

The experience acquired in the use of stillage for the ferti-irrigation of sugarcane in Brasil can be summed up in the following general conclusions:

. The use of stillage to fertilize sugarcane increases the crop sugar yield (pol/ha or real sucrose/ha) (Copersucar, 1977; Stupiello and co-workers, 1977; Magro and Glória, 1977; Rosetto and colleagues, 1977).

. Sugar yield remains practically unaltered through the fourth cutting when the ratoon cane is fertilized with in natura stillage. This evidence is particularly fortunate since the harvest period, which is concurrent to field fertilization, also coincides with distillery operations and, therefore, with stillage availability (Agujaro, 1977).

. Optimal stillage applciation rates are a function of soil fertility, among other parameters, but appear to fall into the 30 to 50 m^3/ha range for cane-molasses stillage and 80 to 200 m^3/ha for cane-juice stillage. Rates substantially higher than these have undesirable effects on the crop while only slightly increasing yield (Gloria, 1978; Gloria and Magro, 1976).

. Adoption of the application rates described above implies the need for a crop area for disposal of the total volume of stillage. This area corresponds to less than 15% of the total area planted for alcohol production (Centro de Tecnologia Promon, 1979b).

. Among the major undesirable effects of stillage fertilization is its ability to increase ash content in sugarcane. This increase is close to that obtained through application of equivalent mineral fertilizers and, for the suggested stillage application rates, jeopardizes neither sugar quality nor industrial yield (Copersucar, 1977; Stupiello and colleagues, 1977; Magro and Gloria, 1977; Rossetto and co-workers, 1975 and 1977; Silva and others, 1976; Rodella and Ferrari, 1977). Increased ash content, which corresponds to increased potassium concentration, can benefit the fermentation yield when cane is used exclusively for alcohol production.

. Stillage application retards sugarcane maturation, noticeably during the first two months of plant growth (Copersucar, 1978; Magro and Gloria, 1977; Rossetto and co-workers, 1975).

. Additional investigation is required for the evaluation of other possible deleterious effects of stillage fertilization, such as increased cane fiber and starch content (Copersucar, 1977; Silva and colleagues, 1976; Cesar, 1977).

. The effects of using mandioca stillage to fertilize mandioca crops are not fully known. The lack of experimental information restricts assessment of this possibility to inferences based on the composition of the mandioca stillage and that plant's requirements. Conclusive results will only be attained through field tests based on the stillage generated at mandioca distilleries that have only recently started operations in Brasil (Brasil Secrataria Tecnol. Ind., 1975; Correa, 1977; Malavolta, 1976).

STILLAGE DISTRIBUTION SYSTEMS

There are two basic methods of applying stillage to the soil: distribution through a furrow system and by mobile equipment such as tank-trucks.

Furrow System. This is the most widely-used and, therefore, the most studied method in Brasil. It involves intermediate storage tanks known as collection boxes, pumps pipes and a network of channels throughout the area where stillage is applied. The land must be prepared to adapt the system to the land topography slopes, so stillage distribution is uniform and erosion caused by excessively fast flows is prevented.

Mobile Equipment. Mobile systems have been used generally to provide for better control of the stillage application rate. However, these systems differ among themselves in terms of areas irrigated per unit of time, soil compaction, operational speed, flexibility, mechanical strength and durability (Stab, 1977).

When tank-trucks are used, stillage is usually carried and applied by the same vehicle. The capacity of the tanks, types of compressors and arrangement of the spray heads are the characteristics that vary the most, depending on manufacturer and user.

The pressurized tractor-drawn tanks operate in conjunction with tank-trucks that carry the stillage from the distillery to the field where it is sprayed. This is known as the Stillage Distribution Vehicle system (SDV); in almost all cases it is more advantageous than the pressurized tank-trucks, inasmuch as the SDV system always requires less vehicles than the tank-truck systems.

Many of stillage's effects on the soil and on the crop can be traced to the type of system and equipment utilized to spread it on the field (Stab, 1977). Table 6 summarizes the advantages and disadvantages of the various stillage application systems (Centro de Tecnologia Promon, 1979b).

TABLE 6

COMPARISON OF SYSTEMS FOR STILLAGE APPLICATION AS FERTILIZER

SYSTEMS	ADVANTAGES	DISADVANTAGES
Tank trucks: Motor-Pump or Pressure-driven (MPTT-PTT)	. 3 to 4 times faster than TTFL(a) . Uniform application over entire area	. Large investment and fuel consumption . Inapplicable on rainy days or on hilly land . Soil is compacted by vehicle, although less than TTFL
Stillage Distribution Vehicle (SDV Unit)	. Less soil compaction . Greater mobility on terrain . May be used on rainy days . Speeds comparable with those of the MPTT or PTT	. High initial capital outlay . Transport requires fleet of tank-trucks
Furrow System	. Low initial capital outlay . Stillage dilution will allow adequate disposal of other distillery waste waters	. High investment in labor to prepare terrain . Soil erosion caused by fast flows . Inaccurate control of the amount applied . High costs for maintenance of piping and pumps

(a) TTFL = Tank-truck with free fall of stillage.

INVESTMENTS

The investments in vehicles and equipment for in natura stillage application throug SDV systems are shown in Figure 4 as economy-of-scale curves covering a range of ca acities (300 to 5,000 m^3 of stillage per day, equivalent to 30 to 360 m^3 of absolut ethanol per day). (Centro de Tecnologia Promon, 1979b.)

Financing costs and charges are not included. Since stillage is an acid waste, all tanks and equipment (mainly pumps) require the use of stainless steel.

In Figure 4, investments in molasses and independent distilleries are depicted together with those required of multiple effect evaporation (MEE) of stillage. MEE is the alternative widely used throughout the world to handle the massive amounts of stillage produced in fermentation ethanol distilleries. It is evident that MEE technology represents a financing barrier for the Brasilian Gasohol program since investments in such units are of the same order of magnitude of those required in the erection of the correspondent distilleries.

Figure 4

ECONOMY OF SCALE OF ABSOLUTE ETHANOL DISTILLERIES, STILLAGE MULTIPLE EFFECT EVAPORATION AND SDV SYSTEMS

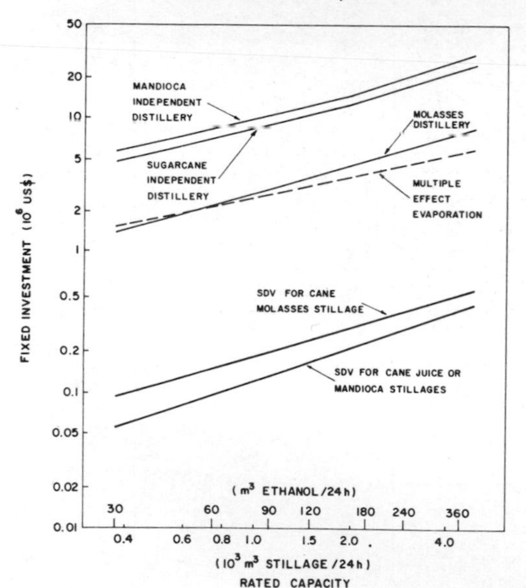

Figure 5

IN-NATURA STILLAGE APPLICATION AS FERTILIZER

-ECONOMIC RADIUS-

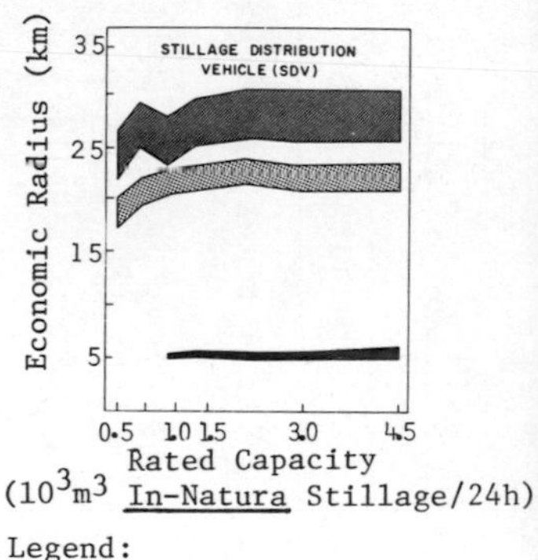

Rated Capacity
($10^3 m^3$ In-Natura Stillage/24h)

Legend:
Type of Stillage:

Cane-Molasses 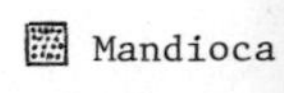 Mandioca

Cane-Juice

Bases: Boundaries Defined for Following Conditions:

- Subsidized financing and own capital schemes
- Capital recovery rates: 5-15% per year.

TRANSPORTATION AND APPLICATION COSTS

The transportation and application costs for the various types of stillage, production capacities and distillery-to-field distances served as the bases for determining the economic stillage transportation and application radius, which was then ascertained by comparing the costs calculated for transporting and applying stillage to the costs associated with conventional NPK fertilization. The economic radius is thus defined as the distillery-to-field distance - in kilometers - for which the US$/hectare cost of transporting and applying stillage is equal to the US$/hectare cost of transporting and applying the conventional fertilizer taken as a reference.

The reference costs (for conventional fertilizer) were determined assuming the same financing scheme and capital recovery rate (CRR) adopted for calculating the stillage fertilization cost and using the Uniform Annual Series of End-of-Year-Payments (UASEP) Method over the 15-year period analyzed (Grant and Ireson, 1970).

Results of this comparative analysis are depicted in Figure 5 where the ranges of values cover the PROALCOHOL financing and own-capital schemes and CRRs of 5 to 15% year (Centro de Tecnologia Promon, 1979b). Inspection of this figure permits the following conclusions:

. the economic radii are a function of the type of stillage and are practically independent of the distillery's daily stillage output;

. the minimum daily cane-juice stillage volume which allows its economical application by SDV systems is 750 m^3;

. the economic radius for mandioca stillage is bigger than that for cane-juice stillage because the annual operating period of the application system (330 days) is longer and the application cost of the equivalent conventional fertilizer is higher. The operating period of sugarcane based distilleries is 200 days per year.

CONCLUSIONS

Based on recent experimental evidence accumulated in the Brasilian sugarcane agroindustry, it is possible to conclude that in natura stillage can be used as a fertilizer for sugarcane. Through this practice, increased cane productivity - in terms of tons of cane per hectare - is obtained without jeopardizing alcohol industrial yield (liter of alcohol/ton of cane).

The SDV system seems to be the most suitable equipment for stillage land distribution due mainly to: its capability of operating under harsh environmental conditions, such as rain and terrain roughness; greater field mobility; better controlled application capability; and lower soil compacting rate.

The effects of applying stillage to the soil may be regarded as resulting from mineral-organic fertilization combined with irrigation. The borders of each of these effects are thus difficult to define, suggesting that stillage is actually a fertilizer with unconventional characteristics.

In comparison with other processing alternatives, the use of in natura stillage as fertilizer presents one of the lowest cost/benefit ratios. The measurable benefit in certain cases can be attributed to the replacement of conventional NPK fertilizer on sugarcane by in natura stillage.

Moreover, this stillage recovery alternative requires low investment for implementation. This investment is one order of magnitude lower than those for industrial stillage processing, such as stillage evaporation through multiple-effect systems.

In case no credit is given to the fertilizer value of stillage, there will be a negative impact on the alcohol selling price on the order of 5%. This figure refers to the molasses stillage application at 30 km from a distillery having a daily production of 30 m^3 absolute ethanol.

The great majority of soils dedicated to sugarcane cultivation in Brasil have characteristics that permit controlled application of in natura stillage as fertilizer. Restrictions would only be the result of topographical characteristics of the terrain that hindered the controlled application of stillage by mobile equipment. Uneven lands require the utilization of a furrow system, in which case it is difficult to control the amount of stillage distributed.

REFERENCES

Agujaro, R. (1977). Uso da vinhaça na Usina Tamoio como ferti-irrigação. Seminário sobre Sistemas de Aplicação da Vinhaca no Solo, STAB/ESALQ, Piracicaba.

Brasil. Secretaria de Tecnologia Industrial (1975). O Etanol como Combustível. Brasília, 94 p.

Centro de Tecnologia Promon (1977). Alcohol From Cassava - Technical-Economic Feasibility, Plant Implementation Distillery: Multiclient Study.

Centro de Tecnologia Promon (1979a). Alcohol: The Brasilian Way Out of The Energy Crisis. Paper submitted to the Unitar Conference on Long-term Energy Resources. Montreal, Canada, November 26/December 7.

Centro de Tecnologia Promon (1979b). Technical-Economic Evaluation of Processes fo Stillage Recovery as Distillery By-product. Multiclient Study.

Cesar, M.A.A. (1977). Níveis de Amido em Cana Tratada com Vinhaça (In print).

Copersucar (1977). Effect of Vinasse Application as Fertilizer in Sugarcane, São Paulo, 28p.

Copersucar (1978). Efeitos da Aplicação de Vinhaça como Fertilizante em Cana-de-Açúcar; Boletim Técnico 7: 9-14.

Correa, H. (1977). A cultura da mandioca como matéria-prima na produção do álcool. Simpósio Estadual do Álcool, 1, Divinópolis, MG, p. 81-100.

Gloria, N.A. (1976). Emprego de Vinhaça para Fertilização. Piracicaba, CODISTIL.

Gloria, N.A. (1978). Private communication.

Gloria, N.A. and J.A. Magro (1976). Utilização agrícola de resíduos da usina de açúcar e destilaria na Usina da Pedra. Seminário Copersucar da Agro-Indústria Açucareira, 4., Águas de Lindóia, SP, p. 163-80.

Grant, E.L. and W.G. Ireson (1970). Principles of Engineering Economy (5th ed.) Ronald Press Co., New York.

Instituto do Açúcar e do Álcool (1977). Planalsucar: Relatório Anual de 1974: 23-3

Magro, J.A. and N.A. Gloria (1977). Adubação de soqueira de cana-de-açúcar com vinhaça: complementação com nitrogênio e fósforo. Brasil Açucareiro 90(6): 363-66.

Malavolta, E. (1976). Manual de Química Agrícola, Nutrição de Plantas e Fertilizan do Solo. Ed. Agronômica Ceres, 527 p.

Rodella, A.A. and S.E. Ferrari (1977). A composição da vinhaça e efeitos da sua aplicação como fertilizante na cana-de-açúcar. Brasil Açucareiro 90 (1): 380-87.

Rossetto, A.J. and others (1975). Utilização da Vinhaça na Usina São Joao: Relatór à Diretoria da Cia. Industrial e Agrícola São João, Araras, SP, 21 p.

Rossetto, A.J. and others (1977). Sistemas de distribuição de vinhaça utilizados pela Usina São João, Araras, SP. Seminário sobre Sistemas de Aplicação da Vinhaç no Solo, STAB/ESALQ, 29p.

Silva, G.M. and others (1976). Comportamento agroindustrial da cana-de-açúcar em solo irrigado e não-irrigado com vinhaça. Seminário Copersucar da Agro-Indústria Açucareira, 4., Águas de Lindóia, p. 107-22.

Spencer, M. (1963). Cane Sugar Handbook. New York, John Wiley & Sons Inc.
Stab (1977). Instituto Zimotécnico da ESALQ. Seminário sobre Sistemas de Applicação de Vinhaça no Solo. Piracicaba.
Stupiello, J.P. and others (1977). Efeitos da aplicação da vinhaça como fertilizante na qualidade da cana-de-açúcar. Brasil Açucareiro 90(3): 185-94.

10
METHANE GENERATION POTENTIAL OF AGRICULTURAL WASTES

J. M. Scharer, M. Fujita and M. Moo-Young

Biochemical Engineering Group, Department of Chemical Engineering, University of Waterloo, Waterloo, Ontario, Canada

ABSTRACT

The dependence of methane production on the composition of agricultural wastes is examined. The carbohydrate and protein content of the raw waste can be related to the methane generation rate. The methane yield from proteinous substances appears to be a function of the retention time, while the yield from carbohydrates depends primarily on the digestibility of the carbohydrate source. On the basis of experimental data, a mechanism for the digestion process is proposed.

Digestion temperature affects the maximum solids loading rate. The rate of hydrolysis of cellulosic materials is particularly temperature dependent. The conversion efficiency, however, remains approximately the same at mesophilic and thermophilic operations.

KEYWORDS

Methane productivity; carbohydrate and protein digestion; temperature effects; retention time; maximum loading.

INTRODUCTION

Environmental and economic considerations require the re-examination of traditional waste management practices. Land disposal of untreated manure can lead to sanitary and odour problems (Welsh and colleagues, 1976). The manure can be stabilized by modern composting methods (Taiganides, 1977) which reduce the odour, pathogenic organisms and insect larvae prior to land disposal. Rising energy costs, however, have given impetus for research into anaerobic digestion of agricultural wastes (Varel and others, 1977; Ashare and colleagues, 1977; Lapp and others, 1975; Chen and Hashimoto, 1978; Scharer and Moo-Young, 1979). Anaerobic digestion of manure offers a promising alternative both as a waste treatment process and as a means of energy (methane) production.

Our interest in anaerobic digestion is twofold. First, we wish to evaluate the methane producing potential of various manure mixtures. Second, we are examining the utilization of the digester effluent either as a direct animal feed supplement or as a basal medium for single cell protein production by aerobic fermentation.

In these respects, the thermophilic operating temperature is favoured, since pathogenic organisms should not survive at the higher temperatures. Studies on thermophilic manure digestion are few. For this reason, the thermophilic digestion is emphasized in the experimental portion of this report.

The Canadian climate presents some serious constraints concerning the successful operation farm-scale anaerobic digestion units, particularly at thermophilic temperatures. For engineering design, an accurate prediction of gas productivity is essential. In this report, we examine the methane generation potential of swine and cattle manures in relation to their fermentable substrate content.

MATERIALS AND METHODS

Swine manure was dried to 5% moisture content and ground to pass a mesh for a particle size of 2mm. This procedure provided a homogeneous feedstock for the digestion studies. The volatile solids and the total Kjeldahl nitrogen content of the dried, ground manure were 71.4% and 4.2%, respectively. A defined quantity of the dried manure was slurried daily to 6% solids content with tap water and allowed to stand overnight at room temperature before feeding the digester. The pH of the slurry averaged 6.6.

A schematic diagram of the experimental anaerobic digester is shown in Fig. 1. This apparatus consisted of an upright cylinder (23 cm diameter) with a conical bottom, a circulation pump, temperature controller, and wet-test gas meter. It was also equipped with a thermometer and a scum breaker. The liquid volume of the digester was 32 L. Steam was used as a controlled heat source and the circulation pump operated simultaneously whenever steam heating was required. The frequency of steam heating and recirculation pumping averaged about 3 times an hour and each operating time was about 15-30 seconds. The daily feeding of the digesters was accomplished in a draw-and-fill manner.

The total Kjeldahl nitrogen (TKN) in both the influent and the effluent were analyzed by Standard Methods (1971) and gas generation was monitored with a wet-test gas meter. Solid-liquid separation was accomplished by centrifugation followed by membrane filtration. Volatile fatty acids in the liquid were analyzed by gas chromatography (Hewlett-Packard, Model 7128A, U.S.A.) using a 0.9 m x 6.2 mm glass column packed with "Porapak GS" (80/100 mesh). The conditions for gas chromatography were as follows: oven temperature = 180°C, flow rate of carrier gas (He) = 2.5 mL/min., temperatures of injection port and detector = 240°C and 230°C, respectively.

The analysis of gas composition was done by gas chromatography using 1.8 m x 6.2mm nickel column packed with 5X Molecular Sieve of 60/80 mesh. The temperature of the injection port was 220°C and the oven temperature was programmed to give a 20°C/min rise to 300°C. Analytical analyses included determinations of total carbohydrate and cellulose (Updegraff, 1969), volatile solids (Standard Methods, 1971), and ammonia (specific ion electrode).

Care must be exercised to initiate the thermophilic digestion successfully (Jewell, 1978). In this study, 10 L of anaerobically digested cattle manure, 6 L of municipal digested sludge and 16L of swine manure slurry (3% solids) were used to fill the experimental digester. The digester temperature was initially at 39°C. The feeding schedule of 1 litre of 6% swine manure slurry was initiated two days after start-up and the temperature of the digester was raised by 2°C each day until 55°C was reached.

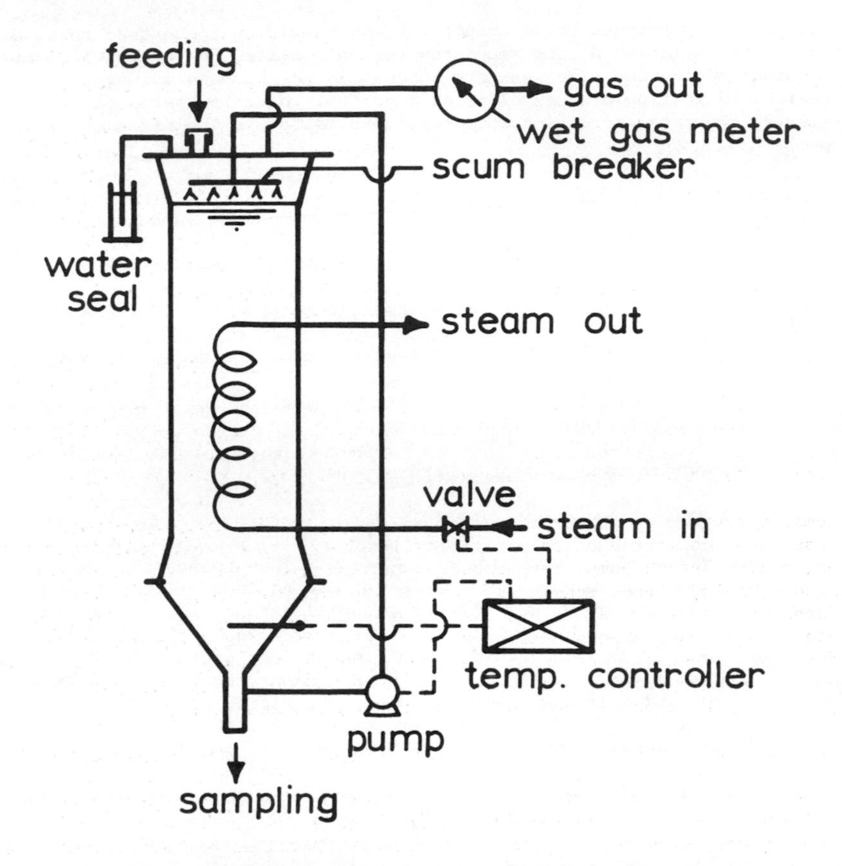

Fig. 1. Schematic diagram of the anaerobic digester

THEORETICAL CONSIDERATIONS

Most researchers recognize three microbiological steps leading to methane formation from complex organic wastes. These include:

1.) Enzymatic hydrolysis of solids to dissolved constituents.

2.) Fermentative processes yielding fatty acids, ethanol, hydrogen, and carbon dioxide.

3.) Methanogenesis from one or two-carbon precursors, i.e. carbon dioxide, formate, acetate, and ethanol.

The first step is catalyzed by extracellular enzymes; the others are intracellular processes. The bioconversion requires the symbiotic association of several microbial species. In manure, both the carbohydrate and the protein fractions are amenable to hydrolysis, hence bioconversion to methane and carbon dioxide.

For efficient conversions, the elementary steps should be concerted, thus, no substantial accumulation of intermediates (organic acids, for example) should occur. Since biological processes are autocatalytic, biomass turnover during the anaerobic digestion process plays an important role. In terms of protein equivalents, the rate expression for proteinous matter in the digester is as follows:

$$\frac{dCp}{dt} = L_p + Y_{x/c} L_c - kCp - \frac{Cp}{\tau} \quad (1)$$

where Cp = concentration of proteinous matter in the digester (kg/m^3)

L_p = specific loading rate of digestible manure protein ($kg/m^3.d$)

L_c = specific loading rate of digestible manure carbohydrate ($kg/m^3.d$)

$Y_{x/c}$ = protein yield from carbohydrate utilization (kg protein produced/ kg carbohydrate utilized)

k = specific rate of protein hydrolysis (d^{-1})

τ = hydraulic retention time (d)

The first term on the right hand side of Equation (1) expresses the protein supply to the digester with the manure feed, while the second term relates protein (biomass) synthesis arising from carbohydrate utilization. Both manure protein and biomass protein are subject to autolysis as expressed by the third term. It is assumed that the rate of hydrolysis of manure feed protein is essentially the same as that of biomass protein. Much of the manure protein is derived from biomass similar to, if not identical with, the microbial flora in the digester. The last term in the equation expresses the rate of hydraulic dilution of the net protein content by the draw and fill schedule.

Stationarity of the digestion process implies no time-dependent trend in "n" observations taken equispaced in time; thus:

$$\frac{1}{n} \Sigma \frac{dCp}{dt} = 0 \quad (2)$$

The constraint expressed by equation (2) is independent of the time scale and must hold on the micro time scale (instantaneous rates) as well as on the macro scale (daily rates, for example). In this case, the protein concentration in the vessel is given by:

$$\overline{Cp} = (\overline{L}_p + Y_{p/c} \overline{L}_c) \left(\frac{\tau}{1 + k\tau}\right) \quad (3)$$

where $\overline{Cp}$, $\overline{L}_p$, and $\overline{L}_c$ represent the arithmetic mean values of the variables as defined previously. The daily mean rate of protein hydrolysis given in equation (1) becomes:

$$k\overline{Cp} = (\overline{L}_p + Y_{p/c} \overline{L}_c) \left(\frac{k\tau}{1 + k\tau}\right) \quad (4)$$

It has been shown previously (Scharer and Moo-Young, 1979) that the breakdown of the carbohydrates (cellulose, hemicellulose, and starches) of manure correlates poorly with the hydraulic retention time. Rather, the rate of carbohydrate hydrolysis and subsequent fermentation to organic acids are solely functions of the specific loading rate of the digestible carbohydrate fraction of manure. The formation of methane from carbohydrates, in turn, is dependent linearly on the loading rate. Combining the methane generation potential of carbohydrate and proteinous resources yields the following equation:

$$G = L\{f_c Y_{G/c} + (f_p + f_c Y_{p/c})(Y_{G/p})(\frac{k\tau}{1 + k\tau})\} \quad (5)$$

where

G = biogas productivity ($m^3/m^3 \cdot d$)
L = volatile solids loading ($kg/m^3 \cdot d$)
f_c = digestible carbohydrate fraction of manure
f_p = digestible protein fraction of manure
$Y_{G/c}$ = biogas yield from carbohydrate (m^3/kg carbohydrate)
$Y_{G/p}$ = biogas yield from protein (m^3/kg protein)
$Y_{p/c}$ = protein yield from carbohydrate (kg protein/kg carbohydrate)
k = specific rate of protein hydrolysis (d^{-1})
τ = retention time (d)

In equation (5), the specific loading rate of the carbohydrate (L_c) and protein (L_p) are expressed as respective fractions of the volatile solids loading. Thus, biogas productivity is the explicit function of two process variables (L and τ), two substrate-dependent variables (f_c and f_p), three yield constants ($Y_{G/c}$, $Y_{G/p}$, and $Y_{p/c}$) and a reaction rate constant (k). Biogas productivity is dependent implicitly on temperature inasmuch as the maximum allowable volatile solids loading and the specific rate of protein hydrolysis are temperature dependent. It is assumed that the pH is maintained at or near the optimum pH range for the anaerobic digestion process (6.8 to 7.8).

If the fermentable carbohydrate and protein fractions of the manure do not vary radically, the yield constants of equation (5) do not need to be established explicitly. For predictive purposes, the equation reduces to:

$$G = L\,[\alpha + \beta(\frac{k\tau}{1 + k\tau})] \quad (6)$$

The combined rate constants α and β express biogas yield from the carbohydrate and the protein fractions of the manure, respectively. Further simplification can be achieved by recognizing that the volatile solids content in the manure feed is usually kept constant, i.e.:

$$L\tau = S_0 \quad (7)$$

where S_0 = concentration of manure volatile solids in the feed.

Substituting equation (7) into equation (6) and rearranging yields:

$$G = \alpha L + \beta\left[\frac{kS_0}{1 + \frac{kS_0}{L}}\right] \quad (8)$$

It should be noted that equation (8) is a function of the loading only and can be used to analyze the results of various other investigators.

RESULTS AND DISCUSSION

Typical results for the swine manure anaerobic digestion are given in Table 1.

TABLE 1

THERMOPHILIC AND MESOPHILIC DIGESTION OF SWINE MANURE

Temperature:-	55°C				39°C
retention time (days)	32	16	8	4	16
loading rate: ($kgTS/m^3/day$)	1.88	3.75	7.50	15.00	3.75
($kgVS/m^3/day$)	1.34	2.68	5.36	10.71	2.68
total solid: inf. (kg/m^3)	60	60	60	60	60
eff. (kg/m^3)	40.80 ± 8.64	42.86 ± 9.70	35.52 ± 5.48	38.02 ± 6.63	40.01 ± 9.
% reduction	32.0	28.6	40.8	36.6	33.2
volatile solid: inf. (kg/m^3)	42.84	42.84	42.84	42.84	42.84
eff. (kg/m^3)	27.39 ± 5.75	29.21 ± 7.10	23.94 ± 3.87	26.09 ± 5.01	27.91 ± 6
% reduction	36.1	31.8	44.1	39.1	34.9
cellulose: inf. (kg/m^3)	6.3	6.3	6.3	6.3	6.3
eff. (kg/m^3)	3.31 ± 0.57	3.57 ± 1.39	2.51 ± 0.63	3.29 ± 0.44	3.04 ± 0.
% reduction	47.3	43.3	60.2	47.8	51.7
cellulose/total solid (%)	8.27 ± 1.01	7.86 ± 1.96	7.15 ± 0.91	7.94 ± 0.26	7.47 ± 1.
TKN: inf. (kg/m^3)	2.52	2.52	2.52	2.52	2.52
eff. (kg/m^3)	1.99 ± 0.34	2.46 ± 0.28	2.22 ± 0.27	2.39 ± 0.13	2.10 ± 0.
NH_4-N eff. (kg/m^3)	0.89 ± 0.19	1.23 ± 0.08	1.01 ± 0.05	1.02 ± 0.07	1.04 ± 0.
Gas Production: ($m^3/m^3/day$)	0.440 ± 0.050	0.676 ± 0.042	1.465 ± 0.061	2.353 ± 0.102	0.708 ±
(m^3/kgVS added)	0.328	0.252	0.274	0.220	0.264
(m^3/kgVS removed)	0.909	0.794	0.620	0.562	0.759
CH_4 (%)	62.7	62.3	66.3	64.9	62.9
volatile acid (as CH_3COOH) kg/m^3	1.42 ± 0.22	0.51 ± 0.13	0.66 ± 0.12	1.00 ± 0.07	
pH	7.38	7.51	7.67	7.56	7.33

inf. = influent. eff. = effluent. TKN = total Kjeldahl nitrogen. NH_4-N = ammonia-nitrogen.

It should be noted that neither the solids reduction nor the cellulose utilizatior efficiency improved markedly by operating the digester under thermophilic conditic The enhanced volumetric gas productivity at shorter retention times was mainly due to the higher loading rate. As predicted by equation (5), the gasification effici ency expressed as (m^3 gas produced)/(kg volatile solids (VS) added) declined stead ly as the volumetric retention time became shorter. In these experiments, the pH ranged from 7.3 to 7.7 and became very stable during stationary state conditions. These pH values are believed to be near optimal for the anaerobic digestion proces Since swine wastes have a favourable carbohydrate carbon-to-nitrogen (ammonia and protein) ratio, no problems with respect to pH control were experienced during the entire series of experiments.

Due to different operating conditions, the comparison of the results with previous studies is often difficult. As indicated by equation (8), the effect of solids loading (expressed as kg VS/m^3.day) on the gas productivity (m^3 gas/m^3 digester volume.day) is one of the few meaningful comparisons. In Fig. 2, our results are compared with the data of van Velsen (1977) and Robertson and others (1975). As predicted by equation (8), gas productivity becomes linearly dependent on the vol

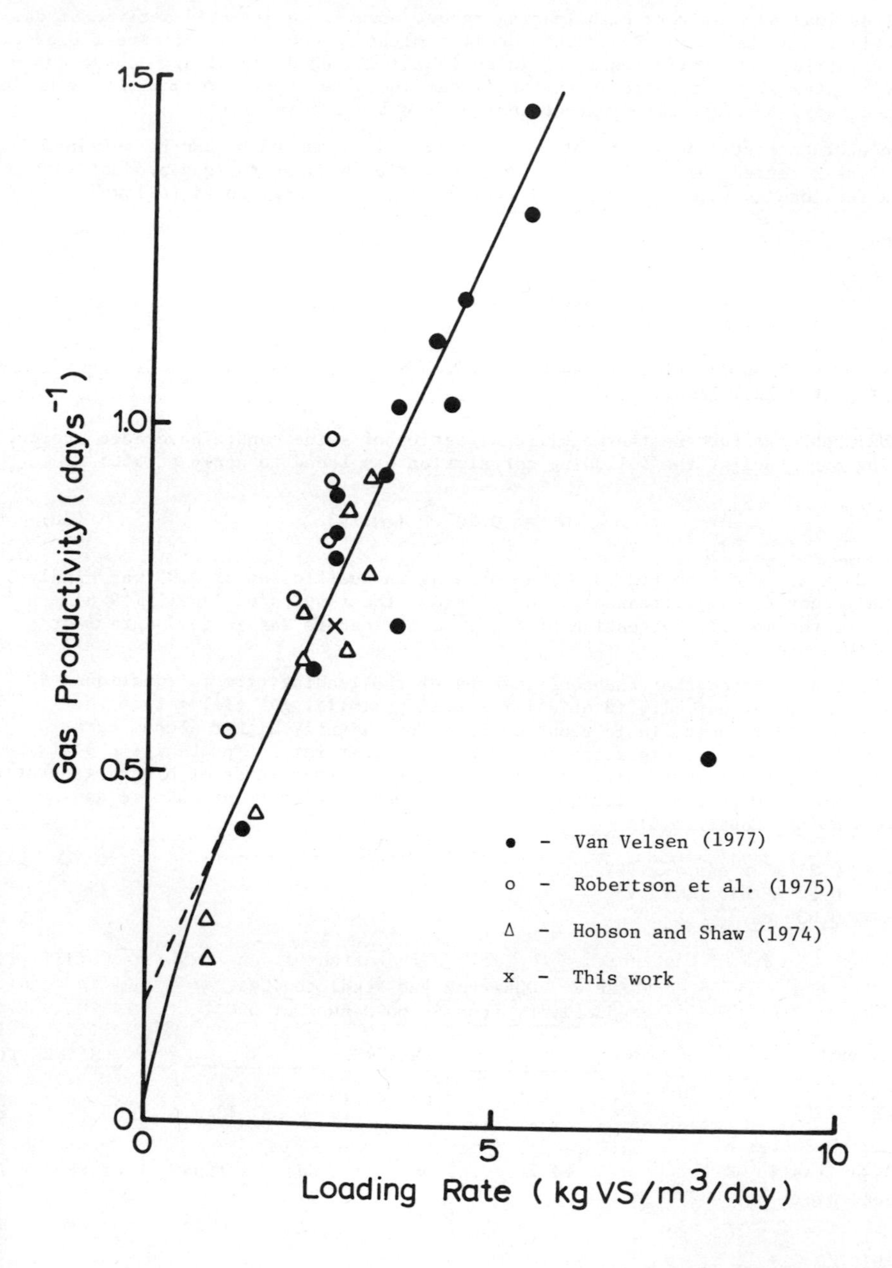

Fig. 2. Relationship between loading and gas productivity in mesophilic (39 C) digestion of swine manure.

solids loading. At very high loading rates, however, a dramatic decline of gas pro ductivity has been observed. This decline might be caused by the excess production of volatile fatty acids resulting in an inhibitory pH depression. Sonoda and other (1965) have also recognized both the linear increase of gas productivity with loadi and a rapid decrease with further increase in the loading rate.

The characteristic constants of equation (8), i.e. α and β can be obtained by linear regression of the data. A correlation between the gas productivity and the loading rate at mesophilic temperatures was obtained as follows:

$$G = 0.224\ L + 0.168 \tag{9}$$

G = gas productivity (m^3 gas/$m^3 \cdot$ day)
L = loading (kg VS / $m^3 \cdot$ day)

The correlation coefficient was 0.93, which is significant at the 5% significance (95% confidence) level.

Published data for the thermophilic digestion of swine manure have been scarce. Using our results, the following correlation was found to apply at 55C:

$$G = 0.207\ L + 0.193 \tag{10}$$

The data are shown on Fig. 3. The correlation coefficient of 0.99 was highly significant (5% significance). In addition, the results of Jewell (1978) for the thermophilic digestion of dairy cattle manure was in good agreement with Equation (10).

It should be noted that the coefficients of the loading term in equations (9) and (10) are essentially identical reflecting similar gas yields from the carbohydrate resource (α in equation 6). The slightly higher second term in case of thermophilic digestion reflects the higher rate of proteolytic activity at the elevated temperature. In Table 2, the observed rates of biogas generation are compared with the predictions. The relevant constants at 55C are as follows (see Equation 6):

α = 0.21 m^3 biogas/kg VS
β = 0.22 m^3 biogas/kg VS
k = 0.033 day^{-1}

TABLE 2 Observed and Predicted Gas Productivity from Swine Manure at 55C

PARAMETER	MANURE I				MANURE II	
RETENTION TIME (DAYS)	4	8	16	32	8	8*
VOLATILE SOLIDS LOADING (kg/$m^3 \cdot$day)	10.71	5.36	2.68	1.34	2.68	7.8
OBSERVED GAS PRODUCTIVITY ($m^3/m^3 \cdot$day)	2.42	1.32	0.74	0.42	0.68	2.[illegible]
PREDICTED GAS PRODUCTIVITY ($m^3/m^3 \cdot$day)	2.52	1.37	0.77	0.43	0.69	2.[illegible]
% ERROR	4.1	1.4	4.1	2.4	1.5	2.4

* 75% Manure, 25% Corn Stover

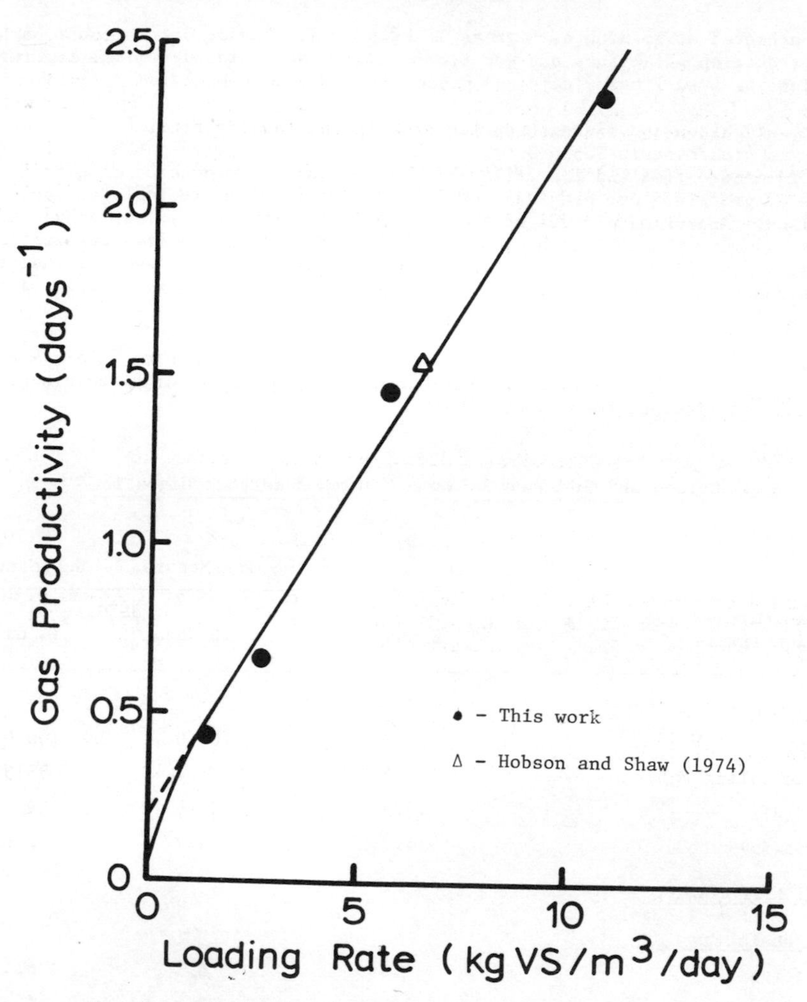

Fig. 3. Relationship between loading and gas productivity in thermophilic (55 C) digestion of swine manure.

The excellent agreement between the predicted and observed gas productivities is evident. The second manure (Manure II) had slightly different carbohydrate composition. In contrast, alternative models (Chen and Hashimoto, 1978) gave up to 60% error in predicted gas productivity.

Comparing mesophilic with thermophilic digestion of swine manure, a major difference appears to be with respect to the loading rate capability. At mesophilic temperatures, the practical limit of volatile solids reduction is approximately 2.5 $kg/m^3 \cdot day$. Thus, above this loading rate the efficiency of solids conversion is expected to decline, although gas productivity may increase to a maximum at near 5 $kg/m^3 \cdot day$. At thermophilic temperatures, the solids reduction efficiency peaks at about 5 $kg/m^3 \cdot day$ loading, but good gas productivity can be achieved at loading as high as 15 $kg/m^3 \cdot day$. Taking into account both solids reduction efficiency and gas productivity, a practical optimal loading rate might be 5 to 7 $kg/m^3 \cdot day$ at thermophilic temperatures

The fate of carbon and nitrogen during thermophilic and mesophilic digestions of manures is presented in Table 3. The table has been prepared on the basis of 100 g dry manure feed and the utilizable carbon was assumed to be either carbohydrate or protein. Our previous results with cow manure are included for comparison. Approximately 20% of the carbon in the manure feed was converted to methane and more than 30% to biogas. Swine manure volatile solids utilization efficiency was 66% at 55 C and somewhat lower (59%) at 39 C. Assuming that the biomass contains 50% carbon, the overall cell yield is 0.28 g biomass/g volatile solids utilized.

Approximately 40% of the particulate nitrogen in the swine manure feed has been converted to ammonia. Generally, 50% of the nitrogen in the digested sludge was in the particulate form, most likely biomass.

TABLE 3

Carbon and Nitrogen Balances During Anaerobic Digestion of Manures

	Swine Manure	Swine Manure	Cow Manure
Temperature:-	55°C	39°C	
Retention Time:-	8 days	16 days	14 days
MANURE FEED			
Total solids (g)	100.0	100.0	100.0
Volatile solids (g)	71.4	71.4	88.5
Carbon (g)	30.9	30.9	29.2
Nitrogen (g)	4.2	4.2	2.6
DIGESTION PRODUCTS			
Carbon basis (as C)			
Methane (g)	7.0	6.3	6.4
Carbon dioxide (g)	3.5	3.8	4.3
Organic acids (g)	0.4	0.4	1.0
Bicarbonate (g)	3.3	2.3	2.0
Biomass (g)	6.2	5.4	3.5
Residual material (g)	10.5	12.7	12.0
Nitrogen basis (as N)			
Ammonia (g)	1.7	1.8	1.6
Dissolved organic N (g)	0.2	0.2	0.3
Particulate organic N (g)	2.0	1.7	0.6

CONCLUSIONS

When swine manure is digested anaerobically, stationary state operation is attainable within 20 days after start-up. The shortest achievable hydraulic retention time is 3 - 4 days at thermophilic and 10 - 12 days at mesophilic temperatures of operation. However, the efficiency of gasification tends to decrease at minimal retention. Also, the daily rate of gas production fluctuates and the system becomes unstable. For these reasons, practical retentions of 8 days at thermophilic and 16 days at mesophilic temperatures are preferable. With regard to retention times, swine manure is similar to other animal wastes.

Good correlations between gas productivity and solids loading were found. The parameters of the linearized equations are virtually the same for mesophilic and thermophilic operations. The major difference appears to be the maximum solids loading rate which is 5 kg vs/m^3·day at mesophilic and in excess of 15 kg vs/m^3·day at thermophilic temperatures. Maximum solids reduction efficiency, however, occurs at substantially lower loadings.

REFERENCES

Ashare, E., D.L. Wise and R.L. Wentworth (1977). Engineering Report COO-2991-10, Dynatech R/D Co., Cambridge, Mass., U.S.A.
Chen, Y.R. and A.G. Hashimoto (1978). Kinetics of methane formation. Symposium on Biotechnology in Energy Production and Conservation, Gatlinburg, Tenn., U.S.A.
Hobson, P.M. and P.G. Shaw (1974). Water Research 8, 507.
Jewell, W.J. (1978). In W.W. Shuster (Ed.) Proceedings: The Second Annual Symposium on Fuels from Biomass, Cornell University Press, Ithaca, N.Y. U.S.A.
Lapp, H.M., D.D. Schulte, A.B. Sparling and I.C. Buchanan (1975). Canadian Agricultural Engineering 17, 97.
Robertson, A.M., G.A. Burnett, P.N. Hobson, S. Bousfield and R. Summers (1975). Proc. of the 3rd International Symposium on Livestock Wastes, American Society of Agricultural Engineers.
Scharer, J.M. and M. Moo-Young (1979). Adv. Biochem. Eng. 11, 85.
Sonada, Y., H. Ono and H. Kobayashi (1965). J. Ferment. Technol. 43, 847.
Standard Methods for the Examination of Water and Wastewater (1971). 13th Ed. APHA, AWWA, WPCF.
Taiganides, E.P. (1977). In E.P. Taiganides (Ed.), Composting of Feedlot Wastes. Applied Science Publishers Ltd., London, England.
Updegraff, D.M. (1969). Anal. Biochem. 42, 420.
Varel, V.H., H.R. Issaacson and M.P. Bryant (1977). Appl. Environ. Microbiol. 33, 298.
Van Velsen, A.F.M. (1977). Neth. J. Agri. Sci. 25, 151.
Welsh, F.W., D.D. Schulte, E.J. Kroeker and H.M. Lapp (1976). Can. Soc. Agricultural Engineering, Annual Meeting. Halifax, N.S., Canada.

11
PYROLYSIS-COMBUSTION PROCESS OF HIGH ASH AGRICULTURAL RESIDUES

M. A. Hamad, H. A. El Abd and L. M. Farag

Pilot Plant Laboratory, National Research Centre, Cairo, Egypt

ABSTRACT

The high ash content of rice hulls and straw (consisting mainly of silica) hinders the complete combustion in case of direct combustion. A residue containing a considerable amount of combustible material is obtained, lowering the calorific value of the agricultural residue. The char obtained after pyrolysis also contains about 50% ash which lowers its value as fuel source.

A pyrolysis-combustion process is proposed for the complete combustion of rice hulls and straw. The proposed process consists of two main stages: the first stage is the pyrolysis of the agricultural residue at a termperature 400-450 C. The second stage is the combustion stage, in which the solid residue from the first stage is ignited in controlled oxygen atmosphere at controlled temperature to obtain the low carbon ash. A material and energy balance of the proposed process is given which clarifies its technical and economical feasibility.

KEYWORDS

Pyrolysis, combustion, energy from waste, rice hulls and straw, material and energy balance.

INTRODUCTION

The declining energy supplies, rapidly increasing consumption, and increasingly severe environmental constraints sharply focus attention on the need for additional amounts of clean energy sources. Conversion of the agricultural wastes into clean energy provides an opportunity to expand the energy resource base while reducing the pollution now associated with waste disposal. By capturing and storing solar energy through carbon fixation, biomass converted to combustible fuels is one of the most convenient sources of energy. Photosynthesis renews daily the biomass by fixation of huge amounts of carbon. The development of conversion methods appears more attractive because the proposed energy resources are renewable.

High ash agricultural residues, such as rice straws and hulls now appear very attractive as raw material, particularly in integrated systems where the raw material is used to produce energy while the resultant residue is marketable at higher prices than the raw material. This concept is especially applicable to rice hulls and straws because of their high ash content which reaches about 20% of the dry material.

Most existing direct combustion processes convert only about two-thirds of the available energy in the rice hulls and straws (Beagel, 1978). This represents mainly the volatizable portion of the material, whereas the remaining heat potentia still exists in the fixed carbon portion. Maheshwari and Ojha (1976) reported that the loss in heat value due to the unburned carbon in refuse constitutes 16.11 - 21.22% in RPEC rice hulls fired furnace. Brink and Charley (1976) proposed pyrolysis - gasification - combustion process for wood. They reported that certain problems inherent in direct combustion have been circumvented by physically separating the stages of conventional combustion and pyrolysis gasification into two steps.

Pyrolysis - combustion process is proposed for the high ash agricultural residues in order to fully utilize the potential energy of the raw material; meanwhile, the complete combustion of fixed carbon-free ash is obtained. The ash of rice hulls and straw consists mainly of silica (about 95%) and is considered a valuable industrial material when obtained in the amorphous form.

PROPERTIES OF RICE HULLS AND STRAWS

Composition

Raw material: Rice hulls and straw have similar chemical composition. The microanalysis of one of the available samples is given on dry basis for both of straw an hulls.

Analysis, wt %	Carbon	hydrogen	nitrogen	oxygen	ash
Rice hulls	37.69	5.23	0.5	36.58	20
Rice straw	38.50	5.30	0.7	37.50	18

Char: The composition of char obtained after thermal decomposition of hulls and straws depends on the temperature of treatment. An example of the obtained char in case of thermal decomposition of rice hulls at 420°C is given in the material balance tables.

Total Volatile Matter

The total volatile matter of rice hulls and straw were determined by heating the sample at 800-850°C for 7 minutes. The loss in weight based on dry matter is defined as the total volatile matter.

Volatile matter of rice hulls = 60.2%
Volatile matter of rice straw = 63.0%

Differential Thermal Analysis

The differential thermal analysis (DTA) is used to clarify the characteristics of the chemical reactions taking place during thermal treatment. The DTA is carried out in atmospheric air and also in argon atmosphere (inert atmosphere) in order to prevent oxidation. The DTA curves of rice hulls is given in Figure 1.

Heat Required for Pyrolysis

The differential scanning calorimetry (DSC) technique was used for evaluation of the heat required for the thermal decomposition of the agricultural residue. The heat required in case of heating rice hulls at a rate of 20°C/min in the range of 100-450°C was 126 cal/g.

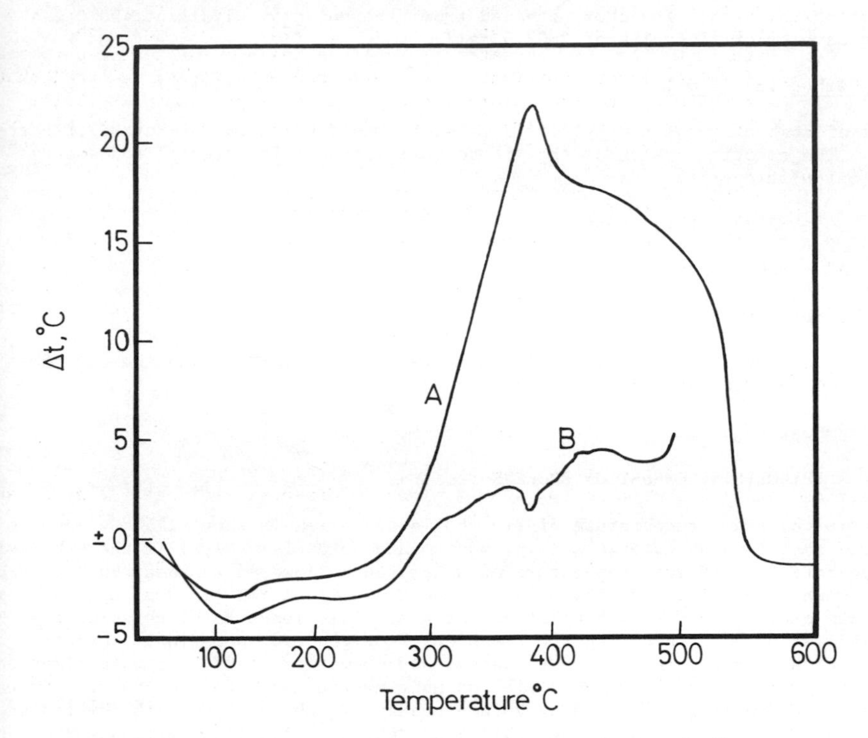

Fig. 1. Differential thermal analysis of rice hulls.
Rate of heating 20°C/min.
A- In air atmosphere , B- In argon atmosphere

Heat of Combustion of Hulls

The heat of combustion of hulls was determined using the emperical Dulong-Berthelot correlation:

$$\text{Cal/g} = 81.37\ (\%C) + 345\ (\%\ H - \frac{\%\ N + \%\ O-1}{8} + 22.2\ (\%\ S)$$

Emperical correlations of this type were originally developed for coal-like fuels. However, Boucher and co-workers (1977) confirmed the applicability of this correlation for determination of the heating values of the agricultural residues, chars and oils.

$$\text{Heat of combn of hulls} = 81.37 \times 33.92 + 345\ (4.7 - \frac{0.45+32.92-1}{8})$$

$$= 2985.6\ \text{cal/g.}$$

Heat of Combustion of Char

The heat of combustion of rice hulls char was determined using the emperical Dulong-

Berthelot correlation. The char obtained from rice hulls pyrolysis at 420°C was found to have calorific value of 3802.5 cal/g.

Heat of Combustion of Oil

The heat of combustion of pyrolytic oil was determined using Dulong-Berthelot correlation. The calorific value of the oil composition used in material and energy balance calculations is:

$$= 81.37 \times 59.8 + 345 \left(5.87 - \frac{0.92 + 34.47 - 1}{8}\right)$$

$$= 5408 \text{ cal/g.}$$

Heat of Combustion of Gases

The heat of combustion of pyrolytic gas mixture was calculated using the gas composition and the standard heats of combustion.

Heat of combustion of gas mixture = 1550.5 cal/g.

PYROLYSIS-COMBUSTION PROCESS

The theoretical flame temperature of rice hulls and straw is about 1850°C. Due to the presence of fluxing materials (e.g. sodium and potassium salts) in the ash resi due components, the fusion temperature of ashes can be lowered to about 900°C. Thu for fast combustion (low heat losses) of high-ash agricultural residues, the bed tem erature can exceed the fusion temperature of ash. This leads to blocking of the intermolecular pores, preventing the transfer of oxygen and, accordingly, hindering the combustion process. According to Bartha and co-workers (1974), crystallization of silica begins at a temperature of 825°C on heating the hulls for one hour. Crysta lization is generally accompanied by volume changes which can affect the diffusion oxygen through the particles of ash. Since amorphous silica is required as the fin product due to its high reactivity, lower temperatures are needed.

The proposed process consists of two main stages. The first stage is the pyrolysis of rice hulls while the second stage is the controlled combustion of the pyrolytic char in presence of air. The volatile matter of the agricultural residue is to be separated by pyrolysis process at temperatures in the range 400-450°C. At such temp eratures, about 50% of the material is decomposed giving combustible gases and liqui The char remaining after pyrolysis is to be ignited in controlled oxygen atmosphere at temperatures of 600-700°C, but not exceeding 825°C.

Pyrolysis process can be carried out in flash, fluidized or rotary type reactors. Combustion process can be carried out in fluidized or rotary type reactors. Part the heat released in the combustion stage is used for heating in the pyrolysis pro while the remaining heat is recovered for steam production or other uses. The flow diagram of the proposed process is given in Fig. 2.

Enthalpies of the material flow streams are designated (H) and heat losses from major equipment items are designated (Q). The material and energy balance of the process are based on 100 kg rice hulls air-dry basis. The oil and gas compositions were taken from the results obtained by Boucher and co-workers (1977) in case of flash pyrolysis of hulls at 427°C, but the amount of oil was corrected to obtain a reasonable balance with carbon and hydrogen. A lot of experimental work is needed but the assumptions of validity of these data help in obtaining approximate evaluation of the proposed process.

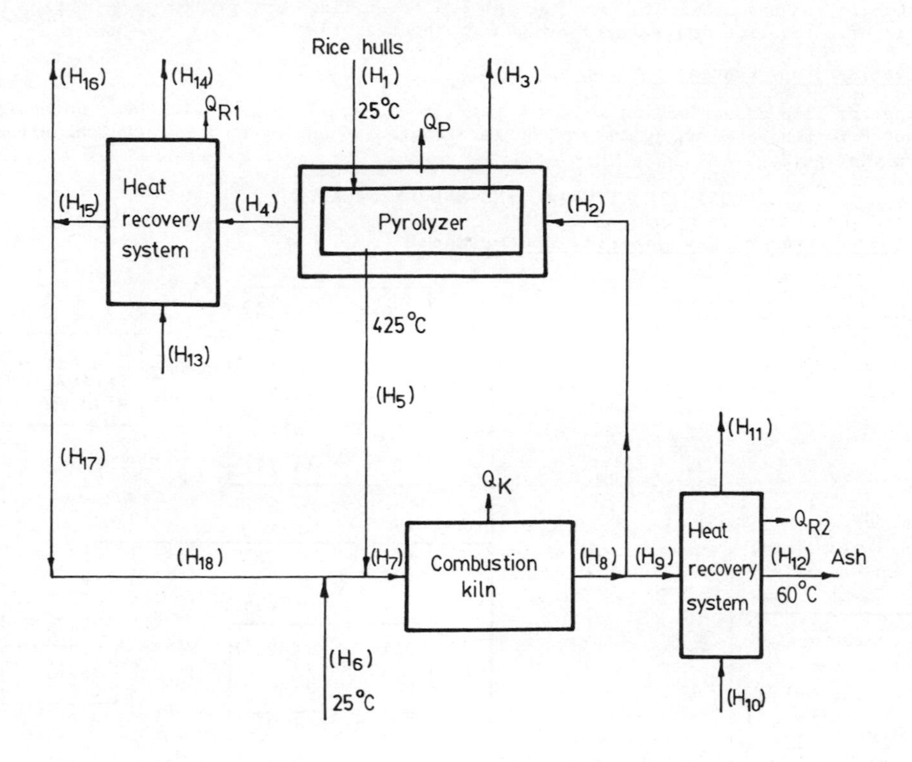

Fig. 2. Pyrolysis-combustion process of high ash agricultural residues.

Pyrolyzer

Raw hulls,as delivered, are introduced to the pyrolyzer, while the rice straw is ground to the proper size. The material is heated indirectly using the hot flue gases from the combustion stage. The temperature is raised to 400-450°C. At this temperature almost all the volatile portion of the material is transformed to gaseous and liquid phases.

Combustion Kiln

The char from the first stage is fed to the combustion kiln. The char is mixed with the required amount of air; 10% excess air was assumed sufficient for the complete combustion of char. The temperature inside the kiln is controlled by circulating the proper ratio of flue gases. Circulation of flue gases decreases the partial pressure of oxygen in the kiln and accordingly decreases the rate of combustion. Thus, a sudden increase in temperature is prevented,specially in the beginning of the combustion process. The kiln lining would be constructed of ceramic material capable of withstanding operating temperatures. The ash is separated from rostatic precipitation. The heat available from hot ash can be utilized for preliminary heating of t air of combustion or recovered using any proper system.

Heat Recovery System from Flue Gases

The hot flue gases leaving the pyrolyzer jacket still contain high heat potential a would be recovered using the appropriate system depending on the technical situatio in the field.

APPROXIMATE MATERIAL AND ENERGY BALANCE

Composition of Raw Material and Products

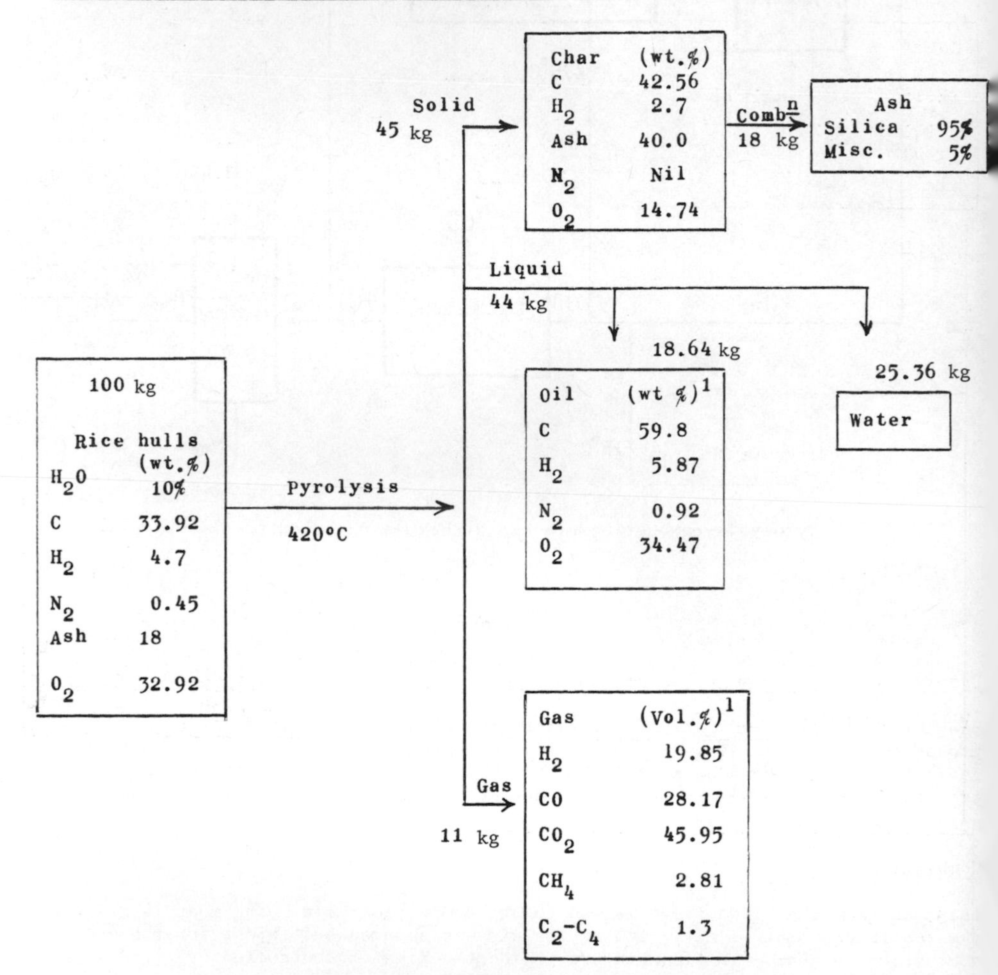

[1] Compositions are taken from Boucher and colleagues (1977)

Material Balance

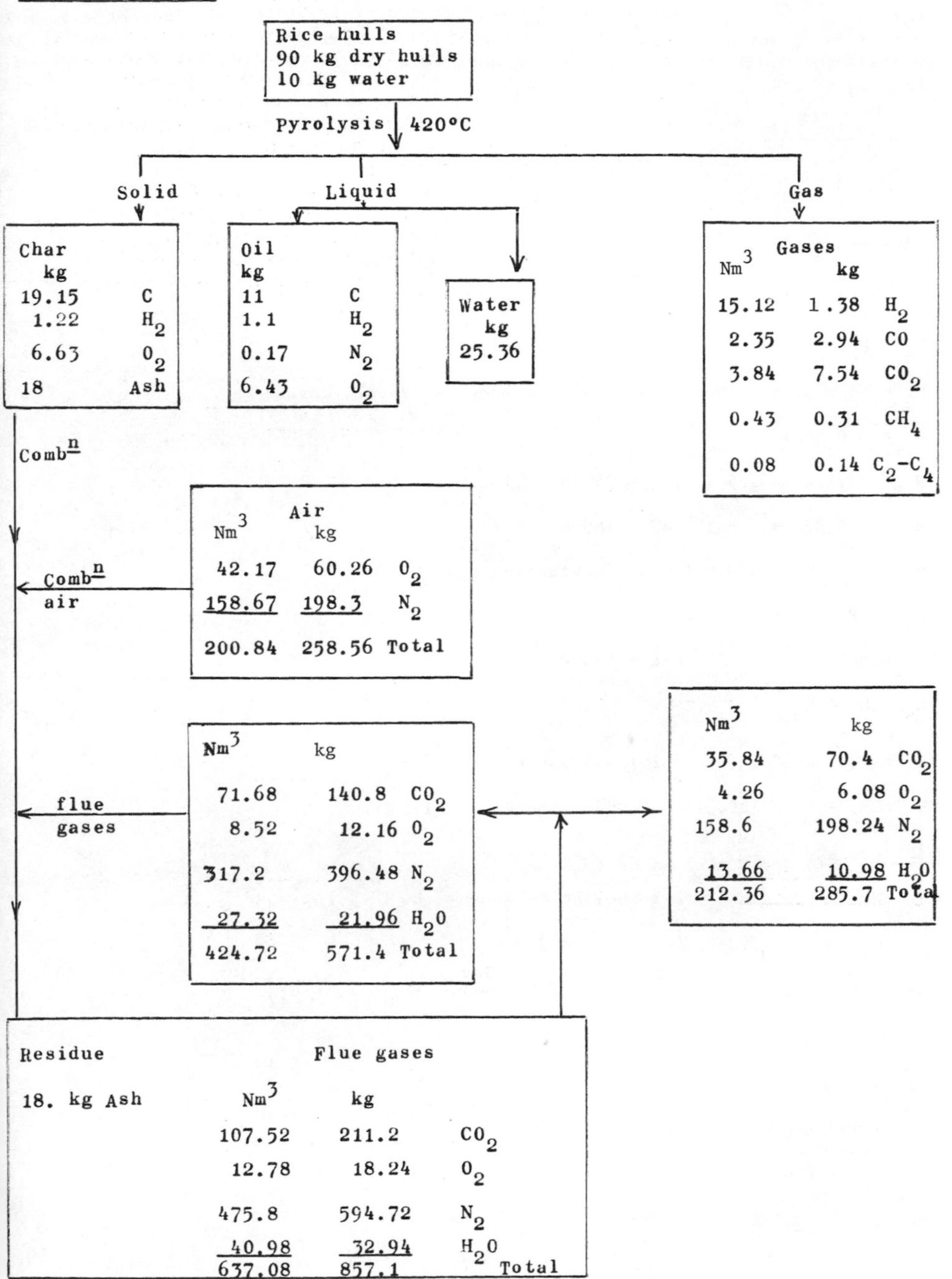

Energy Balance

Due to lack of data, and for simplicity, the heat losses Q from different equipment were first assumed equal to zero and finally an overall value for the whole process was assumed equal to 15% of the total heat recovered. The symbols used are given in Fig. 2.

Combustion kiln

Assuming flow gases temperature inlet = 110°C

$$\text{Flue gases circulation} = \frac{\text{(recirculated flue gases), kg}}{\text{(rec. flue gases + combn. air), kg}} \times 100$$

$$= 68.8\%$$

$H_5 + H_6 + H_{18} = H_7 = H_8 + Q_k$

$H_5 = 45 \times 0.32 \times (425 - 25) + 45 \times 3{,}802.5 = 176{,}872$ kcal

$H_6 = 0.0$ kcal

$H_{18} = 571.4 \times 0.25 \times (110 - 25) = 12{,}142$ kcal

$H_7 = 176{,}872 + 0 + 12{,}142 - 189{,}014$ kcal

$H_8 = H_7 - Q_k = 189{,}014$ kcal (neglecting Q_k)

$H_8 = 189{,}014 = 857.1 \times 0.25 \times \Delta t + 18 \times 0.32 \times \Delta t$

$\Delta t = 860°C$ $\quad\quad t = 885°C$

$H_8 = H_9 + H_2$

$H_9 = 18 \times 0.32 \times (885 - 25) = 4{,}954$ kcal

$H_2 = H_8 - H_9 = 189{,}014 - 4{,}954 = 184{,}060$ kcal

Ash cooler

The ash temperature after cooling is assumed 60°C

$H_{12} = 18 \times 0.32 \times (60 - 25) = 202$ kcal

Heat available for recovery from ash = $H_{11} - H_{10} = H_9 - H_{12}$

$= 4{,}954 - 202 = 4{,}752$ kcal

Pyrolyzer

Heat required inside pyrolyzer
= Sensible heat + heat required for pyrolysis
= $90 \times 0.32 \times (425 - 25) + 10 \times 703 + 90 \times 126 = 29{,}890$ kcal

Pyrolyzer Jacket: $H_2 - H_4 - Q_p = 29{,}890$ kcal

$H_4 = 184{,}060 - 29{,}890 = 154{,}170$ kcal (neglecting Q_p)

Pyrolyzer: $H_1 = 2985.6 \times 100 = 298{,}560$ kcal

$H_1 + 29{,}890 = H_3 + H_5$

$H_3 = 151{,}578$ kcal

Heat Recovery System

$H_4 + H_{13} = H_{14} + H_{15} + Q_{R1}$

Assume flue gases exit temperature = 110^oC and neglect Q_{R1}

$H_{15} = 857.1 \times 0.25 \times (110-25) = 18{,}213$ kcal

Heat available for recovery = $H_{14} - H_{13} = H_4 - H_{15}$

= 154,170 - 182.13 = 135,957 kcal

Total heat available is the sum of heat recovered from hot flue gases, ashes and that contained in the gas and oil stream obtained.

Total heat energy available = $(H_9 - H_{12}) + (H_4 - H_{15}) + H_3$

= 4,752 + 135,957 + 151,578 = 292,287 kcal

Assuming the total heat losses (Q_p, Q_k, Q_{R1}, ,,,) in the process is equal 15%.

Heat energy available = 292,287 x 0.85 = 248,444 kcal

Heat energy available/kg raw hulls = 2,484.4 kcal

$$\% \text{ heat recovery} = \frac{2{,}484.4}{2{,}985.6} = 83.2\%$$

CONCLUSIONS

Pyrolysis - combustion process of high-ash agricultural residues helps in reducing pollution. Complete combustion of char in controlled oxygen atmosphere produces a valuable ash which consists mainly of amorphous silica. The pyrolysis process consists mainly of amorphous silica. The pyrolysis process provides a fuel gas and oil that can be stored, transported and processed. The approximate energy balance of the proposed process clarifies its technical feasibility.

REFERENCES

Bartha, P. and E.A. Huppertz (1974). Structure and crystallization of silica in rice husk. Proc. Rice By-products Conf., Valencia, Spain, 89-98.

Beagle, E.C. (1978). Rice-husk conversion to energy. FAO Agricultural Service Bulletin, No. 31.

Boucher, F.B., E.W. Knell, G.I. Preston and G.M. Mallon (1977). Pyrolysis of Industrial Wastes for Oil and Activated Carbon Recovery. U.S. Environmental Protection Agency EPA-600/2-77-091, May

Brink, D.L. and I.A. Charley (1976). The pyrolysis-gasification-combustion process. Energy consideration and overall processing. Symposium on thermal uses and properties of carbohydrates and lignins. 172nd National Meeting of the American Chemical Society, San Francisco 72-125.

Maheshwari, R.C. and T.P. Ojha (1976). RPEC husk fired furnace. RPEC Reporter, 2, (2) 49-53.

SECTION IV

TOXICITY AND BIOHAZARDS

12
TOXICITY OF LIQUID INDUSTRIAL WASTES TO AQUATIC ORGANISMS

G. R. Craig

Ontario Ministry of the Environment, Water Resources Branch, Limnology and Toxicity Section, P.O. Box 213, Rexdale, Ontario, Canada

ABSTRACT

The application of biological tests using aquatic organisms is outlined with respect to water quality criteria development, industrial waste evaluation and monitoring. Bioaccumulation of organic compounds in fish and a method of estimating bioconcentration factors based on chemical characteristics is described. The opercular/heart rate of fish exposed to industrial waste discharges is being used in an on-line biomonitoring system to identify immediate changes in effluent quality.

KEYWORDS

Fish, LC50, EC50, lethal tests, sublethal tests, reproduction, growth, behaviour, embryo-larval, bioaccumulation, partition coefficient, biomonitoring.

INTRODUCTION

Water quality criteria have been determined for a number of parameters in order to protect the most sensitive use that may be made of receiving waters by aquatic organisms. Legislation in Ontario specifically identifies the need to protect aquatic organisms by stating that "the quality of water---may not become impaired---or---cause injury to any--living thing" (OWRA, 1970), and "---no person shall---discharge a contaminant---into the natural environment that---is likely to cause injury or damage to---plant or animal life" (E.P.A., 1971). Federal legislation (Fisheries Act, 1970) also directs that "no person shall deposit---a deliterious substance of any type in water frequented by fish---". These provincial and federal statues have been embraced by the 1978 Canada-United States Great Lakes Water Quality Agreement that "These waters should be---free from materials---entering the water as a result of human activity that---are toxic or harmful to human, animal or aquatic life---".

The above legislation recognizes that aquatic organisms reflect the health of a river system and that high quality water ensures potable drinking water supplies and a strong base on which to maintain or build fishery and recreational resources.

Industry must also recognize that their future growth and economic role in the community, at a national or international level, are dependent on preserving and improving water quality. Water is not a limitless commodity and must serve a multitude of uses within any one system.

The specific water quality parameters, for which standards have been set, can be monitored singly by analytical means. The biological methods used to determine these criteria are briefly outlined not only to provide an appreciation of how standards are set but to provide insight as to how future regulation might incorporate biological responses into continuous monitoring. Regulatory agencies, having addressed the need to define criteria on an individual parameter basis are now turning their attention to the regulation of mixtures of toxicants. This approach is much more difficult to deal with than assuming that the biological response to one chemical parameter is compounded with the biological response to another parameter, as it does not take into consideration antagonistic or synergistic interactions of the chemicals. The strength of biological monitoring is that an organism exposed to a mixture of toxicants integrates their interactions in its final response. The guesswork can then be taken out of analytical assumptions.

BIOLOGICAL RESPONSE

Measurement of biological response involves estimating how a population of organisms responds to a given stimulus or stress. Some individuals will react to very low levels of stimulus or stress while other individuals require higher levels to elicit the same responses. The population response can be portrayed by normal distribution or the bell curve where, in the case of chemicals, the concentration that produces a response in the greatest number of organisms can be identified as the median response (Fig. 1). Greater confidence can be placed in the estimated median response concentration than in concentrations required to produce a 10% or 90% population response because the data is generated by a greater number of organisms.

Biological response data is usually depicted not as a bell curve but as a dose response relationship where the accumulated population response is plotted against the toxicant or chemical concentration (Fig. 2). Transformation of either response or concentration data may then be completed to promote linearity and improve statistic interpretation. The median response can then be referred to according to the type of response observed. The median lethal concentration of a chemical is referred to as the LC50 (the concentration lethal to ≃50% of the population) while concentratic of compounds that "effect" a 50% reduction in growth,reproduction, behaviour or physiology in a population are referred to as EC50's.

The 50% response estimate is not only important because of the greater number of individuals it represents in a dose response evaluation but it incorporates confidence limits which more fully describe the population. Population responses can then be compared statistically, whether they be different species or families, and different compounds or mixtures can be compared within the same population of organisms.

Methods of determining confidence limits about the median response have been described using probit (Finney, 1971), moving average (Bennett, 1952), Litchfield-Wilcoxon (Litchfield and Wilcoxon, 1949) and Spearman-Karber (Armitage and Allen, 1950) analysis techniques. The relative merits of these methods are discussed by Stephan (1977).

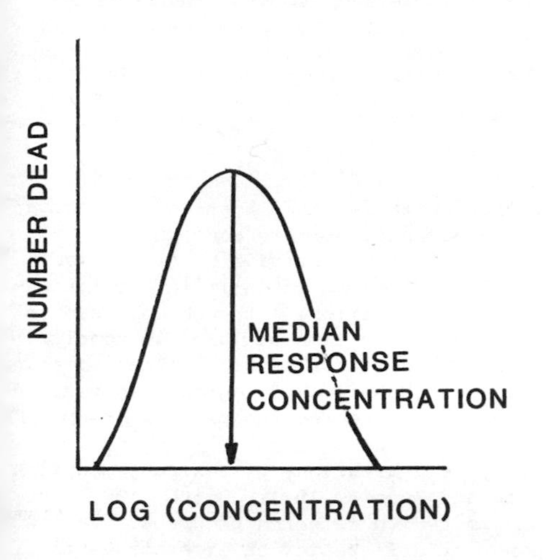

Fig. 1: Frequency distribution of mortality measured in a test population exposed to a toxicant

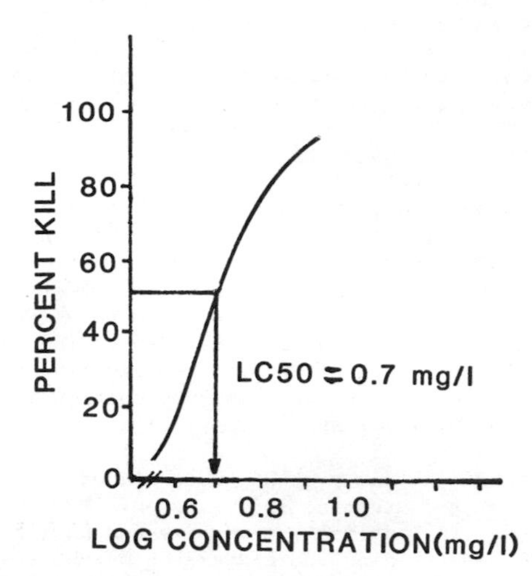

Fig. 2: Dose-response curve to estimate the LC50 of a toxicant to an organism.

LETHAL TESTS

Determination of the lethality of a compound or mixture to aquatic organisms is less difficult and expensive than sublethal evaluations. Lethal evaluations represent the initial order of testing that is applied to chemicals or discharges and a non-lethality requirement provides the first level of protection to the aquatic community.

Fish and daphnia are the common organisms used to determine the lethality of a waste with fish being the most popular historically due to their availability, known stock quality from commercial suppliers and the public's acceptance of their being indicators of good water quality. Test protocols incorporating these organisms have been described by Sprague (1969), A.S.T.M. (1973) and standard methods are soon to be published by the International Standards Organization.

The essence of these tests involves exposing an equal number of organisms to a logarithmic series of effluent or chemical concentrations and recording the percent mortality in each concentration. Tests can be static or continuous flow and are generally of a set duration, 48 or 96 hours. The data are plotted and the LC50 estimated from the mortality response curve. Effluent regulations in Canada (Environment Canada 1974, 1977a, b, c) have required that no more than 50% mortality occur in a 100% waste sample. This allows for the remaining 50% mortality to be accounted by biological, chemical and effluent variation.

Lethal tests need not be limited to regulatory purposes but can be used as investigatory tools to identify toxic agents. Tests can identify trends in effluent quality through time, determine the effectiveness of newly incorporated processes, compare the relative toxic contributions of process streams or performance among similar industries.

Effluents can be treated by altering pH, aeration, removal of specific ions (NH_3,

CN^-, phenols) or filtered with other media prior to lethal testing. Bench scale treatment evaluations and the toxic contribution of individual components to overall lethality can be estimated by this means. Treatment programs can then be designed and implemented to bring effluents into compliance with respect to the lethal test.

SUBLETHAL TESTS

Levels of toxicants in water that do not produce mortality in a population of organisms but impair or inhibit reproduction, growth or behaviour are considered sublethal and are regarded as detrimental to the population's survival. Alteration of one of these components often affects the others because of their interrelationship with each other. The ultimate effect on the population, although more subtle than mortality, can be extinction. The gradual acidification of lakes in poorly buffered waters of Ontario represents a well documented case in point (Beamish and Harvey, 1972), where fish reproduction is impaired (Mount, 1973), growth is reduced (Beamish, 1974) and a number of fish species have disappeared (Beamish and others,197

Emphasis in the past has been placed on life cycle evaluations of toxicants on fish (Mount and Stephan, 1967; Eaton, 1970; McKim and Benoit, 1971; Smith, 1973) but these are long term, expensive and have been limited to research programs. Review of these studies and comparison of the response of different fish life stages to toxicants reported in the literature has led McKim (1977) to suggest that embryo-larval testing protocols, exposing eggs and newly hatched fry, would be more cost effective in determining the potential impact of chemicals and effluents on target organisms.

Embryo-larval toxicity data have been reported for copper (McKim and others,1978), cadmium (Beattie and Pascoe, 1978) and sodium pentachlorophenate (Chapman and Shumway, 1978) using temporate species such as trout, suckers, herring, bass and pike. Eggs from these species are difficult to obtain unless available commercially and then only once or twice a year. Earlier tests of zinc, cyanide, naphthenic acids, potassium, dichromate, detergent (Cairns and colleagues, 1965),phenylmercuric acetate (Kihlstrom and Hulth, 1972), aflatoxin (Abedi and McKinley, 1968) flopet and difolatan (Abedi and Turton, 1968) have used tropical zebra fish, Brachydanio rerio, due to the ease of culture and prolific egg production.

Sublethal levels of toxicants can also affect behaviour in aquatic organisms. Avoidance or preference reactions of fish and invertebrates (Scherer, 1977;Maciorows and co-workers, 1977) can disrupt movement within feeding grounds, inhibit migration between feeding and spawning areas, or attract organisms from sublethal to lethal environments (Ishio, 1965). Swimming speed and endurance, particularly important to salmonids migrating up fast, flowing rivers and streams has been evaluated during exposure to pulpwood fibre (MacLeod and Smith, 1966), alkyl benze sulfonate (Lemeke and Mount, 1963) and laboratory apparatus (Brett, 1964) has been modified for field testing (Jones and colleagues, 1974).

Temperature selection by fish provides a means of regulating metabolism and is critical in the selection of spawning sites to ensure proper incubation of eggs. Exposure of brook trout to sublethal levels of DDT, PCB and phenol has shifted preferred temperatures in brook trout (Miller and Ogilvie, 1975). Waste heat streams from electrical generating plants can change the suitability of local habitats and systems to determine species' thermal preferences have been designed to allow fish to control their environmental temperature (Neill and others, 1972).

Sublethal responses can be measured in the same way as lethal responses using conce tration gradients to estimate the level of chemical required to effect a 50% chang in the population (EC50). The fifty percent sublethal response is often expressed as a fraction of the LC50 for the same chemical or mixture and this factor has been applied to the LC50 of similar compounds in the process of criteria development where no sublethal data is available. In the case of metals, the maximum acceptable tolerance concentration (the highest concentration producing no sublethal effects) for cadmium (Benoit and others, 1976) copper (McKim and Benoit, 1974), chromium (Benoit, 1976), lead (Holcombe and others, 1976), nickel (Pickering, 1974) and zinc (Benoit and Holcombe, 1978) has represented a 0.1 to 0.01 fraction of the LC50. Therefore, in the absence of sublethal information the guidelines for selenium and silver were established at 0.01 of the lowest LC50 for the most sensitive aquatic organism likely to be exposed (E.P.A. 1976). Application factors can be selected for other families of compounds in this way to protect the most sensitive use made of a receiving water.

BIOACCUMULATION

Compounds that are poorly soluble in water tend to be more soluble in organic solvents and when present in the aquatic environment accumulate in the fatty tissue of aquatic organisms. Contaminants that have represented the greatest problem in receiving waters have a low volatility, are highly lipid (fat) soluble, are halogenated for chemical stability and can be structurally similar to physiological chemicals within the organisms. The combination of these properties ensures rapid uptake and storage by the organism, reduced metabolic degradation and limited excretion resulting in persistance. What is observed ultimately, is very low to almost non-detectable levels of chemical in the water and 100 to 10,000 times higher concentrations in the organisms.

The degree to which a chemical concentrates in an animal (Ca) compared to the concentration in the water (Cw) is called the bioconcentration factor (BCF). In a simplistic two compartment model the change in the animal concentration through time is equal to the rate of chemical transport from the water to the animal, k_1, minus the rate of transport from the animal to the water, k_2, (Hamelink, 1977).

$$\frac{d\,Ca}{d\,t} = k_1\,Cw - k_2\,Ca \qquad (1)$$

When d Ca becomes zero, or the animal's body burden reaches equilibrium with the concentration in the water, the rates of uptake (k_1) and clearance (k_2) become important in estimating the bioconcentration factor (BCF) of the chemical.

$$\frac{Ca}{Cw} = \frac{k_1}{k_2} = BCF \qquad (2)$$

Although BCF's have been calculated as a result of laboratory studies it would appear practical to determine BCF's for compounds in the field through fish cage studies by measuring initial rates of uptake (k_1), under contaminated conditions and clearance (k_2), after transfer to uncontaminated conditions (Branson and co-workers, 1975) as illustrated in Fig. 3. This would avoid the difficulty of laboratory control and provide an initial screening evaluation of undocumented compounds identified in receiving waters. Some of the more common historical organic pollutants have the following BCF's (Vieth and others, 1979) in fish and would provide useful frames of reference for field studies where these compounds are also present.

Compound	BCF
PCB (1254)	100,000
pp'DDT	29,400
HCB	18,500
Mirex	18,100

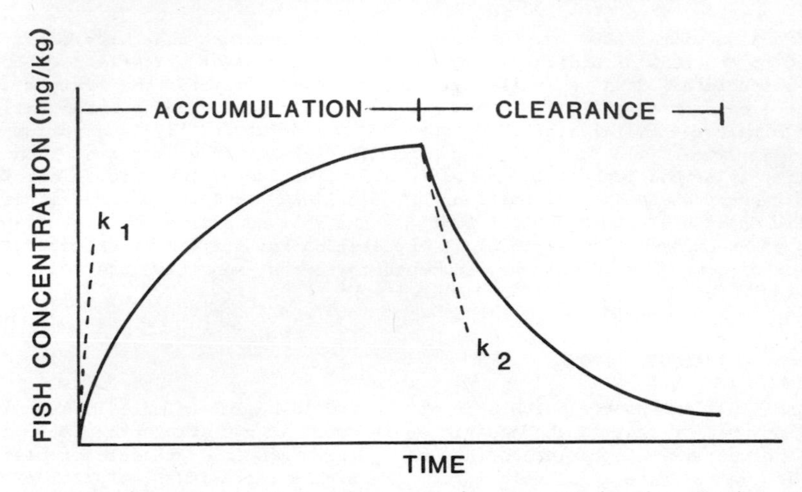

Fig. 3: Accumulation and clearance of an organic compound in fish showing the respective initial rates.

The technical and economic demands of determining BCF's for all organic compounds likely to be discharged into receiving waters would be too great to complete within any reasonable period of time and would outstrip the public's need to know what the accumulation potential of these chemicals is likely to be in aquatic organisms. Consequently physical/chemical characteristics, such as the octanol/water partition coefficient, have been used to estimate the BCF thereby avoiding expensive and time consuming fish studies.

The interface between two immiscible volumes of n-octanol and water has been considered a close approximation of the interface between a biological membrane of an aquatic organism and the surrounding water (Zitko, 1975). By saturating volumes of n-octanol and water in the presence of one another with a specific compound and expressing the concentration of the compound in the n-octanol over that in the water, a coefficient can be determined. Vieth and others (1979) have proposed that the BCF of a compound can be estimated by knowing the partition coefficient (P) based c tests with fathead minnows, rainbow trout and bluegill sunfish according to the following formula:

$$\log BCF = 0.85 \log P - 0.70 \quad (3)$$

The ability of certain compounds to accumulate in fish directly from the water or through the food chain becomes particularly important when the level of contamination and the activity of the compound threatens human health or the health of other wildlife through fish consumption. The bioconcentration effect may have other longer term implications for human health where levels of compound present in fish are similar to levels that produce mutations in bacteria or chromosomal damage in other biological screening systems.

BIOMONITORING

The sublethal biological responses outlined above can be combined to provide a haz assessment of compounds in the development of water quality criteria. They do not, however, lend themselves to monitoring the current quality of a waste as it is discharged. Considerable research has been directed to developing an on line biologi monitoring system that relies on changes in fish opercular (gill) movement and hea beat to detect the presence of toxicants.

Free swimming fish held in aquaria through which diluted or whole effluent is allowed to pass will respond to metabolic or neural toxicants by increasing or decreasing their opercular or heart beat rates. These muscular contractions emit very low level electrical signals, ranging from 1-10 mV, which can be received by external electrodes placed in the aquaria with the fish and amplified for strip chart recording (Gruber and co-workers, 1977). A further refinement of this process includes computer interpretation of the electrical signals (Westlake and van der Schalie, 1977) avoiding visual monitoring of chart output and allows the establishing of alarm limits according to changing rates of fish opercular or heart activity. When alarm conditions are recognized by the computer analytical samples may be collected automatically and plant personnel can be alerted to the change in waste quality.

A number of fish in separate aquaria may be monitored to provide a statistically sound data base and avoid false alarms due to equipment failure, disease outbreaks or feeding responses that can be encountered in single units of this system. Characterization of the effluent by sublethal testing should be completed to relate changes in fish opercular/heart activity to lethal or sublethal responses thereby providing a degree of interpretation to changes in the signals. A biomonitoring system of this type may be used in a permanent installation, a mobile laboratory for field monitoring (Gruber and co-workers, 1979) or to monitor incoming drinking water supplies as is the case in Great Britain (Alabaster pers. comm.) where mixed industrial and municipal uses are made of river systems.

SUMMARY

Water quality criteria are selected on the basis of aquatic toxicity data reported in the literature. The biological tests used to determine levels of lethality or reproduction, growth and behavioural impairment for individual compounds may also be adapted for evaluating mixtures such as industrial wastes. Biological testing data can provide a more complete picture of effluent quality than single parameter chemical testing due to their integration of all the environmental conditions. They are therefore useful for identifying toxic process streams, the relative toxic contribution of effluent constituents and estimating the potential impact of waste discharges on aquatic communities.

Environmental contaminants may also accumulate in aquatic organisms to levels several magnitudes greater than in the water. Comparison of accumulation concentrations with the octanol-water partition coefficient on a number of well studied compounds has led researchers to propose a method of projecting what bioconcentration factor might be expected from recently identified compounds. Caged fish studies might also assist in determining bioconcentration factors based on initial rates of uptake and clearance of compounds analyzed in fish.

Continuous monitoring of opercular and heart rate in fish exposed to an industrial waste discharge can provide an immediate indication of a change in waste quality. Low level electrical signals in fish detected by external electrodes, amplified and interpreted by computer equipment can activate sample collection or personnel alarms when predetermined effluent limits are exceeded. The same technology can be used to monitor drinking water drawn from a multiple use river or lake system.

Biological monitoring and testing systems are not only being used to develop water quality criteria but can be employed as powerful investigatory and surveillance techniques to preserve and improve existing water quality.

ACKNOWLEDGEMENT

The author expresses his gratitude to Dr. T.G. Brydges, J.R. Munro, C. Inniss and D. Wells for their comments and suggestions in the preparation of the manuscript.

REFERENCES

Abedi, Z.H. and W.P. McKinley (1968). Zebra fish eggs and larvae aflatoxin bioassa test organisms. J. of the A.O.A.C. 51, 902-905.

Abedi, Z.H. and D.E. Turton (1968). Note on the response of the zebra fish larvae to folpet and difolatan. J. of the A.O.A.C. 51, 1108-1109.

American Society for Testing and Materials (ASTM) (1973). Biological methods for the assessment of water quality. In J. Cairns Jr. and K.L. Dickson (Eds.). ASTM STP 528, American Society for Testing and Materials, p.256.

Armitage P. and I. Allen (1950). Methods of estimating the LD50 in quantal response data. J. of Hygiene 48, 298-322.

Beamish, R.J. (1974). Loss of fish populations from unexploited lakes in Ontario, Canada, as a consequence of atmospheric fallout of acid. Water Res., 8, 85-95.

Beamish, R.J. and H.H. Harvey (1972). Acidification of La Cloche Mountain Lakes, Ontario and resulting fish mortalities. J. Fish. Res. Bd. Can., 29, 1131-1143.

Beamish, R.J., W.L. Lockhart, L.C. van Loon and H.H. Harvey (1975). Long-term acidification of a lake and resulting effects on fishes. Ambio 4, 98-102.

Beattie, J.H. and D. Pascoe (1978). Cadmium uptake by rainbow trout, Salmo gairdneri eggs and alevins. J. Fish. Biol. 13, 631-637.

Bennett, B.M. (1952). Estimation of LC50 by moving averages. J. of Hygiene, 50, 157-164.

Benoit, D.A. (1976). Chronic effects of hexavalent chromium on brook trout (Salvelinus fontinalis) and rainbow trout (Salmo gairdneri). Water Res., 10, 497-500.

Benoit, D.A., E.N. Leonard, G.M. Christensen and E.P. Hunt (1976). Toxic effects of cadmium on three generations of brook trout (Salvelinus fontinalis). Trans. Am. Fish. Soc. 105, 550-560.

Benoit, D.A. and G.W. Holcombe (1978). Toxic effects of zinc on fathead minnows (Pimephales promelas) in soft water. J. Fish. Biol., 13, 701-708.

Branson, D.R., G.E. Blair, H.C. Alexander and W.B. Neely (1975). Bioconcentration of 2, 2', 4, 4' - tetrachlorobiphenyl in rainbow trout as measured by an accelerated test. Trans. Am. Fish. Soc., 4, 785-792.

Brett, J.R. (1964). The respiratory metabolism and swimming performance of young sockeye salmon. J. Fish. Res. Bd. Can., 21, 1183-1226.

Cairns, J. Jr., A. Scheier and J.J. Loos (1965). A comparison of the sensitivity to certain chemicals of adult zebra danios Brachydanio rerio (Hamilton-Buchanan) and zebra danio eggs with that of adult bluegill sunfish Leponis macrochirus raf Notulae Naturae 383, 109.

Chapman, G.A. and D.L. Shumway (1978). Effects of sodium pentachlorophenate on survival and energy metabolism of embryonic and larval steelhead trout. In. K.R. Rao (Ed.) Pentachlorophenol. Plenum Publishing Corp., New York, N.Y.

Eaton, J.G. (1970). Chronic malathion toxicity to the bluegill (Leponis macrochiru Rafinesque). Water Res., 4, 673-684.

Environment Canada (1974). Petroleum Refinery Effluent Regulations and Guidelines. Report EPS 1-WP-74-1. p.29.

Environment Canada (1977a). Metal Mining Liquid Effluent Regulations and Guideline Report EPS 1-WP-77-1.

Environment Canada (1977b). Meat and Poultry Products Plant Liquid Effluent Regula tions and Guidelines. Report EPS 1-WP-77-2. p.48.

Environment Canada (1977c). Potato Processing Plant Liquid Effluent Regulations ar Guidelines. Report EPS 1-WP-77-4. p.33.

Environmental Protection Act (EPA) (1971). Queen's Printer for the Province of Ontario, Canada.

Environmental Protection Agency (EPA) (1976). Quality Criteria for Water. U.S. Environmental Protection Agency, Washington, D.C.

Finney, D.J. (1971). Probit Analysis. 3rd ed. Cambridge University Press.

Fisheries Act (1970). Fisheries Act. R.S. e.119, S.1. Fisheries and Environment Canada.

Gruber, D., J. Cairns Jr., K.L. Dickson, R. Hummel III, A. Maciorowski and W.J. van der Schalie (1977). An inexpensive, noise-immune amplifier designed for computer monitoring of ventilatory movements of fish and other biological events. Trans. Am. Fish. Soc., 106, 497-499.

Gruber, D., J. Cairns Jr., K.L. Dickson, A.C. Henricks and W.R. Miller III (1979). Initial testing of a recent biological monitoring concept. J. Water Poll. Control Fed., 51, 2744-2751.

Hamelink, J.L. (1977). Current bioconcentration test methods and theory. In F.L. Moyer and J.L. Hamelink (Eds.), Aquatic Toxicology and Hazard Evaluation, American Society for Testing and Materials STP, 634, p.149-161.

Holcombe, G.W., D.A. Benoit, E.N. Leonard and J.M. McKim (1976). Long-term effects of lead exposure on three generations of brook trout (Salvelinus fontinalis). J. Fish. Res. Bd. Can., 33, 1731-1741.

Isio, S. (1965). Behaviour of fish exposed to toxic substances. Adv. in Water Poll. Res., 2, 19-40.

Jones, D.R., J.W. Kiceniuk and O.S. Bamford (1974). Evaluation of the swimming performance of several fish species from the Mackenzie River. J. Fish. Res. Bd. Can., 31, 1641-1647.

Kihlstrom, J.F. and L. Hulth (1972). The effect of phenylmercuric acetate upon the frequency of hatching eggs from the zebra fish. Bull. Environ. Contamin. Toxicol., 7, 111-114.

Lemke, A.E. and D.I. Mount (1963). Some effects of alkyl benzene sulfonate on the bluegill (Leponis macrochirus). Trans. Am. Fish. Soc., 92, 372-378.

Litchfield, J.T. Jr. and F. Wilcoxon (1949). A simplified method of evaluating dose-effect experiments. J. of Pharmacology and Experimental Therapeutic, 96, 99-113.

Maciorowski, H.D., R. McV. Clarke and E. Scherer (1977). The use of avoidance-preference bioassays with aquatic invertebrates. In. W.R. Parker, E. Pessah, P.G. Wells and G.F. Westlake (Eds.). Proceedings of the Third Annual Toxicity Workshop, EPS-5-AR-77-1, p.49-58.

MacLeod, J.C. and G.L. Smith (1966). Effect of pulpwood fibre on oxygen consumption and swimming endurance of the fathead minnow, Pimephales promelas. Trans. Am. Fish. Soc., 95, 71-84.

McKim, J.M. (1977). Evaluation of tests with early life stages of fish for predicting long-term toxicity. J. Fish. Res. Bd. Can., 34, 1148-1154.

McKim, J.M. and D.A. Benoit (1971). Effects of long term exposure to copper on the survival, growth and reproduction of brook trout. J. Fish. Res. Bd. Can., 28, 655-662.

McKim, J.M., J.G. Eaton and G.W. Holocombe (1978). Metal toxicity to embryos and larvae of eight species of freshwater fish - II Copper. Bull. Environ. Contam. Toxicol., 19, 608-616.

Miller, D.L. and D.M. Ogilvie (1975). Temperature selection in brook trout (Salvelinus fontinalis) following exposure to DDT, PCB and Phenol. Bull. Environ, Contam. Toxicol., 14, 545-555.

Mount, D.I. (1973). Chronic effect of low pH on fathead minnow survival, growth and reproduction. Water Res., 1, 987-993.

Mount, D.I. and C.E. Stephan (1967). A method for establishing acceptable toxicant limits for fish - malathion and butoxyethanol ester of 2,4-D. Trans. Am. Fish. Soc., 96, 185-193.

Neill, W.H., J.J. Magnuson and G.D. Chipman (1972). Behavioural thermoregulation by fishes: A new experimental approach. Science, 176, 1443-1445.

O.W.R.A. (1970). The Ontario Water Resources Act, Queen's Printer for the Province of Ontario, Canada.
Pickering, Q.H. (1974). Chronic toxicity of nickel to the fathead minnow. J. Wat. Pollut. Control Fed., 46, 760-765.
Scherer, E. (1977). Behavioural assays - principles, results and problems. In W.R. Parker, E. Pessah, P.G. Wells and G.F. Westlake (Ed.) Proc. Third Aquatic Toxicity Workshop, EPS-5-AR-77-1. p.33-40
Smith, W.E. (1973). A cyprinodontid fish, Jordanella floridae, as reference animal for rapid chronic bioassays. J. Fish. Res. Bd. Can., 39, 329-330.
Sprague, J.B. (1969). Measurement of pollution toxicity to fish - I Bioassay methods for acute toxicity. Water Res., 3, 793-821.
Stephan, C.E. (1977). Methods for calculating an LC50. In F.L. Nayer and J.L. Hamelink (Eds.), Aquatic Toxicology and Hazard Evaluation, American Society for Testing and Materials STP 634, p.65-84.
Vieth, G.D., D.L. DeFoe and B.V. Bergstedt (1979). Measuring and estimating the bioconcentration factor of chemicals in fish. J. Fish. Res. Bd. Can., 36, 1040-1048.
Westlake, G.F. and W.H. van der Schalie (1977). Evaluation of an automated biological monitoring system at an industrial site. In John Cairns Jr., K.L. Dickson and G.F. Westlake (Eds.). Biological Monitoring of Water and Effluent Quality, American Society for Testing and Materials STP, 607, p.30-37.
Zitko, V. (1975). Structure activity relationships in fish toxicology. In G.D. Vieth and D.E. Konasewich (Eds.). Structure-Activity Correlations in Studies of Toxicity and Bioconcentration with Aquatic Organisms. International Joint Commission Research Advisory Board, Windsor, Ontario.

13
ELIMINATION OF THE MUTAGENICITY OF BLEACH PLANT EFFLUENTS

M. A. Nazar and W. H. Rapson

Dept. of Chemical Engineering and Applied Chemistry, University of Toronto, Toronto, Canada

ABSTRACT

The mutagenicity of the aqueous effluents from the bleaching of kraft pulp has been assayed by the Ames Salmonella/Microsome Mutagenicity Test. Confirming the results of Swedish investigators (Ander and co-workers, 1977; Eriksson and co-workers, 1979), only the chlorination stage filtrate produced a significant response, but this decreased almost linearly with increasing substitution of chlorine dioxide for equivalent chlorine. The component of pulp mainly responsible for mutagenicity produced by chlorination is shown to be lignin. Methods of decreasing or eliminating the mutagenicity as well as variations in the bleaching process which may increase the mutagenicity are discussed. In particular raising the pH to 7-8 destroys much of the mutagenicity in a relatively short time. Since effluents are often neutralized with lime before being discharged to secondary treatment or to the receiving water much of the mutagenicity would be destroyed.

KEYWORDS

Pulp bleaching; mutagenicity; chlorination; bleach plant effluents; chlorine dioxide; destruction of mutagens.

INTRODUCTION

This is the second of a series of reports of extensive research into the production of mutagens in the bleaching of pulp, the nature of the mutagenic substances, the mechanism of their formation, methods of avoiding their formation and methods of destroying them (Rapson, Nazar and Butsky, 1980). The previous work in this field reported by Ander and co-workers (1977) and Eriksson and co-workers (1979) is confirmed and extended. Methods of substantially decreasing or even eliminating the mutagenicity of effluents from the bleaching of chemical pulps are presented.

The effluents from pulp and paper mills have been known for many years to contain substances toxic to aquatic life. Most of the BOD, COD and colour and much of the toxicity is found in the filtrates from the first two stages of the bleaching process (Das and co-workers, 1969; Leach and Thakore, 1975). Sophisticated analyses of effluents from the first two stages have identified many of the organic constituents, some of which are acutely toxic (Lindstrom and Nordin, 1976; CPAR, 1976). See Appendix for more detail.

The toxicity of substances to aquatic life has been measured traditionally by a short term test procedure for acute toxicity to fish or other test species (CPAR, 1976). Such rapid tests are useful indices of the toxicity of the effluent and the efficacy of any process modification or effluent treatment to decrease the hazard to the environment (CPAR, 1976 and 1977). Toxicologists have known for decades that low-level chronic exposure to many chemicals can lead to harmful effects (Casarett, 1975), particularly chemicals that are stored in animal bodies, which are consumed by other animals up the ecosystem, a process termed bioaccumulation. Chronic toxicity of bleach plant effluents has not been studied due in part to the large investment of time and money required.

The toxic effect of a chemical may involve an inheritable change in the genetic code of living cells (a mutation) or a change in function of an unaltered gene (a change of expression). Many investigators believe that although most genetic aberrations do not lead to malignancy, either or both of these aberrations may be the primary cause of cancer (Bridges, 1976; Brusick, 1978; Setlow, 1978). If it happens to be a cell involved in reproduction that is mutated, inheritable genetic disorders may be produced. This type of effect is even more difficult to assess, since a study over several generations is necessary.

Recently, several short term tests for mutagenicity have been developed and evaluated for their ability to detect mutation caused by chemicals of known carcinogenicity in laboratory animals (Brusick, 1978; Purchase and co-workers, 1976). In terms of cost, rapidity and ease of testing, the Ames Salmonella/Microsome Mutagenicity Test has an advantage over other assays. It also has a very high success rate in correctly detecting carcinogens (85%) and non-carcinogens (90%) according to Ames and co-workers (1975) and McCann and co-workers (1975). Although carcinogenicity produced by chemicals in test animals in the laboratory does not have an immediate or direct correlation with carcinogenicity in humans, almost all organic chemicals known to cause cancer in humans do cause cancer in animals and are positive on the Ames test (McCann and Ames, 1976).

TESTING PROCEDURES

The Ames test has been described in detail by Ames and co-workers (1975). Testing was carried out initially at the Ontario Cancer Institute[1] in Toronto until our own facility was installed in an isolated room dedicated to mutagenesis testing.

The parent strains of Salmonella typhimurium bacteria used are both genetically defective relative to wild type bacteria in that they can no longer synthesize essential histidine from other materials with which they are in contact. The strains used in this work are TA98 and TA100, in which further alterations have been made in the DNA of the parent bacteria. These strains are very sensitive to mutagens in general and can revert through mutation to a form which synthesizes its own histidine and can grow into an observable colony. Each

[1] We are deeply indebted to Prof. W.R. Bruce of the Ontario Cancer Institute for the use of his laboratory for our initial work and for the advice we have received throughout this project.

strain has a different type of histidine mutation·

a) TA98 is a frameshift mutation which involves the addition or deletion of base-pair(s) from DNA. Base-pairs are the building blocks for DNA and carry the genetic code.

b) TA100 is a base-pair substitution mutation in which one base-pair has been replaced by the other. (There are only four bases in virtually all DNA in all species and they are always found paired in a unique way to give only two base-pairs.)

The procedure for assaying is as follows: To a petri dish containing solidified agar and several necessary nutrients, is added 2.0 mL of molten agar which contains a specified amount of the chemical or solution to be tested, plus 0.1 mLof $\sim 10^9$ bacteria/mL suspension and a trace of two more nutrients, histidine and biotin. The molten agar is quickly poured onto the petri dish, allowed to solidify and incubated for 48 hours (TA100) or 72 hours (TA98) at 37^{0}C. The trace of histidine added is sufficient to allow the bacteria to undergo only a few cell divisions until that histidine is exhausted. Most often, genetic damage is done at the DNA duplication stage during cell division. If a particular mutation occurs, namely the one which restores the ability to synthesize histidine from base ingredients, then the revertant, through multiple cell divisions, forms a colony. At the end of the incubation period the colonies are counted using a colony counter. Generally, dose-response curves are determined by applying decreasing volumes of a solution, or applying a constant volume of solutions of decreasing concentration. It should be noted that only the degree of reversion to histidine independence can be measured in this test.

Not all cell replications are exactly faithful and a few spontaneous reversions do occur without the intentional addition of mutagens. These are thought to arise from naturally occurring uncorrected errors in transcription of DNA during cell division as well as from trace mutagens in the environment. In each experiment a known mutagen is added to two blank plates to confirm the reversion properties of the bacterial culture. For TA100, N-methyl-N'-nitro-N-nitrosoguanidine (MNNG) was used; for TA98, 4-nitro-o-phenylenediamine (4-NPD). The spontaneous revertants are counted on blank test plates to determine the extent of natural mutation.

In some cases the test chemicals or filtrates are toxic to the bacteria and may or may not be mutagenic. A low spontaneous reversion rate or a decrease in the number of microcolonies formed in the beginning due to the trace of histidine present (known as the "background lawn", viewed under a microscope) signals a toxic effect. Normally, dilution of the test sample in a repeat experiment will decrease the toxicity and show up the mutagenicity.

The majority of carcinogens are in fact not direct acting but require activation by enzyme systems to induce cancer (Weisburger, 1975). Following the method of Ames and co-workers (1975), rat liver microsomes which contain a proportionately large concentration of enzymes can be added to the molten agar to activate the test chemicals.

Bacterial strains used in this work have been kindly supplied by Dr. B.N. Ames. They have been tested for their integrity and maintained for us by Dr. C.R. Fuerst, Dept. of Medical Genetics, Faculty of Medicine, University of Toronto. Agar slants of the culture are prepared for us monthly as a source of innoculum for the suspensions which are grown in our own laboratory weekly.

Plates and solutions for the Ames test are supplied by Central Services in the Faculty of

Medicine or by the Media Preparation Department in the Ontario Cancer Institute.

PREPARATION OF BLEACH PLANT FILTRATES

An unbleached spruce-jackpine kraft pulp of known Kappa Number was chlorinated at 3.5% consistency in sealed preserving jars. The Kappa Number is a measure of the lignin content of pulp. The total available chlorine applied was kept constant, except where otherwise indicated. After the chlorine and/or chlorine-chlorine dioxide solutions were added, the contents were vigorously shaken and left at 25°C for 30 minutes. For higher temperatures, pulp and make-up water were preheated, chemical was added and the jar was placed in a water bath at 60°C for 30 minutes. After filtration the solutions were placed in stoppered bottles and kept at 4°C until tested.

Then, in the extraction stage, each well-washed pulp was treated at 12% consistency at 70° for 2 hours with the required sodium hydroxide to pulp ratio.

RESULTS

In Fig. 1 the upper line shows the number of TA100 revertants vs volume of filtrate applied per plate for a softwood kraft pulp treated with pure chlorine. Up to about 0.4 mL, the dose response curve was linear. Since chlorination filtrate is the most mutagenic of filtrates tested, 0.4 mL of filtrate has been chosen as a standard volume applied for mutagenicity testing.

The middle line shows the mutagenicity for the same volumes of filtrate for the same pulp when 70% of the available chlorine was replaced by equivalent chlorine dioxide, D/C 70/30. There was a dramatic decrease in mutagenic activity. With pure ClO_2 the mutagenicity of the filtrate (lower line) was barely above the spontaneous reversion rate indicated on the ordinate scale.

This important observation is shown more effectively in Fig. 2 where the replacement of chlorine by equivalent chlorine dioxide was varied from 0 to 100%. Decreasing mutagenic with increasing replacement with ClO_2 is clearly evident, not only on TA100 but also on TA98. These two strains have different histidine point mutations as described above and revert at different rates, so the scales are different. The qualitative trend in both strains is very similar and shows that both frameshift and base pair substitution mutagens are decreased as ClO_2 substitution increases. An inflection at D/C 10/90 has been reproduced several times. These data agree well with those of Eriksson and co-workers (1979).

No metabolic activation by rat liver enzymes was required by these untreated first stage filtrates to produce mutagenicity. Therefore the mutagens are direct acting. However direct acting carcinogens (and presumably mutagens) are chemically reactive entities (Weisburger, 1975) and can often be rendered less harmful as shown below.

In order to decrease effluent volumes many mills partially recycle C-stage filtrate with a consequent rise in temperature in the chlorination tower. The effect of temperature on the mutagenicity of first stage filtrate is shown in Fig. 3. Decreasing mutagenicity with increasing ClO_2 substitution is clearly evident at both 25°C and 60°C.

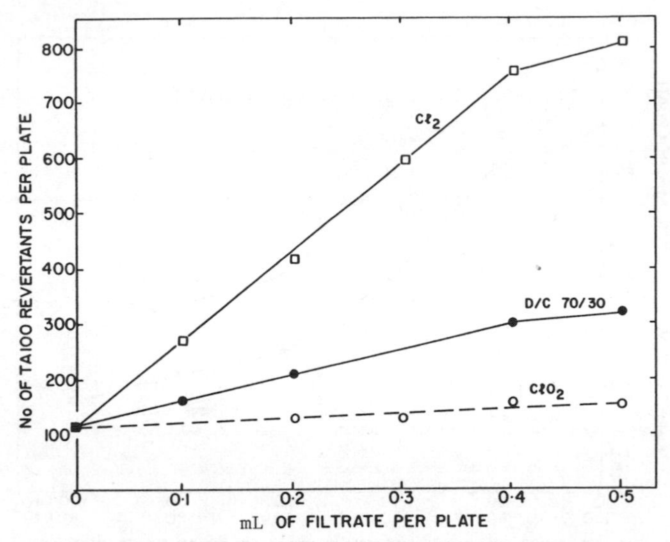

Fig. 1. Plot of TA100 revertants per plate versus volume of filtrate applied for bleaching softwood kraft pulp of Kappa No. 34.3 with 7.6% total available chlorine as pure chlorine (□), D/C 70/30 (●) and pure chlorine dioxide (○).

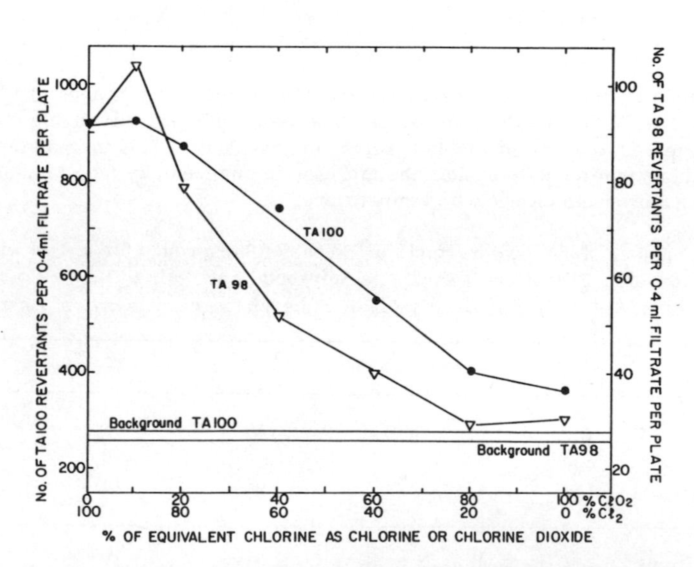

Fig. 2. Plot of TA100 (●) and TA98 (▽) revertants versus per cent of total available chlorine as chlorine dioxide for 0.4 ml of filtrate from bleaching softwood kraft pulp of Kappa No. 34.2 with 7.6% total available chlorine.

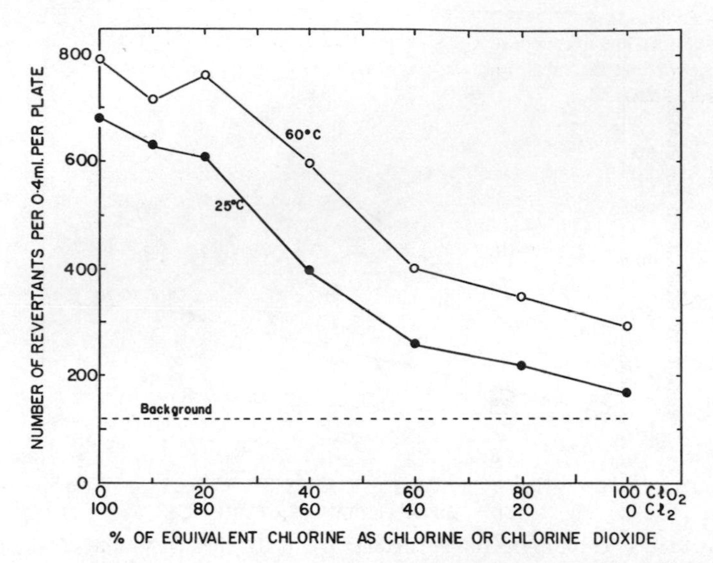

Fig. 3. Plot of TA100 revertants versus per cent of total available chlorine as chlorine or chlorine dioxide for 0.4 mL of filtrate from bleaching softwood kraft pulp of Kappa No. 26.7 with 5.8% total available chlorine at 25^0C (●) and 60^0C (○).

Residual available chlorine was found only in the 25^0C series. It was about 0.1% available chlorine up to D/C 60/40 and then increased. Therefore only 5.7% available chlorine was consumed compared to 5.8% for the 60^0C series (no residual). This increase in chlorine consumption is not sufficient to explain the increase in mutagenicity (see Fig. 5). It appea that the reaction products change with temperature.

TABLE 1 Number of TA100 Revertants in Excess of Background for 0.4 mL of Filtrate for Each Filtrate in Bleaching a Softwood Kraft Pulp of Kappa No. 34.2 in Five Stages $C_{7.6}E_{4.3}D_{1.0}E_{0.5}C_{0.4}$. Subscripts are % by wt. of Chemical on Pu

D/C Ratio	C	E_1	D_1	E_2	D_2
0/100 at 25^0C	450	- 8	- 9	-16	-47
0/100 at 60^0C	436	-28	-10	- 9	-66
70/30 at 60^0C	181	28[x]	9	-14	-23

[x] only 3.0% NaOH on pulp was required.

In Table 1 are shown the results of tests done on the filtrates from five stages of bleaching a softwood kraft pulp for three sets of bleaching conditions. Only the filtrates from the fi stage showed significant mutagenic activity. Since most of the organic carbon content of bleach plant filtrates is found in the first two stages of bleaching it would be expected that mutagenic chemicals would be found in either or both of these stages. However, extraction stage filtrate was not mutagenic, even after metabolic activation with rat liver homogenate (prepared according to Ames and co-workers (1975) by Litton Bionetics Inc., Maryland, U.S.A.) (see Table 2).

Table 2 Comparison of the Number of Revertants in Excess of Background for E_1 Filtrate With and Without Liver Enzyme (S9) Activation. Softwood Kraft Pulp Kappa No. 34.2 was Treated with 7.6% Available Chlorine On Pulp.

D/C Ratio	% NaOH	TA100 Revertants		TA98 Revertants
		-S9[+]	+S9[x]	-S9[+]
0/100	4.2	8	0	-7
10/90	4.1	- 3	- 5	-1
20/80	3.9	-23	- 9	3
40/60	3.6	- 1	9	-2
60/40	3.3	53	10	3
80/20	3.0	32	8	2
100/0	2.7	54	19	4

[+] 0.1 mL of E_1 filtrate used.
[x] 0.4 mL of E_1 filtrate used.

In the cases where no significant response was seen, the background lawn was checked for the possibility of a lethal effect of the filtrate on the bacteria. No decrease in the background lawn was observed. Since a positive available chlorine residual may be toxic to the bacteria, any residual chlorine is normally reduced with thiosulphate. The wide variation in the pH of the filtrates does not have a serious effect on the Ames test. This was demonstrated by determining the pH of the molten top agar to which either 0.4 mL of C stage or E_1 stage filtrate from a pure chlorine bleaching sequence had been added. Solutions of HCl or NaOH were added at sufficient concentration to change the pH of the top agar to the same values when either C stage or E_1 stage filtrate was used. No change in the spontaneous reversion rate was found due to a change in pH alone.

ELIMINATION OF MUTAGENICITY

Substitution of Equivalent ClO_2 for Cl_2

It has been shown in this work and previously by Eriksson and co-workers (1979) (on our recommendation) that substantial substitution of ClO_2 for equivalent Cl_2 greatly decreases and almost eliminates mutagenicity in the Ames test. This has been shown for three Salmonella bacteria strains, TA98 and TA100 in this work and TA1535 by Eriksson and co-workers (1979).

The benefits in the quality of bleach plant effluent by using chlorine dioxide instead of chlorine in the first stage have been known for some time (Rapson and co-workers, 1966 and 1977). BOD, COD, TOC, colour and acute toxicity of the effluent are substantially decreased as more of the available chlorine is supplied as chlorine dioxide. The decrease in mutagenicity is a further environmental benefit. In addition, the bleached pulp produced with chlorine dioxide is cleaner, stronger, more stable towards yellowing with age and shows an increase in yield of pulp from the wood. The use of ClO_2 for bleaching has been shown to be cost competitive with Cl_2 (Reeve and Rapson, 1980).

Elevation of pH Decreases Mutagenicity

The acidity of C-stage filtrate is normally neutralized with extraction stage filtrate, lime or by dilution in the receiving body. The effect on mutagenicity of this change of pH on chlorinated softwood kraft pulp filtrate was investigated by raising the filtrate pH with NaOH. After periods of 1 hour or 24 hours, aliquots were reacidified to pH 2 to sterilize them. In Fig. 4 is shown the decrease in mutagenicity as a function of elevation of pH. At the highest pH the mutagenicity is almost destroyed and shows only a small difference as a function of time. The results shown in Fig. 4 confirm the work of Eriksson and co-workers (1979) and may in part explain the decreased mutagenicity of effluents seen by others (Bjorseth and co-workers, 1979).

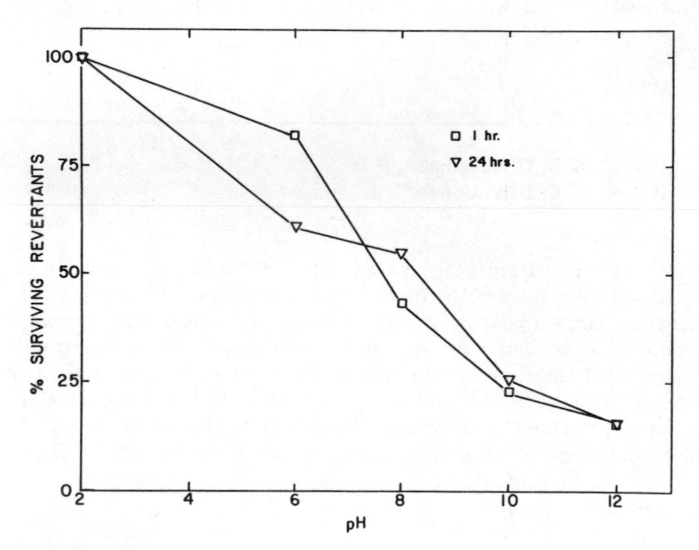

Fig. 4. Per cent of original number of revertants in excess of background remaining after raising the pH of chlorination filtrate with NaOH to 6, 8 10 and 12 for the times shown and then reacidifying to pH 2. The filtrate is from bleaching softwood kraft pulp of Kappa No. 34.2 with 7.6% chlorine.

Closed Cycle Operation

There are several other methods to decrease (but not totally eliminate) the mutagenicity of bleach plant effluent (Eriksson and co-workers, 1979). A mill in which all water streams are returned to process contributes no BOD, COD, toxicity, colour or mutagenicity to any receiving body. In the effluent-free mill (Rapson, 1967 and 1968), first stage seal tank filtrate is recycled to dilute incoming stock instead of fresh water. This decreases the volume of effluent discharged many-fold and thereby increases the concentration of organic matter and therefore the mutagens approximately the same number of times. However, the total mutagenicity per ton of pulp should not be changed. The mutagenicity has no effect on the environment if the filtrate is returned to the process.

ENHANCEMENT OF MUTAGENICITY

The variation in bleaching conditions experienced in the mill may alter the mutagenicity of filtrates. Investigations have been made into the effects of poor practices in the first stage.

The chemical to pulp ratio may vary if the chemical addition is poorly controlled or pulp mixing is inefficient. The effect of changing the chemical to pulp ratio is shown in Fig. 5 where mutagenicity versus applied available chlorine is plotted for pure chlorine (C), pure chlorine dioxide (D) and a mixture in which 70% of the total available chlorine is supplied by chlorine dioxide (D/C 70/30). When pure chlorine dioxide is applied, even at 15% available chlorine on pulp, no significant increase in the mutagenicity is observed over the spontaneous revertants.

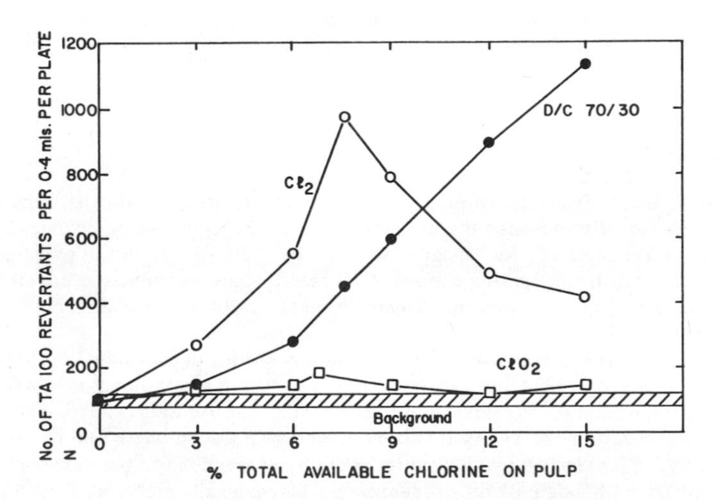

Fig. 5. Plot of TA100 revertants versus the applied total available chlorine for the filtrate from bleaching softwood kraft pulp of Kappa No. 30.7 with pure chlorine (○), pure chlorine dioxide (□) and D/C 70/30 (●).

To reach an E_1 Kappa number of 6.0, a normal target in industrial pulp bleaching, the necessary chlorine to pulp ratio is 6.8% for this pulp. In the event of overchlorination for both pure chlorine and D/C 70/30 bleaching, the mutagenic activity increases. The greatest activity is seen at 7.6% applied chlorine for pure chlorine bleaching. Similar behaviour is observed for all other kraft pulps tested with only slight shifts in the peak position. The application of chlorine beyond the peak decreases the mutagenic activity, but subsequent experiments have shown that this may be due to an increased concentration of reaction products which are lethal to the bacteria, thereby destroying many of the bacteria and reducing the number of colonies.

A mixture of chlorine and chlorine dioxide in which 70% of the chlorine is replaced by equivalent chlorine dioxide (D/C 70/30) shows moderate mutagenicity when applied at 6.8%. However as the total available chlorine is increased, so is the mutagenicity. Chlorine dioxide itself produces no significant mutagenicity when used for bleaching pulp, and the increase in activity when D/C 70/30 is used is not solely due to the presence of elemental chlorine. For example, at 15% total available chlorine in a D/C 70/30 mixture, 4.5% is from elemental chlorine. By comparison of the mutagenicity of filtrates produced from D/C 70/30 at 15% total available chlorine and pure Cl_2 at 4.5% there is greater than a 3 fold higher mutagenicity for the mixture. The effect of chlorine dioxide in D/C 70/30 mixtures is synergistic. It is possible that chlorine dioxide ruptures the aromatic rings and chlorination of the reaction products produces the increased mutagenicity. This was shown by adding chlorine to a non-mutagenic filtrate from treatment of pulp with pure chlorine dioxide which produced substantial mutagenicity.

A further example of synergism although not as pronounced has been observed when, instead of mixtures, the chlorine dioxide is added before the chlorine, called serial bleaching (Reev and co-workers, 1979). The total applied available chlorine remains the same. Although this method of bleaching has substantial advantages, laboratory experiments have shown that serial bleaching at D/C 60/40 produces about 30% more revertant colonies than the equivalent mixture.

Black Liquor Carryover

Both weak black liquor from the pulping operation and E_1 filtrate from the first extraction stage do not significantly increase the number of revertants over background. This is shown in Tables 3 and 4 respectively for 0% applied chlorine. However, in the event of black liquo carryover or E_1 filtrate penetrating the sheet on the C-stage washer, the possibility exists that after acidification and chlorination these filtrates may become mutagenic.

Accordingly, weak black liquor from a kraft mill was acidified to pH 2 with 1N H_2SO_4 and treated with D/C 0/100, 70/30 and 100/0 mixtures in increasing amounts. Water was added before the chemical solution to maintain a uniform dilution (25 fold). After sitting for 1/2 hour at room temperature, the residual was removed with thiosulphate and the solution teste for mutagenicity. The results are shown in Table 3, corrected for the varying volumes of thiosulphate added. Although all three treatments increase the mutagenicity, 70/30 mixture appear to be the most effective. This may be due to the synergistic effects of mixtures or due to the increased toxicity of filtrates when pure chlorine is applied in excess compared t equivalent chlorine dioxide or D/C 70/30.

Chlorination of Caustic Extraction Stage Filtrate

Chlorinated and oxidized lignin remaining in the pulp after the first stage is extracted in the next hot alkaline stage. The breakdown products of lignin might be expected to produce mutagens if they are subsequently chlorinated at low pH.

TABLE 3 Number of TA100 Revertants in Excess of Background for Available Chlorine Added to Acidified Weak Black Liquor (Unoxidized) from Kraft Pulping of Softwood (80% Jack Pine 20% Spruce). Liquor was diluted 25 Fold with Make-up Water and Chemical.

Available Chlorine % By Weight On Undiluted Liquor	D/C		
	0/100	70/30	100/0
0.0	31	42	35
0.2	95	59	22
2.0	86	448	140
20.0	14	14	14

After treatment with a D/C 10/90 mixture in the first stage at 7.6% available chlorine, a softwood kraft pulp was extracted with 4.0% NaOH on pulp, at 70°C at 12% consistency, for 2 hours. After filtering, the solution was acidified to pH 2 with HCl and treated with increasing amounts of aqueous chlorine in one set of experiments and chlorine dioxide in another, and submitted to the Ames test. After correction for dilution, the number of revertants per 0.4 ml of filtrate is shown in Table 4.

TABLE 4 Number of TA100 Revertants in Excess of Background for Cl_2 or equivalent ClO_2 Added to Acidified E_1 Filtrate from Extracting at 12% Consistency a Softwood Kraft Pulp Which Had Been Treated With 7.6% By Weight Available Chlorine as D/C 10/90. Unbleached Pulp Kappa No. 34.2

% Available Chlorine+	Cl_2	ClO_2
0	23	23
10	105	26
50	388	112
100	830	29

+ expressed as % by weight on pulp

Treatment of the caustic extraction filtrate with chlorine produced increasing mutagenicity with increasing application of chlorine, but chlorine dioxide produced very little mutagenicity In terms of the ratio of chemical to pulp from which the filtrate arose, the chlorine applied to the filtrate was far more than would ever be applied in a commercial bleach plant.

Chlorination of Lignin

The organic material in pulp which produces mutagens during chlorination was identified by bleaching Alaska yellow cedar milled wood lignin kindly supplied to us by Professor W. G. Glasser of Virginia Polytechnic Institute. Suspensions of lignin in water (equivalent to 2% by weight of lignin in pulp at 3.5% bleaching consistency) were treated with increasing per cent by weight chlorine in one series of experiments or equivalent chlorine dioxide in anothe The filtrates produced gave results as shown in Fig. 6 in the Ames test.

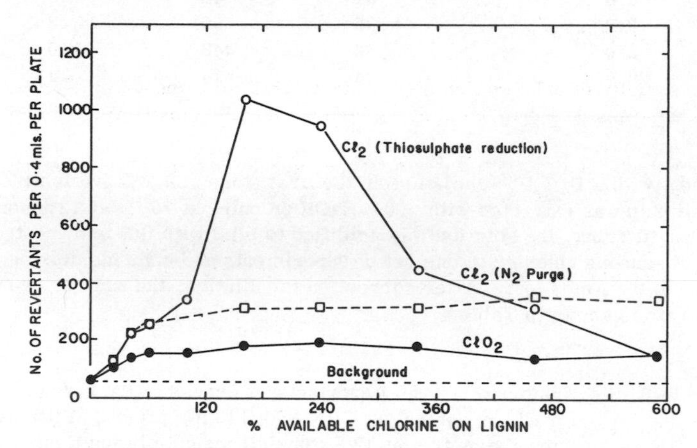

Fig. 6. Plot of TA100 revertants versus per cent by weight total available chlorine on lignin treated with pure chlorine or chlorine dioxide (●). The residual from reaction with chlorine was removed by thiosulphate (○) or purged with nitrogen (□). The residual from reaction with chlorine dioxide was removed by purging with nitrogen. A subsequent test for residual available chlorine with KI was negative.

Reacting chlorine dioxide with lignin produced a marginal mutagenic response over the rang of chemical application studied. However, increasing application of chlorine on lignin produced a plot of mutagenicity qualitatively similar to pulp bleaching as seen in Fig. 5.

The active chlorine residual in the experiment with chlorine was removed from one half of the filtrate volume by reduction with thiosulphate and from the other half by purging with dr nitrogen. The filtrates reduced with thiosulphate are both more mutagenic and more toxic for the same 0.4 mL test dose than are the nitrogen stripped samples. This implies that th thiosulphate alters the organic compounds in the filtrate. Above 100% by weight chlorine o lignin a toxic effect was observed, indicated by the decreased per cent bacterial survival relative to controls. Up to the peak, the mutagenicity outweighs the toxicity.

The experiment which produced mutagenicity on chlorinating the non-mutagenic filtrate from treatment of pulp with chlorine dioxide, was repeated using purified lignin. The lignin was treated with 160% of its weight of available chlorine as chlorine dioxide at tenfold higher concentration than in Fig. 6, and the filtrate was neither mutagenic nor toxic. Aliquots of this filtrate were treated with increasing chlorine to lignin ratio. When residual chlorine was removed with dry nitrogen all the filtrates were mutagenic but not toxic. When the residual chlorine was reduced with thiosulphate the solutions were more mutagenic but toxic only when larger amounts of chlorine were applied.

The filtrates from chlorination of other sources of lignin and lignin-like materials including groundwood and high yield sulphite have been tested and all are mutagenic. Therefore mutagenicity arises as a result of chlorination of any lignin and is not limited to chlorination during the bleaching of kraft pulp.

CONCLUSIONS

1. The mutagenicity of chlorination stage filtrate on TA100 and TA98 decreases almost to zero with increasing substitution of equivalent chlorine dioxide for chlorine.

2. No other filtrate in a CEDED sequence (including E_1 filtrate with liver homogenate activation) is mutagenic.

3. Raising the pH of chlorination filtrate to 7-8 destroys much of the mutagenicity in a relatively short time. This is normal practice in most mills before discharging.

4. The change in mutagenicity with changing chemical to pulp ratio for D/C 0/100, 70/30 and 100/0 has been investigated. Even a large excess of pure chlorine dioxide on pulp produces very little mutagenicity, if any.

5. The chlorination of weak black liquor and E_1 filtrate produces mutagenicity.

6. The chlorination of pure lignin as well as groundwood, kraft and sulphite pulp produces mutagenicity. Therefore lignin in wood or lignin altered by pulping produces mutagens on treatment with chlorine. In all experiments carried out with pure chlorine dioxide no significant mutagenicity was produced.

ACKNOWLEDGEMENTS

We would like to express our appreciation to the people in our research group who over the past thirty months have contributed to the results in this and subsequent papers: M. Brook, V. Butsky, A. Hoefler, S. Ing, A. Magued, S. May, G.B. Strumila. This work was supported by grants from the Natural Sciences and Engineering Research Council of Canada and from Erco Envirotech Limited.

This paper is presented with permission of the Canadian Pulp and Paper Association.

REFERENCES

Ames, B.N., J. McCann and E.Yamasaki (1975). Methods for detecting carcinogens and mutagens with the Salmonella/Mammalian-Microsome mutagenicity test. Mutation Research, 31, 347-364.

Ander, P., K. Eriksson, M. Kolar and K. Kringstad (1977). Studies on the mutagenic properties of bleaching effluents. Svensk. Papper. 80, 454-459.

Bjorseth, A., G.E. Carlberg, and M. Moller (1979). Determination of halogenated organic compounds and mutagenicity of spent bleaching liquors. The Science of the Total Environment, 11, 197-211.

Bridges, B.A. (1976). Short term screening tests for carcinogens, Nature, 261, 195-200.

Brusick, D.J. (1978). The role of short term testing in carcinogen testing. Chemosphere, 5, 403-417.

Casarett, L.J. (1975). In Casarett, L.J. and Doull, J. (ed.), Toxicology, Macmillan Publishing Co., New York. Chap. 2, pp. 13.

CPAR (1976). Identification of the toxic constituents in kraft mill bleach plant effluents. Cooperative Pollution Abatement Research Programme (CPAR) Report No. 245, Canadian Forestry Service, Dept. of the Environment.

CPAR (1977). Biodegradability of various toxic compounds in pulp and paper mill effluents. Cooperative Pollution Abatement Research Programme (CPAR) Report No. 408, Fisheri and Environment Canada, Environmental Protection Service.

Das, B.S., S.G. Reid, J.L. Betts and K. Partick (1969). Tetrachloro-o-benzoquinone as a component in bleached kraft chlorination effluent toxic to young salmon, J. Fish. Res. Board Can. 26, 3055-3067.

Eriksson, K., M. Kolar and K. Kringstad (1979). Studies on the mutagenic properties of bleaching effluents. II. Svensk. Papper. 82, 95-104.

Leach, J.M. and A.M. Thakore (1975). Isolation and identification of constituents toxic to juvenile rainbow trout in caustic extraction effluent from kraft pulp mill bleach plants. J. Fish. Res. Board Can. 32, 1249-1257.

Lindstrom, K. and J. Nordin (1976). Gas chromatography-mass spectrometry of chlorophenols in spent bleach liquors. J. Chromatogr. 128, 13-26.

McCann, J. and B.N. Ames (1976). The Salmonella/Microsome Mutagenicity Test: predictive value for animal carcinogenicity. In Hiatt, H.H., Watson, J.D. and Winsten, J.A. (ed.), Origins of Human Cancer, Cold Spring Harbour Laboratory, New York (Conference 1976) p. 1436.

McCann, J., E. Choi, E. Yamasaki and B.N. Ames (1975). Detection of carcinogens as mutagens in the Salmonella/Microsome test. Proc. Nat'l. Acad. Sci. U.S.A., 72, 5135-5139.

Purchase, I.F.H., E. Longstaff, J. Ashby, J.A. Styles, D. Anderson, P.A. Lefevre and F.R. Westwood (1976). Evaluation of six short term tests for detecting organic chemic carcinogens and recommendations for their use. Nature, 264, 624-627.

Rapson, W.H. (1967). The feasibility of recovery of bleach plant effluent to eliminate water pollution by kraft pulp mills. Pulp and Paper Mag. Can., 68, (12), T635-640.

Rapson, W.H. (1968). New concepts for stream improvement by recovery of bleach plant liquors from kraft pulp mills. Pulp and Paper Mag. Can., 69, (6), T161-166.

Rapson, W.H. and C.B. Anderson (1966). Mixtures of chlorine dioxide and chlorine in the chlorination stage of pulp bleaching. Pulp and Paper Mag. Can. 67, (1), T47-T55.

Rapson, W.H., C.B. Anderson and D.W. Reeve (1977). The effluent-free bleached kraft pulp mill, Part VI. Substantial substitution of chlorine dioxide for chlorine in the first stage of bleaching. Pulp and Paper Can., 78, (6), T137-T148.

Rapson, W.H., M.A. Nazar and V.V. Butsky (1980). Mutagenicity produced by aqueous chlorination of organic compounds. Bull. Environm. Contam. Toxicol., 24, 590-596.

Reeve, D.W., C.B. Anderson and W.H. Rapson (1979). The effluent-free bleached kraft pulp mill. Part X. The effect of dissolved organic matter on the first bleaching stage. Presented at the 1979 International Pulp Bleaching Conference, Toronto, Canada, June 1979.

Reeve, D.W. and W.H. Rapson (1980). Developments in chlorine dioxide bleaching. Proceedings of TAPPI Pulping Conference, 403-409.

Setlow, R.B. (1978). Repair deficient human disorders and cancer. Nature. 271, 713-717.

Weisburger, J.H. (1975). In Casarett, L.J. and Doull, J. (ed.), Toxicology, Macmillan Publishing Co., New York, Chap 15, pp. 334.

APPENDIX

The bleaching of wood pulp is carried out in several stages in order to minimize chemical costs. Each stage consists of a retention tower, washer and filtrate seal tank. The residual lignin in the pulp remaining after the pulping operation contains the chromophores which are oxidized and dissolved in the bleaching process.

Generally the first stage, chlorination (C), involves chlorine or chlorine-chlorine dioxide mixtures to oxidize and dissolve the lignin. The second stage is an alkaline extraction (E_1) which dissolves almost all the remaining lignin. The final stages in pulp bleaching are used to whiten the pulp to the high brightness of market pulp. One of the more common sequences is $D_1E_2D_2$: chlorine dioxide (D_1) followed by caustic extraction (E_2) and a final chlorine dioxide stage (D_2).

In the past, these effluents were discharged into receiving bodies of water with little if any treatment except perhaps control of the combined effluent pH. Due to more recent environmental constraints, the volume of effluent has been decreased by internal recycle of the filtrates produced from some of the stages. Chlorination stage filtrate has been used to replace part of the fresh water used to dilute pulp to the proper bleaching consistency. Consequently the concentration of organic matter (and mutagens) dissolved in the filtrate builds up as the filtrate is recycled.

Since the effluent volume is decreased due to recycling, comparison of the mutagenicity produced with non recycled filtrate is made on a per ton pulp basis.

14
THE GENETIC HAZARD OF SMALL DOSES OF NITROSO COMPOUNDS IN THE ENVIRONMENT

N. B. Akhmatullina and M. Ch. Shigaeva

Institute of Microbiology and Virology, Academy of Science of the Hazakh SSR, Alma-Ata, USSR

ABSTRACT

Our data showed that small doses of nitroso compounds have mutagenic activity. The use of nitroso compounds in some areas of the national economy and their resulting wide distribution in nature increase the degree of their genetic hazard. Micro-organisms contribute to the spreading of these compounds in nature as well by their metabolic products which include nitrosamines.

KEYWORDS

Nitroso compounds; small doses; genetic effects; stimulation and mutation; mutagen.

INTRODUCTION

The achievements in the field of chemical mutagenesis have allowed study of genetic effects of environmental pollution. Various aspects of the field have included assays for mutagenic activity of natural and synthetic chemical compounds to animals, plants and microorganisms, the study of potential mutagenic hazards of the environment to man, and to predict what new compounds, for medical and commercial use, might be genetically active. A general scheme for testing different compounds has been developed to show their action on the genetic structure of living organisms. A series of test systems is recommended for these purposes, using micro-organisms in the first stage. The latter are useful not only as primary tests but as models for studying general aspects of mutagenesis. This includes the analysis of chronic, multiple and joint actions of environmental contaminants which may differ in their chemical and mechanistic natures.

One more aspect of chemical mutagenesis by factors in the environment is no less important, i.e., the establishment of minimal doses with mutagenic activity. Until now, little attention has been paid to solving this question. However, the study of supermutagens, with nitroso compounds in particular, indicates the advisability of restricting the study of genetic effects to a narrow dose range.

Nitroso compounds are widely spread in nature; they are found in air, water, soil and food. In natural substrates, nitrosamines are accumulated as products of interactions of nitrites and secondary amines, formed by microbiological conversion of ammonium and nitrates. Nitrosamines also are found in metabolic by-products. Some bacterial species (*E. coli*, *Streptococcus epidermidis*, *Pr. vulgaris*, *Mortierella porvispora, Asp. oryzae,* etc.) were found to be capable *in vitro* of forming dimethylnitrosamines in a culture medium in the presence of nitrites and dimethyl-mine.

The greatest amount of dimethylamine has been reported to be found at 36 h of growth corresponding to 0.29, 0.62 and 2.2 mcg/mL. In the microbe-free medium, the dimethylamine content is less than 0.1 mcg/mL, an amount also found in human urea infected with *Pr. vulgaris* (Ayanaba and others, 1973).

Many chemical mutagens are also carc inogenic; or for example, N-dialkylnitrosamines N-alkylnitrosourethans and N-alkylnitrosourea are the strongest carcinogens in the class of nitroso compounds. These also accumulate in the environment as a result of microbial activity. Thus, bacteria of the *Clostridium* genus synthesized from 12 to 800 mcg of 3,4-benzo(a)pyrene per unit dry weight under laboratory conditions (Lot, 1969). There are reports concerning carcinogenic effects in combination with the action of oncogenic viruses and mutagens (Golubev and others, 1972), all again suggesting the action of small doses of chemical substances in the environment.

The purpose of the present investigation was to determine the genetic effects of low concentrations of some nitroso compounds and diazoalkanes on microorganisms with different organization levels.

MATERIALS AND METHODS

Act. olivaceus, Act. roseoflavus var. roseofungini, *Ps. aeruginosa,* and fowl plague virus were the objects of this investigation. Among the mutagenic compounds tested were nitrosomethylurea (NMU), nitrosodimethylurea (NDMU) with the dose range 0.125 to 0.5 mcg/mL, and 1,4-bisdiazoacetylbutane (DAB) at 2.5 to 25 mcg/mL. Actinomycete spores, bacterials cells (24 h culture) and extracellular virions were treated for 2 h.

The production of forward mutations, differing in their morphological properties, and the capacity for antibiotic and vitamin formation, served as criteria of the mutagenic action of small doses of nitroso compounds in actinomycetes. The criteri of the genetic effect in bacteria was phage induction. The genetic effect on virus of vertebrates was indicated by data showing their increased reproducibility and induction of different mutations.

RESULTS AND DISCUSSION

In the present study, treatment of spores of *Act. roseoflavus* var. roseofungini with NMU at a dose of 0.25 mcg/mL produced different visible mutations with a high frequency (up to 66%), which did not differ in their spectrum from mutations induce by higher concentrations, including the change of antibiotic formation. The range of variability was greater than that of the control population as far as this latte feature was concerned. A mean antibiotic activity value of the actinomycete popula-tion, which was obtained after NMU action, was 1450 $\pm$ 635 mcg/L, while the activity value of the control population was 1242 $\pm$ 606 mcg/L. The increase of the mean was associated with the production of highly active variants which produced the antibiotic roseofungini up to 3000 mcg/L in such populations.

Similar data were obtained after NDMU action on *Act. olivaceus*. At low concentra-tions, the mutagen increased the range of mutational variability, due to its vitamin formation feature, and allowed selection of mutants which formed twice the amount of vitamin B_{12} as the parent culture. Small doses of NDMU also induced rare muta-tions in *Act. olivaceus*. With a frequency of 0.003, for example, mutants were foun that were capable of using agar-agar as a source of carbon. They differed in colon morphology, had the white acrialmycelium, grew well on a mineral Czapek's medium without carbon sources, and formed zones of deepening and thinning agar (2-4 mm diameter). Such mutants appeared phenotypically with considerable delay, i.e., the nature of their origin seemed to be associated with the phenomenon of delayed mutagenesis or expression.

The experiments with the influenza virus also confirmed the mutagenic action of small doses. NMU at a dose of 0.05%, without inducing virus inactivation, produced a noticeable mutagenic effect, which was determined by a number of genetic markers (thermosensitivity, pathogenicity for mice, plaque size). Under growing conditions of fowl plague virus in the presence of low concentrations of DAB in the culture medium, there was a marked shift of particles from S^+ to S^-.

The virus-cell system proved to be more sensitive for displaying biological effects of chemical wastes, responding to doses which are lower by an order of magnitude in comparison with the stimulating ones. In the experiments with the lysogenic culture of *Ps. aeruginosa*, it was established that NMU at doses of 1-200 mcg/mL induced mature phage. The discovery of the virus inducibility indicates another aspect of the action of small doses of the mutagenic factors.

Thus, the data given in Table 1 show that the mutagens, even with low concentrations, produce the following genetic effects:

TABLE 1. Genetic Effects of Mutagens on Test Organisms

Mutagens	Organism	Genetic effect
NMU	*Act. rimosus*	Reverse mutations
	Act. roseoflavus var. roseofungini	Increase in antibiotic activity
	Ps. aeruginosa	Phage induction
	Influenza virus	Forward mutations
NDMU	*Act. olivaceus*	Forward mutations
NBU	*Act. rimosus*	Reverse mutations
NEU	*Pen. chrysogenum*	Morphological mutations
DAB	*Act. antibioticus*	Reverse mutations

Their wide distribution and high biological activity allow reference to nitroso compounds as hazardous environmental wastes. In connection with this, the investigations of the effects of low concentrations of these compounds on different organisms are of particular importance. One of the manifestations of the action of small concentrations of nitroso-mutagens is a monumental stimulation of growth and development of plants and some microorganisms (Golubev and others, 1972; Rapoport, 1966; Rapoport, 1968; Rapoport and others, 1971). We have demonstrated previously a stimulation effect of nitroso compounds on actinomycetes, bacteria and viruses (Shigaeva and Akhmatullina, 1979). The stimulation was expressed by the initiation of spore formation of actinomycetes, increase in bacterial division, increases of infectious virus titer, and by bacteriophage induction. The stimulation effect, as a mutagenic one, was found to depend upon the stage of organism development. Our experiments with actinomycetes and bacteria demonstrated that small doses neither caused any stimulation nor increased the microbial growth and development in the constant presence of the mutagens in a nutrient medium (Rapoport and others, 1980; Shigaeva and Akhmatullina, 1979).

All these data indicate the plurality of the actions of the so-called "small doses" of mutagens. The number of such investigations is not great, but the results are similar. Marquardt and others (1963) found that nitrosomethylacetateamide did not produce any lethal effect at a dose of 0.3 mcg, but it induced mutations from auxotrophy to prototrophy in adenine-limited *S. cerevisiae*. Rapoport and co-workers (1971) detected the mutagenic effect of stimulating doses of a number of compounds (NMU, NEU, NBU, DAB). The concentrations of the mutagens, which gave over 100%

survival, increased the yields of reverse mutations in *Act. rimosus* by a factor as high as 7 times. In *Pen. chrysogenum* NEU, used at maximally stimulating doses, up to 78% morphological mutation frequencies were induced. Reduced effects were found in *Act. streptomicini* and *Act. antibioticus*.

The present data cannot suggest directly the hazards of small doses of mutagens to man, and the problem of their human potential needs further investigation. Investigation of the mutagenic action of small doses may be important, not only of nitroso compounds, but for other chemical wastes as well. Moreover, the consequences of the actions of microquantities of the chemical mutagens on the natural microbial population in the sense of probable conversion of potential damages into fixed mutations are still unknown. The possibility of the increase in toxicity and virulence of microorganisms for plants, animals and men, and the formation of new pathogenic types of bacteria and viruses could be the most hazardous aspects.

The diversity of genetic effects of the chemical agents is of particular interest. It can be similar to the action of high concentrations or can have its own peculiarities. The results reported by Schendel and co-workers (1978), obtained during the study of the action of weak concentrations of ethyl methanesulfonate and nitrosoguanidine on *E. coli*, points to the difference between the mechanisms of induction of mutations with weak and strong concentrations of these mutagens. Similar investigations on the nitrosomutagens used in this study were not conducted.

REFERENCES

Aynaba A. and M. Alexander (1973). *Appl. Microbiol.*, *25*, 862-868.

Golubev, D.B. and M.A. Shlanketich (1972). *Modern Aspects of Virus Theory of Malignant Tumor Growth.* Medical Publ. House, Leningrad. (in Russian)

Lot, F. (1969). *Science*, *3416*, 491-492.

Marquardt, H., F. Zimmerman and R. Schweier (1963). *Naturwissenschaften*, *50*, *419*, 625-629.

Rapoport, I.A. (1966). *Chemical Mutagenesis.* Moscow. Znanie. (in Russian)

Rapoport, I.A. (1968), In: I.A. Rapoport (Ed.) *Mutation Selection.* Moscow. Nauka, p. 230-249. (in Russian)

Rapoport, I.A., A.G. Domrachewa, E.S. Lebedy, D.V. Keskinova and I.W. Kovalenko (1971). In: I.A. Rapoport (Ed.) *Chemical Mutagenesis and Selection.* Moscow. Nauka, p. 29-36. (in Russian)

Rapoport, I.A., M.H. Shigaeva and N.B. Akhmatullina (1980). *Chemical Mutagenesis.* Alma-Ata, Nauka. (in Russian)

Shendel, P.F., M. Defais, F. Veggo, Z. Samson and J. Caizns (1978). *J. Bacteriol.*, *135*, 466-475.

Shigaeva, M.H. and N.B. Akhmatullina (1979). *Vestnik Acad., Sci Kaz. SSR*, *I*, 26-30. (in Russian)

15
GROWTH OF MONO AND MIXED CULTURES IN SALINE ENVIRONMENT

A. A. Esener, N. W. F. Kossen and J. A. Roels

Biotechnology Group, Dept. of Chemical Eng., Delft University of Technology, Jaffalaan 9, Delft, The Netherlands

ABSTRACT

The effects of salinity on the kinetics and energetics of mono cultures were studied in batch.Results were compared with those reported for activated sludge. It was shown that the response of both mono and mixed cultures to increased salinity followed a similar pattern but the magnitudes of the effects differed significantly. Salinity was shown to be an important parameter influencing the kinetics and energetics of biological systems. Practical implications of these findings are discussed.

KEYWORDS

Salinity; mono culture; mixed culture; energetics; kinetics; biological waste water treatment.

INTRODUCTION

There are several processes that produce brines. These processes include distillation of sea water, production of salts, water softening by ion exchange, reverse osmosis, electrodialysis, pickling, canning, cheese and fishmeal manufacturing etc. Some of the wastes from these processes and/or combination of them with domestic waste waters present a special case for the conventional biological waste water treatment facilities. The major load of saline waste water, however, arise from the use of sea water for domestic purposes and as a carrier of domestic and industrial waste water at coastal locations with limited supply of fresh water. Saline wastes are also generated and has to be treated on board of marine vessels and off-shore installations. Discharge of waste water into seas and salt lakes is also common. Study of microbial growth in saline environment is therefore, of practical importance.

The effects of the presence of inorganic salts on microbial growth have first been studied by microbiologists, particularly for the formulation of suitable growth media for exacting organisms and mammalian cells (Ingram, 1939; Pirt and Thackeray, 1964; Scott, 1953; Wodzinski and Frazier, 1960). From these studies it can be concluded that addition of salts, in most cases NaCl, increased the respiration rate up to a specific salt concentration, thereafter a decrease was observed. Ingram

(1939) concluded that cation of the salt was of importance in determining the respiration rate, but this finding was not confirmed by others. It was generally agreed that maximum adaptation was obtained when the salinity of the medium was raised gradually.

In waste water field the effect of NaCl on activated sludge and trickling filtration processes have been studied (Burnett, 1974; Iami, Endoh and Kobayashi, 1979a, 1979b; Kessick and Manchen, 1976; Kincannon and Gaudy, 1966, 1968; Kincannon, Gaudy and Gaudy, 1966; Lawton and Eggert, 1957; Ludzack and Noran, 1965; Tokuz and Eckenfelder, 1979). The results of most of these studies confirm that high concentrations and/or shocks of NaCl have adverse effects on the performance of treatment plants. Small shocks or gradual increases in salt levels had little effect on the system performance and usually the system retained its activity after some adaptation time. Studies with mixed cultures, however, all suffered from the same phenomena; adaptation, selection and predominance of species prevented the experimentor to draw pure cause-response relationships from their experiments. e.g. Kincannon and Gaudy (1968) have observed a tremendous increase in biomass yield (75 %) in the presence of 8 kg/m^3 NaCl but were not able to identify whether this increase was due to a change in the efficiency of microbial metabolism or selection of salt tolerant species. Moreover, there is little data on the kinetics and energetics of microbial growth in saline waters.

A more fundamental approach has, therefore, been adopted in this study. First, the growth behaviour of *Klebsiella pneumoniae (aerogenes)*; a bacterium commonly present in soil and waste waters, was studied in batch cultures. With this approach the pure response of one of the species present in activated sludge communities was determined. The results obtained were then compared with those reported for activated sludge growing under similar environmental conditions (Imai, Endoh and Kobayashi, 1979a). It is shown that although the responses of mono and mixed cultures to increasing NaCl levels followed a similar pattern, the magnitude of responses differed significantly. Prediction of the amount of biomass which is to be expected due to breakdown of organic matter and the corresponding oxygen demand is of great importance in the design and operation of treatment plants. Therefore, various yields were evaluated as functions of NaCl concentration in the growth environment. Practical implications of these findings are discussed.

MATERIALS AND METHODS

Klebsiella pneumoniae (aerogenes) NCTC 418 was chosen as a representative bacterium existing in mixed cultures developed in waste water treatment facilities. The organism was cultivated aerobically in simple salts medium with glycerol being the only carbon source. All experiments were performed in a 11 x 10^{-3} m^3 working volume bioreactor in batch mode. Temperature and pH were set and controlled at 308^0K and 6.8 respectively. Inoculum was provided by an exponentially growing culture in NaCl free medium. Special attention was paid for the accurate determination of gas flow and concentrations. Elemental composition of biomass was determined by an element analyser. Oxygen content was found by difference. Ash content was determined separately. No carbon containing product other than biomass and carbon dioxide could be detected at significant quantities. Amount of biomass containing 12 grams of carbon was defined as 1 mole of biomass. Results were treated and corrected by a statistical analysis as reported elsewhere (de Kok and Roels, 1980).

RESULTS

Experimental results of mono cultures were obtained in this laboratory. Results of activated sludge, named as mixed culture hereafter, were obtained from a publication of Imai, Endoh and Kobayashi (1979a). These authors studied the response of activated sludge to increasing concentrations of NaCl in a respirometer. They used chloride concentration as a parameter, therefore, their results were converted by us in order to obtain a NaCl dependence. The experiments were performed in 0 - 40 kg/m^3 NaCl range which also covered the sea water case (approx. 33 kg/m^3 NaCl).

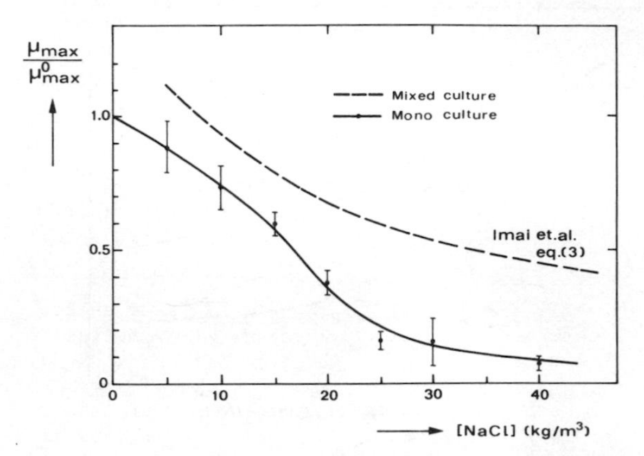

Fig. 1. Maximum specific growth rate vs. NaCl concentration.

In Fig. 1, the normalized μ_{max}, the maximum specific growth rate as defined by Monod kinetics, is shown as a function of NaCl concentration in the growth medium together with 95 % confidence levels (the superscript 0 denotes the value obtained in NaCl free medium). μ_{max} was determined from dry weight data by nonlinear regression. Imai, Endoh and Kobayashi observed a maximum μ_{max} in the low NaCl range. Such a maximum was not observed in this study. However, it must be noted that the first two data points have overlapping confidence levels.

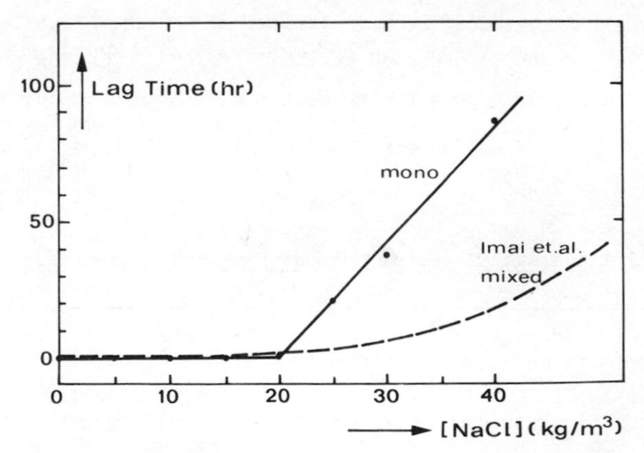

Fig. 2. Observed lag time vs. NaCl concentration.

In Fig. 2, lag time as defined by Dean and Hinshelwood (1966) is shown as a function of NaCl concentration. The inoculum concentration was not determined for each experiment but was of the same order of magnitude. The observed lag times have, therefore been assumed to be only a function of the NaCl concentration. The same assumption was also adopted by Imai, Endoh and Kobayashi (1979a).

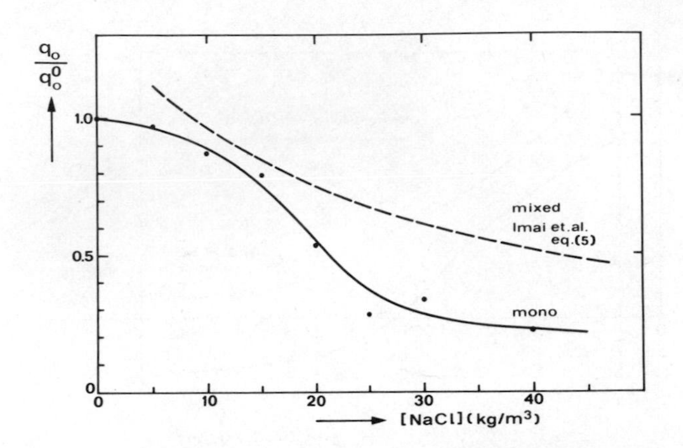

Fig. 3. Oxygen uptake rate vs. NaCl concentration.

Respiratory activity as the total oxygen uptake rate, q_o, is plotted in Fig. 3. A maximum q_o reported by Imai, Endoh and Kobayashi (1979a) was not observed in this study.

No significant change in the elemental composition of biomass could be detected in the experimental range and an average formula as shown in Table 1, was used for yield calculations.

Yields of biomass on substrate, Y_{sx}, oxygen, Y_{ox}, and carbon dioxide, Y_{cx}, are presented in Fig. 4. All yields are expressed as mole/mole. Slightly distinct optima in yield values can be observed at about 5 kg/m^3 NaCl. The presence of a corresponding minimum in the respiratory quotient, RQ, ensures the existence of an optimal yield range (Fig. 5). It therefore, becomes evident that the optimal yield is achieved not at the highest growth rate but at about $0.9 \cdot \mu_{max}$, where the NaCl concentration is 5 kg/m^3.

TABLE 1 Average Elemental Composition of *K. pneumoniae*

C	H	N	O	Formula
49.68	6.71	13.45	30.16	$C H_{1.62} N_{0.23} O_{0.46}$

% ash free dry weight. Average ash content 8.63 %.

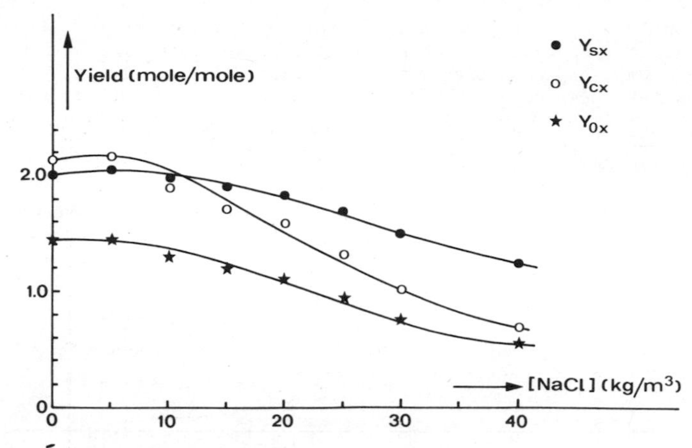

Fig. 4. Influence of NaCl concentration on yields of biomass.

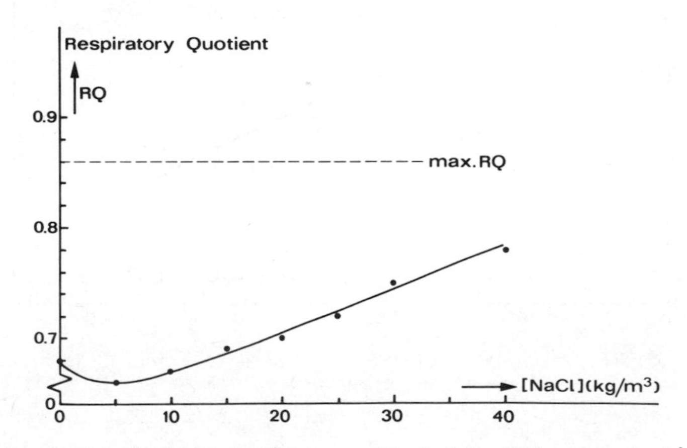

Fig. 5. Respiratory quotient as effected by NaCl concentration.

DISCUSSION

Comparison of the Behaviour of Mono and Mixed Cultures

From data presented in Figs. 1, 2 and 3, it can be concluded that both mono and mixed cultures show similar behaviour in saline environment. Mono cultures, however, are shown to be much more sensitive to increasing NaCl concentration. From 0 to 40 kg/m^3 NaCl, μ_{max} of the mono culture decreased by about 10 times, whereas the mixed culture by 5. Lag time and respiration rate data also showed a similar pattern. The functions adversely effected by the presence of NaCl showed a smooth loss of activity in mixed cultures, throughout the experimental range. Mono cultures, however, seemed to have a critical salinity range above which the culture activity slowed down drastically (15 - 20 kg/m^3). Possibly, above this level significant damage is done to the cell membrane by the osmotic pressure of its exterior. The relative success of the mixed culture in saline environment is most probably due to adaptation and predominance of salt tolerant species. This possibility has already been pointed at by Kincannon and Gaudy (1968). These authors have reported an increase in biomass yield (about 75 %) at 8 kg/m^3 NaCl. However, in this work with mono cultures only a slight increase in the yield was observed (<5 %) at the same NaCl range. Based on this observation it is unlikely that any other species could increase its yield at such amounts. This means that increase in yield observed by Kincannon and Gaudy (1968) was mainly due to changes in the predominance of the species.

Energetic Considerations

In Fig. 6, the thermodynamic efficiency of the growth process, η , is shown as a function of NaCl concentration in the medium.

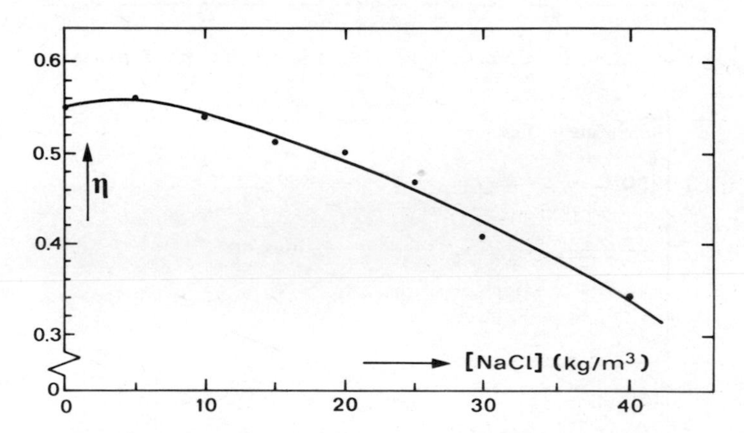

Fig. 6. Thermodynamic efficiency as a function of NaCl concentration.

η is given by Roels (1980) as:

$$\eta = Y_{sx} / Y'_f \quad (1)$$

where Y_{sx} is the biomass yield on substrate (mole/mole) and Y'_f is the maximum value of Y_{sx} consistent with the second law of thermodynamics. η describes the efficiency of the growth process by considering its irreversibility.It is important to note that knowledge of η allows straightforward estimation of the oxygen demand by:

$$Y_{ox} = (4 / \gamma_x) \cdot \eta / (1 - \eta) \quad (2)$$

where γ_x is the degree of reduction of biomass (Erickson, Minkevich and Eroshin 1978; Roels, 1980).

From the data presented it is clear that biosynthesis gets less efficient at high NaCl levels. Although detailed biochemical and physiological reasons for this decrease are still not clear, it is believed that the cells in saline environment have to do extra work to maintain concentration gradients and viability. This type of energy expenditure is currently accounted for, by the so-called "maintenance energy" requirement; a term which includes energy expenditure for other functions too. Significant increase in maintenance requirements in highly saline media has already been reported in literature (Stouthamer and Bettenhaussen, 1973; Watson, 1970). Unfortunately the current state of microbial energetics does not allow a precise assesment of the direct contribution of the energy spent for osmotic work, to maintenance requirements. Therefore, only a rough analysis will be reported here.

Assuming the law of linear substrate consumption is valid, q_s (mole/mole/hr), specific substrate consumption rate, is given by:

$$q_s = \mu / Y_{sx}^{max} + m_s \quad (3)$$

where Y_{sx}^{max} is the maximal yield on substrate (mole/mole) and m_s is the maintenance coefficient (mole/mole/hr). Here, both of these parameters may be influenced by salinity. If they are assumed as constants, statistical analysis of the experimental data yields the values of 2.09 (1.97 - 2.21) and 0.032 (0.023 - 0.040) for Y_{sx}^{max} and m_s , respectively (95 % confidence limits in parenthesis). The maintenance value determined is about 50 % higher than that reported by Herbert (1958) for the same organism in salt free medium. Although m_s is expected to be a function of NaCl concentration the present data does not allow the rejection of constant m_s hypothesis. However, it must be noted that the value of m_s obtained in this analysis is influenced most by data collected at low μ i.e. at high salinities.
Rewriting equation (3) as:

$$r_s / r_x = 1 / Y_{sx}^{max} + m_s / \mu \quad (4)$$

where r_s and r_x are the net flows of substrate and biomass to and from the system,

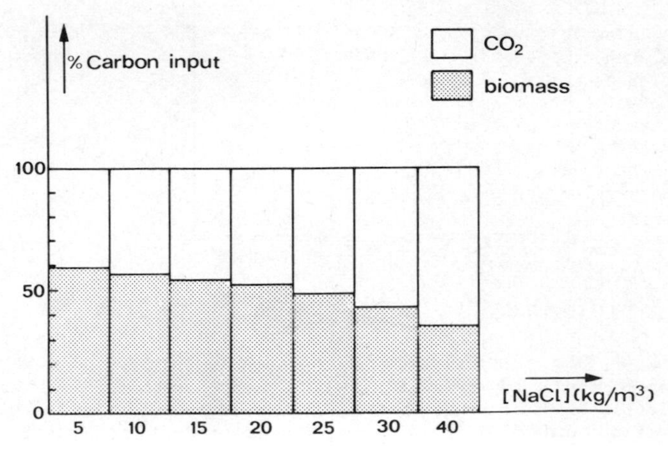

Fig. 7. Effect of NaCl level on the distribution of substrate carbon.

respectively, one can see the effect of increasing NaCl level on the distribution of substrate input. If Y_{sx}^{max} is constant, elevation of NaCl content increases m_s and decreases μ; thus the portion of substrate used for non-growth associated functions gets higher. This is also apparent from Fig. 7, where the distribution of substrate carbon in the system is shown.

Practical Aspects

The significant change in the substrate carbon distribution with increasing salinity, has an important consequence in relating the two commonly used parameters, COD and BOD, to each other. COD is defined here as the amount of oxygen required for the complete combustion of organic substrate. BOD is the amount of oxygen necessary for the biological conversion of organic substrate to carbon dioxide and biomass. The value of BOD, therefore depends on the efficiency of the biological growth process which is most conveniently described by η as shown in Fig. 6. Utilizing the recently developed regularities of fermentation processes and macro balancing principles (Erickson, Minkevich and Eroshin, 1978; Roels, 1980) a simple expression can be derived for the ratio, COD/BOD by noting that:

$$COD = (1/4) \cdot \gamma_s \text{ and } BOD = (\gamma_s/4) \cdot (1-\eta)$$

where γ_s is the degree of reduction of substrate. Here COD and BOD are expressed as moles oxygen required for the breakdown of substrate containing 12 grams of carbon. The ratio is then simply given by:

$$COD/BOD = 1 / (1 - \eta) \qquad (5)$$

In Fig. 8, dependence of COD/BOD to η as influenced by the presence of NaCl concentration is shown for this experimental system.

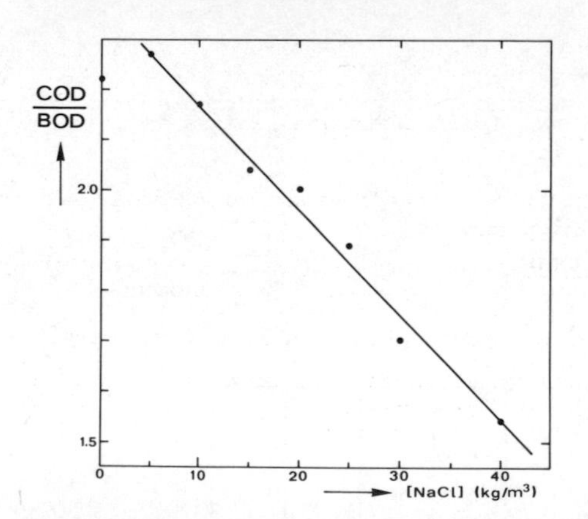

Fig. 8. Influence of NaCl level on the ratio, COD/BOD.

Generally high sludge production is undesirable in treatment installations. Remova[l] and disposal of sludge may present financial and environmental problems. It is, therefore, desirable to operate an activated sludge process at low μ as to increas[e] maintenance requirements of the culture, thereby reducing the amount of sludge formed. From this point of view, high salinity is a desirable quality. However, fr[om]

Fig. 6, one must remember that the thermodynamic efficiency is greatly reduced at high salinities. A consequent decrease in Y_{ox}, as predicted by equation (2), will then call for higher aeration capacity. Thus the engineer will be faced with an optimization problem, solution of which of course, depends on the relative costs involved. In Fig. 9, Y_{sx} shows the amount of sludge formed per mole substrate consumed and Y_{sx} / Y_{ox}, is the amount of oxygen taken up per mole substrate consumed. Therefore it is desirable to minimize both.

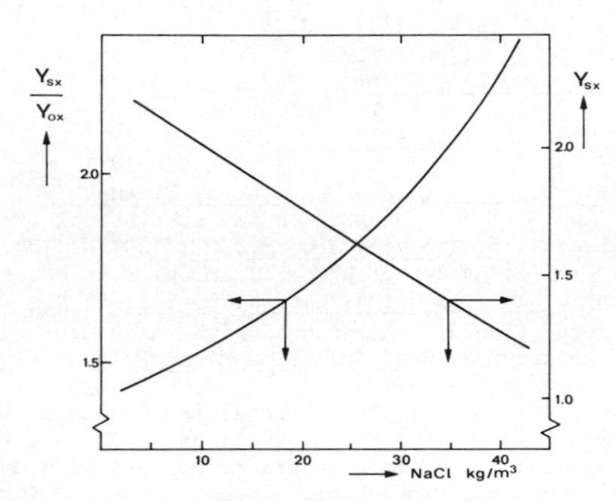

Fig. 9. Graphical representation of the optimization problem.

CONCLUSIONS

1. Mono and mixed cultures show similar response to increase in salinity in their environments. Mono cultures, however, are much more sensitive and therefore less efficient in such environments.
2. Salinity has significant effects on the efficiency and kinetics of growth process and therefore is an important parameter.
3. Design calculations based on data obtained from mono cultures will provide overestimates.
4. COD/BOD ratio can not be assumed constant, for a system with varying salinities.
5. For efficient operation an optimization must be carried out to minimize sludge production while maximizing yield on oxygen.

REFERENCES

Burnett, W. (1974). The effect of salinity variations of the activated sludge process. Wat. Sewage Wks., 121(3), 37 - 55.

Dean, A. C. R., and C. Hinshelwood (1966). Growth Function and Regulation in Bacterial Cells., Clarendon Press, Oxford. pp. 55 - 68.

Erickson, L. E., I. G. Minkevich, and V. K. Eroshin (1978). Application of mass and energy balance regularities in fermentation. Biotechnol. Bioeng., 20, 1595 - 1621.

Herbert, D. (1958). Some principles of continuous culture. In G. Tunewall (Ed), Recent Pregress in Microbiology, Almquist and Wiksell, Stockholm, p. 381.

Imai, H., K. Endoh, and C. Kobayashi (1979a). Effects of high salinity on the respiration characteristics of activated sludge. J. Ferment. Technol., 57(4), 333 - 340.

Imai, H., K. Endoh, and C. Kobayashi (1979b). Respiratory activity and sludge volume index of activated sludge during acclimation to saline water. J. Ferment. Technol., 57(4), 453 - 459.

Ingram, M. (1939). The endogeneous respiration of bacillus cereus. J. Bacteriol., 38, 613 - 629.

Kessick, M. A., and K. L. Manchen (1976). Salt water domestic waste treatment. J. Wat. Pollut. Control. Fed., 48(9), 2131 - 2136.

Kincannon, D. F., and A. G. Gaudy (1966). Some effects of high salt concentrations on activated sludge. J. Wat. Pollut. Control. Fed., 38(7), 1148 - 1159.

Kincannon, D. F., and A. F. Gaudy, and E. T. Gaudy (1966). Sequential substrate removal by activated sludge. Biotechnol. Bioeng., 8, 371 - 378.

Kincannon, D. F., and A. F. Gaudy (1968). Response of biological waste treatment systems to changes in salt concentrations. Biotechnol. Bioeng., 10, 483 - 496.

de Kok, H. E., and J. A. Roels (1980). Method for the statistical treatment of elemental and energy balances. Biotechnol. Bioeng., 22, 1097 - 1104.

Lawton, G. W., and Eggert (1957). Effect of high sodium chloride concentrations on Trickling filter slimes. Sewage and Ind. Wastes, 29(11), 1228.

Ludzack, F. J., and D. K. Noran (1965). Tolerance of high salinities by conventional waste water treatment processes. J. Wat. Pollut. Control Fed., 37(10), 1404 - 1416.

Pirt, S. J., and E. J. Thackeray (1964). Environmental influences on the growth of erk mammalian cells in monolayer cultures. Exp. Cell Res., 33, 396 - 405.

Roels, J. A. (1980). The application of macroscopic principles to microbial metabolism, Bioengineering Report. Biotechnol. Bioeng., Accepted for publication

Scott, W. J. (1953). Water relations of staphylococcus aeureus at 30°C. Aust. J. Biol. Sci., 6, 549 - 564.

Stouthamer, A. H., and C. Bettenhaussen (1973). Utilization of energy for growth and maintenance in continuous and batch cultures of microorganisms. Biochim. Biophys. Acta., 301, 54 - 69.

Tokuz, R. Y., and W. W. Eckenfelder (1979). The effect of inorganic salts on the activated sludge process performance. Water Res., 13, 99 - 104.

Watson, T. G. (1970). Effects of sodium chloride on steady state growth and metabolism of S. cerevisiae. J. Gen. Microbiol, 64, 91.

Wodzinski, R. J., and W. C. Frazier (1960). Moisture requirements of bacteria. J. Bacteriol., 79, 572 - 578.

SECTION V

BIOLOGICAL TREATMENT OF WASTES

16
KINETICS APPLIED IN PROCESS DESIGN FOR BIOLOGICAL WASTE WATER TREATMENT

A. Moser

Institute for Biotechnology, Microbiology and Waste Technology, Technical University Graz, Austria

ABSTRACT

In recent years a systematic approach to bioprocess design has become more important. New types of treatment plants,e.g.,plug flow reactors with concentration gradients and also gradientless reactors, must be developed more rapidly and with less risk. The correct determination of kinetics needs a special strategy. This paper intends to generalize the evaluation methods for the kinetics of homogeneous and heterogeneous bioprocessing. Integral analysis becomes impossible in the case of complex kinetics, and at the same time plug flow systems need differential evaluation of kinetics from batch data due to "kinetic similarity". The differential method of process kinetic analysis is summarized for Monod kinetics modified by lag and stationary phase, endogenous metabolism, S-inhibition, sludge adsorption, multi-S-reactions and the influence of mass transfer on floc and film geometry. To faciliate the handling of process design of truly heterogeneous 3-phase systems, a definition of "pseudohomogeneity" is outlined with quantitative criteria, which are proposed to be used to check bioprocesses.

KEYWORDS

Process kinetic analysis; differential evaluation; plug flow reactors; integral reactors; kinetic similarity; pseudohomogeneity.

INTRODUCTION

The design of waste water treatment is often carried out empirically. Due to the complex nature of the process, a scientific approach seems to be too tedious for practice. Modern process design, however, will show less risk and more rapid development. The use of mathematical models is recommended, as models are "invented" by man for situations which are too complex for a complete scientific elucidation. Working with models, however, requires a certain strategy and mode of thinking to be successful(A.Moser, 1981). There are different kinds of models, as shown in Table 1. Mechanistic models are based on reaction mechanisms. These models are presently being developed in fermentation processing for control and optimization purposes (Moreira, van Dedem and Moo-Young, 1979), but are too sophisticated for reaction engineering calculations. Models to be used in process design must be simple enough to handle with a minimum of parameters. An alternative, therefore, is the use of numerical models, i.e. purely mathematical functions, represented

TABLE 1. Different Types of Kinetic Models

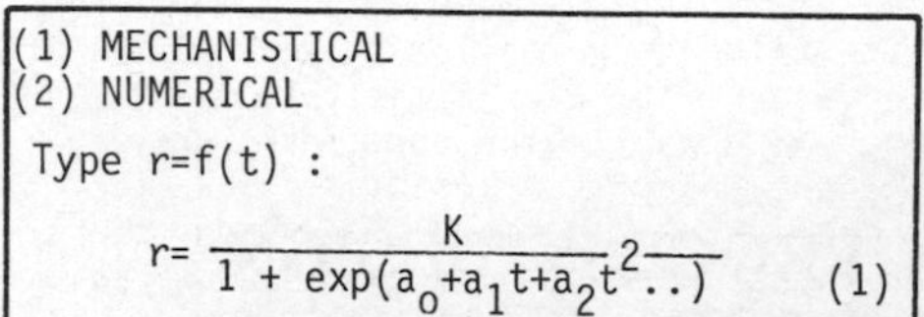

(1) MECHANISTICAL

(2) NUMERICAL

Type $r=f(t)$:

$$r = \frac{K}{1 + \exp(a_o + a_1 t + a_2 t^2 ..)} \qquad (1)$$

$$r = k.\exp[-a_3(t-t_o)] \qquad (2)$$

Type $r=f(c)$: $r = a_o + a_1 c + a_2 c^2 + ...$ (3)

(3) FORMAL KINETICS from ANALOGIES

$$r = k(T).f(c) \qquad (4)$$

Type $k(T)$: Arrhenius (5)

Type $f(c)$: $r = k.c^n$ (6)

$$r = \frac{k_1.c}{1 + k_2.c} \qquad (7a)$$

$$-\frac{dS}{dt} = r_s = r_{max}\frac{S}{K_m + S} \qquad (7b)$$

$$-\frac{1}{X}\cdot\frac{dS}{dt} = \sigma = \sigma_{max}\frac{S}{K_s + S} \qquad (7c)$$

$$\frac{1}{X}\cdot\frac{dX}{dt} = \mu = \mu_{max}\frac{S}{K_s + S} \qquad (7d)$$

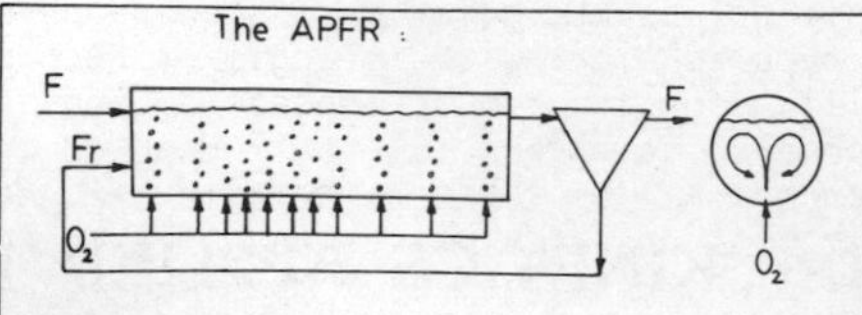

FIG.1 : The AERATED PLUG FLOW REACTOR and CHARACTERISTICS.

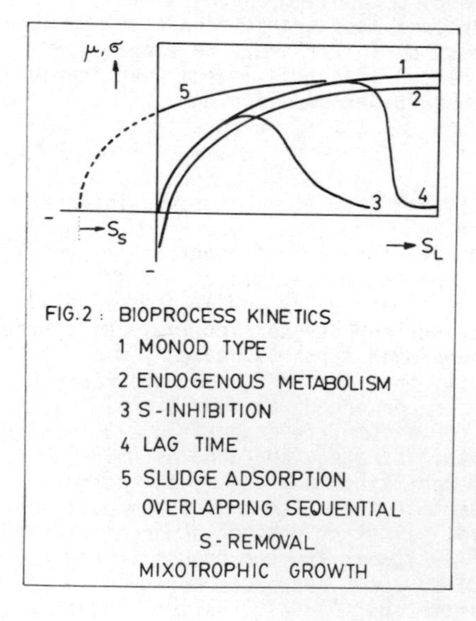

FIG.2 : BIOPROCESS KINETICS
1 MONOD TYPE
2 ENDOGENOUS METABOLISM
3 S-INHIBITION
4 LAG TIME
5 SLUDGE ADSORPTION
OVERLAPPING SEQUENTIAL
S-REMOVAL
MIXOTROPHIC GROWTH

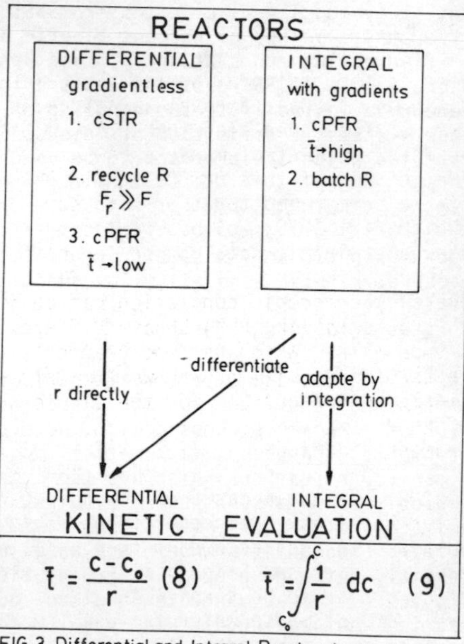

FIG.3 Differential and Integral Reactor Analysis

by eqn.(1)-(3) in Table 1. Even when they do form a workable solution, it is not satisfying, as the parameters show no physical or biological meaning. Nevertheless, this type is often used in engineering practice. The only realistic approach in modelling for the aim of process design is the use of formal kinetics. See eqn.(4) for general setup, eqn.(5) for T-dependence and eqn.(6) and (7) for concentration dependence. All of these mathematical functions are taken from analogies (Moser, 1978). The research philosophy of this approach is that the model has to be "verified", i.e. identified by comparison to experiments, where the parameters are adjusted to their values. This modelling is adaptive (Boyle and Berthouex, 1974), and it is necessary to modify and/or extend the basic model function.
Analogies for waste water treatment are taken from fermentation and chemical kinetics, but due to the complex nature it is questionable whether simple models are satisfactory. The main differences between normal fermentation processes and waste water plants can be summarized: 1) the influents are more complex by a sum of unknown components, fluctuating in quality and quantity and with changing conditions like T and pH and 2) only global methods of measurements for concentrations (BOD, COD, TOC) are available. Due to the complex situation of bioprocessing and the interaction between kinetics and mass transport, the determination of kinetic model parameters is influenced by the type of reactor chosen in kinetic analysis. This fact is illustrated in this paper in the case of a plug flow reactor, which was developed in Austria during the last decade. This new type of treatment unit seems to be unconventional not only in the normal sense (PFR are uncommonly used) but also mainly by the fact that discrepancies with well established techniques on the basis of the stirred tank appear in their use. The discrepancy shown in this paper, the sludge adsorption phenomenon, concerns the heterogeneous nature of bioprocesses and can be observed primarily in reactors with concentration profiles. For better understanding of these facts, a general concept for process kinetic analysis is reviewed in this paper.

REACTION ENGINEERING PROBLEMS

New Plug Flow System

Related to the design of agitated tubular fermenters for aerobic processes (A.Moser, 1973, 1977a, 1980a), a novel aerated plug flow reactor (APFR) was developed in Austria and patended also in Germany (F.Moser,1972). This unit is now used in pilot scale for treatment of municipal waste (F.Moser,Wolfbauer and Taibinger, 1977). A schematic view of this new design is given in Fig. 1, where the advantages of this plant are summarized together with some problems (F.Moser,1977; F.Moser, Theophilou and Wolfbauer, 1979; Wolfbauer, Klettner and F.Moser,1978). Numerical values of reactor and kinetic data can be taken from these articles.

A general macroscopic conclusion can be stated: reactors with concentration profiles (with gradients like PFR) behave differently compared to reactors of the stirred tank type (STR), which have no gradients (gradientless or "lumped"). Quantitative concepts of microbial growth were mainly developed from specialized kinds of studies in single continuous STR for the single culture-single substrate case.

Bioprocess kinetics

In Fig. 2, realistic cases of homogeneous biokinetics are summarized in plots of reaction rates vs. S-concentrations exhibiting drastic deviations from the simple Monod-type kinetics according to eqn.(7). Curves 2-5 illustrate the cases, where endogenous metabolism, S-inhibition, lag phase and sludge adsorption with multi-S-kinetics appear to be significant for a special process. During process kinetic analysis of any waste water process, the type of adequate model becomes evident by comparison to experimental data. Parameter estimation follows this model identification, where different hypotheses (cf. Fig. 2) in integral or differential analysis (shown in Tables 3 and 4) can be used.

PROCESS KINETIC ANALYSIS

General

The proper use of reactors in kinetic analysis is based on reaction engineering fundamentals, with reactor types of PFR and STR and operational modes of dis-, semi- and continuous. The basic behaviour of these situations is characterized by their concentration/time (c/t) and concentration distance (c/z) profile. When using the various reactors for kinetic measurements, the classification of reactor as with or without gradients is important. As shown in Fig. 3, gradientless reactors behave like differential reactors: small differences in conversion are measur so that reaction rate, r, is directly related (cf. eqn. 8). Integral reactors show gradients, so that high differences between inlet and outlet concentration exist and no direct connection to r is given. The evaluation of r is only possibl after adaptation of the model function to the experiment by integration (cf. eqn.9

Even though there is a lack of complete understanding of bioreactor operation, a crude classification with the aid of the criteria steady or unsteady state, with or without gradients and balanced or unbalanced growth can be made (Table 2).

TABLE 2 . Bioreactor Operation Techniques and Biological Behaviour

reactor operation \ biological growth		"balanced"	"unbalanced"
gradient-less	steady state	CSTR	
	quasi steady state	"extended culture"	
	unsteady state	←---- CSTR, SCSTR --------→	
	"periodic"	←-"transient operation techniques"-→	
with gradients	steady state	←----- CPFR,NCSTR ----→	
	unsteady state	←--- BSTR,CPFR,NCSTR ----→	
	"periodic"	←-"transient operation techniques"-→	

The CSTR and also a special case of semi-CSTR, the "extended culture", are the on ones with true balanced growth, while in BSTR and CPFR it is not certain that balanced growth really occurs. In NCSTR and in other configurations of SCSTR, growth unbalanced. Further investigations are needed, perhaps on the basis of a concep using a ratio between the characteristic time of the perturbation and the biologi response (Bailey, 1973).

Homogeneous Systems - Integral Evaluation

To apply the integral method of kinetic evaluation to bioprocessing, we start wit case 1 as the simpliest analogy from enzyme kinetics from eqn.(7b)in Table 1. Af integration, the so-called Henry equation (eqn.10a,Table 3) is yielded, which can linearized on a Walker diagramm (Walker and Schmidt, 1944). This procedure, ofte used in waste water treatment, is only valid, however, in the case where no growt occurs. Therefore, this solution of case 1 can only be applied where $Y_{X/S} = 0$ and represents a special case of the general case 2. Both S-utilization and biomass

TABLE 3. Integral Method of Kinetic Evaluation

Case 1: Substrate utilization with enzyme kinetics: Eqn(7b) with r_{max} (g/L.h)

integration (Henry equation)

$$r_{max}.t = K_m.\ln S_o/S + (S_o-S) \quad (10a)$$

linearization (Walker plot)

$$(S_o-S)/t = -K_m \frac{1}{t} \ln S_o/S + r_{max} \quad (10b)$$

Case 2: Substrate utilization and biological growth

$$\text{Eqn. (7d) and (7c) with } -\sigma = (1/Y)\,\mu \quad (11)$$

integration

$$\sigma_{max}.X_{max}.t = K_S.\ln S_o/S + (\frac{1}{Y} X_{max} + K_S) \ln \frac{X_{max}-Y.S}{X_o} \quad (12a)$$

linearization (Gates plot)

$$\frac{1}{t} \ln \frac{X_{max}-Y.S}{X_o} = \frac{\sigma_{max}.Y.X_{max}}{X_{max} + Y.K_S} - \frac{Y.K_S}{X_{max} + Y.K_S} . \frac{1}{t} \ln \frac{S_o}{S} \quad (12b)$$

TABLE 4. Differential Method of Kinetic Evaluation.

Case 1: Eqn. (7c) and (7d): Linearizations:

Lineweaver-Burk plot $\quad 1/\sigma = (K_S/\sigma_{max})(1/S) + 1/\sigma_{max}$ (13a)

Eadie-Hofstee plot $\quad \sigma = \sigma_{max} - K_S\,(\sigma/S)$ (13b)

Langmuir plot $\quad S/\sigma = K_S/\sigma_{max} + S/\sigma_{max}$ (13c)

Case 2: S-Inhibition

$$\mu = \mu_{max}\,(1 + K_S/S + S/K_I)^{-1} \quad (14a)$$

linearization only if $S \gg K_S$: $\quad 1/\mu = (K_I/\mu_{max})(1/S) + 1/\mu_{max}$ (14b)

Case 3: Lag-time $\mu(S,t) = \mu(S).(1 - e^{-t/t_L})$ with $\ln X/X_o = \mu(t-t_L)$ (15a,b)

Case 4: Stationary growth phase $\quad r_x = \mu_{max}.X\,(1 - X/X_{max})$ (16a)

for CSTR $\quad \bar{t} = \left[\mu_{max}\,(1 - X/X_{max})\right]^{-1}$ (16b)

Case 5: Endogenous metabolism

$$\mu = Y.\sigma - k_d \quad \text{with Eqn.(7d)} \quad (17)$$

$$\text{or} \quad -\sigma = (1/Y)\,\mu + m \quad \text{with Eqn.(7c)} \quad (18)$$

$$\text{or} \quad \mu = \mu(S) - \mu_d(S) \quad \text{with Eqn. (8)} \quad (19a)$$

$$\text{with} \quad \mu_d(S) = \mu_{d,max}\,(1 - \frac{S}{K_d+S}) \quad (19b)$$

growth are considered together (note that $\sigma = r_s/X, \sigma_{max} = r_{max}/X$). In this case of variable sludge concentration, X, integration and also linearization according to Gates and Marlar (1968) are still possible (eqn. 12). However, a graphical trial and error method for the determination of the kinetic parameters has to be carried out due to the complex equation system. No unique line appears as in the Walker plot, as different conditions (S_o, X_o, X_{max})yield different lines. As a result of these considerations, it can be stated that the integral method of kinetic evaluation is restricted to simple kinetics.

On the other hand, in designing a CPFR, which is an integral reactor, one needs data from reactors with gradients. More attention, therefore, should be given to the examination of integral reactors, but the differential methods of evaluation are to be applied for more convenience. Based on reaction engineering fundamentals, the similar profiles in BSTR and CPFR result in a "kinetic similarity", which should be taken into consideration for process kinetic analysis (Moser, 1978). In agreement with this fact, a CPFR can be designed properly with kinetic parameters which are measured in BSTR.

Homogeneous Systems - Differential Evaluation

In Table 4, the differential methods of kinetic evaluation are summarized for a variety of kinetic effects from Fig. 2. The best plot, only rarely used for evaluation of the Monod parameters, μ_{max} and K_s, is given by the Langmuir linearization of eqn.(13c)(Langmuir, 1918). This plot exhibits the narrowest confidence region. Equations (14-19) represent possible functions to describe the formal kinetics of S-inhibition (Humphrey, 1978), lag time (Bergter and Knorre, 1972; Pirt, 1975), stationary growth phase (Roels and Kossen, 1978; La Motta, 1976a) and endogenous metabolism (Pirt, 1965; A.Moser and Steiner, 1975; Humphrey, 1978). The kinetic parameters in all cases can be evaluated using forms given in the subsequently-written equations b. Note that in all cases of formal kinetics, alternative functions can be used instead of the given equations.

In the case of non-stationary behaviour due to perturbations in inflows, a simple but adequate method was recently described, using the basic Monod relationship but with kinetic parameters depending on organic loading (Mona, Dunn and Bourne, 1979). The model equations (20) - (21) developed and successfully used are given in Fig. 4. The time term, τ , takes into account the retardation observed between perturbation (dotted step-line) and biological response and is defined according to Fig. 4.

Sludge Adsorption

Finally, a specific behaviour of activated sludge, called sludge adsorption, is demonstrated in Fig. 5. Examining the kinetics of S-elimination together with O_2 consumption, a time difference in both utilization rates was observed by Theophilo, Wolfbauer and Moser (1978). The COD as a measure of substrates shows that they are eliminated from the liquid phase much more quickly than the utilization of them in the sludge as indicated by the BOD. This can be interpreted as sludge adsorption (Walters, 1966; Jones, 1971), which takes place in the first period of contact between sludge and waste water. The adsorption rate in the presence of S-elimination is represented by Fig. 5b. As a consequence of this effect, kinetic studies with activated sludge often show zero order kinetics (Wuhrmann, van Beust and Ghose, 1958). This effect is demonstrated in Fig. 2 (cf. curve 5). This deviation is quite clear as all measurements are carried out in the liquid phase; nevertheless the reaction takes place in the solid phase. The real S-utilization rate in the solid phase represented by S_s could show the dotted curve indicated in Fig. 2 whi the rate of dependence of S_L ^{S}follows zero order. Normally, this effect is not recognized during fermentations, as substrate adsorption in these cases is very sma compared to waste water sludge, so that "pseudohomogeneity" , i.e. no concentrati

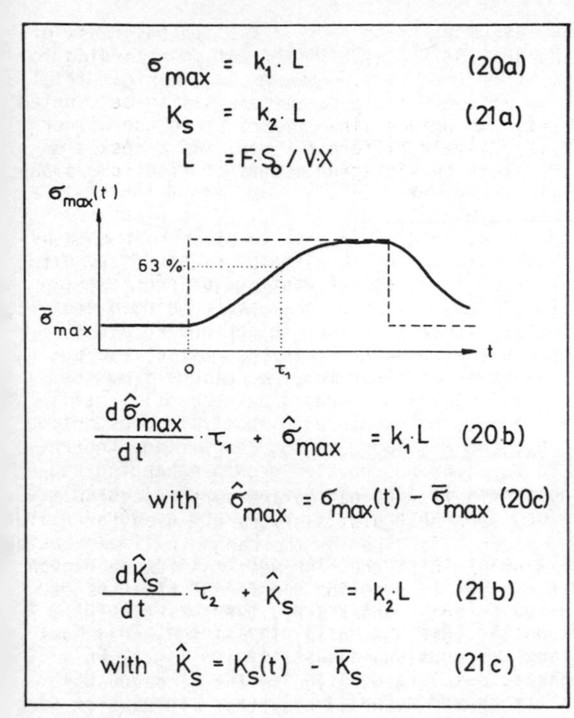

Fig. 4 : Non-stationary kinetics due to inflow perturbation

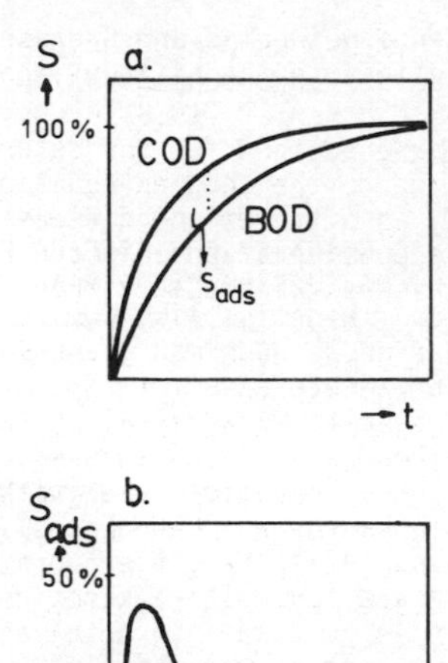

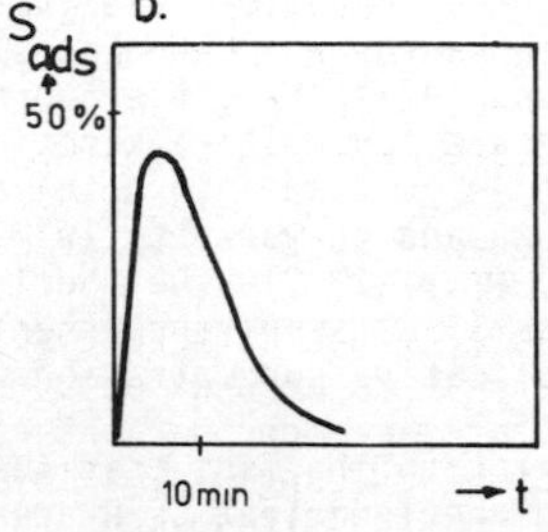

Fig. 5 : S-elimination from liquid phase (COD) and S-utilization in the sludge (BOD) showing sludge-adsorption (S_{ads}).

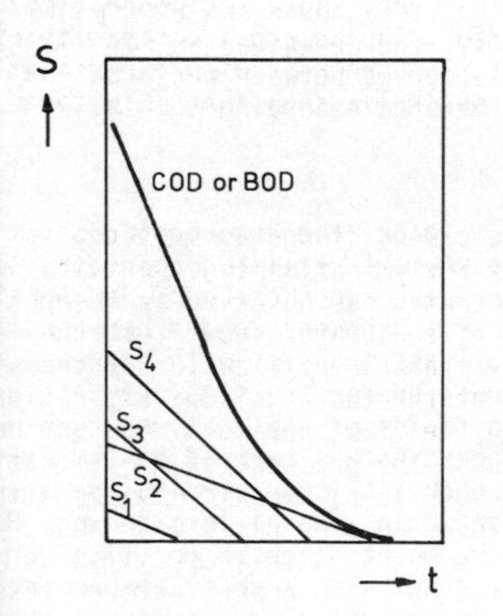

$$r_{tot} = \sum r_i \qquad (22)$$

$$\mu(S_1, S_2) = \mu(S_1) + \mu(S_2) \cdot f_1 \qquad (23)$$

$$\mu_{tot} = \left[\mu_{max,o} + \sum \mu(S_i)\right] \prod \mu(S_E) \qquad (24)$$

Fig.6 : Simultaneous substrate removal following Wuhrmann kinetics.

gradients between L and S phase can be assumed. As CPFR operate with smaller concentrations in the effluent than CSTR, this effect was observed in this reactor.

Multi-S-Removal

The problem of multi-S kinetics is widely studied in the field of bioengineering. From the reaction engineering point of view, however, the consequences are mostly overlooked. Recently, Wolfbauer, Klettner and Moser (1978) illustrated the effects of global kinetics like COD or BOD, representing a sum of single zero order reactions (qn.22, Wuhrmann,van Beust and Ghose, 1958).This problem is illustrated by Fig. 6 , where a set of zero order reactions, obviously affected by sludge adsorption, results in an overall kinetic order of 1. In actual waste water processes, global methods are often used. Therefore, many investigators have concluded that first order kinetics occur with respect to substrate (Eckenfelder and Ford, 1970; Grau, Dohanyos and Chudoba, 1975; Joschek and co-workers, 1975; Krötzsch and co-workers, 1977). Fig. 6 elaborates on a number of other model equations from the literature for multi-S kinetics (eqn. 23-24), valid to describe sequential utilization (e.g. diauxic growth) and simultaneous or overlapping substrate consumption (Imanaka and co-workers, 1972; van Dedem and Moo-Young, 1975; Bergter and Knorre, 1972; A.Moser,1978). The model given in eqn.(24)incorporates growth-enhancing substrates S_i as a sum and essential substrates S_E like glucose and oxygen, which are multiplicative concentration terms (Tsao and Hanson, 1975; Megee and co-workers,197

Returning to the fact that the overall reaction order often can be taken as 1, one could conclude that CPFR are superior to CSTR according to reaction engineering laws (Levenspiel, 1972). This conclusion is only partly true, however, according to results given in Fig.1 from the work on the APFR with activated sludge. This fact obviously must be connected with sludge adsorption, illustrated on Fig.5. Only 10% better removal was observed when separate settling devices for the CPFR and CSTR were used (F.Moser,1977;Wolfbauer, Klettner and Moser, 1978). As a consequence of this behaviour, sludge contact stabilization is recommended where the liquid and solid phases are separated first and the sludge with the adsorbed substrates is the aerated in a separate aeration vessel, so that the substrates can be metabolized by the sludge. In this way, the purified liquid phase, following the COD curve in Fig. 5a, can be released from the first aeration basin in a very short residence time. In the second aeration vessel the sludge is "unloaded". The adsorption capacity of sludge is responsible at the same time for the fact, in agreement with experiments, that CPFR are not more sensitive to perturbations (shock-loading) than CSTR types.

Heterogeneous Systems

All kinetic models mentioned in the previous section ignore the heterogeneous properties of the 3-phase reaction of bioprocessing. A more realistic approach, therefore, would be a true heterogeneous model of process kinetics, considering al differential equations of components and all transport phenomena in and between th 3 phases. In Table 5, an overview of the problems of mass transfer with biochemica reactions is shown, summarizing the set-up of an equation for the simplest case of 1-dimensional steady state (eqn. 25a) together with fields of applications. Most o the work on internal transport limitation in bioprocessing was carried out by Atki son and his group; this work is reffered to in his book (Atkinson, 1974). The solu tion with given boundary conditions (eqn.25b,c) is shown in general form in eqn.(26 This heterogeneous model, compared to the homogeneous model given in eqn.(7a),contains not only the same kinetic parameters k_1 and k_2, but also a physical paramete k_3, taking into account the transport in the solid phase (D_s). The meaning of the three parameters of the biological rate equation (eqn. 26a) is given in eqn.(26 b- The exact mathematical solution is too complex to be described and used here, but can be demonstrated clearly in a graph, as shown in Fig. 7. The specific substrate consumption rate, σ , dependent on S-concentration, is drawn in Fig. 7a as a functi

TABLE 5 . Mass Transfer with Reaction.

$$D\frac{d^2c}{dz^2} - r = 0 \qquad (25a)$$

BOUNDARY CONDITIONS

1. $z = 0 \;\ldots\; c = c^*$ (25b)
2. $z = \delta \;\ldots\; dc/dz = o$
   or $\ldots\; r = Tr$ (25c)

APPLICATIONS :

(1) TRANSPORT LIMITATIONS

1.1 "EXTERNAL" G/L , L/S

1.2 "INTERNAL" in S

(2) TRANSPORT ENHANCEMENT G/L

SOLUTION for BIOPROCESS in case 1.2 :

$$r_S = f(k_1, k_2, k_3, d, S) \qquad (26a)$$

$$\text{with } k_1 = a.\sigma_{max}/K_S \qquad (26b)$$

$$k_3 = (k_1/D_S)^{1/2} \qquad (26c)$$

$$k_2 = 1/K_S \qquad (26d)$$

TABLE 6. Formal Kinetics of Heterogeneous Systems

General concept : effectiveness factor η

$$\frac{r_{eff}}{r_{max}} = \eta_r = f\left(\frac{k_r}{k_{Tr}}\right) \qquad (27a)$$

in case 1.2 : $(0 \leqslant \eta_r \leqslant 1)$:

$$\eta_r = f(Da_{II})^{1/2} = f(\Phi) \qquad (27b)$$

$$\text{with } \Phi = d\,(k_r.c^{n-1}/D_s)^{1/2} \qquad (27c)$$

in case 2 : $(\eta_{Tr} \geqslant 1)$:

$$\frac{r_{eff}}{r_{max}} = \eta_{Tr} = f\left(\frac{k_r}{k_{Tr}}\right) = f(Ha) \qquad (27d)$$

ATKINSON 1974 :

$$\sigma = \sigma_{max}\cdot\frac{S}{K_S + S}\cdot\eta_r \qquad (28a)$$

$$\text{with } \eta_r = f(k_2.S\,,\,k_3.d) \qquad (28b)$$

POWELL 1967 :

$$\sigma = \sigma_{max}\frac{S}{K_{eff} + S} \qquad (29a)$$

$$\text{with } K_{eff} = K_S.R \qquad (29b)$$

HARREMOES 1978 :

$$r_S = k_{1/2}.X.S^{1/2} \qquad (3\ )$$

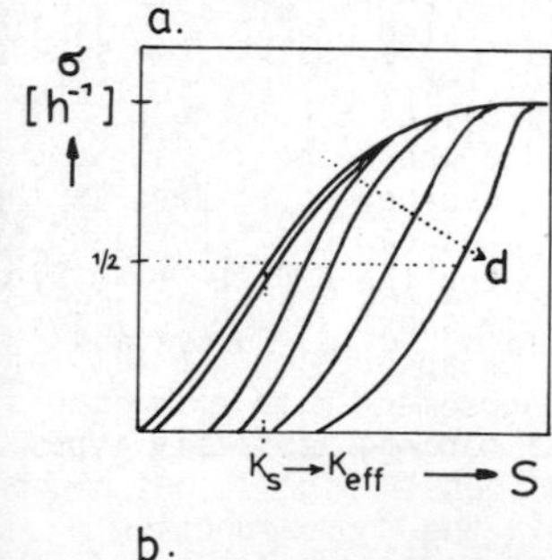

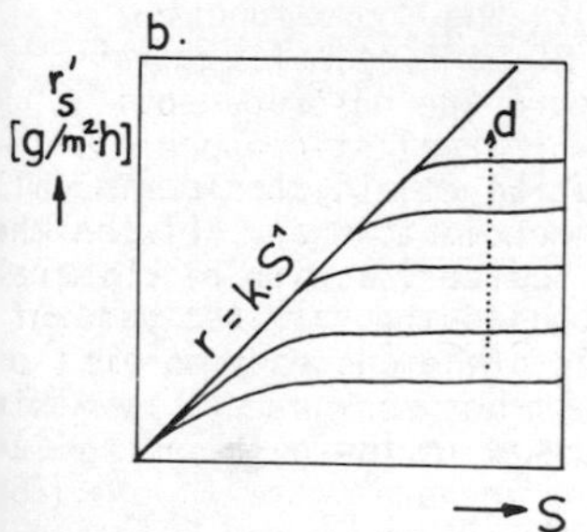

Fig. 7: Graphical reapresentation of the biological rate equation (eqn.26) for flocs (a) and films (b) with

$$r_S' = \sigma\cdot\varrho\cdot V/A \qquad (31)$$

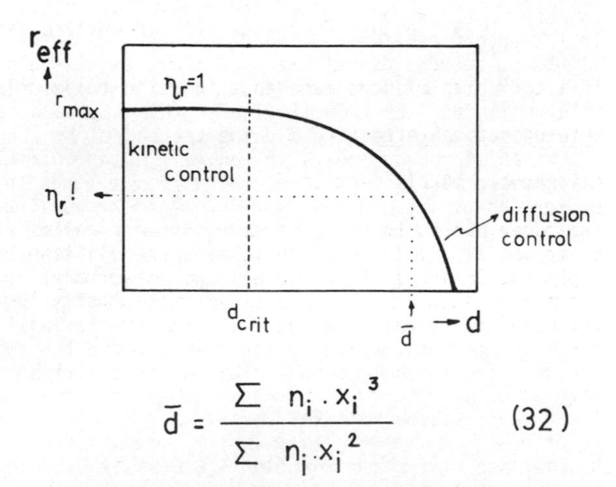

$$\bar{d} = \frac{\sum n_i\cdot x_i^{\,3}}{\sum n_i\cdot x_i^{\,2}} \qquad (32)$$

Fig. 8 : Determination of effectiveness factor η_r according eqn.(27) for the use in eqn.(28) from the mean diameter d .

of the thickness, d, of biological mass. At small values of d, the simple Monod-type relationship appears, while with greater thicknesses, large deviations occur. This graph is valid principally for floc and film geometry. In the case of films, however, a more reliable magnitude for measuring consumption rate is the rate related to the surface of microbial mass, r_s'. Such a plot is shown in Fig. 7b, again as a function of the thickness, d, of the film. The limiting case of first order kinetics, already mentioned previously, appears in this diagram.

In Table 6, some known equations are compiled which can be used for a formal kinetic description of heterogeneous systems in a pseudohomogeneous approach. Derived from the truly heterogeneous model according to Atkinson (1974), a pseudohomogeneous equation is shown as eqn.(28a). The concept of an effectiveness factor for the reaction rate, η_r, is defined in eqn.(28b). The use of an effectiveness factor is a general concept in engineering, the physical meaning of which is presented in eqn.(27a). An effectiveness factor compares an effective rate to the maximal rate and can be written for reactions (η_r) or transport (η_{TR} in eqn. 27d). In all cases of application, η is dependent on the ratio of the rate constants of the reaction and of transport. While the Thiele modulus, Φ (which is identical with the root of the Damköhler number of second kind, Da_{II}) is used in the case of internal transport limitation (eqn. 27b,c), a Hatta number, Ha, is applied in the case of transport enhancement (cf. eqn. 27d and 34b), all of which relate k_r to k_{TR} at different reaction orders. For convenience, the result is given in graphical form as a plot of r_{eff} vs. diameter, d, instead of η_r vs. Φ or Da_{II}, with a typical shape given in Fig. 8. The regions of kinetic control and diffusion control can be distinguished. This graphical representation of the theoretical solution of heterogeneous models can be used to demonstrate the usefulness of the η concept. According to Atkinson and Ur-Rahman (1979), the distribution function of floc sizes in a bioreactor can be represented by a mean value, $\bar{d}$, calculated from eqn.(32) as the surface mean, which closely corresponds to the mean floc size. This value $\bar{d}$ indicates the magnitude of η_r in the single situation, shown in Fig. 8.

Similar solutions of more complex cases for product and substrate inhibition can be achieved only by numerical methods (Moo-Young and Kobayashi, 1972). A solution in the case of the occurence of endogenous metabolism was derived by Howell and co-workers (1979). Double substrate limitations are discussed by Fujie, Furuya and Kubota (1979).

An alternative pseudohomogeneous function to describe the heterogeneous situation is given in eqn. 29 (Powell, 1967). This approach can be seen in Fig. 7a. At higher thicknesses, an effective K value instead of K_s appears in the Monod function. A similar approach appears in eqn.(30), using an analogy of half-reaction order (Harremoës, 1978).

Even when pseudokinetic parameters offer a workable solution in singular cases, their general use is dangerous, as extrapolations to other conditions would create inaccurate preconditions (Howell and co-workers, 1979). However, as simple kinetics are prefered for better practicability, mostly the homogeneous models are used. The paper presented intends to accentuate the application of homogeneous models for bioprocessing, but only for cases where the real heterogeneous nature can be treated as pseudohomogeneous. This problem will be discussed in the next section.

PSEUDOHOMOGENEITY

In the case of 3-phase reactions (G/L/S), the concentration profile of a component penetrating from the G through L into the S phase , where the reaction takes place is shown in Fig. 9. Concentration gradients occur especially at each interface where the transport coefficients k_G, k_{L1}, k_{L2} and k_S are related to the thickness of layers δ_G, δ_{L1}, δ_{L2} and d (d_s), which represent the transport resistance according

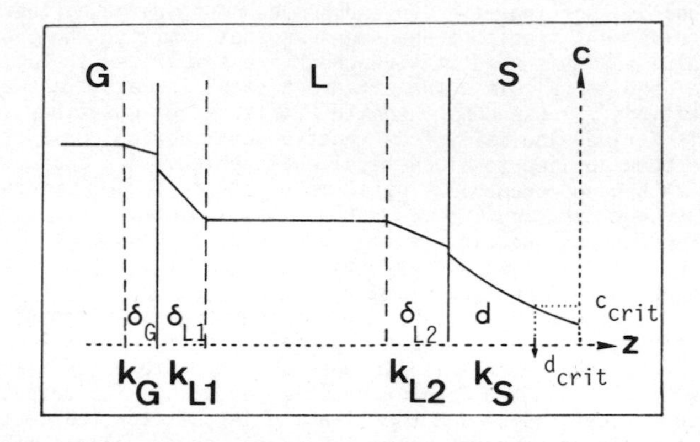

Fig. 9: Concentration profile in heterogeneous 3-phase systems with different mass transport coefficients k.

TABLE 7. Quantitative Tests for Psuedohomogeneity of 3-phase Systems.

PROBLEMS	IN THE CASE OF	CRITERIA
1. G/L :	high soluble gases	k_G
(k_{L1})	aerobic processes	$k_{L1}.a(O^*-O) \gg \sigma_O.X$ (gas dynamics!) (33)
η_{Tr}	$d \ll \delta_{L1}$ flocs only	$0.3 \leqslant Ha \leqslant 1 \rightarrow \eta_{Tr} > 1$ (34a)
		with $Ha = \frac{1}{k_{L1}}(\frac{2}{n+1}.k_r.O^{*n-1}.D)^{1/2}$ (34b)
2. L-phase :	BSTR and CSTR	$t_m \leqslant 0.1\ t_r$ (35a)
micromixing	Recycle reactors	$t_c \leqslant t_r$ (35b)
macromixing	CSTR CPFR	$N \rightarrow 1$ $Bo \rightarrow \infty$ (36)
3. L/S :		calculate (analogy to chemical processes)
(k_{L2})	flocs	$Sh_{L2} = 2 + 0.4\ Re_\varepsilon^{1/4}.Sc^{1/3}$ (37a)
	films	$k_{L2} \sim v_L^{0.7}$ (37b)
4. S-phase:	flocs and films	$d \leqslant d_{crit}$ (38a)
(D_S)	1-S-limited	calculate $d_{crit} = \frac{c^*}{K_S} \cdot \frac{(1+2.c^*/K_S)^{1/2}}{1 + c^*/K_S} \cdot \left(\frac{Y.D_s.K_S}{\sigma_{max}.\rho}\right)^{1/2}$ (38b)

to the 2-film theory. Working with a pseudohomogeneous approach, therefore, requires that the different transport phenomena are not limiting the process rate. The criteria useful for this quantitative check are summarized in Table 7. These equations can be used to calculate the different steps in mass transfer and to control the preconditions for pseudohomogeneity. This set of equations represents a proposal which is derived logically from reaction engineering considerations. Part of them only has been applied to biochemical engineering. A 3-phase process can be treated with a pseudohomogeneous kinetic model if eqn.(33)-(38) are fulfilled.

In most cases of processes occuring in liquid phases, pseudohomogeneity should be achieved if mixing time is less than any reaction time, t_r. Good mixing is related to good aeration and high turbulence in δ_{L2} and also reduces the floc size. The enhancement of OTR can occur with flocs only if they are small enough and in the case of fast respiring populations, have high cell density at low k_{L1} values (A.Moser,1980b). This enhancement is also possible in waste water treatment (Nagel, Kürten and Heger, 1977). In the case of recycle reactors, like deep shaft fermenter (Hines and co-workers, 1975) and the plunging jet aerator (Lafferty and co-workers, 1978), the mixing time concept is often no longer valid so that cycling time, t_c, has to be compared to the reaction time. The value for t_r can be taken from the diagram of the dynamic measurement of OTR with a P_{O2} electrode, where it represents the time where the concentration drops down to a critical O_2 concentration, at which point respiration starts to be rate limiting. The weak point in the verification of pseudohomogeneity, however, lies in the problem of measuring k_{L2}. As a substitute, values can be estimated with the aid of the Kolmogoroff theory of local isotropic turbulence or with the theory of terminal velocity-slip velocity. The approach according to eqn.(37a) is an analogy to chemical engineering, where k_{L2} can be calculated from the Reynolds-number based on energy dissipation ε in the reactor. The author believes that this analogy holds for bioprocessing, even if no experimental verification is known yet. ε can be calculated from volumetric power input in reactor. Different researchers have elaborated relationships which can be unified in a single equation given by eqn.(37a) (Shah, 1979; Calderbank and Moo-Young, 1961; Sano, Yamaguchi and Adachi, 1974; Boon-Long, Laguerie and Condere, 1978). In the case of films, eqn.(37b) was elaborated (La Motta, 1976b). According to eqn.(37b), the L/S-mass transfer coefficient k_{L2} is directly related to the relative velocity v_L between L- and S-phase. The relative velocity in the case of rotating biofilms is proportional to rotational speed so that these reactors offer an opportunity to study this effect. The magnitude of the critical thickness of biological mass can be approximated with the aid of eqn.(38b) for single substrate limitation (Atkinson and Knight, 1975) and for dual substrate limitation according to Howell and Atkinson (1976). They show the appearance of anaerobic zones when the ratio of O_2 to glucose concentration is low. In deeper parts of the film, S is exhausted and endogenous respiration causes lysis of cells, weakening the adhesive bond so that sloughing will occur. The critical thickness of films is reported to be 100 μm (Wuhrmann, 1963), and in the case of flocs, this critical diameter can rise up to 1 mm if O_2 concentration is near saturation (8 ppm). The real size of films is 2-5 mm and flocs are 0.2-1 mm.

CONCLUSIONS

Due to discrepancies (e.g., sludge adsorption) observed in the design of "unconventional" reactors (plug flow systems) on the basis of "conventional" measured kinetics, more effort should be made in process kinetic analysis. Differential evaluation in reactors with gradients, e.g., batch stirred tanks on bench scale, as an expression of "kinetic similarity" must be preferred in this case. This fact is the consequence of a general research strategy, where kinetics of the biological reactions and mass transfers in the bioreactor configurations are always coupled (A.Moser, 1981). Due to the truly heterogeneous character of bioprocessing, the influence of transport phenomena should not be disregarded. For easier handling of modeling, a pseudohomogeneous approach is preferred. Therefore, more attention should be given to checking the validity of "pseudohomogeneity" in experiments where kinetic parameters are to be evaluated.

NOMENCLATURE

a	cm^2/cm^3	specific interfacial area
a_0, a_1, etc.	-	empirical constants
Bo	-	Bodenstein number $Bo = \frac{v.d}{D_L}$
c	g/L	concentration
D	cm^2/sec	diffusion coefficient
d	cm	diameter, thickness, characteristic length
Da_{II}	-	Damköhler number of second kind (cf. eqn. 27c) $Da_{II} = \frac{d^2.r}{c.D}$
f	-	mathematical function
F	m^3/h	flow rate
Ha	-	Hatta number $Ha = \frac{1}{k_L} \left(\frac{2}{n+1} k.c^{*n-1}.D\right)^{1/2}$
k	-	rate coefficient
K	g/L	equilibrium constant
K_I	g/L	inhibition constant
K_m	g/L	Michaelis constant
K_s	g/L	Monod constant
L	h^{-1}	organic loading ($L = F.S_0/V.X$)
n	-	reaction order
n_i	-	number
N	-	number of equivalent tanks in series
O	g/L	O_2 concentration
r	g/L	reaction rate
r'	g/m^2h	reaction rate related to surface
R	g/L	formal volumetric coefficient for transport resistance
Re_ε	-	Reynolds number based on energy dissipation $Re_\varepsilon = \frac{\varepsilon.d_p^4}{\nu_L^3}$
S	g/L	substrate concentration
Sh	-	Sherwood number $Sh = \frac{k_L.d}{D}$
Sc	-	Schmidt number $Sc = \frac{\nu}{D}$
T	°C	temperature
t	sec	time
t_c	sec	cycle time
t_L	sec	lag time
t_r	sec	reaction time
t_m	sec	mixing time
v	cm/sec	velocity
V	m^3	volume
X	g/L	biomass or sludge dry weight concentration

Y	-	yield coefficient
z	-	coordinate

Greek Letters

α_i	-	empirical coefficients
δ	cm	thickness of layers in 2-film theory
ε	cm^2/sec^3	energy dissipation
η	-	effectiveness factor
μ	h^{-1}	specific growth rate
ρ	g/cm^3	density
σ	h^{-1}	specific substrate utilization rate
σ_0	h^{-1}	specific O_2 utilization rate
τ	sec	retardation time

Indices

crit	critical
d	dead
e	endogenous
eff	effective
ex	exit
G	gas
L	liquid
max	maximal
o	inlet
p	particle
r	reaction
s	solid or substrate
TR	transport
*	saturation value
-	mean value
^	deviation value

Abbreviations

BSTR	batch stirred tank reactor
CPFR	continuous plug flow reactor
CSTR	continuous stirred tank reactor
NCSTR	continuous stirred tank cascade with N reactors
OTR	O_2-transfer rate
PFR	plug flow reactor
SCSTR	semicontinuous stirred tank reactor
SVI	sludge volume index

REFERENCES

Atkinson, B. (1974) Biochemical Reactors, Pion Ltd., London.
Atkinson, B. and F. Ur-Rahman (1979) Biotechnol. Bioeng., 21, 221.
Atkinson, B. and A.J. Knight (1975) Biotechnol. Bioeng., 17, 1245.
Bailey, J.E. (1973) Chem. Eng. Commun., 1, 111.
Bergter, F. and W.A. Knorre (1972) Zschr. Allgem. Mikrob., 12, 613.
Boon-Long, S., C. Laguerie and J.P. Condere (1978) Chem. Eng. Sci., 33, 813.
Boyle, W.C. and P.M. Berthouex (1974) Biotechnol. Bioeng., 16, 1139.
Calderbank, P.H. and M. Moo-Young (1961) Chem. Eng. Sci., 16, 39.

Eckenfelder, W.W. and D.L. Ford (1970) Water Pollution Control, Pemberton Press, Austin, Texas.
Fujie, K., T. Furuya and H. Kubota (1979) J. Ferment. Technol., 57, 99.
Gates, W.E. and J.T. Marler (1968). J. Wtr. Poll. Con. Fed., 40, 469.
Grau, P., M. Dohanyos and J. Chudoba (1975) Water Research, 9, 637.
Harremoës, P. (1978) In Water Pollution Microbiology, Vol. 2, Chap. 4, p.71.
Hines, D.A., M. Bailey, J.C. Onsby and F.C. Roesler (1975) I. Chem. E. Symp. Series, 41, D1.
Howell, J.A. and B. Atkinson (1976) Biotechnol. Bioeng., 18, 15.
Howell, J.A., M.G. Jones, M.K. Klu and A.R. Khan (1979) In M. Moo-Young and G.J. Farquhar (Eds.), Waste Treatment and Utilization, Pergamon Press, Oxford, p.395.
Humphrey, A.E. (1978) Amer. Chem. Soc. Symp. Series, 72, 1.
Imanaka, T., T. Kaieada, K. Sato and H. Taguchi (1972) J. Ferment. Technol., 50, 633.
Jones, P.H. (1971) Adv. Water Poll. Res., 1, Pergamon Press, Oxford.
Joschek, H.I., J. Dehler, W. Koch, H. Engelhardt and W. Geiger (1975) Chem. Ing. Techn., 47, 422.
Krötzsch, P., H. Kürten, H. Daucher and K.H. Popp (1977) Ger. Chem. Eng., 1, 15.
Lafferty, R.M., A. Moser, W. Steiner, A. Saria and J. Weber (1978) Chem.Ing. Techn., 50, 401.
La Motta, E.J. (1976a) Biotechnol. Bioeng., 18, 1029.
La Motta, E.J. (1976b) Biotechnol. Bioeng., 18, 1359.
Langmuir, I. (1918). J. Amer. Chem. Soc., 40, 1361.
Levenspiel, O. (1972) Chemical Reaction Engineering, J. Wiley & Sons, New York.
Megee, R.D., J.F. Drake, A.G. Fredrickson and H.M. Tsuchiya (1972) Can. J. Microbiol. 18, 1733.
Mona, R., I.J. Dunn and J.R. Bourne (1979) Biotechnol. Bioeng., 21, 1561.
Moo-Young, M. and T. Kobayashi (1972) Can. J. Chem. Eng., 50, 162.
Moreira, A.R., G. van Dedem and M. Moo-Young (1979) Biotechnol. Bioeng. Symp.,9,179.
Moser, A. (1973) Biotechnol. Bioeng. Symp., 4, 399.
Moser, A. (1977) Chem. Ing. Techn., 49, 612.
Moser, A. (1978) In Proc. 1st Europ. Congress on Biotechnology, Interlaken, Switzerland, part I, 88.
Moser, A. (1980a) 6th Internat. Ferment. Symp. London, Ontario, Canada, 20-25 July.
Moser, A. (1980b) 2nd Internat. Symp. on Bioconversion and Biochemical Engineering, IIT, New Delhi, India, March 3-6.
Moser, A. (1981) Bioprozesstechnik, Springer Verlag, Wien, New York.
Moser, A. and W. Steiner (1975) Europ. J. Appl. Microb., 1, 281.
Moser, F. (1972) FRG Patent P2239 158.8-25.
Moser, F. (1977) Verfahrenstechnik, 11, 670.
Moser, F., J. Theophilou and O. Wolfbauer (1979) Österr. Abwasser-Rundschau, 24, 83.
Moser, F., O. Wolfbauer and P. Taibinger (1977) Prog. Wat. Techn., 8 (6), 235.
Nagel, O., H. Kürten and B. Hegner (1977) Symposium für Biotechnologie in Tutzing (März) Dechema Monographie, 81.
Pirt, S.J. (1965) Proc. Roy. Soc., B163, 224.
Pirt, S.J. (1975) Principles of Microbe and Cell Cultivation, Blackwell Scientific Publ., Oxford, London.
Powell, E.O. (1967) In E.O. Powell, C.G.T. Evans, R.E. Strange and D.W. Tempest (Eds.), Microbial Physiology and Continuous Culture, Her Majesty's Stationery Office, p. 34.
Roels, J.A. and N.W.F. Kossen (1978) In M.J. Bull (Ed.), Prog. Ind. Microb., 14, 95. Elsevier Publ.
Sano, Y., N. Yamaguchi and T. Adachi (1974) J. Chem. Eng. (Japan) 7 (4), 255.
Shah, Y.T. (1979) Gas-Liquid-Solid Reactor Design, McGraw Hill, New York.
Theophilou, J., O. Wolfbauer and F. Moser (1978) Gas-Wasser-Fach, Wasser/Abwasser 119, 135.
Tsao, G.T. and Th.P. Hanson (1975) Biotechnol. Bioeng., 17, 1591.
van Dedem, G. and M. Moo-Young (1975) Biotechnol. Bioeng., 17, 1301.
Walters, C.F. (1966) Ph.D. Dissertation, University of Illinois, Urbana.

Walker, A.C. and C.L.A. Schmidt (1944) Arch. Biochem., 5, 445.
Wilderer, P. (1976) In L. Hartmann (Ed.), Karlsruher Berichte zur Ingenieurbiologie, 8. Universität Karlsruhe.
Wolfbauer, O., H. Klettner and F. Moser (1978) Chem. Eng. Sci., 33, 953.
Wuhrmann, K. (1963) In W.W. Eckenfelder and J. McCabe (Eds.), Advances in Biological Waste Treatment, Pergamon Press, New York, p.27.
Wuhrmann, K., F. van Beust and T.K. Ghose (1958) Schweizer Zschr. Hydrol., 20, 284.

17

EVALUATION OF LIGNITE LIQUEFACTION WASTEWATER TREATMENT BY ACTIVATED SLUDGE PROCESS

Yung-Tse Hung

Department of Civil Engineering, University of North Dakota, Grand Forks, North Dakota, USA

ABSTRACT

A laboratory study was undertaken to determine the feasibility of using an activated sludge process in treating lignite liquefaction wastewaters. Raw wastewaters contained high concentrations of COD, phenols, and ammonia. Pretreatment by lime precipitation and air stripping was necessary for effective biological waste treatment. The pretreated wastewaters had 16,800 to 45,000 mg/L COD and 3,290 to 11,700 mg/L phenol. Bench-scale reactors were operated to evaluate the effects of feed strength, aeration tank hydraulic detention time, and sludge age on the reactor performance. The wastewater was diluted to 3.3 to 10.0% of its original strength. The aeration tank hydraulic detention time and the sludge age ranged from 1.88 to 8.77, and 5.81 to 16.3 days, respectively. Results indicated that increasing the feed strength decreased treatment efficiencies. When the hydraulic detention time as well as the sludge age were increased, reactor performance improved in terms of COD and phenol removals. Percent removal rates for ammonia, phenol, and COD were 63 to 99%, 64 to 99%, and 60 to 91%, respectively. Organic loading rates were from 0.33 to 1.88 mg COD applied/mg MLVSS-day. Kinetic constants of oxygen utilization and sludge production were also determined in this study.

KEYWORDS

Coal conversion waste; coal liquefaction waste; activated sludge process; phenol removal.

INTRODUTION

Since a large percentage of the petroleum supplies are being imported in the United States, there is an increasing interest in the development of coal conversion processes to meet the energy demand. A great deal of preliminary investigations are underway in the operation and design of the coal conversion facilities. Additional research will also be necessary to characterize the wastewater streams produced, to evaluate the liquid waste clean up efficiencies, and to determine the environmental effects relating to the discharge of coal conversion wastewaters. In this investigation, treatment of wastewaters generated through liquefaction of North Dakota lignites using activated sludge processes was studied. The main objective was to examine conditions necessary to achieve optimum treatment of lignite liquefaction wastewaters. This included altering hydraulic detention times and feed strengths. Performance characteristics were measured in terms of phenol and COD (chemical oxygen demand) removal rates. The scope of this study included wastewater characterization, wastewater pretreatment, and activated sludge treatment of pretreated wastewaters. The processes involved in the conversion of lignite generate a waste stream which is about 20% of the original weight of the raw lignite. The lignite conversion wastewaters contain high levels of hydrocarbons and inorganic substances. If not properly treated, these wastewaters can exert adverse effects on the environment. In a recent study (Singer, Pfaender and Lamb, 1977) regarding characterization of coal conversion wastewaters, it was reported that approximately 60 to 80% of the total organic carbon appeared to be phenolic in nature, consisting of monohydric, dihydric, and polyphenols. The remainder of the organics consisted of nitrogen-containing aromatics, oxygen-and-sulfur-containing heterocyclics, polynuclear aromatic hydrocarbons, and simple aliphatic acids. At the concentrations present in untreated wastewaters, the discharge of these wastes would have an adverse impact on aquatic life. Biological treatment of coal conversion wastewaters to minimize environmental impact has proven to be very useful. The biological treatability of synthetic coal conversion wastewater was examined by Lamb and co-workers (1979). Substantial TOC (total organic carbon) removals were observed for chemostats operating at hydraulic detention times of 5, 10, and 20 days. Undesirable environmental impact was reduced as the effluent quality was consistently improved with increasing detention time. Reap and co-workers (1977) studied the treatment of wastewaters generated from liquefaction of coal using the H-Coal process. Before pretreatment, the wastewaters had COD, phenol, NH_3-N, and sulfide concentrations of 88,600, 6,800, 14,400 and 29,300 mg/L, respectively. The major portion of organic constituents were phenolic compounds. Pretreatment was used to reduce NH_3-N, oil and grease. The pretreated wastewaters were then diluted to 22% strength to reduce toxic effects to the microorganisms used in the activated sludge process. The sludge age varied from 17 to more than 100 days and the average sludge yield coefficient was 0.48 mg VSS (volatile suspended solids) produced/mg BOD (biochemical oxygen demand) removed. Nitrification was indicated by reductions in TKN (total Kjeldahl nitrogen) and NH_3-N. It was concluded that with proper pretreatment successful biological oxidation would produce an effluent acceptable for discharge on a COD or BOD basis. Studies were conducted by Stamoudis and Luthy (1980) to review the efficiency of activated sludge treatment for removal of organic contaminants from coal gasification process effluents. It was concluded that about 99% of removable and chromatographable organic material was present in the acidic portion of the gasification wastewaters. Activated sludge processes were effective in removing a majority of the organic constituents. Removal rates were also high in the alkaline fraction of gasification wastewaters.

MATERIALS AND METHODS

Wastewater Source

The wastewaters used in this study were obtained from the Project Lignite process development unit at the University of North Dakota (UND), Grand Forks, North Dakota U.S.A. The purpose of Project Lignite was to upgrade lignite to a more convenient and less polluting fuel. The continuous process originally envisioned was the two-stage conversion of lignite to fuel liquids, with solvent refined lignite (SRL) as an intermediate and useful fuel product. Figure 1 depicts the first stage of lignite liquefaction for a 50 lb/hr processing development unit that was developed and used at UND until June 1978. The unit used a water gas (CO and H_2) mixture to convert lignite to SRL. This occurred in the presence of a continually regenerated hydrogen donor solvent. Powdered lignite was introduced to a high pressure reactor where operating conditions were approximately 2000 psi and 350°C. Table 1

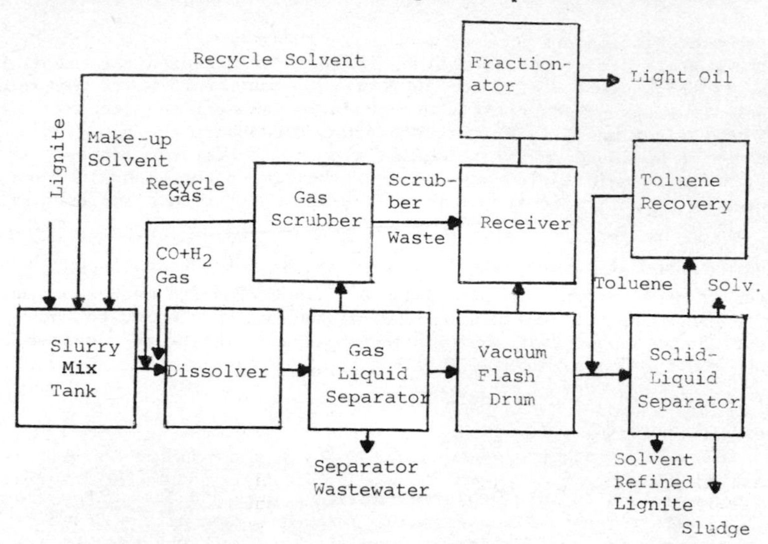

Fig. 1. Flow diagram of Project Lignite 50 lb/hr coal liquefaction process development unit.

TABLE 1 Percent Material Balance for Typical Lignite Liquefaction

Compound	% Starting Materials	% Product
MAF Lignite	45.4	-
Ash	4.2	3.9
Water	24.2	22.6
CO	24.3	17.2
H_2	1.9	1.8
SRL	-	24.0
Liquid Organics	-	5.9
Unconverted Lignite	-	3.5
Organic Gases	-	2.4
CO_2	-	18.5

ıows starting material and product composition summaries for lignite liquefaction. ıe wastewater generated at this stage of operation was called separator wasteıter and was used in this study. Large quantities of wastewater of about 100 ıllons were collected in an attempt to maintain a feed of similar characteristics ır biological treatment. Preservation was achieved by freezing the wastewater in gallon plastic containers.

ıe second stage of lignite liquefaction utilized catalytic hydrogenation to convert ıfined lignite to premium fuels. A typical sample used in this conversion conıined carbon, hydrogen and nitrogen concentrations of 40, 6.7, and 0.3%, respecıvely. Sulfur, oxygen, and ash levels were determined to be 0.9, 44, and 9.0%.

Pretreatment of Wastewaters

One of the most important considerations to be made in biological treatment is to determine if the waste contains any inhibitory constituents which may render the biological treatment system totally or partially inoperative. According to preliminary investigations, lignite liquefaction wastewaters are composed of contaminants which would suppress biological growth. Therefore, pretreatment was necessary. Lime precipitation was used for the removal of high alkalinity. It also proved an effective treatment for removing SO_4^-, organic nitrogen, SCN^-, TOC, and TSS. Lime precipitation also prepared the wastewater for NH_3 removal. Air stripping was used to reduce high levels of NH_3-N. Air was diffused through the wastewater at an elevated pH. Dissolved ammonia is favored over ammonium ion at an elevated pH. The pretreated undiluted wastewater had 16,800 to 45,000 mg/L COD, and 3,290 to 11,700 mg/L phenol. Nutrient additions were necessary to maintain a BOD : N : P ratio of 100 : 5 : 1. Solutions of 40 gm/L Na_2HPO_4 plus 24.2 gm/L $NaH_2PO_4 \cdot H_2O$, and 297 gm/L NH_4Cl were used as phosphorous and nitrogen additions, respectively.

Microorganism Acclimation

Microorganism acclimation was achieved through a fill and draw batch operation. The initial culture was obtained from the aerated lagoon of the Grand Forks municipal wastewater treatment plant at Grand Forks, North Dakota, U.S.A. A water extract of soil underlying a coal pile of North Dakota lignite was also added to the sewage seed. Acclimation procedures began by introducing the seed to very dilute, NH_3-stripped lignite liquefaction wastewaters. The feed concentration was increased gradually once the microorganisms were adapted to the feed concentratio Oxygen uptake rate measurements were used to evaluate microorganism reactions to various loading conditions. Batch reactors of acclimated seed were maintained to provide a source of acclimated microorganisms for the continuous activated sludge reactors.

Activated Sludge Reactors

Bench scale, completely mixed, continuous flow activated sludge reactors were used in this investigation. The aeration tank and settling tank were separated by a removable baffle. Filtered compressed air was supplied to the aeration tank at a rate adequate to insure thorough mixing and maintenance of aerobic conditions at all times. Masterflex tubing pumps were used to provide continuous feed to the activated sludge reactors. Samples were taken regularly from reactor feed tanks and effluent tanks in order to determine performance. Chemical analyses included effluent total suspended solids (TSS), effluent volatile suspended solids (VSS), COD, phenol, NH_3-N, alkalinity, oxygen uptake, sludge volume index (SVI), zone settling velocity (ZSV), conductivity, alkalinity, and pH. Mixed liquor samples, taken after the baffle was removed, were also analyzed for mixed liquor suspended solids (MLSS) and mixed liquor volatile suspended solids (MLVS The combination of the aeration tank and settling tank with the baffle removed was defined as the reactor tank (RT) in this report. The chemical analyses were performed in accordance with Standard Methods (American Public Health Associatio 1976). Phenol concentrations were determined by the direct photometric method. Reactors were run for a period of 2 to 3 weeks to reach steady state conditions which were evidenced by constant COD and phenol removal efficiencies. After steady state conditions were reached, the reactors were maintained for an additi al period of 6 weeks for data collection. Two periods of reactor run, each of 8 to 9 weeks duration, were involved in this experiment in order to examine the effects of altering the feed strengths and the hydraulic detention times on the reactor performance. In period 1, the pretreated lignite liquefaction wastewate diluted to 10, 5 and 3.3% of original strength was used as feed for reactors 1, 2, and 3, respectively. The hydraulic detention time averaged 2.6 days. In period 2, 10% strength wastewater was used as feed for all reactors. The hydraulic detention times for reactors 1, 2, and 3 were 1.88, 5.70 and 8.77 days, respectively. Table 2 exhibits the run protocol for both periods.

TABLE 2 Run Protocol for Periods 1 and 2

	Period 1			Period 2		
	Reactor 1	2	3	1	2	3
Hydraulic Detention Time (day)	2.57	2.43	2.78	1.88	5.71	8.77
% Pretreated Liquefaction Wastewater	10.00	5.00	3.33	10.00	10.00	10.00
Total Volume of Feed (L/day)	0.87	0.94	0.84	0.70	0.67	0.70

RESULTS AND DISCUSSION

Wastewater Characterization

The chemical analysis of the raw liquefaction wastewaters was divided into organic and inorganic portions. Table 3 shows an organic compound summary for lignite liquefaction wastewaters before pretreatment (Baltisberger and Bartak, 1979).

TABLE 3 Organic Compound Summary for Liquefaction Wastewaters Before Pretreatment

Run Number	M-31	L-1
Compound*		
Methanol	3,000	4,400
Ethanol	327	160
Acetonitrile	82	53
Isopropanol	240	N.D.
Acetone	3,500	6,300
Acetic Acid	13,200	N.D.
Phenol	3,400	3,600
O-cresol	1,944**	2,050**

*Concentration in mg/L
**Total content includes cresol and xylene isomers
N.D. - Not Determined

Acetic acid was the major organic compound found in the wastewater. It was present at over 13,000 mg/L. Acetone, phenol and methanol also made up significant portions of waste contaminants. The major inorganic compounds present in the wastewaters were NH_3 and bicarbonate. Major ion concentrations are presented in Table 4. The relative amounts of bicarbonate and acetate varied, but their total concentration (mole/L) was found approximately equal to that of ammonia (mole/L). Raw wastewater analyses for calcium, magnesium, iron, and aluminum indicated total levels were less than 100 ppm. Light oil, solvent, and heavy oil concentrations in the raw wastewaters averaged 92,000, 75,000 and 5,800 mg/L, respectively. A summary of chemical characteristics of the pretreated and diluted wastewater used as reactor feed is shown in Table 5. The raw wastewater was pretreated and diluted to

TABLE 4 Major Ion Concentrations in Liquefaction Wastewaters

Run Number	M-28	M-29	M-30	M-31
Compound*				
NH_4^+ (ammonium)	5,400	4,500	5,600	4,500
CH_3COO^- (acetate)	12,400	12,400	10,600	13,200
HCO_3^- (bicarbonate)	6,470	3,050	7,930	2,440

*Concentration in mg/L

TABLE 5 Characterization of Pretreated 10% Strength Liquefaction Wastewaters

Parameter*	Period 1	Period 2
COD	3,391	3,509
Phenol	575	980
NH_3-N	37	66
Alkalinity	1,189	932

*Concentration in mg/L

10% strength in order to achieve effective biological waste treatment.

In period 1, the feed COD and phenol concentrations were about 3400 and 540 mg/L, respectively. Ammonia-nitrogen was relatively low at about 40 mg/L. Approximatel 1,000 mg/L alkalinity was observed. In period 2, the organic level was slightly higher. The feed COD and phenol levels were 3500 and 500 mg/L, respectively. Ammonia-nitrogen and alkalinity concentrations were 66 and 930 mg/L, respectively.

Performance of Activated Sludge Reactors

Organics removal. Reactor operating conditions and performance for period 1 and are summarized in Tables 6 and 7. The major variables in the operating condition included altering the feed strength of the lignite liquefaction wastewater, the hydraulic detention time and the sludge age.

Generally speaking, as the hydraulic detention time and sludge age increased, and the feed strength decreased, the activated sludge reactor performance efficiencie improved. In period 1, organic removal rates were not appreciably enhanced by decreasing the feed strength from 5% to 3.3%. The maximum COD removal rate was 79%. This occurred at a 5% feed strength or wastewater to tap water ratio of 1:1 During this period, phenol removal rates increased the most in going from reactor to 2 where removal efficiency increased from 65 to 86%. In reactor 3 the phenol rate increased to only 87%. Figure 2A shows a graphical relationship of COD and phenol removal rates versus feed strength in period 1 at an aeration tank hydraulic detention time of 2.6 days and a sludge age of 14.7 days.
When the feed strength was increased to 10%, there was a significant drop in

TABLE 6 Summary of Average Steady State Performance. Data for Period 1, Feed Strength; Reactor 1, 10%; Reactor 2, 5%; Reactor 3, 3.3%

Reactor	Hydraulic Detention Time (day)	MLSS	MLVSS	Oxygen Uptake Rate (mg/L/day)	COD Inf.	COD Eff.	COD % Removal	Phenol Inf.	Phenol Eff.	Phenol % Removal
1	2.57	9637	2457	2147	3441	1390	59.6	536	191	64.6
2	2.43	7641	2259	1351	1773	372	79.0	321	44	86.3
3	2.78	3150	1229	1016	1062	261	75.4	182	24	86.8

Reactor	NH_3-N Inf.	NH_3-N Eff.	NH_3-N % Removal	Alkalinity Inf.	Alkalinity Eff	Alkalinity % Removal	pH RT+	Sludge Age (day)	F/M	q
1	37.8	8.7	77.0	1092	461	57.8	7.7	13.05	0.64	0.32
2	21.6	3.6	83.3	624	233	62.7	7.7	14.74	0.39	0.26
3	10.5	0.8	92.4	409	200	51.1	7.8	16.27	0.38	0.23

All units are in mg/L unless stated otherwise
RT+ denotes reactor tank (aeration tank and settling tank combined)

TABLE 7 Summary of Average Steady State Performance Data for Period 2, 10% Feed Strength for all Reactors

Reactor	Hydraulic Detention Time (day)	MLSS	MLVSS	Oxygen Uptake Rate (mg/L/day)	COD Inf.	COD Eff	COD % Removal	Phenol Inf.	Phenol Eff.	Phenol % Removal
1	1.88	4670	1279	1179	3538	1415	60.0	-	-	-
2	5.71	3831	1639	794	3510	622	82.3	506	5.75	98.9
3	8.77	3051	1521	229	3480	308	91.1	475	3.38	99.3

Reactor	NH_3-N Inf.	NH_3-N Eff.	NH_3-N % Removal	Alkalinity Inf.	Alkalinity Eff.	Alkalinity % Removal	pH RT+	Sludge Age (day)	F/M	q
1	69.9	26.2	62.5	928	384	58.6	7.8	5.81	1.88	0.88
2	65.2	0.8	98.8	1001	140	86.0	7.6	7.15	0.50	0.31
3	64.2	0.4	99.4	867	139	84.0	7.7	7.86	0.33	0.24

All units are in mg/L unless stated otherwise

RT+ denotes reactor tank (aeration tank and settling tank combined)

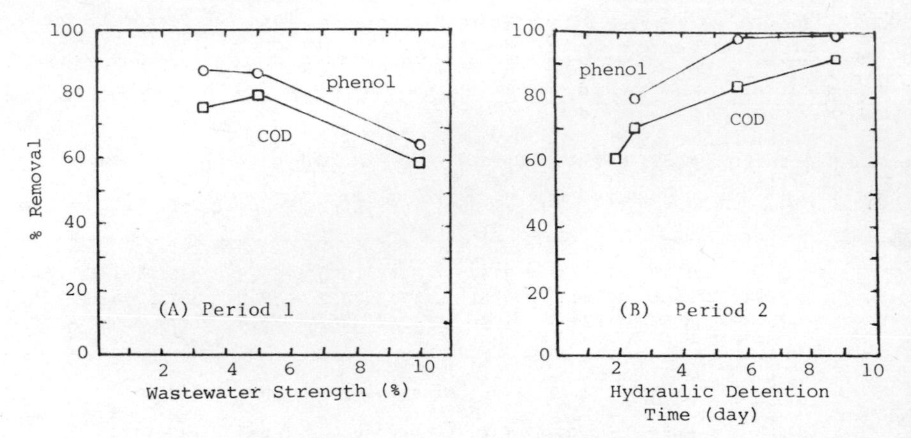

Fig. 2. COD and phenol removal efficiencies vs wastewater strength and hydraulic detention time.

both COD and phenol removal efficiencies, possibly due to the higher levels of toxic substances present. In period 2, the aeration tank hydraulic detention time was the most significant variable. The hydraulic detention times for reactors 1, 2, and 3, were 1.88, 5.71, and 8.77 days, respectively. Influent flow rates were maintained between 0.67 and 0.70 L/day. With similar flow rates, the reactor detention times were increased by using aeration tanks with increasing volumes. As the hydraulic detention time was increased, the organics removal also increased as shown in Fig. 2B. At a feed strength of 10% and an average sludge age of 6.9 days, the COD removal increased from 60% for reactor 1 at 1.9 days hydraulic detention time to 91% for reactor 3 at 8.8 days hydraulic detention time. A significant portion of the COD seems to be unaffected by biological treatment. It is also interesting to note that removal efficiencies do not increase significantly beyond a hydraulic detention time of 6 days. This indicates that a longer detention time would probably be unpractical from an economic standpoint.

The MLVSS was used as an indication of level of the biological mass in the reactor. In period 1, the MLVSS concentration was highest in reactor 1. This was related to the highest organic loading. In period 2, reactor 1 also had the highest organic loading, but had a lower concentration of MLVSS. This could be the result of a shorter hydraulic detention time and a shorter sludge age.

Nitrification. The conversion of NH_3-N to NO_3^- and NO_2^- by nitrifying bacteria is known as nitrification. This oxidation process is carried out in two steps. First the *Nitrosomonas* oxidize NH_3 to NO_2^-, and then the *Nitrobacter* oxidize the NO_2^- to NO_3^-. This reaction is fairly complete so there is relatively little nitrite found in the waste effluent in comparison to nitrate. In period 1, with an average hydraulic detention time of 2.6 days and a sludge age of 14.7 days, the degree of nitrification decreased with increasing feed strength. Reactor 1 had the lowest percentage of NH_3-N removed. A similar trend was observed in reactors 2 and 3, only to a lesser degree. These results are typical for wastewaters containing substances toxic to biological growth. In period 2, the feed was 10% strength in all reactors. The sludge age averaged 6.9 days. The NH_3-N removal efficiency was lowest in reactor 1 due to the shortest hydraulic detention time. In reactors 2 and 3 the nitrification rates were similar. The NH_3-N oxidation in these reactors exceeded 98%. Nitrification appears to be favored by decreasing feed strength and increasing hydraulic detention time.

O_2 requirement. The COD removal rate (q) was needed in order to determine the O_2 requirement for the activated sludge process. It was calculated using the following equation (Adams and Eckenfelder, 1974):

$$q \text{ (mg COD removed/mg MLVSS-day)} = \frac{\text{Infl. COD (mg/L)} - \text{Eff. COD (mg/L)}}{\text{MLVSS (mg/L)} \cdot t \text{ (day)}} \tag{1}$$

where t = hydraulic detention time in aeration tank (day).

Wastewater contaminant removal rates versus COD removal rates (q) are shown in Fig. 3A and 3B. Increasing q reduced treatment efficiencies. The COD removal rate

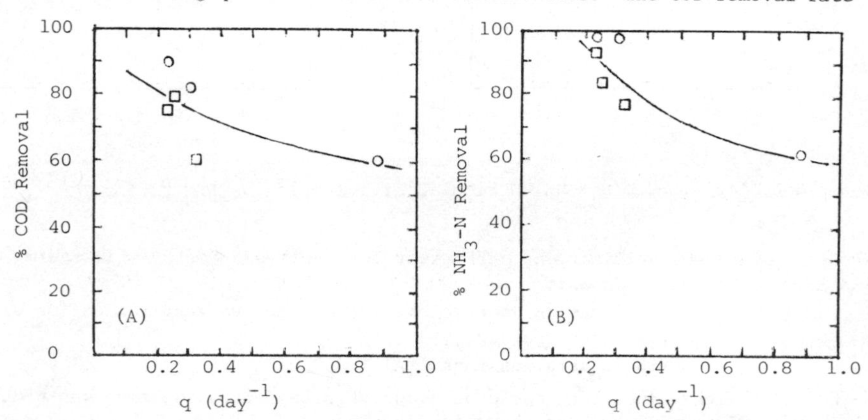

Fig. 3. COD and NH_3-N removal efficiencies vs COD removal rates

followed the trends of the oxygen uptake rates. Therefore, the COD removal rate was directly related to the biological activity.

Organic loading. The organic loading is also defined as F/M ratio.

$$\frac{F}{M} = \frac{\text{Influent COD rate (mg/day)}}{\text{MLVSS (mg in aeration tank)}} \tag{2}$$

In period 1, the organic loading was most affected by the influent COD concentration. Reactor 1 had the highest influent COD and the highest food to microorganism ratio. In period 2, the hydraulic detention time had a significant effect on the organic loading rate. Figure 4A shows the relationship between the COD removal rates and the food to microorganism ratio. Increasing organic loading inhibited organic removal. These same trends were observed for NH_3-N removal in Fig. 4B. The sludge age may also influence these trends. Lower sludge ages were associated with higher COD removal rates. By correlating the COD removal rates with the specific oxygen utilization rate, the O_2 utilization coefficients, a' and b', could be determined. The model used to represent this relationship was (Adams and Eckenfelder, 1974):

$$\frac{R_r}{X} = a'(q) + b' \tag{3}$$

where:

R_r = O_2 uptake rate (mg/day)

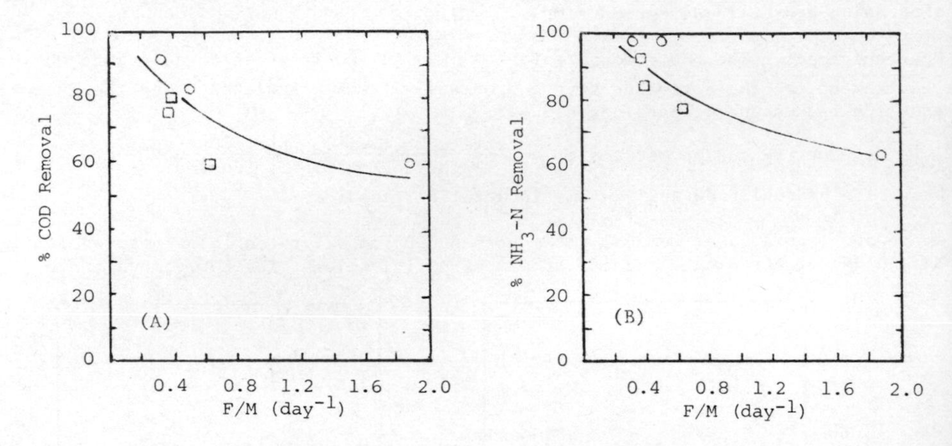

Fig. 4. COD and NH_3-N removal efficiencies vs F/M ratios.

a' = O_2 utilization coefficient for synthesis (mg O_2 utilized/mg COD removed)

b' = O_2 utilization coefficient for endogenous respiration (mg O_2 utilized/mg MLVSS-day)

X = MLVSS in aeration tank (mg)

One of the straight lines that could be drawn when oxygen utilization was plotted against the COD removal rate is shown in Fig. 5A.

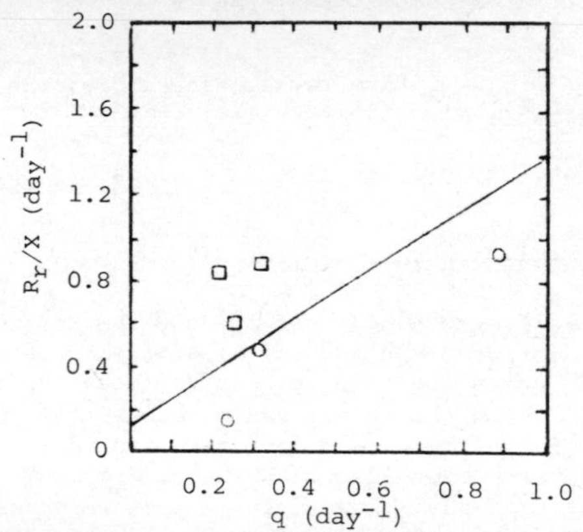

Fig. 5A. Kinetic coefficients of O_2 utilization.

The experimental data collected in this study does not seem to fit this model, possibly due to the inhibitory effect of toxic constituents in the liquefaction wastewaters. The slope of the line was defined to be a' and the y-intercept was equal to b'. The values for a' and b' were found to be 1.3 and 0.1 day^{-1}, respectively.

Sludge production. The rate of sludge production was determined according to the following equation:

$$\frac{\Delta X}{X} = \frac{\text{Sludge production (mg/day)}}{\text{MLVSS (mg/L). Vol. of aeration tank (L)}} \quad (4)$$

Sludge production may be correlated with COD removal rates according to the following equation (Adams and Eckenfelder, 1974):

$$\frac{\Delta X}{X} = a(q) - b \quad (5)$$

Where:

a = sludge synthesis coefficient (mg MLVSS produced/mg COD removed)
b = sludge auto-oxidation coefficient (mg MLVSS destroyed/mg MLVSS in aeration tank - day)

The coefficients of sludge production were determined through graphic analysis. When the rates of sludge production were plotted against the COD removal rates a straight line can be drawn which might indicate the general trend of experimental data. As in the case of O_2 utilization, the scattering of data might result from the inhibitory effect of toxic substances present in the liquefaction wastewaters. The slope of the line was defined as a and the y-intercept was equal to b. Figure 5B illustrates this relationship. Values for a and b were found to be 0.298 and 0.02 day^{-1}.

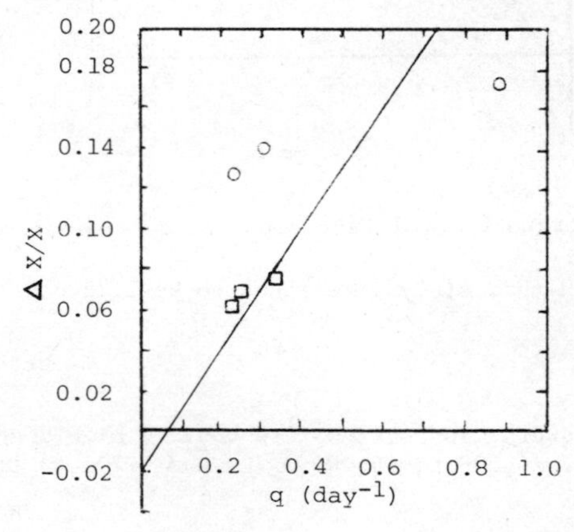

Fig. 5B. Kinetic coefficients of sludge production.

Substrate removal rate. The following equation was used to determine the substrate removal rate (Adams and Eckenfelder, 1974):

$$S_o(S_o - S_e) / (Xt) = K(S_e - Y) \quad (6)$$

Where:

K = substrate removal rate constant (day^{-1})
S_o = influent COD concentration (mg/L)
S_e = effluent COD concentration (mg/L)
Y = non-biodegradable portion (mg/L)
X = MLVSS in aeration tank (mg/L)
t = hydraulic detention time in the aeration tank (day)

The substrate removal determination is graphically represented in Fig. 6.

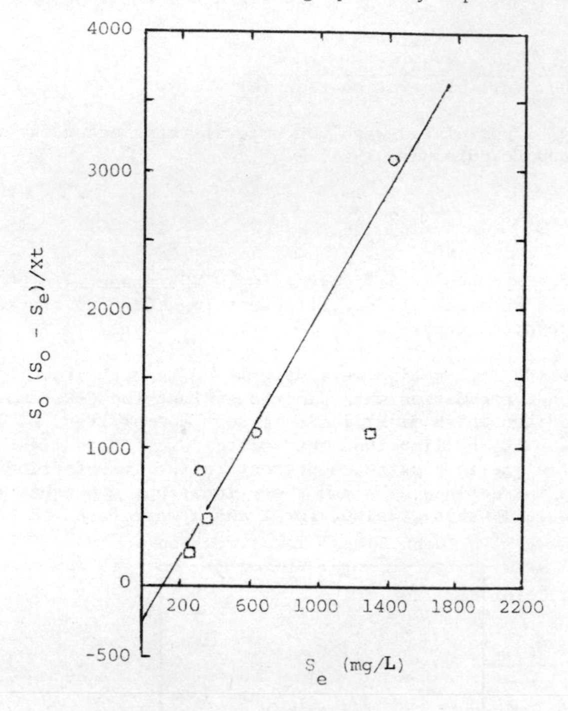

Fig. 6. Substrate removal rate constant determination.

The values for K and Y in this study were found to be 2.28 day^{-1}, and 109.7 mg/L, respectively.

CONCLUSIONS

1. Pretreated lignite liquefaction wastewaters contain high concentrations of COD and phenol in the range of 16,800 to 46,000 mg/L, and 3,200 to 11,700 mg/L, respectively.

2. Pretreatment methods such as lime precipitation, ammonia stripping, and diluti are necessary in order to successfully treat lignite liquefaction wastewaters.

3. Biological treatment of lignite liquefaction wastewaters was found to be inhibited by excessive organic loading rates.

4. For lignite liquefaction wastewater, feed strength and hydraulic detention tim appear to have a significant effect on the treatment efficiencies. Pretreated wastewater at 5% feed strength and a hydraulic detention time of 6 days seemed to be necessary for successful treatment of lignite liquefaction wastewaters.

5. In this study the COD removal rates and organic loading rates varied from 0.24 to 0.88 mg COD removed/mg MLVSS-day, and 0.33 to 1.88 mg COD applied/mg MLVSS-day, respectively.

6. Kinetic constants for oxygen utilization, a' and b' were 1.3 and 0.1 day^{-1}, respectively. The sludge yield and decay coefficients were 0.30, and 0.02 day^{-1}, respectively.

REFERENCES

Adams, C. E. and W. W. Eckenfelder, Jr. (1974). Process Design Techniques for Industrial Waste Treatment. Robert J. Young Co., Nashville, Tennessee. pp. 51-78.

American Public Health Association. (1976). Standard Methods for the Examination of Water and Wastewater. 14th Ed., American Public Health Association, Washington, D.C.

Baltisberger, R. J. and D. R. Bartak (1979). Lignite Liquefaction Wastewater Suitability for Reuse and Cleanup for Discharge into the Environment. Dept. of Chemistry, University of North Dakota, Grand Forks, North Dakota, prepared for U.S. Dept. of Interior, Office of Water Research and Technology.

Lamb III, J. C., P. C. Singer, R. D. Rader, D. A. Reckhow and G. E. Speitel (1979). Characterization and biological treatability of coal conversion wastewaters Annual Purdue Industrial Waste Conference. ESE Publication No. 149.

Reap, E. J., G. M. Davis, J. H. Koon and M. J. Duffy (1977). Wastewater characteristics and treatment technology for the liquefaction of coal using the H-Coal Process. 32nd Purdue Indust. Waste Conference. pp. 929-943.

Singer, P. C., J. C. Pfaender and J. C. Lamb III (1977). Composition and biodegradability of organics in coal conversion wastewaters. 3rd Symposium of Environmental Aspects of Fuel Conversion Technology. Industrial Environmental Research Lab. EPA.

Stamoudis, V. C. and R. G. Luthy (1980). Biological Removal of Organic Constituents in High-BTU Coal Gasification Wastewaters. Energy and Environmental Systems Division. Argonne Natl. Lab., Argonne, Ill.

18

PRELIMINARY SCREENING FOR THE BIOLOGICAL TREATABILITY OF COAL GASIFICATION WASTEWATERS

Yung-Tse Hung*, G. O. Fossum*, L. E. Paulson and W. G. Willson****

**University of North Dakota, Grand Forks, North Dakota, USA*
***Grand Forks Energy Technology Center, U.S. DOE, Grand Forks, North Dakota, USA*

ABSTRACT

A treatability study was made on three wastewaters generated during gasification of lignite in the Grand Forks Energy Technology Center's slagging fixed-bed gasifier. The raw wastewaters contained 5,500 to 7,260 mg/L of phenols, 27,500 to 30,200 mg/L of COD, and 10,500 to 11,500 mg/L of TOC. Pretreatment by lime precipitation, ammonia stripping, coagulation and solvent extraction removed a significant quantity of various chemical constituents. Respirometry studies using Warburg or E/BOD techniques were conducted to determine the effects of pretreatment and dilution on oxygen consumption of acclimated microorganisms. On a batch basis, raw gas liquor and ammonia stripped gas liquor can be biologically treated at 1:25 to 1:50 dilutions. However, in a completely mixed continuous reactor much higher influent concentration could be tolerated. Wastewaters which had been solvent extracted could be biologically treated without dilution. A seven day batch biodegradation study was performed on a 1:50 diluted raw gas liquor and showed that phenols, COD, and SCN could be reduced by more than 96, 73, and 70%, respectively. These biodegradation tests showed no reductions in ammonia.

KEYWORDS

Coal gasification wastewaters; biological treatment; Warburg respirometry; E/BOD; O_2 uptake rate; pretreatment.

INTRODUCTION

The ever-increasing demand for energy and the abundance of this country's coal resources, have prompted intensive investigations on the part of industry and the federal government to develop processes for converting coal into gaseous and liquid fuels. Coal conversion processes produce a complex wastewater stream which must be treated in order to minimize detrimental environmental impact. The wastewater is composed of a variety of contaminants including phenolic compounds, ammonia, polynuclear aromatic hydrocarbons and numerous other organic and inorganic pollutants. Wide variations exist in the composition of the coal conversion wastewaters depending on the type of conversion processes, operating conditions and the nature of the coal used. Generally speaking, there are consistently high concentrations of phenol and ammonia.

In the coking industry, which produces an effluent similar to coal conversion wastewaters, biological treatment has been used successfully for many years in conjunction with other physical-chemical treatment processes.

Pretreatment methods, including dilution, were found to be necessary by Singer and others (1979) in treating synthetic coal conversion wastewater. Results showed tha a minimum sludge age of 10 days may be necessary to achieve a reasonable degree of treatment. Cytotoxicity was shown to decrease with increasing degrees of biological waste treatment.

The bench scale activated sludge treatment of Synthane process wastewaters were examined by Johnson and collegues (1977), and Neufeld and co-workers (1978). Resul indicated that Synthane process wastewater could be successfully treated at 15% strength with a hydraulic detention time of about 1 day and a sludge age of 10 days Ammonia-nitrogen had been stripped to about 500 mg/L in this investigation.

Luthy and Tallon (1978) evaluated the biological treatment characteristics of Hygas coal gasification process pilot plant wastewaters. COD and phenol removal efficiencies were approximately 80 and 99%,respectively. It was concluded that about 400 mg/L COD was non-biodegradable. Nitrification was suspected to be inhibited by toxic organic compounds and high boron concentrations. Processing conditions for ammonia-stripped wastewaters included a hydraulic detention time of 2 to 3 days and a mean cell residence time of 10 to 40 days.

A study of the applicability of available wastewater technology to wastewaters from coal gasification plants was conducted by Muela (1978). It was concluded that zero discharge appeared to be achievable with available wastewater treatment technology, particularly in the locations where evaporation ponds could be utilized as an ultimate disposal process.

Klein and Lee (1978) examined the biological treatability of liquid wastes from coal conversion processes. Biological oxidation showed a high potential for application to hydrocarbonization wastewater. The principal organic waste, phenol, had efficient removal rates, averaging greater than 99.5%. The polynuclear aromatic hydrocarbon and xylenol concentrations were also decreased to some extent, depending on the chemical species involved and the reactor residence time.

The objectives of this study were to determine the characterization of raw lignite gasification wastewater and pretreated effluents, and to determine the biological treatability of gas liquor using Warburg and E/BOD techniques.

MATERIALS AND METHODS

Wastewater Source

The slagging fixed-bed gasifier at the U. S. DOE Grand Forks Energy Technology Center (GFETC) is the only unit of its type in the USA. A slagging fixed-bed gasifier (SFBG) differs from the conventional dry-ash fixed bed gasifier in that only the stoichiometric amount of steam is supplied for gasification. In a dry-ash unit, steam is also supplied to solidify the ash and cool the grate. Some advantages of a slagging unit are that about one-quarter the amount of steam is used, much less wastewater is produced, and gas production is increased as much as three or four times.

A process flow diagram of slagging fixed bed pilot plant is shown in Fig. 1. Coal is fed through a lock hopper and gravity fed into the gasifier. As the coal enters the gasifier, it is heated, dried and carbonized by hot rising gases.

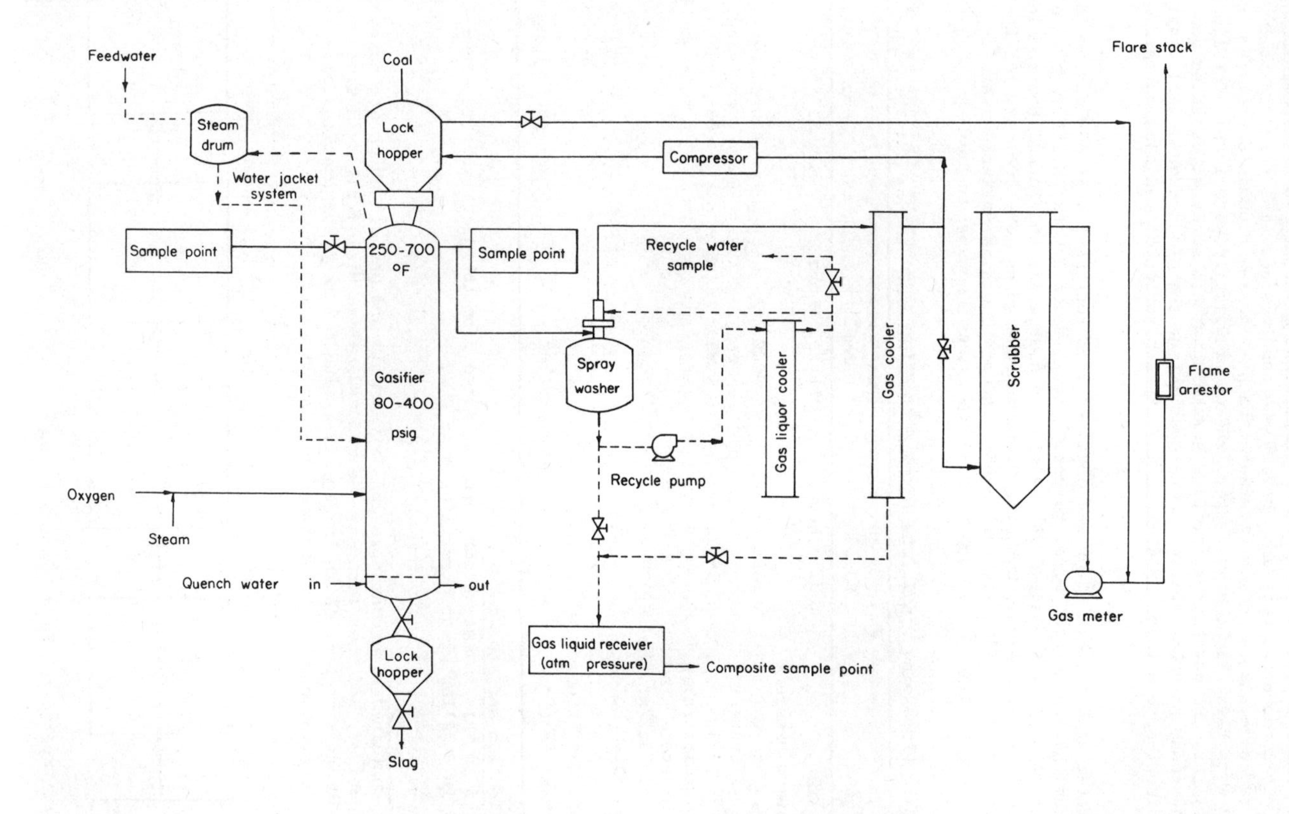

Fig. 1. - Process flow diagram for the GFETC slagging gasification pilot plant. Configuration used until September 1978.

Carbonized coal (char) is contacted with an oxygen-steam mixture just above the hearth and combustion occurs. A hearth temperature of about 3100°F is hot enough to completely consume the coal and convert the ash to slag. The slag drains through a tap hole and is cooled in a water bath. The raw product gas stream is contaminanted with tars, light oils, water, and other volatile materials. The light oil, water, tar, and dust are removed by scrubbing with a spray of recycled gas liquor. A second scrubber is used to further refine the product gas. All wastewater is directed to a receiver where a gravity separation of low boiling tars occurs.

The wastewater used in this study was obtained from runs RA-52, RA-63 and RA-65. Lignite from the Indian Head mine (Mercer County), North Dakota, U.S.A. was gasified in RA-52. In RA-63 and 65 the lignite source was the Baukol Noonan mine (Burke County), North Dakota, U.S.A.. Operating pressures were 300 psig for RA-52, 200 psig for RA-63 and 400 psig for RA-65. Other processing variables were essentially constant. A list of processing conditions is shown in Table 1.

TABLE 1 Processing Conditions for Gasification Tests RA-52, 63, and 65.

Parameter	Run Number RA-52	 RA-63	 RA-65
Coal Type	Indian Head	Baukol Noonan	Baukol Noonan
Operating Pressure (psig)	300	200	400
Moisture Content (pct)	29.11	31.3	32.3
Test Conditions:			
Oxygen Rate (std. cu. ft./hr.)	6,000	6,000	6,000
Oxygen/Steam Molar Ratio	1.0	1.0	1.0
Product Gas Rate (std. cu. ft./hr.)	33,032	32,624	32,720

Sample Collection and Storage

Samples used in this investigation were collected from the gas liquor receiver and stored in 5 gallon plastic containers at 0°C.

Pretreatment Procedure

Gas liquor contained high concentrations of ammonia and phenols which are toxic to microorganisms, and therefore must be reduced before biological oxidation. The pretreatment sequence, shown in Fig. 2, included lime precipitation, air stripping, coagulation, and solvent extraction.

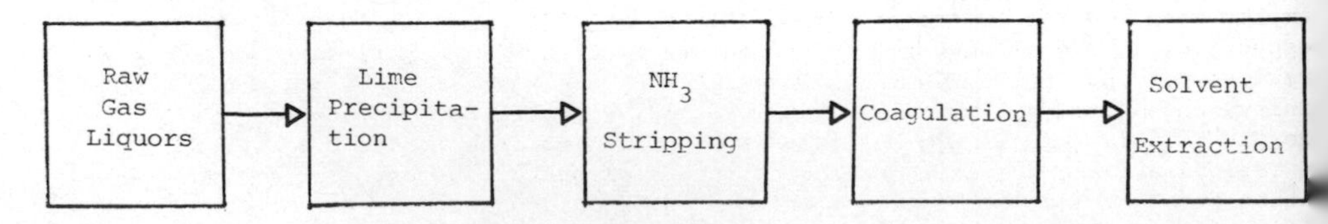

Fig. 2. Pretreatment scheme for gas liquors.

Wastewater was prepared for ammonia stripping by addition of lime which reduced alkalinity and increased pH to 12. Lime and gas liquor were mixed for 10 minutes then allowed to settle for several hours. The supernatant was air stripped until ammonia nitrogen concentration was reduced to less than 500 mg/L.

Gasifier wastewater contains suspended and colloidal materials. Coagulants used to reduce these contaminants included ferric chloride ($FeCl_3 \cdot 6H_2O$), ferric nitrate ($Fe(NO_3)_3 \cdot 9H_2O$), aluminum sulfate ($Al_2(SO_4)_3 \cdot 18H_2O$), and ferrous sulfate ($FeSO_4 \cdot 7H_2O$). Testing procedures consisted of adding coagulants and 100 mL of wastewater sample to plastic beakers and mixing for 3 minutes at 100 rpm followed by mixing for 30 minutes at 20 rpm. Mixing was done with a Phipps and Bird six-place laboratory stirrer. The flocculant material was allowed to settle for at least 30 minutes before supernatants were taken and analysed for percent transmittance with a Bausch and Lomb Spectronic 20.

The wastewater contained a significant quantity of phenols and other organic contaminants. These dissolved organics are toxic to bacteria and are present in high enough concentrations to be economically recovered by solvent extraction. In this investigation the solvents examined were n-hexane, cyclohexane, toluene, and xylene.

Warburg and E/BOD respirometry Procedure

Warburg and E/BOD respirometry were used to determine the O_2 demand of gas liquors. The principle and techniques of Warburg respirometry were described by Umbreit and co-workers (1964), and Hung and Eckenfelder (1978). The instrument was manufactured by Gilson Medical Electronics. The theory of operation and procedures for the E/BOD respirometer are listed in the procedure manual by Oceanography International Corporation (1978).

Batch Degradation Study Procedure

Respiration rates of microorganisms, as determined by E/BOD, were used to measure biodegradation. Gas liquor samples were mixed and maintained at 20°C for seven days in an oxygen rich atmosphere. Comparative tests were made between seeded and unseeded samples in order to differentiate between chemical and biological oxygen demand. Samples of the liquor were taken before and after testing for COD, phenol, NH_3-N, and SCN analysis. Removal of these components would indicate the effectiveness of biological or chemical oxidation of these compounds.

Analytic Procedures

All chemical analytic procedures were performed in accordance with "Standard Methods for the Examination of Water and Wastewater" (APHA, 1976).

RESULTS AND DISCUSSION

Waste Characterization

Table 2 summarizes the chemical analysis data for raw gas liquors from the three gasification runs. Significant levels of organics were observed in all gas liquor samples with TOC and COD ranging from 10,500 to 11,500 and 27,500 to 30,200 mg/L, respectively. Phenol appeared to be the greatest organic constituent with concentrations varying from 5,570 to 7,260 mg/L. The concentrations of NH_3-N and alkalinity ranged from 5,520 to 7,020 mg/L and 12,700 to 21,800 mg/L, respectively. The gas liquors were highly buffered with pH ranging from 8.3 to 8.7. Also listed in Table 2 are the operating pressures and the lignite sources. Increasing gasification pressure had little effect on the amount of organics present but had significant influence on the level of NH_3-N in solution. Evidence is inconclusive on the effect of coal source on composition of gas liquor.

TABLE 2 Chemical Summary of Raw Gas Liquors

Test*	RA-52	RA-63	RA-65
TSS	650	212	140
Phenol	5,570	7,260	6,900
SCN	305	-	248
NH_3-N	5,520	5,140	7,020
Organic-N	779	54	313
Sulfide	160	-	-
Sulfate	515	60	261
COD	27,500	30,200	29,600
Alkalinity	21,800	14,600	12,700
pH	8.3	8.7	8.5
Cyanide	2.8	5.0	-
TOC	10,500	11,500	11,300
NO_3-N	0.9	2.4	1.5

* Values recorded in mg/L

Tables 3, 4 and 5 show chemical analysis summaries of the coal gasification wastewaters after various degrees of pretreatment. In general, the chemical constituents showed an appreciable decrease with increasing degrees of pretreatment.

TABLE 3 Chemical Summary of Pretreated RA-52 Gas Liquors

Test	Raw Gas Liquor	Lime Precipitated Gas Liquor	NH_3 Stripped Gas Liquor	Coagulated Gas Liquor	Solvent Extracted Gas Liquor
TSS	650	368	231	50	46
Phenol	5,570	5,300	5,100	4,800	1,800
SCN	305	395	317	258	165
NH_3-N	5,520	4,050	360	360	350
Organic-N	779	687	621	634	433
NO_3-N	0.9	0.9	0.4	0.1	0.1
Sulfide	160	25	-	-	-
Sulfate	515	264	85	57	59
COD	27,500	26,400	26,000	24,200	12,000
Alkalinity	21,800	835	187	75	77
pH	8.3	11.4	9.8	9.2	9.8
Cyanide	2.8	2.4	14.0	2.6	1.6
TOC	10,500	9,500	9,160	8,000	3,900

* All values in mg/L

Lime precipitation destroyed the buffer capacity of gas liquors and significantly reduced alkalinity and increased pH. A portion of the free NH_3 formed during lime addition escaped to the surrounding atmosphere,thus reducing the NH_3-N level in the gas liquors. Air stripping was found to be effective in the removal of NH_3-N,with efficiencies reaching 98% for RA-52 gas liquors.

Coagulated gas liquor samples showed rather limited reductions in sulfate, alkali nity and organic constituents. This was possibly due to prior pretreatment by lime precipitation in which lime served as coagulant and had removed certain amount of these constituents.

TABLE 4 Chemical Summary of Pretreated RA-63 Gas Liquors

Test*	Raw Gas Liquor	Lime Precipitated Gas Liquor	NH_3 Stripped Gas Liquor	Coagulated Gas Liquor	Solvent Extracted Gas Liquor
TSS	212	184	140	96	56
Phenol	7,260	6,600	6,450	6,310	670
NH_3-N	5,140	4,460	415	355	340
Organic-N	54	53	48	42	39
NO_3-N	2.4	1.2	1.2	0.8	0.8
Sulfate	60	37	24	22	21
COD	30,200	28,800	28,300	27,300	9,100
Alkalinity	14,600	6,660	4,600	1,290	730
pH	8.7	12.4	12.1	7.5	7.9
Cyanide	5	-	-	-	-
TOC	11,500	10,300	10,100	9,080	2,300

* All values in mg/L

TABLE 5 Chemical Summary of Pretreated RA-65 Gas Liquors

Test*	Raw Gas Liquor	Lime Precipitated Gas Liquor	NH_3 Stripped Gas Liquor	Coagulated Gas Liquor	Solvent Extracted Gas Liquor
TSS	140	99	35	24	19
Phenol	6,900	6,370	6,140	5,860	780
SCN	248	160	150	140	120
NH_3-N	7,020	5,820	380	380	375
Organic-N	313	220	192	142	76
NO_3-N	1.5	1.4	1.4	1.0	0.5
Sulfate	261	195	132	113	112
COD	29,600	28,100	27,800	26,300	8,900
Alkalinity	12,700	7,210	6,950	1,860	1,530
pH	8.5	11.5	11.0	7.9	8.0
TOC	11,300	9,900	9,650	8,500	2,700

* All values are in mg/L

Solvent extraction significantly reduced phenol concentrations. As shown in the tables, 89% and 87% of the phenol was removed from RA-63, and 65 wastewaters in two stages of extraction by xylene. Investigations on the partition coefficients of solvents and the best candidate solvent are still in progress. In addition to phenols,other dissolved organics are removed to a lesser extent.

Respirometric Study

Respirometric techniques were used to determine the effect of pretreatment and dilution on the O_2 consumption of gas liquors. Although several respirometric methods were available, only the Warburg respirometry and E/BOD respirometry were used in this investigation.

Figure 3 shows the O_2 consumption of four different dilutions of RA-52 raw gas liquor using E/BOD techniques with microorganisms acclimated with diluted ammonia-stripped gas liquor.

Raw gas liquor and those diluted 1:5 (20% strength) and 1:10 (10% strength) showed only chemical oxygen demand. The highest O_2 uptake rates were with the raw gas liquor. Chemical oxygen demand appeared to be proportional to the gas liquor concentrations. Dilution of the wastewaters between the range of 1 to 25 and 1 to 50 will give an oxygen demand which is primarily from the respiration of bacteria. Further dilution to 0% strength will probably result in the reduction of O_2 consumption as feed concentrations diminish. Based on chemical oxidation alone, the O_2 demand for a 1:50 dilution would be insignificant.

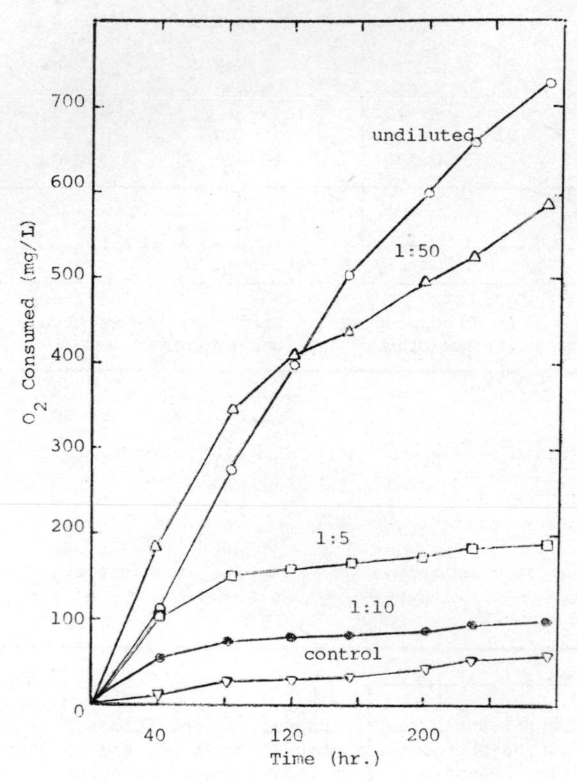

Fig. 3. O_2 consumption of seeded RA-52 raw gas liquor.

Figure 4 shows the O_2 consumption for 90% strength raw gas liquors of runs RA-63 and RA-65 for both seeded and unseeded conditions. For both cases, gas liquor with addition of acclimated microorganisms had almost the same amount of O_2 consumption as the gas liquors without addition of acclimated microorganisms. This implied that at 90% strength the gas liquors were too toxic for biological oxidation, and all the O_2 demand was due to chemical oxidation. This further substantiated the explanation for O_2 demand for undiluted gas liquors of RA-52.

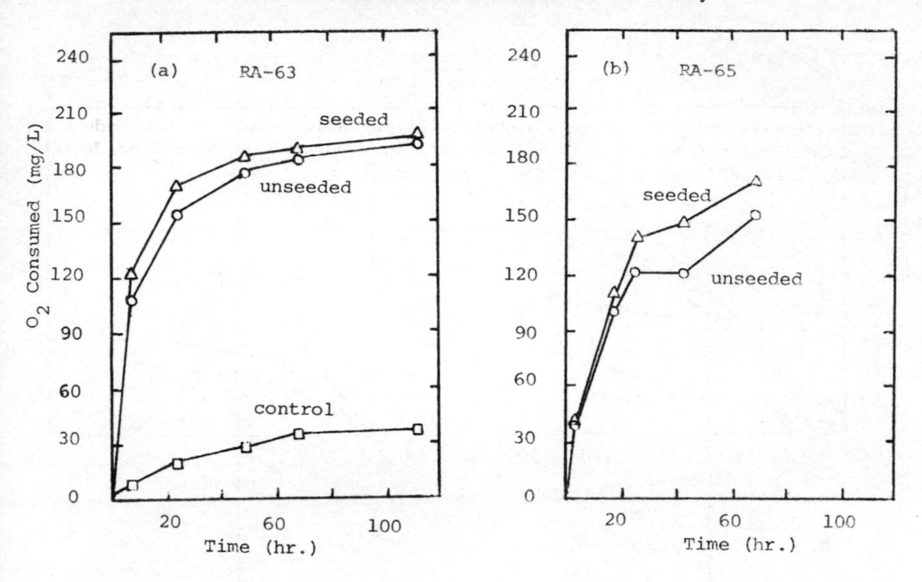

Fig. 4. O_2 consumption of 90% strength of RA-63 and 65 raw gas liquors.

The O_2 consumption for NH_3-stripped gas liquors with addition of acclimated seed is depicted in Fig. 5. For RA-52 gas liquor the 1:50 dilution had higher O_2 demand than either 1:25 or 1:75 dilution. For RA-63 and RA-65 gas liquors the 1:25 dilutions had the highest O_2 demand among the three dilutions tested.

Gas liquor that was solvent extracted could be biologically treated at almost full strength. Results as shown in Fig. 6 indicate that over 90% of the O_2 consumption is due to biological demand.

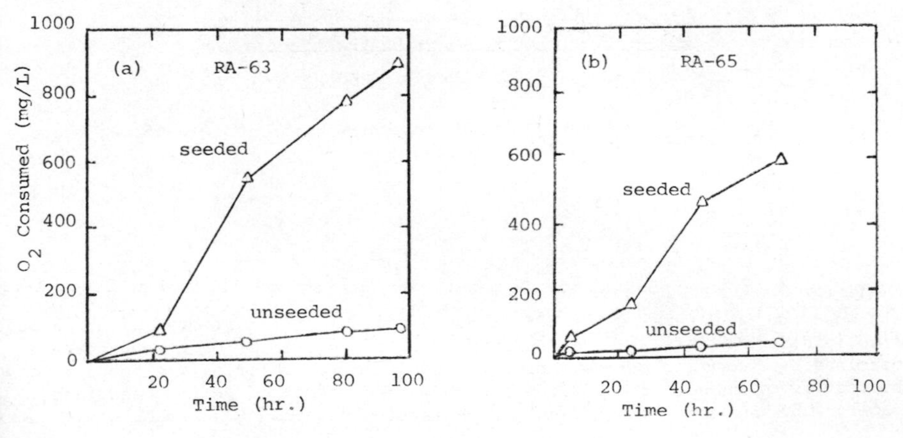

Fig. 6. O_2 consumption of 90% strength solvent extracted RA-63 and 65 gas liquors.

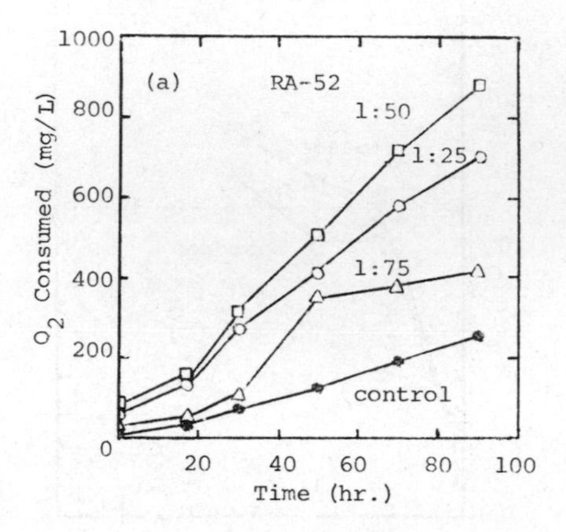

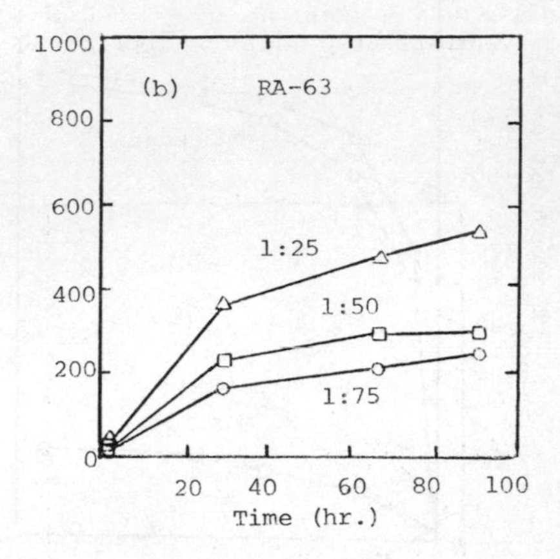

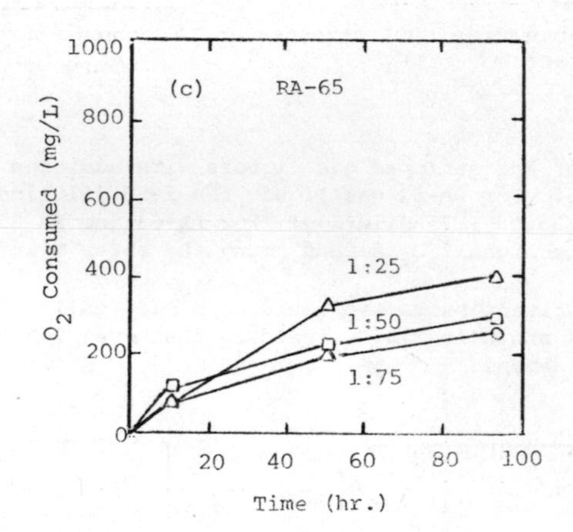

Fig. 5. O_2 consumption of NH_3-stripped RA-52, 63 and 65 gas liquors at various dilution ratios.

<u>Coagulation Test</u>

Tests were run to determine the efficiency of coagulants in removal of suspended and colloidal materials from wastewaters. The effectiveness of coagulants were determined by measuring percent transmittance. Ferric chloride was found to give best results of those tested. Figure 7 shows the effect of pH and coagulant dosage on the percent transmittance of NH_3-stripped RA-52, 63, and 65 gas liquors.

Generally both high pH and high coagulant dosage are required to achieve a high percent transmittance. The optimum pH ranged from 10.3 to 12.3, while the optimum coagulant dosage varied from 6,000 to 8,000 mg/L. As shown in Tables 3, 4 and 5, only a small amount of COD and TOC were removed by coagulation. This pretreatment step would be impractical because of the high dosage requirement.

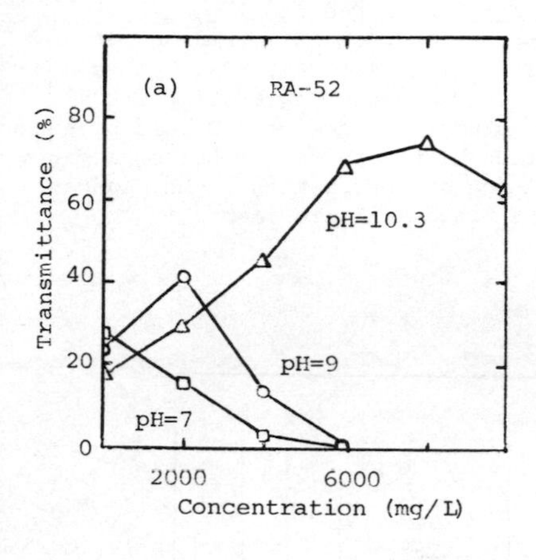

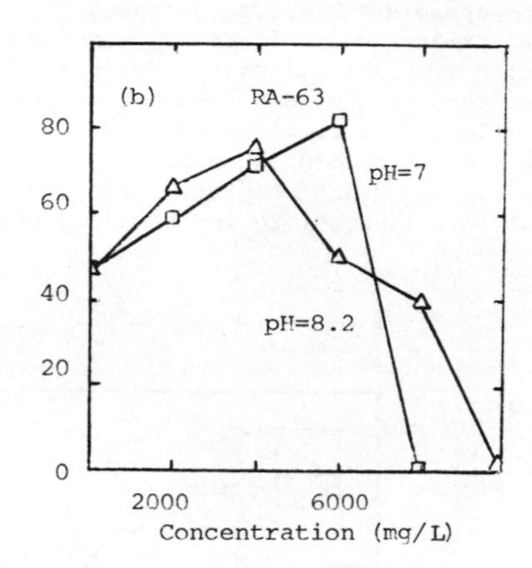

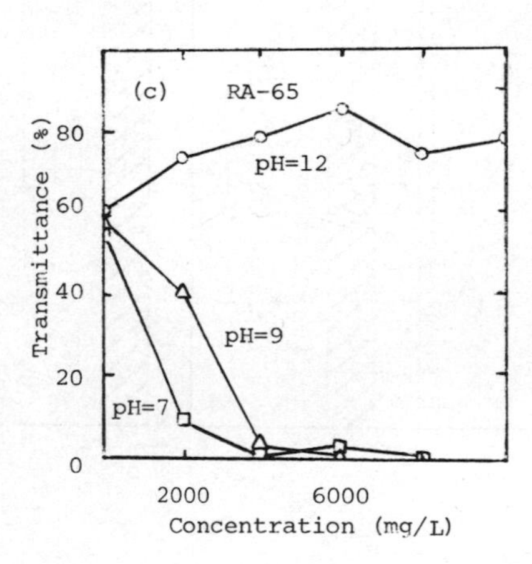

Fig. 7. Percent transmittance of NH_3-stripped RA-52, 63 and 65 gas liquors at various pH and $FeCl_3 \cdot 6H_2O$ dosages.

Batch Degradation Study

Figure 8 shows the breakdown of selected gas liquor constituents in the unseeded RA-52 raw gas liquor. After 7 days of aeration little reduction was observed for all constituents tested. Breakdown of the components of unseeded raw gas liquors was due only to chemical oxidation. Figure 9 depicts the breakdown of compounds in a 1:50 diluted RA-52 raw gas liquor. After 7 days of aeration, little breakdown in components was noted for unseeded gas liquors in which chemical oxidation was responsible for the constituent removal. For the seeded gas liquors, biological oxidation significantly removed COD, phenol and SCN while the degree of NH_3 reduction was insignificant. The removal efficiencies were 96%, 73%, 70% and 5%, for phenol, COD, SCN and NH_3, respectively. The batch degradation study substantiates the respirometer results in this study. Effective COD and phenol removal can be accomplished for the 1:50 dilution RA-52 raw gas liquor on a batch basis. In using the complete-mix activated sludge process, a much higher concentration will be possible in providing effective biological treatment.

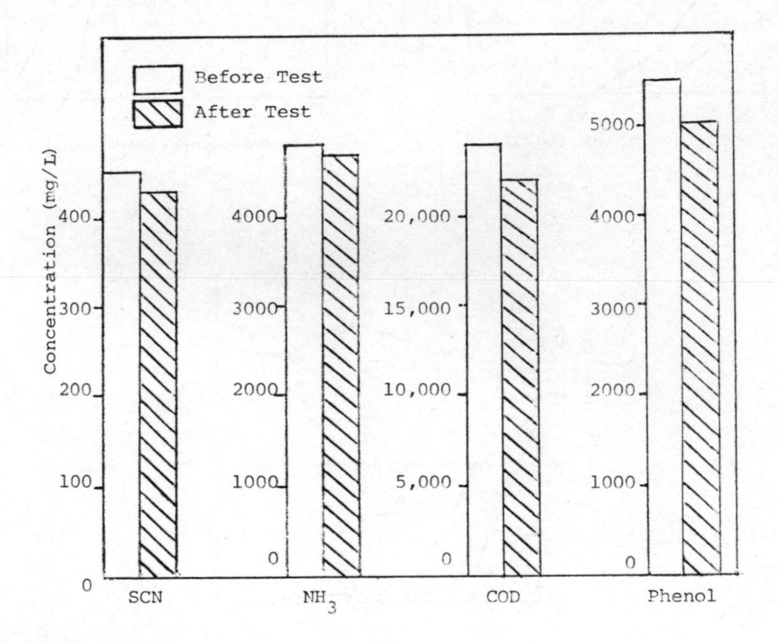

Fig. 8. Chemical breakdown of compounds in undiluted raw gas liquors in a 7-day E/BOD test.

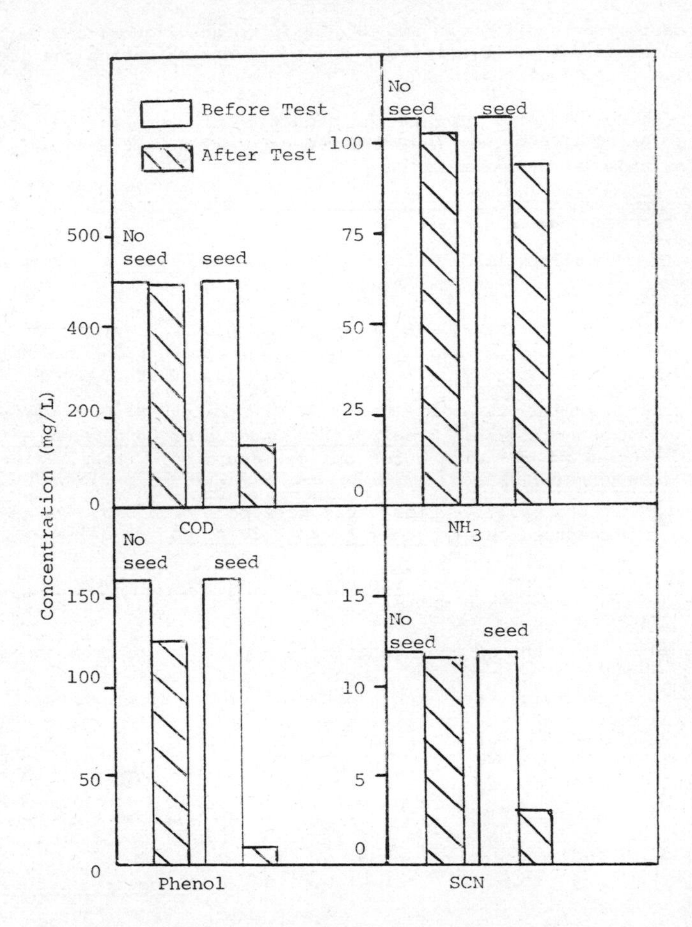

'ig. 9. Chemical breakdown of compounds in 2% strength raw gas liquors in a 7-day E/BOD test.

CONCLUSIONS

. Wastewaters produced by a slagging fixed bed gasifier, gasifying 30% moisture lignite, contained 5,500 to 7,260 mg/L of phenols, 27,500 to 30,200 mg/L of COD, and 10,500 to 11,500 mg/L of TOC.

. Pretreatment by lime precipitation, air stripping, coagulation and solvent extraction were effective in reducing alkalinity, NH_3-N, suspended solids, and phenols.

3. On a batch basis, raw gas liquors and NH_3-stripped gas liquors can be biologically treated at 1:25 to 1:50 dilution, while solvent extracted gas liquors can be treated at full strength.

4. Biological treatment is effective in the reduction of phenols, COD and SCN. A 7-day batch degradation study showed a reduction of 96, 73 and 73%, respectively for these three parameters.

REFERENCES

American Public Health Association (1976). Standard Methods for the Examination of Water and Wastewater. 14th Ed. American Public Health Association, Washington, D.C.

Hung, Y. T. and W. W. Eckenfelder (1978). Application of the Warburg and reaeration techniques in the determination of oxygen uptake rate constants. Proceedings of the N.D. Academy of Science. Vol. 29. Part V, 30-39.

Johnson, G. E., R. D. Neufeld, C. J. Drummond, J. P. Strakey, W. P. Haynes, J. D. Mack and T. J. Valiknac (1977). Treatability Studies of Condensate Water from Synthane Coal Gasification. Pittsburgh Energy Technology Center, Pittsburgh, Pa. Report No. PERC/RI-77/13.

Klein, J. A. and D. D. Lee (1978). Biological treatment of aqueous wastes from coal conversion processes. Biotechnol. Bioeng. Symp. No. 8, 379-390.

Luthy, R. G. and J. T. Tallon (1978). Biological Treatment of Hygas Coal Gasification Wastewater. Carnegie-Mellon University, Department of Civil Engineering. Bulletin No. FE-2496.

Muela, C. A. (1978). Treatment of coal gasification plant wastewater. Coal Processing Technology. Vol. IV, Radian Corp. Austin, Texas.

Neufeld, R. D., C. J. Drummond and G. E. Johnson (1978). Biokinetics of activated sludge treatment of Synthane fluidized bed gasification wastewater. Proceedings 33d Purdue Industrial Waste Conference, 278-285

Oceanography International Corp. (1978). E/BOD Respirometer System Operating Procedures. College Station, Texas.

Singer, P. C., J. C. Lamb III, F. K. Pfaender, R. G. Goodman and R. Jones (1979). Evaluation of coal conversion wastewater treatability. Symposium on Environmental Aspects of Fuel Conversion Technology, IV. USEPA.

Umbreit, W. W., R. H. Burris and J. F. Stauffer (1964). Manometric Techniques. 4th Ed. Burgess Publishing Company.

19

PROCESSING OF OPERATIONAL INFORMATION ON AN INDUSTRIAL WASTEWATER TREATMENT

J. Ganczarczyk and S. W. Hermanowicz

Department of Civil Engineering, University of Toronto, Toronto, Canada

ABSTRACT

Following previous work on operational performance of a full-scale extended aeration treatment of coke-plant effluents, a more detailed analysis of the process was made for a subsequent period of over nine months. The study covered 21 parameters, describing influent and effluent quality and the operational regime of the treatment. All data collected were coded, stored on perforated cards, and edited to allow for a time-series analysis.

The most significant difference in the pattern of plant operation is in the mixed liquor recycle rate and this was used to distinguish four different periods during the operation of the plant. For the long-term analysis of the plant operation the average data, their standard deviation and probability density functions for particular parameters were analyzed and their goodness-of-fit was tested. This analysis proved to be of only limited importance, and therefore more effort was directed to the short-term performance analysis using, for several parameters, autocorrelation and cross-correlation functions and their Fourier transformations (power density spectra functions).

KEYWORDS

Coke-plant effluent treatment; statistical analysis; time-series analysis; thiocyanate decomposition; nitrification inhibition.

INTRODUCTION

There is a feeling that in many cases extensive and frequent measurements and numerous laboratory analyses for the control of activated sludge treatment plants have only very limited practical significance. Often these are used only to fill the monthly and annual reports. Yet at the same time, for immediate plant control, only a few simplified tests are utilized. The application of this simplified approach is justified only if the particular plant has been previously studied to a reasonable extent and unexpected effects are unlikely.

For such plant operation studies, conventional statistical processing of the plant parameter data can be effectively used, as was demonstrated previously (Ganczarczyk, 1969, 1974; Ganczarczyk and Elion, 1978). However, the data processing techniques applied in these studies proved to be not completely satisfactory, and many questions remained unanswered.

This work attempted to look more deeply into the short-term performance of the plant by applying some more sophisticated techniques of time-series analysis, made available by the use of computerized data processing. As it is anticipated that computer storage and processing of wastewater treatment plant operational data soon will become a relatively common procedure, this work appears to be a practical step rather than an exercise of an academic interest only.

PREVIOUS WORK ON COKE-PLANT EFFLUENTS TREATMENT

Ganczarczyk and Elion (1978) studied the performance of a full-scale extended aeration treatment of coke-plant effluents at Dominion Foundries and Steel Limited (DOFASCO) in Hamilton, Ontario. At that time, only a preliminary attempt was made to use a comparison of the types of data distribution as a measure of wastewater characteristics and a treatment stability. In particular, types of distributions (usually normal or log-normal) were taken into consideration, as well as standard deviation values (slopes of lines on distribution graphs). The type of distribution reflects the absence or presence of skew in the data which *per se* is a measure of process stability; the standard deviation value determines the relative size of the dispersion range of data in question.

NEW DATA COLLECTION AND INITIAL EDITING

New operational data for the above activated sludge treatment plant was collected for a period of 273 days, from January 1, 1976 to September 29, 1976. These daily data included flow rate, influent and effluent characteristics: pH, temperature, concentrations of ammonia, phenolics, thiocyanates and suspended solids, and aeration tank parameters for both tanks: mixed liquor suspended solids (MLSS), mixed liquor volatile suspended solids (MLVSS), recycle rate and temperature.

Altogether, 21 process parameters were monitored and coded on perforated cards. During weekends, a limited number of parameters were determined (generally influent characteristics were not recorded).

The missing values for time-series analyses of data were approximated using linear interpolation between two neighbouring points. This technique is relatively simple, but is based on the assumption that no sharp changes occurred within the period of interpolation. This might not be true; however, daily sampling did not allow detecting of changes with periods shorter than two days. It seems, therefore that this method can be acceptable, at least at the first approach to the problem.

DIFFERENCES IN OPERATIONAL PATTERNS

During the course of the study it was possible to distinguish four operational periods at a different recycle rate for both aeration tanks. The average values of the recycle rate are shown in Fig. 1 and the characteristics of corresponding periods are presented in Table 1.

During each period the recycle rate was relatively stable with the standard deviation generally less than 10 percent. The plant performance data within each of the periods were compared in terms of the long-term stability of the plant influent and effluent. Means and standard deviations of all parameters were calculated for all four periods and for the whole time of the study. Observed frequency distributions of all parameters were compared with normal probability distribution using the chi-square goodness-of-fit test based on equiprobable class

TABLE 1. Study Periods of Different Recycle Rate

Period #	Begins Date	Begins Day of Study	Ends Date	Ends Day of Study	Duration (days)
1	1.01.1976	1	2.04.1976	93	93
2	3.04.1976	94	9.07.1976	191	98
3	10.07.1976	192	16.08.1976	229	38
4	17.08.1976	230	29.09.1976	273	44

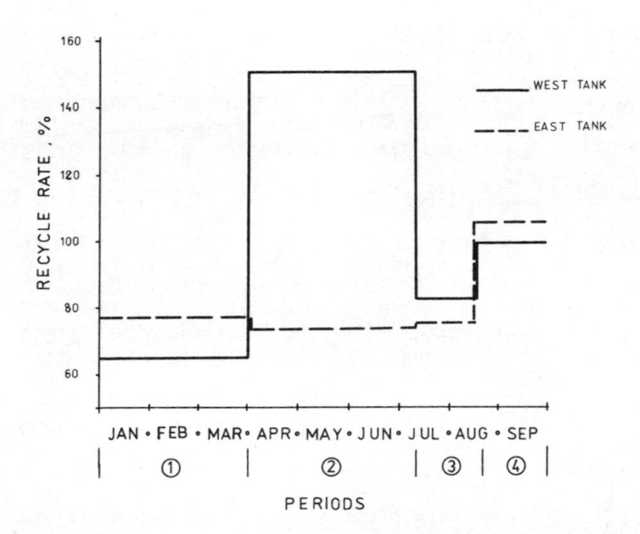

Fig. 1. Average recycle rate

(Kendall and Stuart, 1961). The test was performed at two significance levels, 0.05 and 0.01. For calculations,the FORTRAN subroutine GFIT of International Mathematical and Statistical Library (IMSL, 1979) was used. The same procedure was followed for log values of the initial data and corresponding log-normal distributions were evaluated. The means and standard deviations for selected parameters and their log values are presented in Table 2, and the results of the goodness-of-fit test are shown in Table 3.

LONG-TERM ANALYSIS

The visual inspection of the gathered data did not reveal any significant trends, except for MLSS and MLVSS because of changes in the recycle rate.

The calculated means of influent and effluent characteristics were compared, using the Student t-test. For log-normally distributed parameters the test was also performed for the means of log values and in all cases it confirmed the results for corresponding untransformed data. The results of the above

TABLE 2. Means and Standard Deviations of Selected Parameters

Parameter	Unit	All Data	Periods 1	2	3	4
Flow rate	m^3/h	34.1±4.3	24.2±4.4	34.5±3.8	33.4±4.4	33.4±4.8
Influent concentration of:						
ammonia	mg/L	219±231	191±183	228±243	249±288	233±242
phenolics	mg/L	286±61	235±47	296±36	313±41	345±69
thiocyanates	mg/L	224±80	272±109	213±44	197±29	169±26
Effluent concentration of:						
ammonia	mg/L	249±213	224±206	307±209	216±251	200±172
phenolics	mg/L	1.52±.88	1.53±.95	1.22±.64	2.02±1.0	1.78±.80
thiocyanates	mg/L	65.3± 55.1	59.4± 67.0	63.4± 51.1	80.6± 49.6	69.1± 35.4
MLVSS						
west tank	mg/L	8580 ±3750	7100 ±2310	10400 ±3060	11200 ±5180	5360 ±2120
east tank	mg/L	7970 ±3060	6130 ±1650	9490 ±2610	10260 ±3400	6520 ±2960
Temperature						
west tank	°C	31.8±2.1	31.9±1.7	31.2±1.6	21.7±3.5	32.8±2.0
east tank	°C	30.8±2.1	30.0±2.2	31.6±1.5	21.3±2.6	30.4±1.7
Effluent pH		7.80±.25	7.86±.23	7.79±.22	7.80±.28	7.67±.23

analysis can be summarized as follows:

- no statistically significant long-term trend was found for ammonia concentration in the influent and effluent, or for ammonia load;
- the concentration of phenolics in the influent increased slightly during the whole study period, but the increase was very slow and the difference between any of two neighbouring periods was statistically insignificant;
- the concentration of phenolics in the effluent was stable over the period of investigation;
- the concentration of thiocyanate in the influent decreased over the 273-day period but, as in the case of phenolics, changes were very slow;

- the concentration of thiocyanate in the effluent remained stable in the long-term sense.

This long-term stability of plant performance did not, however, indicate that steady conditions of operation existed on a day-to-day basis. Some measures of the variations were presented as relatively large values of standard deviation (Table 2), which pictured the substantial changes of the process parameters which seem to oscillate around their stable means. The probability distribution functions were another reflection of this behaviour. The normal distribution could be fitted to a few parameters only, namely: MLVSS in the west tank and the phenolics and thiocyanate concentrations in the influent and effluent during periods 3 and 4 and the MLVSS in the east tank during all operational periods.

TABLE 3. Goodness-of-Fit Test Log-Normal Distribution

Parameter	All Data	Periods			
		1	2	3	4
Flow rate	-	-	-	-	-
Influent concentration of:					
ammonia	+	+	+	+	+
phenolics	-	-	-	+	-
thiocyanates	-	+	-	+	±
Effluent concentration of:					
ammonia	+	+	+	+	+
phenolics	-	±	-	-	+
thiocyanates	-	-	-	-	+
MLVSS					
west tank	±	-	-	-	-
east tank	±	+	-	-	±

Note: + fit at significance level 0.05

± fit at significance level 0.01

- lack of fit

Log-normal distribution (Table 3) could be fitted to more parameters because the initial data were quite often skewed towards the larger values as a result of operational upsets and, in some cases, a result of the fluctuating efficiency of the plant.

For some parameters (phenolics, thiocyanates), neither normal nor log-normal distribution were appropriate during all operating periods.

THE APPLIED TECHNIQUES OF SHORT-TERM ANALYSIS

Probability distribution and related statistical parameters are based on relatively long periods of observations and, in some sense, average the pattern of changes of the variables. Besides this, they do not describe sequences in which given values appeared. Therefore, their usefulness for short-term analysis is limited. To investigate these day-to-day changes in the process parameters, time series analysis was used. It included the examination of auto- and cross-correlation functions for selected parameters and corresponding spectra and cross-spectra.

Auto-correlation function and power density spectrum (cross-correlation and cross-spectrum, respectively) are theoretically equivalent as the spectrum is a Fourier transformate of the correlation function. Generally speaking, spectrum shows how much power is associated with the oscillations of a particular frequency which are present in observed data. Unlike the correlation function, values of spectrum for neighbouring frequencies are virtually independent and interpretation of spectrum may be sometimes less difficult than that of the corresponding correlation function. However, there are some problems related to the estimation of spectrum from a sample of finite length. The most important factor in the estimation is the so-called "truncation point" (length of estimated correlation function used for computation of spectral functions) or corresponding "window bandwidth". By narrowing bandwidth, more details can be revealed in spectrum but, unfortunately, the estimating function becomes unstable. The process of finding the optimum truncation point is called "window closing" and consists of gradually narrowing the bandwidth and comparing the details and stability of estimation and interpretation of correlation function and spectrum as described in detail by Jenkins and Watts (1968). In the present study, it was found that the optimum truncation point for analyzed data was close to the value of 32. Computations of correlation functions were performed using subroutine FTCROS (IMSL, 1979) and spectra were estimated by subroutine FTFPS (IMSL, 1979) using the Parzen spectral window.

RESULTS OF SHORT-TERM ANALYSIS

Ammonia. As mentioned previously, ammonia concentration in the plant influent varied very significantly during all the periods, as indicated by very large standard deviations. Auto-correlation function of influent ammonia concentration is presented in Fig. 2. The function is close to zero at very short lags. Some correlation exists between consecutive days, but for the lags longer than 2 days is not significant and sharp changes within 3 or more days could be expected. This instability was probably caused by malfunction of the ammonia still which preceded the aeration tanks. The fate of ammonia in the plant can be described by the cross-correlation function of the influent and effluent ammonia concentrations (Fig. 3). The function displays a large peak at lag of one day. This means that the concentration of ammonia in the effluent is very significantly correlated to the influent with a delay of one day which corresponds to the wastewater retention time in the plant. The treatment plant had very limited buffer capacity with regard to ammonia and any change in the influent was reflected in due time in the quality of the effluent. It should be noted that the average concentration of ammonia in the effluent was usually higher than that in the influent due to the decomposition of thiocyanates. The above observations confirmed previous ones (Ganczarczyk and Elion, 1978) that nitrification was practically inhibited in the plant and that ammonia was merely passing through the plant with some increase due to degradation of thiocyanates.

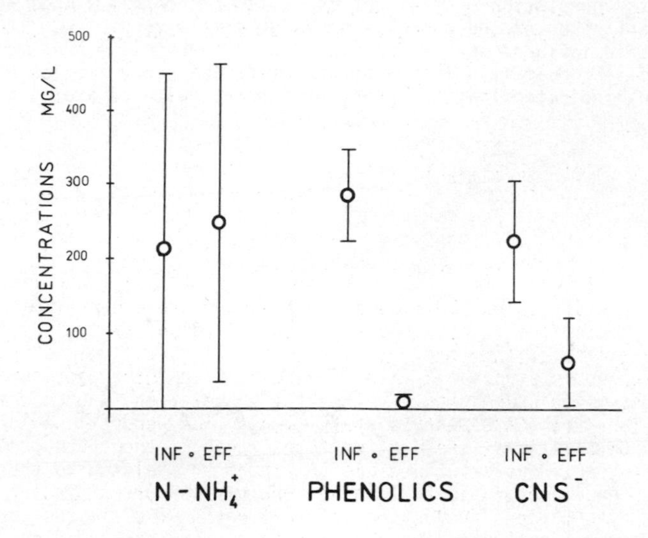

Fig. 2 Autocorrelation function of influent ammonia concentration

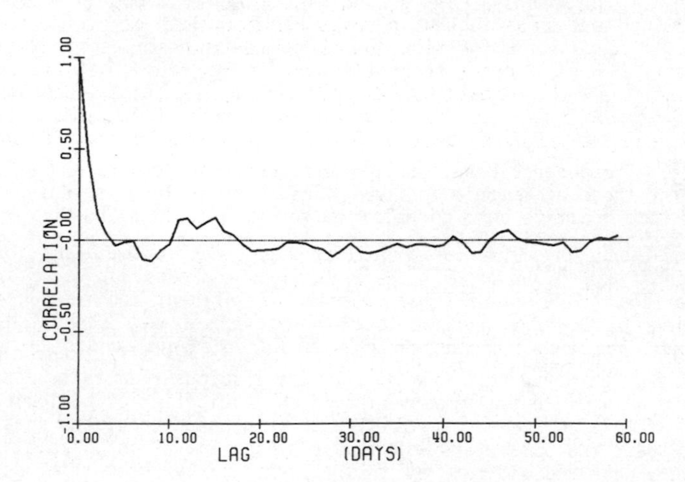

Fig. 3 Cross-correlation function of influent and effluent ammonia concentration

henolics. Influent phenolics concentration was more stable than that of ammonia. uto-correlation function of influent phenolics concentration is shown in Fig. 4. he correlation between concentrations in subsequent days was relatively strong. his indicates that the changes in influent were gradual and "smooth". Unlike ammonia, cross-correlation of influent and effluent phenolics does not show any significant peak and the correlation is generally very weak. It means that shock changes in influent were effectively damped during the treatment and were no affecting effluent quality. This high buffer capacity and the low effluent concentrations indicated that the plant performance with regard to phenolics was good.

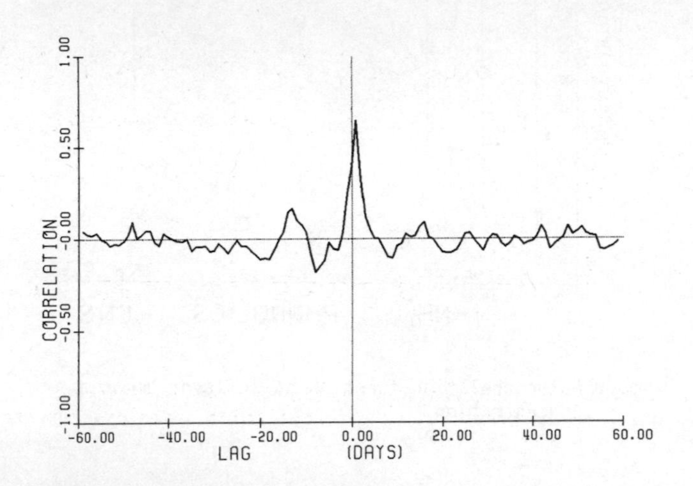

Fig. 4 Autocorrelation function of influent phenolics concentration

Thiocyanates. The general behaviour pattern for the thiocyanates was somewhere in between for those of ammonia and phenolics. The influent thiocyanate concentra tion showed some positive auto-correlation for shorter lags, however not so stron as in the case of phenolics. Cross-correlation revealed a minor but well-defined peak at a lag of one day. It could mean that some part of thiocyanates was not affected by the process but passed through the plant in a similar way to ammonia

Interaction of ammonia and thiocyanates. It was previously reported (Ganczarczyk and Elion, 1978) that high concentrations of ammonia may affect the biological decomposition of thiocyanate. However, it was also observed that a shock loading of ammonia did not always result in the inhibition of biological degradation of this compound. To clarify this inconsistency, the cross-correlati function of these two parameters was studied (Fig. 5). This correlation was generally very weak and did not reveal any particularly interesting details. Nex cross-spectrum was estimated and also coherency cross-spectrum (Fig. 6). Coherency spectrum is the result of the division of cross-spectrum over a product of auto-spectra of involved parameters and, therefore, is "normalized" ir the same sense as a cross-correlation function is a "normalized" cross-covariance The spectrum presented showed that large power is connected with higher frequenci above $0.4 d^{-1}$(period of oscillations less than 2.5 days). It seemed that perhaps even more power could be related to frequencies above 0.5 (period less than 2 days), but with daily sampling frequency regime, the corresponding delay (or phase shift) was equal 0. There was also a flat peak with the maximum at about $0.2\ d^{-1}$ (period = 5 days) with a phase shift corresponding phase spectrum

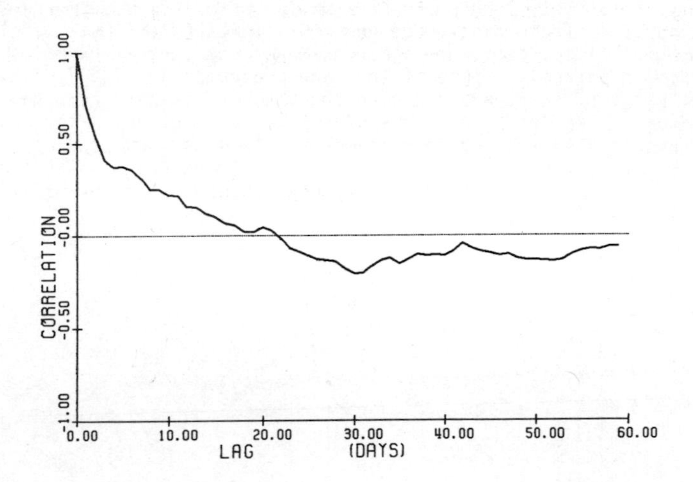

Fig. 5 Cross correlation function of influent ammonia concentration and effluent thiocyanate concentration

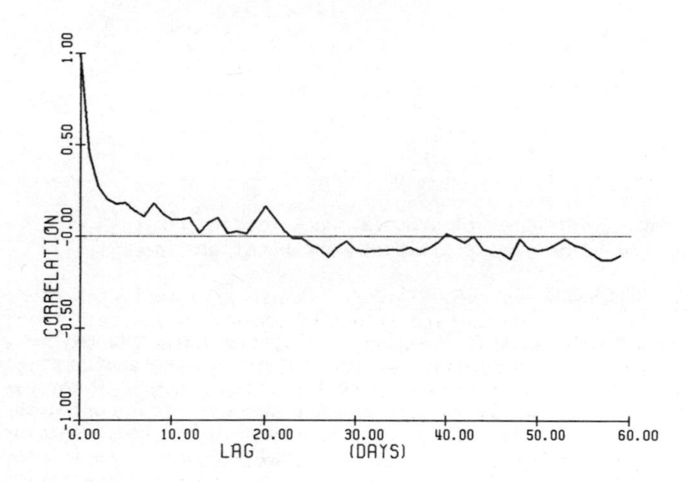

Fig. 6 Coherency cross-spectrum of influent ammonia concentration and effluent thiocyanate concentration

f approximately 3 to 4 days. Relatively small power was associated with frequen-ies lower than, say 0.14 - 0.17 d^{-1} (period 6 to 7 days). From the above analysis, t could be concluded that very steep changes in ammonia concentrations (peaks horter in duration than 2.5 days) should result in peaks in the thiocyanate concentration in the effluent with the maximum at the same time. Longer peaks (4 to 5 days long) should produce thiocyanate peaks delayed by 3 to 4 days, and, finally, relatively long peaks of ammonia (6 to 7 days or more) should not significantly affect thiocyanate decomposition. After these conclusions had been drawn, the initial data were again examined and several examples were found to support the hypothesis. Some of them are presented in Fig. 7. The first peak in May was very sharp and the corresponding thiocyanate peak had its maximum in the same day. Two peaks in February/March of a length of 4 to 5 days and the thiocyanate peaks were indeed shifted by 3 to 4 days as predicted by spectrum analysis. Finally, the very large and relatively long peak in May (longer than 6 days) did not seem to influence the effluent thiocyanate concentration significantly.

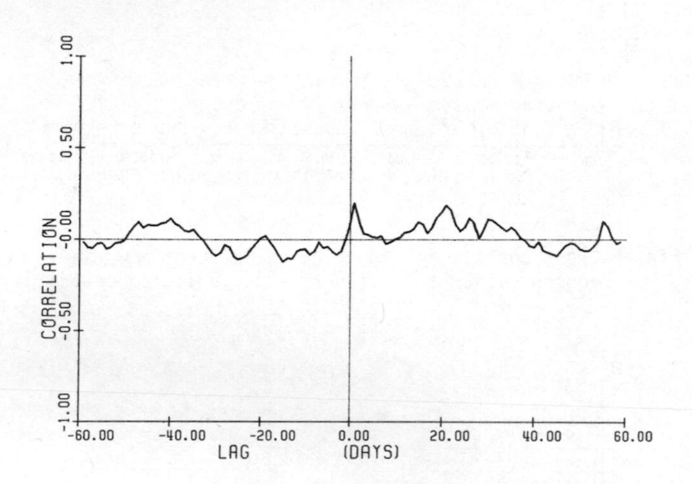

Fig. 7 Influence of ammonia shock loading on thiocyanate effluent concentration

CONCLUSIONS

1. The conventional statistical analysis of operational data may be useful in the evaluation of the long-term performance of the wastewater treatment plant, but has some limitation, especially when the data do not show any particular long-term consistency. To learn more about plant behaviour during shorter periods (day-to-day), the time-series analysis methods should be used. Interpretation of correlation functions should be carried on simultaneously with interpretation of spectral functions which help to reveal all details of the time series.

2. The examined plant operated during the period of study with large variations of influent and effluent quality; however, the average characteristics of influent and effluent were virtually stable.

3. The plant showed very little buffering capacity in regard to ammonia concentrations, but removal of phenolics was always very efficient and changes in the concentration of phenolics in the influent were easily damped and did not

affect the effluent concentrations.

4. The ammonia concentration in the influent influenced bio-decomposition of thiocyanate, but the inhibition of the bio-degradation appeared to be caused rather by steep changes of ammonia concentrations (concentration gradient) than by the concentration itself. However, this hypothesis is drawn on the basis of only a relatively small sample and is not supported by any investigations of metabolism of micro-organisms. Therefore, further studies are needed to examine this phenomenon.

ACKNOWLEDGEMENTS AND CREDITS

The authors are indebted to Mr. M.S. Greenfield and Mr. A. Kruzins of the DOFASCO management for releasing the information used in this paper and for permission to publish it. Special thanks are also due to Drs. B.J. Adams and J.G. Henry for their internal review of this paper.

This work was initiated as a student team project on industrial wastewater treatment data processing in the University of Toronto graduate course "Industrial Pollution Control" taught by Professor Ganczarczyk. Subsequently, a more extended statistical analysis was performed by S. Hermanowicz, a doctoral student at the University of Toronto. At the time of this study he was on leave of absence from Warsaw Technical University, Poland.

REFERENCES

Ganczarczyk, J. (1969). Performance studies of the unbleached kraft pulp mill effluent treatment plant in Ostrolenka, Water Research (Brit.), 3, 519-529.

Ganczarczyk, J. (1974). Evaluation of activated sludge treatment plant performance. In J. Pratt (Ed.) Statistical and Mathematical Aspects of Pollution Problems, Chapter 22, pp. 339-352, Marcel Dekker, New York.

Ganczarczyk, J. and D. Elion (1978). Extended aeration of coke-plant effluents, Proceedings, 33rd Industrial Waste Conference, Purdue University, pp. 895-902.

IMSL (1979). International Mathematical and Statistical Libraries, Ed. 7, IMSL Inc., Houston, Texas.

Jenkins, G.M. and D.G. Watts (1968). Spectral Analysis and its Applications, Chapter 6-9, pp. 209-420, Holden-Day, San Francisco.

Kendall, M.G. and A. Stuart (1961). The Advanced Theory of Statistics, Vol. 2, Chapter 30, Hafner Publishing Co., New York.

20

BIOLOGICAL EXCESS PHOSPHORUS REMOVAL IN THE ACTIVATED SLUDGE PROCESS IN WARM TEMPERATE CLIMATES

I. Siebritz, G. A. Ekama and G. v. R. Marais

Water Resources and Public Health Engineering, University of Cape Town, Rondebosch 7700, Cape, South Africa

ABSTRACT

A review of the prerequisites for achieving biological excess phosphorus removal in the single sludge nitrification-denitrification activated sludge process indicates that these prerequisites lack quantitative definition principally because the basic parameters are not yet fully identified. A basic parameter appears to be the redox-potential, but because a reliable measure of this parameter is not yet possible in the activated sludge process, an alternative parameter, the anaerobic capacity is proposed. The anaerobic capacity is defined as the nitrate concentration per litre influent that an unaerated reactor could remove by denitrification if the nitrate were made available to it. Experimental investigations indicate that at a specific influent COD concentration, there is an associated anaerobic capacity that clearly distinguishes between an anaerobic state which induces biological excess phosphorus removal and one which does not. Utilizing the anaerobic capacity and the nitrification-denitrification theory of Ekama, van Haandel and Marais (1979), an analysis of the phosphorus and nitrogen removal propensities of the Phoredox process indicates that the process (1) can establish the required anaerobic capacity to induce excess biological phosphorus removal only at relatively low influent TKN/COD concentration ratios and (2) has little operational flexibility to accommodate influent TKN/COD ratios higher than that assumed in its design. A new process configuration is presented which (1) can establish the required anaerobic capacity at high influent TKN/COD ratios and (2) has operational flexibility to accommodate influent TKN/COD concentration ratios other than that assumed in its design.

INTRODUCTION

In 1974, Barnard, from pilot scale studies, noted that excess biological phosphorus removal (i.e.,phosphorus removal in excess of that attributable to normal metabolic requirements), is induced in the Bardenpho process (Fig. 1), if at some point in the configuration the mixed liquor is subjected to an anaerobic state* such that phosphorus is released by the sludge mass to the bulk liquid. To obtain this state efficiently, an "anaerobic" reactor was introduced ahead of the primary anoxic reactor. This modification is called the Phoredox process (Fig. 2). Barnard (1976) suggested that because the Bardenpho process has the propensity for virtually complete denitrification, the anaerobic reactor of the Phoredox process would have little nitrate recycled to it via the sludge underflow, and the establishment of the

---

* An environment in which no dissolved oxygen, nitrate and nitrite concentrations are present.

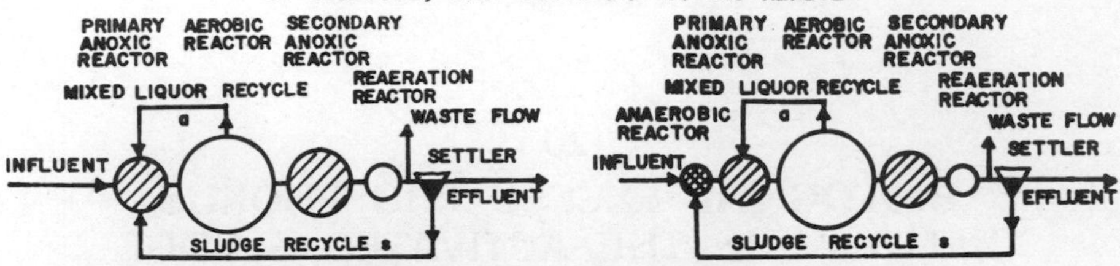

Fig.1. *The Bardenpho configuration for biological denitrification.*

Fig.2. *The Phoredox configuration for nitrogen and phosphorus removal.*

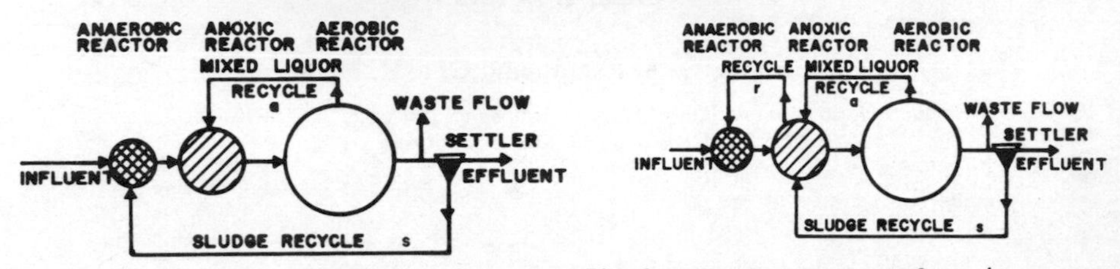

Fig.3. *The Modified Phoredox Process for nitrogen and phosphorus removal.*

Fig.4. *The UCT Process for nitrogen and phosphorus removal.*

anaerobic condition is unlikely to be jeopardized by nitrate in the sludge recycle. Nicholls (1975), McLaren and Wood (1976) and Simpkins and McLaren (1978), following up Barnard's suggestion, succeeded in inducing excess biological phosphorus removal in the Modified Phoredox proces, i.e., a Phoredox process excluding the secondary stages (Fig. 3).

Performance of the Phoredox processes has not always been satisfactory, to an extent that it is doubtful whether at the design stage excess phosphorus removal can be guaranteed. The unpredictable behaviour can be attributed to inadequacy in design arising from (1) a deficiency in the models describing nitrification and denitrification with the result that the nitrate concentrations in the reactors of the processes are not accurately predicted and (2) inadequate definition of the prerequisites for excess biological phosphorus removal.

(1) In the Phoredox processes, the establishment of the anaerobic condition in the anaerobic reactor is directly linked to the nitrification and denitrification performance because the sludge underflow recycle is discharged to the anaerobic reactor; if complete denitrification is not achieved, there will be nitrate in the effluent, and hence in the underflow recycle. If the nitrate concentration exceeds a certain value, the anaerobic condition required for excess phosphorus removal cannot be established. The Phoredox processes, once designed, have little operational flexibility to reduce the nitrate mass recycled to the anaerobic reactor. It can be reduced effectively only by reducing the underflow recycle ratio. However, this is of limited scope because (1) the increase in sludge mass in the settler often induces denitrification with associated loss of sludge over the effluent weirs by flotation, (2) the settling tank may fail in its thickening function*. Consequently, at the design stage, the nitrification and denitrification performance of the process needs to be accurately known to ensure the required anaerobic state for excess phosphorus removal. This, however, usually is not possible; nitrification and denitrification are intimately bound up with the

* Settling properties of activated sludges in Phoredox processes have been observed to be very poor (Osborn, Pitman and Venter, 1979).

influent sewage characteristics, i.e., TKN and COD. Even if the nitrification-denitrification theory is reliable, the response of the plant may be wrongly assessed if the sewage characteristics are different from that assumed in the design.

(2) There is increasing evidence that phosphorus release in an anaerobic reactor is required in order to obtain consistent biological phosphorus removal (Barnard, 1974, 1976; McLaren and Wood, 1976; Nicholls, 1975 and Simpkins and McLaren, 1978). However, it is not yet understood whether the phosphorus release *per se* is a prerequisite for excess biological phosphorus removal or whether it indicates that a necessary condition has been established. Barnard (1976) states that it is not the phosphorus release *per se* that triggers off the excess biological phosphorus removal mechanism, but that it indicates that a certain minimum redox potential has been achieved, which triggers off the excess biological phosphorus removal mechanism. He suggests that because the redox potential cannot be reliably measured in the activated sludge process, the degree of phosphorus release can serve as a substitute parameter for assessing the intensity of the anaerobic condition. This suggestion however, cannot serve as criterion for process design because although it describes the behaviour that appears to be associated with excess phosphorus removal, it provides no guidance at the design stage as to how it can be ensured under different conditions. The best that could be done at the design stage was to specify a nominal hydraulic retention time for the anaerobic reactor of about 1 hour based on average daily flow. There is sufficient evidence that this rule is not adequate due to the behaviour described in (1) above.

Ekama, van Haandel and Marais (1979) presented a model wherein the denitrification kinetics are integrated with the general activated sludge kinetics including nitrification. This model, *inter alia*, describes the nitrification-denitrification behaviour of any configuration of reactors connected by any system of recycles and any cyclic loading conditions. Under constant flow and load conditions, it allows the nitrification in the aerobic zone and the denitrification in the primary and secondary anoxic zones to be accurately expressed by simple functional relationships. These relationships have formed the basis for the development of two parameters: (1) the maximum mass of nitrate which an unaerated reactor can denitrify, called denitrification capacity, and (2) the maximum mass of nitrate that is generated in the aerobic reactors, called nitrification capacity. These two parameters led to a new approach in investigating the prerequisites for excess phosphorus removal. Ekama and co-workers (1979) stated that if the mass of nitrate entering an unaerated reactor is less than its denitrification capacity, an "anaerobic capacity" is set up in the reactor and that this capacity can perhaps serve to quantify the anaerobic state.

In this paper, the parameters nitrification capacity and denitrification capacity and their application to nitrogen removal processes are briefly reviewed. The concept of anaerobic capacity is developed and applied to phosphorus and nitrogen removal processes. A brief description of the experimental investigation into the parameter anaerobic capacity and its implications is given.

ANALYSIS OF NITROGEN REMOVAL PROCESSES

To evaluate nitrogen removal processes, four parameters are introduced: (1) nitrification capacity, (2) maximum unaerated sludge mass fraction, (3) denitrification capacity, (4) TKN/COD concentration ratio of the influent. It must be noted that the formulations of these parameters are valid only under constant flow and load conditions.

Nitrification Capacity

The nitrification capacity is defined as the mass of nitrate per unit influent flow that the process can produce through nitrification of the influent TKN concentration.

The nitrification capacity (N_c) of a process is given by the difference between the influent TKN concentration (N_{ti}) and the sum of the effluent TKN concentration (N_{te}) and the nitrogen required for net sludge production (N_s), i.e.,

$$N_c = N_{ti} - N_s - N_{te} \tag{1}$$

The influent nitrogen concentration required for net sludge production (N_s) is given by Marais and Ekama (1976) as:

$$N_s = S_{ti}\{[Y_h(1 - f_{us} - f_{up})(1 + fb_{hT}R_s)/(1 + b_{hT}R_s)] + f_{up}/P\}f_n \tag{2}$$

The effluent TKN concentration (N_{te}) is given by the sum of the ammonia (N_{ae}) and organic nitrogen (N_{oe}) concentrations, i.e.

$$N_{te} = N_{ae} + N_{oe} \tag{3}$$

The N_{ae} from a process with a total unaerated sludge mass fraction of (f_{xt}) is given by Ekama and colleagues (1979) as:

$$N_{ae} = K_{nT}(b_{nT} + 1/R_s)/\{(1 - f_{xt})\mu_{nmT} - (b_{nT} + 1/R_s)\} \tag{4}$$

Temperature dependencies of K_{nT}, b_{nT} and μ_{nmT} are given in Table 1.

When dealing with municipal wastewaters, N_{oe} at the long sludge ages common to nutrient removal processes can be taken to be less than 3 mg N/ℓ (Ekama and Marais, 1978).

In general, under constant flow and load conditions, the nitrification capacity is independent of the process configuration provided the aerobic sludge mass fraction is sufficiently large to allow stable nitrification.

Maximum Unaerated Sludge Mass Fraction

At a fixed R_s, if f_{xt} is increased above a certain maximum value f_{xm}, nitrification ceases and N_{ae} increases up to the influent ammonia concentration (N_{ai}). Substituting N_{ai} for N_{ae} and f_{xm} for f_{xt} into Eq. (4) and solving for f_{xm} yields the maximum allowable unaerated sludge mass fraction to ensure nitrification, i.e.

$$f_{xm} = 1 - (K_{nT}/N_{ai} + 1)(b_{nT} + 1/R_s)/\mu_{nmT}$$

but as $N_{ai} >> K_{nT}$, $K_{nT}/N_{ai} \simeq 0$ and hence

$$f_{xm} \simeq 1 - (b_{nT} + 1/R_s)/\mu_{nmT} \tag{5}$$

A diagrammatic representation of Eq. (5) is given in Fig. 5 for μ_{nm20} = 0.65 and 0.33 per day, respectively. These are the approximate extremes of the range of values reported in the literature. Figure 5 shows that f_{xm} is very sensitive to μ_{nmT}; unless a sufficiently large aerobic sludge mass fraction is provided, nitrification will not take place and consequently no nitrogen removal by denitrification is possible.

In design, f_{xt} should always be significantly lower than f_{xm} given by Eq. (5) for the following reasons:

(1) Nitrification becomes unstable as f_{xt} approaches f_{xm}.

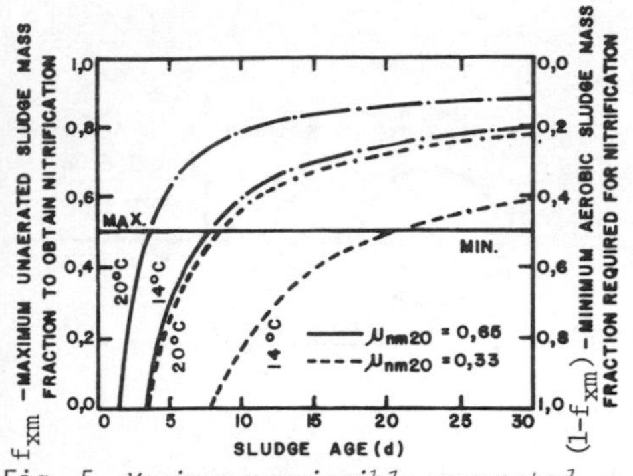

Fig. 5. *Maximum permissible unaerated fraction or minimum aerated sludge mass fraction required to sustain nitrification versus sludge age for different maximum specific growth rate constants of the nitrifiers at 20°C and 14° C.*

TABLE 1. Temperature dependency of kinetic constants of the activated sludge process.

Kinetic Constant	Ref.
$b_{hT} = 0.24(1.029)^{(T-20)}$	Marais & Ekama
$\mu_{nmT} = \mu_{nm20}(1.123)^{(T-20)}$	Ekama (1978-79)
$\mu_{nm20} = 0.20 - 0.65/d$	
$K_{nT} = 1.0(1.123)^{(T-20)}$	Ekama (1978-79)
$b_{nT} = 0.04(1.029)^{(T-20)}$	Ekama & Marais
$K_{2T} = 0.1008(1.080)^{(T-20)}$	Ekama et al Stern
$K_{3T} = 0.072(1.029)^{(T-20)}$	Van Haandel Wilson

(2) The value of μ_{nm20} is difficult to assess. It is very sensitive to toxic or inhibiting substances in the influent and is sharply reduced in processes treating municipal wastewaters where these contain an appreciable industrial fraction.(It varies not only with different sewages but also with different batches of the same sewage). A high value of μ_{nm20} is about 0.65, and a low value 0.20/day. Although the μ_{nm20} value is a fundamentally important kinetic constant, its sensitivity in processes treating different municipal wastewaters requires it to be considered as a sewage characteristic.

(3) The presence of unaerated sludge mass fractions has two effects on the nitrifiers:

(i) Nitrifiers grow only in the aerobic zone but undergo "endogenous mass loss" in both aerobic and unaerated zones. It is this behaviour that led to the formulation of Eq. (5) (Ekama and co-workers, 1979).

(ii) As f_{xt} approaches f_{xm}, the <u>activity</u> of the nitrifiers is reduced. After having been subjected to anoxic conditions, the nitrifiers require a period of adjustment in the aerobic zone before nitrification can proceed at the normal rate (Downing, Painter and Knowles, 1964). This period of adjustment has the effect of reducing the μ_{nm20} value, the degree of reduction depending on the unaerated sludge mass fraction.

(4) The severity of the daily cyclic variations in the load and flow, in particular the variation in the flow, decreases the nitrification efficiency. For the same cyclic flow variation, this effect is reduced as R_s increases beyond the minimum steady state value required for nitrification (R_{sm}).

To accommodate these effects, it is recommended that the minimum *aerobic* sludge mass fraction $(1 - f_{xm})$ is increased by 1.25 to 1.50 times and that the design f_{xm} is always less than 0.50* in the temperature range 12°C to 24°C. Hence Eq. (5) becomes

$$f_{xm} = 1 - S_f(b_{nT} + 1/R_s)/\mu_{nmT} \quad \text{and} \quad f_{xm} \leq 0.50 \tag{6}$$

where S_f = factor of safety: between 1.25 and 1.50.

---

* Apart from the effect that $f_{xm} > 0.40$ has on nitrification, the settling characteristics of the sludge are also severely reduced at $f_{xm} > 0.40$ to an extent that it becomes uneconomical to design plants with $f_{xm} > 0.50$.

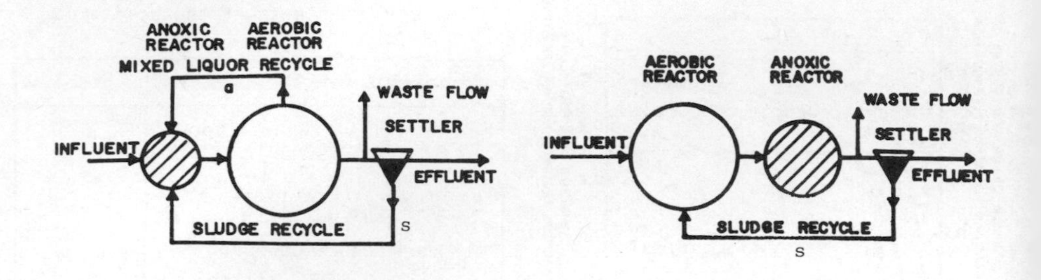

Fig. 6. *The Modified Ludzack-Ettinger (MLE) configuration for biological denitrification.*

Fig. 7. *The Wuhrmann configuration for biological denitrification.*

Denitrification Capacity

The denitrification capacity of a process is defined as the concentration of nitrate per unit influent flow that the process can denitrify. The denitrification capacity of a reactor (under constant flow and load) was formulated by Ekama and colleagues (1979) as:

$$D_{cp} = (1-f_{us}-f_{up})S_{ti}\{[f_{bs}(1-PY_h)/2.86] + [Y_h R_s K_2 f_{xp}/(1+b_{hT}R_s)]\} \quad (7a)$$

$$D_{cs} = (1 - f_{us} - f_{up})S_{ti}\, Y_h R_s K_3 f_{xs}/(1 + b_{hT}R_s) \quad (7b)$$

In Eq. (7a), the first and second terms refer to the first and second rates of denitrification, respectively, in the primary anoxic reactor. The first term is attributable to the utilization of the rapidly biodegradable COD fraction which is complete in a very short time - about 10 minutes at 14°C for R_s 12 to 20 days; hence it can be taken as complete in any practical design. Eq. (7b) describes the rate in the secondary anoxic reactor. Equations (7a and 7b) give closely the respon calculated by the general activated sludge model (Van Haandel, Ekama and Marais, 1981). The D_c of a modified Ludzack-Ettinger (MLE) process (Fig. 6) is given by Eq. (7a), that for the Wuhrmann process (Fig. 7) by Eq. (7b) and that for the Bardenpho process (Fig. 1) by the sum of Eqs. (7a and 7b). If D_c of a reactor is greater than the nitrate recycled to it, the nitrate concentration in its outflow will be zero but its denitrification performance is less than D_c. If D_c is less th or equal to the nitrate recycled to it, there will be nitrate in the outflow and it denitrification performance is equal to D_c. From Eqs. (7), it is evident that the location of an anoxic reactor in a process affects its D_c; the energy source for denitrification in a primary anoxic reactor is the two fractions of influent COD, whereas in the secondary anoxic reactor it is self-generated through "endogenous mass loss" - the different sources of energy result in different denitrification rates (K_2 and K_3). Consequently, the D_c of a process comprising of primary and secondary anoxic reactors depends on the fractional division of f_{xt} between f_{xp} anc f_{xs}.

Influent TKN/COD Concentration Ratio

The TKN/COD of the influent is defined as the ratio of the TKN concentration to the total COD concentration. A high influent TKN concentration will generate a high concentration of nitrate in the aerobic zone, the mass of nitrate generated being virtually completely independent of the influent COD concentration. In contrast, the D_c for a specific process configuration, temperature and R_s is proportional to the influent COD concentration (see Eqs.7). Consequently, the influent TKN/COD provides an approximate relative measure of N_c to D_c.

Nitrogen Removal Processes

Designs for steady state conditions can be based on the four parameters discussed above. From the sewage characteristics,(i.e., the influent TKN/COD ratio, μ_{nm20}, temperature), f_{xm} can be calculated for a given R_s (Eq. 6). The f_{xm} is then divided into f_{xp} and f_{xs} and this division fixes the D_c of the reactors. When considering the subdivision of f_{xm} into f_{xp} and f_{xs}, the following should be noted:

(1) For the same f_x, D_c of a primary anoxic reactor is always greater than that of a secondary anoxic reactor.
(2) Sufficient nitrate must be recycled to a primary anoxic reactor via the mixed liquor recycle to fully utilize its D_c.
(3) If the combined D_c of the anoxic reactors is sufficient to obtain complete denitrification, this is possible only in a configuration including a secondary anoxic reactor.
(4) If the combined D_c is insufficient to obtain complete denitrification for any division of f_{xt} between f_{xp} and f_{xs}, then from (1) above, the highest nitrogen removal is obtained if f_{xt} is utilized as effectively as possible, i.e., as f_{xp}.

In the Bardenpho process (Fig. 1) at 25 days and 14°C, if the influent TKN/COD ratio is low (< 0.08 mg N/mg COD*), the D_c of the process is sufficiently high to allow complete denitrification (provided the mixed liquor(a) recycle is such that D_{cp} is fully utilized). Consequently, the nitrate concentration in the effluent (N_{ne}) is zero. If the TKN/COD ratio is high (> 0.09 mg N/mg COD),* not all the nitrate can be removed by denitrification. Under these circumstances it is disadvantageous to incorporate a secondary anoxic reactor; it is preferable to utilize f_{xt} as a primary anoxic reactor to remove the maximum concentration of nitrate.

It might be thought that incomplete denitrification may be resolved by increasing R_s for this would allow a higher f_{xm} thus producing a higher D_c. However, f_{xm} is limited at a maximum of 0.50 and once this limit is reached, further increases in R_s become counter-productive; both D_c and N_c increase (N_c because less sludge is wasted), but the increase in D_c no longer compensates even for the increase in N_c. Hence, once an upper limit to f_{xm} is fixed and attained in design, there is also an upper limit to R_s beyond which N_{ne} is no longer reduced. At 14°C, f_{xm} = 0.50 and μ_{nm20} = 0.35/d, the maximum R_s is 30 days and the Bardenpho process can achieve complete denitrification only if the TKN/COD ratio < 0.09 mg N/mg COD*. Consequently, from (4) above, if the TKN/COD ratio > 0.09mg N/mg COD*, a lower N_{ne} will be achieved with the MLE process than with the Bardenpho. For $f_{xm} \leq 0.50$, plotting the maximum TKN/COD ratio with which the Bardenpho process can achieve complete denitrification versus μ_{nm20} at 40° and 20°, defines the "success area" of this process in terms of the 3 sewage characteristics (Fig. 8). For TKN/COD ratios outside this area, the MLE process will be more efficient in nitrogen removal than the Bardenpho.

ANALYSIS OF NITROGEN AND PHOSPHORUS REMOVAL PROCESSES

Despite the difficulties in identifying the prerequisites for excess biological phosphorus removal, it has been found that when excess biological phosphorus removal takes place, the removal can be quantified. Martin and Marais (1975), Marsden and Marais (1977) and Hoffmann and Marais (1977) concluded that under constant flow and load conditions the combined effect of normal metabolic phosphorus requirements and excess phosphorus uptake is limited and can be expressed in terms of the steady state activated sludge process theory as developed by Marais and Ekama (1976), i.e.,

$$P_s = S_{ti}\{[(1 - f_{us} - f_{up})Y_h(\alpha + f_p f b_{hT} R_s)/(1 + b_{hT} R_s)] + f_p f_{up}/P\} \quad (8)$$

* Refers to total COD concentration (S_{ti})

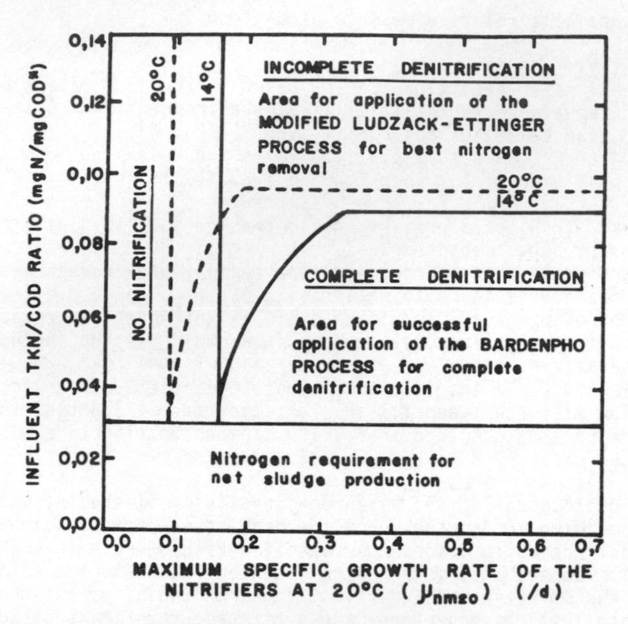

Fig. 8. *Diagram showing combination of sewage characteristics with which the Bardenpho process can achieve complete denitrification at 14 C° and 20 C°.*

The value of α was found to be highly variable ranging from 0.025, a value associated with no excess phosphorus removal, to 0.15 mg P/mg VASS. For laboratory scale processes exhibiting excess phosphorus removal they proposed α = 0.10 mg P/mg VASS and f_p = 0.025 mg P/mg VSS as reasonable values for design. The varying value of α was due to the inability to specify the prerequisites for excess phosphorus removal correctly.

Anaerobic Capacity

In an endeavour to find a criterion by means of which the prerequisites for excess biological phosphorus removal can be predicted, it was decided to look for a parameter which shows the same behavioural tendency as the redox potential but which is quantifiable in the activated sludge process. The parameter selected is the "anaerobic capacity" (A_c). The formulation of A_c was developed as follows: by means of Eqs.(7), D_c of an unaerated reactor can be calculated. By deducting the nitrate discharged to the reactor (in terms of mg N per litre influent flow) from D_c, the nitrate concentration deficit or anaerobic capacity, A_c, is found.

It was hypothesized that (1) if A_c has a high positive value, it is likely that the redox potential concomitantly will be low and (2) excess phosphorus removal will be induced if A_c exceeds a certain minimum value A_{cm}. Hence

$$A_c = D_{ca} - s.N_{ne} \quad \text{for the Phoredox Process} \tag{9}$$

$$A_c = D_{ca} - r.N_{nx} \quad \text{for the UCT Process (described below)} \tag{10}$$

It will be shown below that when A_c is greater than A_{cm} in an anaerobic reactor, phosphorus release is almost invariable observed. For the present, therefore, the hypothesis that $A_c > A_{cm}$ needs to be established to initiate phosphorus release and its subsequent removal in excess will be accepted. Implications of this on design are as follows:

(1) The anaerobic capacity is necessarily an unutilized denitrification capacity.

(2) The D_c of the anaerobic reactor must be sufficiently high to denitrify all the nitrate discharged to it <u>and</u> establish $A_c > A_{cm}$.

Phosphorus Removal Processes

In the Phoredox process (Fig. 2), a certain part, f_{xa}, of f_{xt} is set aside as an anaerobic reactor to establish $A_c > A_{cm}$ required for excess biological phosphorus removal. The remaining *anoxic* sludge mass fraction ($f_{xt} - f_{xa}$) is available for denitrification. If f_{xa} is estimated on the basis that zero nitrate is recycled to the anaerobic reactor via the underflow (s) recycle, then complete denitrification must be achieved in the anoxic sludge mass fraction. The denitrification behaviour of the *anoxic* sludge fraction of the Phoredox process is the same as that for the Bardenpho process (Fig. 1), provided there is no nitrate in the underflow (s) recycle. Consequently, for the same conditions, the Phoredox process has a lower unaerated sludge mass fraction available for denitrification than the Bardenpho process. Hence,in order to obtain complete denitrification in the Phoredox process, the sewage characteristics must necessarily be more favourable than those with which the Bardenpho process can produce complete denitrification: If A_{cm} = 10 mg N/ℓ influent#,then at 14°C,R_s = 25 days, f_{xt} = 0.50 and S_{ti} = 500 mg COD/ℓ with a biodegradable COD fraction of 80 percent, the maximum TKN/COD ratio with which excess phosphorus removal can be achieved is 0·08 mg N/mg COD*. When the TKN/COD ratio is such that complete denitrification cannot be obtained, there will be nitrate in the effluent and underflow recycle. The nitrate recycled to the anaerobic reactor via the underflow reduces A_c, thereby reducing the possibility of obtaining excess phosphorus removal. This conclusion is supported by reports on the behaviour of the Phoredox process - excess phosphorus removal was only observed when the effluent nitrate concentration was low, which usually occurs when the influent TKN/COD ratio was low (Nicholls, 1975; Osborn, Pitman and Venter, 1979).

In the Modified Phoredox process† (Fig. 3),complete denitrification is impossible so that always there will be nitrate in the effluent and underflow recycle. The denitrification capacity of the anaerobic reactor must be sufficiently high to denitrify the nitrate entering it as well as establish $A_c > A_{cm}$. However, the nitrate denitrified in the anaerobic reactor reduces D_c of the anoxic reactor due to COD consumption during denitrification. As the TKN/COD ratio increases, the effluent and underflow recycle nitrate concentrations also increase.Hence,in order to maintain $A_c > A_{cm}$, D_c of the anaerobic reactor must increase concomitantly by increasing f_{xa}. However, because f_{xt} is limited by f_{xm}, the increase in f_{xa} can take place only at the expense of f_{xp}; accordingly, in order not to load the smaller anoxic reactor with nitrate above its denitrification capacity, the mixed liquor (a) recycle ratio needs to be reduced.ØIn this fashion, at a certain TKN/COD ratio, $f_{xa} = f_{xt} = f_{xm}$, and at this stage f_{xp} and the(a) recycle both will be zero (i.e., the process configuration will be reduced to two reactors, an anaerobic and an aerobic linked only by the underflow (s) recycle). At this TKN/COD ratio,D_c of the

---

See Experimental Investigations section

† The conclusion above that the secondary anoxic reactor in nitrogen removal processes no longer serves a useful purpose when the sewage characteristics are such that complete denitrification cannot be obtained also applies to the phosphorus and nitrogen removal processes.

Ø This is not essential because the reactor will not denitrify more nitrate than its D_c. However, a reduction in the(a) recycle reduces the mass of dissolved oxygen discharged to the reactor, thereby leaving a larger fraction of D_c for denitrification.

anaerobic reactor is just sufficient to denitrify all the nitrate recycled to it via the recycle and establish $A_c = A_{cm}$. Any further increase in TKN/COD ratio will increase the effluent and underflow nitrate concentrations. The additional nitrate load on the anaerobic reactor will be denitrified, but at the expense of the anaerobic capacity. Consequently, even in the Modified Phoredox process, certain TKN/COD ratios make it impossible for the process to establish $A_c > A_{cm}$: If A_{cm} = 10 mg N/ℓ, then at 14°C, R_s = 25 days, f_{xm} = 0.50, S_{ti} = 500 mg COD/ℓ with a biodegradable COD fraction of 80 percent, the maximum TKN/COD ratio with which excess phosphorus removal can be achieved is 0.12 mg N/mg COD.*

The failure of the Phoredox processes to establish $A_c > A_{cm}$ at high TKN/COD ratios, stems principally from their inflexibility: once the required degree of denitrification cannot be achieved, very little can be done by way of process operation to re-establish $A_c > A_{cm}$. The situation would be improved if A_c were less dependent on nitrate concentration in the sludge underflow; e.g., the sludge underflow (s) recycle should not be discharged to the anaerobic reactor. A configuration called the UCT process was found to satisfy this requirement.

In the UCT process (Fig. 4), the sludge underflow (s) recycle is discharged to the anoxic reactor and an additional mixed liquor (r) recycle from the anoxic to the anaerobic reactor is introduced. The nitrate recycled to the anoxic reactor can be controlled by appropriately adjusting the a recycle such that the nitrate concentration in the anoxic reactor (N_{nx}) is approximately zero. Hence the (r) recycle is not compromised by excessive nitrate input. The sludge concentration in the anaerobic reactor is lower than that in the other two reactors - by a fraction $r/(r + 1)$ - which reduces A_c. However, the reduction in A_c is not in proportion to the reduction in sludge concentration because the principal contributor to the D_{ca} is the rapidly biodegradable COD fraction of the influent (1st term, Eq. 7a) which theoretically is independent of the sludge concentration except at very low anaerobic hydraulic retention times.

The UCT process has greater flexibility than the Phoredox processes to accommodate changes in the TKN/COD ratio: if the TKN/COD ratio increases, then for a fixed underflow (s) recycle ratio, the mixed liquor (a) recycle ratio can be reduced until the nitrate load on the anoxic reactor again is approximately equal to its D_c. In this fashion, the anaerobic reactor may be protected by appropriately adjusting only the (a) recycle ratio. The effluent nitrate concentration (N_{ne}) will increase as the influent TKN/COD ratio increases, but A_c is now independent of it. However, at a certain TKN/COD ratio, N_{ne} will be so high that the nitrate input via the underflow (s) recycle fully loads the anoxic reactor to its D_c. In that event, the (a) recycle ratio will be reduced to zero to maintain a low N_{nx}. Any further increase in TKN/COD ratio will result in an increase in N_{nx} which, when recycled to the anaerobic reactor via the (r) recycle, is likely to result in $A_c < A_{cm}$. Hence, although the UCT process has a considerable degree of operational flexibility to cater for different TKN/COD ratios, there is a maximum TKN/COD ratio above which this process cannot establish $A_c > A_{cm}$. If A_c = 10 mg N/ℓ influent flow, then at 14°C, R_s = 25 days, f_{xm} = 0.50 and S_{ti} = 500 mg COD/ℓ with a biodegradable COD fraction of 80 percent, the maximum TKN/COD ratio is 0.15 mg N/mg COD.* This maximum TKN/COD ratio is almost double that which causes failure of the Phoredox process and is a value high enough to encompass the TKN/COD ratios of most of the municipal wastewaters.

Success or failure of phosphorus removal - nitrification-denitrification processes can be summarized in a success diagram similar to that for the Bardenpho process for nitrification-complete denitrification (Fig. 8). By plotting the maximum allowable TKN/COD ratio versus μ_{nm20} at a given temperature, the boundaries of the success area of the different processes for achieving excess biological phosphorus removal are defined (Fig. 9).

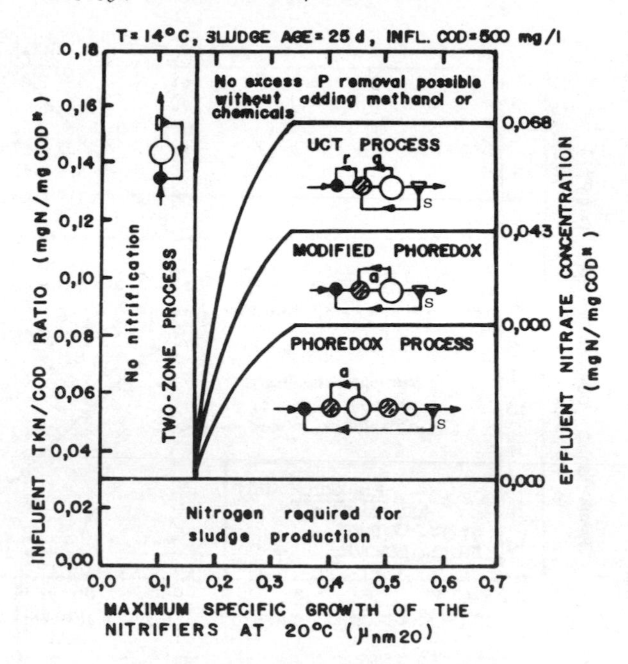

Fig. 9. Diagram showing combination of sewage characteristics with which the different activated sludge process configurations can achieve excess phosphorus removal at 14°C.

EXPERIMENTAL INVESTIGATIONS

iebritz and Marais (1980) investigated the effects of the influent TKN/COD ratio n anaerobic capacity, biological excess phosphorus removal and effluent nitrate oncentration. They operated a Modified Phoredox process and a UCT process (Fig. 3 nd 4), both having the same volumetric dimensions for the anaerobic, anoxic and erobic reactors, with a total unaerated volume fraction of 0.33 and sludge age 0 days under constant flow and load conditions, (Table 2). Over a period of about year, raw sewage influent TKN/COD ratios ranging from 0.065 to 0.105 mg N/mg COD* ere tested.

he unaerated *volume* fractions of the processes were selected at 0.33 rather than he more usual value of 0.40 - 0.50. This was done deliberately because at lower naerated volume fractions (or f_{xt}), D_c of the unaerated reactors is reduced, re-ulting in higher concentrations of nitrate in the processes and effluent which re-uce the possibility of establishing $A_c > A_{cm}$ for excess biological phosphorus re-oval. Consequently, the configurations selected were considered to provide a evere test on the phosphorus removal propensity of the two processes and therefore ore likely to show up differences in behaviour.

ffect of Anaerobic Capacity on Phosphorus Removal

 Fig. 10 the average data are shown plotted: (i) system phosphorus release in the aerobic reactor, (ii) system phosphorus uptake and (iii) system P removal versus aerobic capacity for both processes. Evidently, irrespective of the process, when > 9 to 10 mg N/ℓ influent flow, a qualitative change in the behaviour took place: osphorus removal was obtained. When A_c < 9 mg N/ℓ, but greater than zero (i.e., aerobic conditions in the reactor), excess phosphorus removal was still observed,

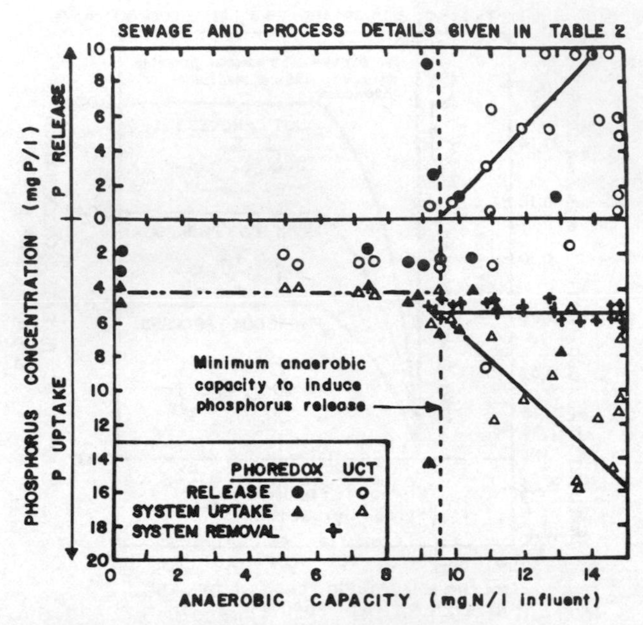

Fig.10. Phosphorus release in the anaerobic reactor, process phosphorus uptake and process phosphorus removal versus anaerobic capacity observed in the Modified Phoredox and UCT processes.

TABLE 2. Experimental Units' Sewage and Process Parameters utilized by Siebritz and Marais (1980).

1. Sewage Characteristics		2. Process Parameters		
		Process	Phoredox (Fig. 3)	UCT (Fig.4)
Temperature °C	19°			
Total COD (mg COD/ℓ)	500	Sludge age (d)	20	20
Total TKN (mg N/ℓ)	33-55	Mixed liquor pH	~7.3	~7.4
TKN/COD ratio	0.065-0.110	Reactor vol.,vol.frac. and mass frac. -		
Sewage Type	raw	Anaerobic (ℓ; %; %)	2;11;11	2;11;6
Unbiodegradable particulate COD fraction (f_{up})	~0.13	Anoxic (ℓ; %; %)	2;22;22	4;22;24
		Aerobic (ℓ; %; %)	12;67;67	12;67;70
Unbiodegradable soluble COD fraction (f_{us})	~0.07	Unaerated (ℓ; %; %)	6;33;33	6;33;30
		Recycle ratios -		
Influent flow constant (ℓ/d)	35	Underflow (s)	1:1	1:1
		Mixed liquor (a)	2:1	2:1
		Mixed liquor (r)	-	1:1

but at a lower magnitude. Excess phosphorus removal at low anaerobic capacities has been observed a number of times in other investigations (Hoffmann and Marais, 1977) and cannot be explained at present. However, excess removal under these circumstances often showed instability and consequently cannot be depended upon. From Fig. 10, once the phosphorus release is obtained, the removal is constant, i.e.,the extent of phosphorus release does not appear to affect the extent of the removal, and the extent of excess biological phosphorus removal appears to be limited. In terms of the phosphorus removal equation (Eq. 8), the processes achieved a coefficient of excess phosphorus removal (α) of 0.12 mg P/mg VASS if the phosphorus content of the inert sludge (f_p) is taken to be 0.015 mg P/mg VSS.†

Effect of Influent TKN/COD Ratio on Anaerobic Capacity and Effluent Nitrate Concentration

In Figs. 11 and 12, the average data are plotted: anaerobic capacity and effluent nitrate concentration versus influent TKN/COD ratio for the Modified Phoredox and UCT processes, respectively. Also shown are the theoretical predictions calculated from Eqs. (1 - 9) above. To incorporate the effect of dissolved oxygen in the recycles, the calculations were repeated including 2 mg / ℓ dissolved oxygen in the mixed liquor (a) and underflow (s) recycles and also plotted in Figs. 11 and 12. The theoretical anaerobic capacity for the Modified Phoredox process (Fig. 11a) compares well with the values calculated from the experimental data,# whereas that for the UCT process (Fig. 12a) appears to be over-predicted. In both processes the effluent nitrate concentration tends to be slightly under-predicted (Figs. 11b and 12b). The difference between the theoretical and experimental data can be reconciled to a considerable degree by reducing the second denitrification rate (K_2 in Eq. 7a), but was not done due to insufficient evidence for accepting a lower value. A lower value is not unacceptable as the source data (Stern and Marais, 1974; Marsden and Marais, 1977 and Wilson and Marais, 1976) from which Van Haandel, Ekama and Marais (1980) recalculated the second denitrification rate in the primary anoxic reactor (K_2) show a fairly wide spread.

Effect of Effluent Nitrate Concentration on Anaerobic Capacity

To assess the relationship between the anaerobic capacity and effluent nitrate concentration in the Modified Phoredox and UCT processes, the average data are plotted anaerobic capacity versus effluent nitrate concentration in Fig. 13. Also shown are the theoretical predictions for the processes calculated from Eqs. (1 - 9) above. Figure 13 clearly shows the UCT process is less sensitive to effluent nitrate concentration than the Modified Phoredox process and, under the same conditions, the UCT process can establish the minimum anaerobic capacity for excess phosphorus removal at higher effluent nitrate concentrations than the Modified Phoredox process.

DISCUSSION

A comparison of the results of the Modified Phoredox and UCT processes (Figs. 11, 12 and 13) shows that with the same unaerated *volume* fraction, it is possible to establish the required minimum anaerobic capacity (to induce excess phosphorus removal) with a higher influent TKN/COD ratio (and, hence, a higher effluent nitrate concentration) in the UCT process than in the Modified Phoredox process. In general, the UCT process provides greater surety in obtaining excess biological phosphorus

---

The value of 0.015 mg P/mg VSS for the phosphorus content of the inert fractions as obtained after an analysis of the effect of variation of inert material from atch to batch of sewage. This value appears to lead to greater consistency in redicting phosphorus removal. With this value for f_p, the α value of a process ot exhibiting excess phosphorus removal is approximately 0.05 mg P/mg VASS.

It should be noted that the "experimental" anaerobic capacity requires the use of q. (7a) to estimate the denitrification capacity of the anaerobic reactor.

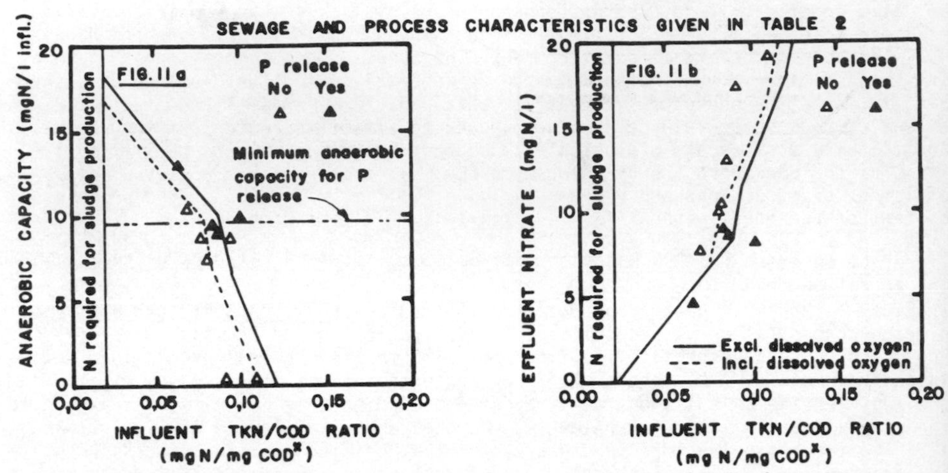

Fig. 11. Experimental and theoretical anaerobic capacity and effluent nitrate concentration versus influent TKN/COD ratio for Modified Phoredox Process.

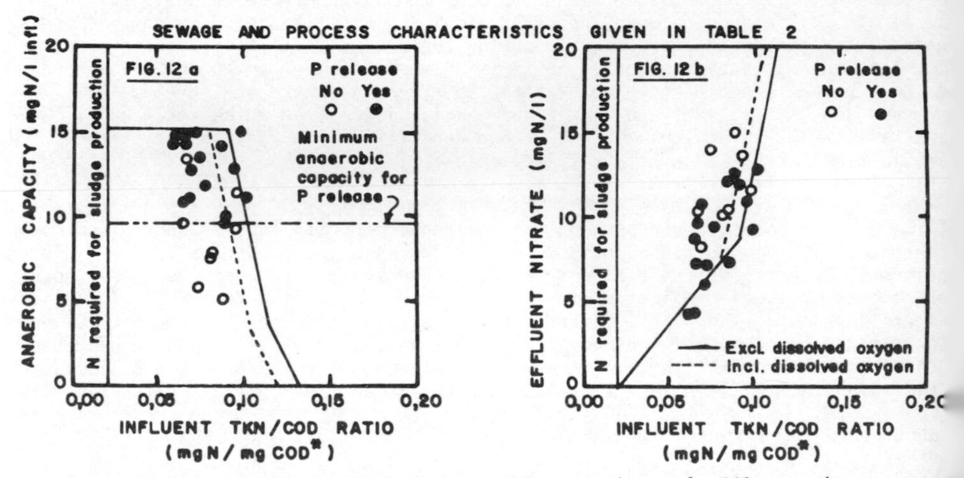

Fig. 12. Experimental and theoretical anaerobic capacity and effluent nitrate concentration versus influent TKN/COD ratio for the UCT Process.

removal than the Phoredox processes. The limitations set on the two Phoredox processes by the influent TKN/COD ratio considerably reduce their application in design in situations where the influent TKN/COD ratio is not well established prior to process design; if the actual influent TKN/COD ratio is higher than provided for in design, then, due to inflexibility of these processes, they are unlikely t achieve excess phosphorus removal. In contrast, a design based on the UCT proces can accommodate higher actual influent TKN/COD ratios than it was designed for by reducing the mixed liquor (a) recycle to the anoxic reactor. This corrective procedure can be applied up to the point where the (a) recycle is reduced to zero, after which the UCT process also becomes inflexible. With regard to the Phoredox Process, in general the influent TKN/COD ratio of many municipal wastewaters is such that the Phoredox process is unlikely to be successful;it should be consider

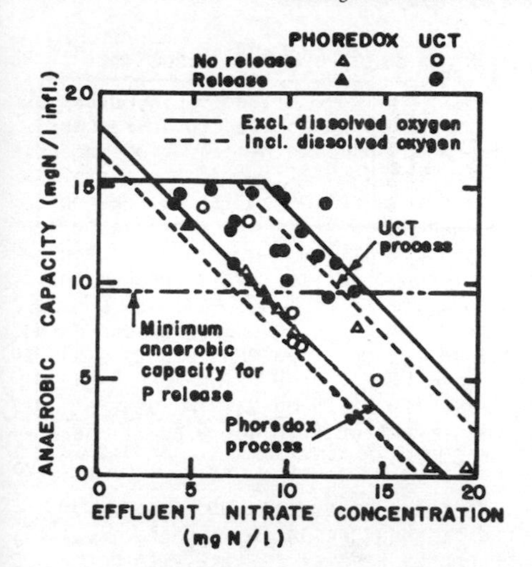

Fig. 13. Experimental and theoretical anaerobic capacity versus effluent nitrate concentration for the Modified Phoredox and UCT processes.

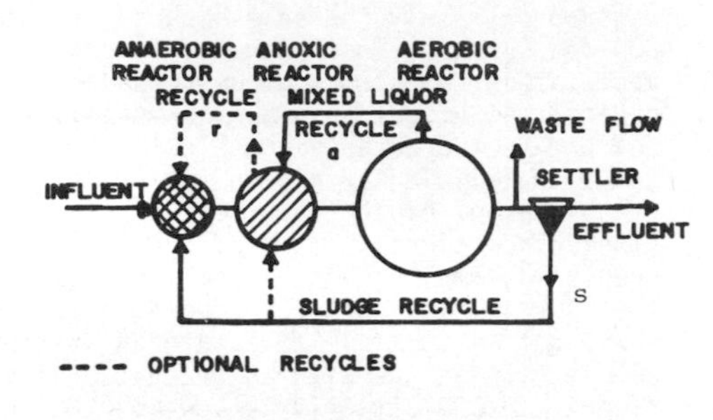

Fig.14. Recommended design configuration to incorporate both the Modified Phoredox and UCT processes.

only under circumstances where the influent TKN/COD ratio is well defined and low. With regard to the Modified Phoredox process, this process can accommodate higher influent TKN/COD ratios, but even in cases where this process appears acceptable, it is recommended to include the relatively minor additions which allow operation as a UCT process, Fig. 14.

The anaerobic capacity was specifically formulated as an intensive parameter, i.e., as a concentration, mg N/ℓ influent flow, instead of an extensive parameter, i.e., as a mass, mg N/d. This distinction is very important: when the processes reported above were fed the influent sewage at half the COD concentration but at twice the influent flow, (i.e. the same mass of COD per day), the masses in the processes and the mass changes were approximately the same, but the concentrations of the soluble constituents were halved. Consequently, the anaerobic capacity also was halved (to about 6 mg N/ℓ influent flow) and no phosphorus release in the anaerobic reactor or excess phosphorus removal was observed. Consequently, the anaerobic capacity must be analogous to an intensive parameter. It is probably related in some fashion to the redox potential, although the relationship is not likely to be straightforward. From the experimental data in Fig. 10, a further characteristic of the anaerobic capacity is evident. In the Modified Phoredox process, the biodegradable COD discharged to the anaerobic reactor is principally that in the influent flow - the underflow recycle is unlikely to contain a significant concentration of biodegradable COD. In contrast, in the UCT process, the mass of biodegradable COD discharged to the anaerobic reactor is that in the influent flow as well as that in the mixed liquor (r) recycle from the anoxic reactor - a significant concentration of biodegradable COD is likely to be present in the anoxic reactor because only limited oxidation has taken place by denitrification. The difference in the masses of biodegradable COD discharged to the anaerobic reactors of the two processes should have resulted in the UCT process achieving phosphorus release at a lower anaerobic capacity (as formulated by Eq. 9) than the Modified Phoredox process. However, Fig. 10 demonstrates that this did not happen, i.e., both the processes commenced to release phosphorus at the same anaerobic capacity of 10 mg N/ℓ influent. A possible explanation is that the anaerobic capacity is principally established by the readily biodegradable COD fraction in the influent. In both processes, the mass of readily biodegradable COD discharged to the anaerobic reactor was approximately the same because in the UCT process all the readily biodegradable COD was consumed in the anoxic reactor by denitrification. It would therefore appear that the slowly biodegradable COD fraction does not establish an anaerobic capacity as effectively as the rapidly biodegradable COD fraction. This is perhaps not unexpected, because the slowly biodegradable COD fraction requires adsorption and extracellular breakdown prior to transfer through the cell wall, so that the establishment of the anaerobic capacity is perhaps restricted by the rates of the kinetic reactions. If these conclusions on the nature of the anaerobic capacity are valid, implications in design of municipal wastewater treatment plants would be that for low influent COD concentrations (< 300 mg COD/ℓ, i.e. ≃ 150 mg BOD/ℓ), it would not be possible to obtain an anaerobic capacity of 10 mg N/ℓ influent at 14°C, 20 days sludge age with an unaerated sludge mass fraction of 0.40, even if the influent TKN/COD ratio is as low as 0.08 mg N/mg COD.*

From the experimental investigation, it is evident that the anaerobic capacity, as formulated by Eq. (9), can be utilized to predict when phosphorus release is likely to occur and hence when excess biological phosphorus removal is likely to be achieved. However, experimentally, the magnitude of the anaerobic capacity to achieve phosphorus release has been measured only at an influent COD concentration of 500 mg COD/ℓ, 19°C and 20 days sludge age. It will be necessary to repeat the investigation at different influent COD concentrations, temperatures and sludge ages before it can serve as a general design parameter.

CONCLUSIONS

1. Phosphorus release appears to be a prerequisite for consistent excess biological phosphorus removal.
2. Phosphorus release is probably induced if the redox potential is reduced to below some critical value. In the absence of a reliable means for measuring the redox potential in activated sludge processes, a suitable parameter, the anaerobic capacity, is proposed to predict when phosphorus release is likely to take place. The anaerobic capacity is defined under constant flow and load conditions as the difference between the nitrate concentration (per litre influent flow) that an unaerated reactor can remove and that which it actually removes.
3. From experimental evidence, for a sludge age of 20 days, temperature 19°C and total influent COD of 500 mg COD/ℓ, the establishment of an anaerobic capacity of 10 mg N/ℓ influent, or greater, appears to induce phosphorus release.
4. The extent of excess biological phosphorus uptake is limited. A formulation linking excess uptake with the process parameters is proposed which shows that under excess uptake, the phosphorus content in the active volatile mass is 0.12 mg P/mg VASS while that in the influent inert material and endogenous residue generated is 0.015 mg P/mg VSS.
5. In general, the establishment of the required anaerobic capacity becomes increasingly difficult as (i) the influent COD concentration decreases and/or (ii) the influent TKN/COD ratio increases.
6. The Phoredox process can establish the minimum anaerobic capacity only with influents that have relatively low TKN/COD ratios, of 0.08 mg N/mg COD* or less at 14°C; the Modified Phoredox (without secondary anoxic zone) with ratios less than 0.11 mg N/mg COD* at 14°C.
7. Both Phoredox processes have little operational flexibility to achieve the required anaerobic capacity if the TKN/COD ratio is higher than that for

which the process was designed.

8. A new process, called the UCT process is proposed which "protects" the anaerobic capacity in the anaerobic reactor from the effluent nitrate concentration. This process can be designed to give excess biological phosphorus removal for influent TKN/COD ratios up to about 0.15 mg N/mg COD*. If designed for a particular influent TKN/COD ratio and the actual influent TKN/COD ratio is higher (but less than 0.15 mg N/mg COD*), the process has a large measure of operational flexibility to re-establish excess biological phosphorus removal.

9. The influent TKN/COD ratio of many municipal wastewaters is such that the Phoredox process has limited application. The Modified Phoredox process has a wider range of application, but even in situations where it is the appropriate process, provision should always be made at the design stage to allow operation also as a UCT process.

ACKNOWLEDGEMENTS

This research was carried out under contract with the Water Research Commission of South Africa. The authors wish to thank the Commission for permission to publish this paper.

NOMENCLATURE

a = mixed liquor recycle ratio from aerobic to anoxic reactor.

A_c = anaerobic capacity (mg N/ℓ influent). Additional subscript m refers to the minimum value to induce phosphorus release.

b_n# = nitrifier endogenous mass loss rate (/d).

b_h# = heterotrophic organism endogenous mass loss rate (/d).

D_c* = denitrification capacity (mg N/ℓ). Additional subscripts a, p and s refer respectively to anaerobic, primary anoxic and secondary anoxic reactors.

f = endogenous residue fraction = 0.20 (mg VSS/mg VASS).

f_{bs} = rapidly biodegradable COD fraction of the influent biodegradable COD = 0.24 mg COD/mg COD for average municipal wastewaters.

f_n = fraction of nitrogen in the sludge mass = 0.10 mg N/mg VSS.

f_p = fraction of phosphorus in the sludge mass = 0.015 mg P/mg VSS.

f_x = unaerated sludge mass fraction. Additional subscripts a, p, s, m and t refer respectively to anaerobic, primary anoxic, secondary anoxic, maximum and total fractions.

f_{up} = particulate unbiodegradable COD fraction of the influent (mg COD/mg COD).

f_{us} = soluble unbiodegradable COD fraction of the influent (mg COD/mg COD).

K_n# = Monod half saturation nitrification coefficient (mg N/ℓ).

K_2# = second denitrification rate in the primary anoxic reactor (mg No_3-N/mg VASS/d).

K_3# = denitrification rate in the secondary anoxic reactor (mg NO_3-N/mg VASS/d)

N = general symbol for nitrogen (mg N/ℓ). Subscripts a, o, n and t refer respectively to ammonia, organic nitrogen, nitrate and TKN concentrations. Additional subscript i, e or x refers respectively to influent and effluent and anoxic reactor concentrations.

N_c = nitrification capacity (mg N/ℓ influent).

N_s = nitrogen required for sludge production (mg N/ℓ influent).

P = COD/VSS ratio = 1.48 mg COD/mg VSS.

R_s = sludge age (d).

r = mixed liquor recycle rate from anoxic to anaerobic reactor.

S_{ti} = total influent COD concentration (mg COD/ℓ).

s = underflow recycle ratio.

Y_h = heterotrophic yield coefficient.

α = coefficient of excess phosphorus removal, i.e. fraction of phosphorus in active mass (mg P/mg VASS).

$\mu_{nm}^{\#}$ = maximum specific growth rate of the nitrifiers.

REFERENCES

Barnard, J.L. (1974) Cut P and N without chemicals. Water and Wastes Eng., 11, 7, 33-36.

Barnard, J.L. (1976) A review of biological phosphorus removal in the activated sludge process. Water S.A., 2, 3, 136-144.

Downing, A.L., H.A. Painter and G. Knowles (1964). Nitrification in the activated sludge process. J. Proc. Inst. Sew. Purif., 64, 2, 130-158.

Ekama, G.A. and G.v.R. Marais (1978). The dynamic behaviour of the activated sludge process. Res. Recpt. No. W. 27, Dept. of Civil Eng., Univ. of Cape Town.

Ekama, G.A., A.C. van Haandel and G.v.R. Marais (1979). The present status of research on nitrogen removal: a model for the modified activated sludge process. Presented at the Symposium on Nutrient Removal, S.A. Branch of I.W.P.C., Pretoria, May, 1979. Proceedings to be published by the Water Research Commission.

Hoffmann, R.J. and G.v.R. Marais (1977). Phosphorus removal in the modified activated sludge process. Res. Rept. No. W. 22, Dept. of Civil Eng., Univ. of Cape Town.

Marais, G.v.R. and G.A. Ekama (1976). The activated sludge process Part 1 - steady state behaviour. Water S.A., 2, 4, 163-200.

Marsden, M.G. and G.v.R. Marais (1977). The role of the primary anoxic reactor in denitrification and biological phosphorus removal. Res. Rept. No. W. 19, Dept. of Civil Eng., Univ. of Cape Town.

Margin, K.A.C. and G.v.R. Marais (1975). Kinetics of enhanced phosphorus removal in the activated sludge process. Res. Rept. No. W. 14, Dept. of Civil Eng., Univ. of Cape Town.

McLaren, A.R. and R.J. Wood (1976). Effective phosphorus removal from sewage by biological means. Water S.A., 2, 1, 47-50.

Nicholls, H.A. (1975). Full scale experimentation on the new Johannesburg extended aeration plant. Water S.A., 1, 3, 121-132.

Osborn, D.W., A.R. Pitman and S.L. Venter (1979). Design and operation experience with nutrient removing activated sludge plants in Johannesburg. Presented at the Symposium on Nutrient Removal, S.A. Branch of I.W.P.C., Pretoria, May 1979. Proceedings to be published by the Water Research Commission of S.A.

Temperature dependency given in Table 1. Additional subscript T or 20 refers to value at T or 20°C.

Rabinowitz, B. and G.v.R. Marais (1980). Chemical and biological phosphorus removal in the activated sludge process. Res. Rept. No. W. 32, Dept. of Civil Eng., Univ. of Cape Town.
Siebritz, I. and G.v.R. Marais (1980). Chemical and biological phosphorus removal in the activated sludge process. Res. Rept. No. W. 34, Dept. of Civil Eng., Univ. of Cape Town.
Simpkins, M.J. and R.A. McLaren (1978). Consistent biological phosphate and nitrate removal in an activated sludge plant. Proc. Wat. Tech., 10, 5/6, 433-442.
Stern, L.B. and G.v.R. Marais (1974). Sewage as the electron donor in biological denitrification. Res. Rept. No. W. 7, Dept. of Civil Eng., Univ. of Cape Town.
Van Haandel, A.C., G.A. Ekama and G.v.R. Marais (1981). The activated sludge process Part 3 - application of the general kinetic model to single sludge denitrification. To be published in Water Research.
Wilson, D.E. and G.v.R. Marais (1976). Adsorption phase in biological denitrification. Res. Rept. No. W. 11, Dept. of Civil Eng., Univ. of Cape Town.

21
PRINCIPLES AND APPLICATIONS OF THE ROTATING BIOLOGICAL CONTACTOR

D. J. L. Forgie

Dept. of Civil Engineering, University of Saskatchewan, Saskatoon, Saskatchwan, Canada

ABSTRACT

The Rotating Biological Contactor (RBC) is a relatively new and promising method of biologically treating a variety of wastewater types.

To acquaint the reader with RBC's the basic process principles and history of the RBC are reviewed. RBC operation and process design considerations, e.g. hydraulic loading, organic loading, media surface area, effects of temperature, and mathematical RBC models, are discussed in greater detail.

Comparisons between RBC's and other, more conventional, biological wastewater treatment methods are made from applications, operation and energy use points of view. An overview of current RBC applications including removal efficiencies, operational problems and cost effectiveness is presented. Advantages and disadvantages of using RBC's are discussed and future RBC application potential and research directions outlined.

KEYWORDS

Secondary wastewater treatment; biological; fixed-film, rotating biological contactor; RBC; principles; applications; cost effectiveness.

BASIC PROCESS PRINCIPLES

The rotating biological contactor (RBC) is a secondary, fixed-film, biological wastewater treatment method.

It consists of some type of media (nominally closely spaced discs) fixed on a shaft and partially submerged (e.g. 30 to 45%) in the wastewater to be treated. Subsequent rotation of this assembly leads to the development of a natural, microbiological film or "biomass" on the media.

This rotating biomass receives two-phase contact with the wastewater: the first while submerged in the bulk of the wastewater and the second with the liquid film which adheres to the biomass as it is rotated out of bulk of the wastewater, into the atmosphere. This vigorous and repetitive exposure to both the wastewater (sub-

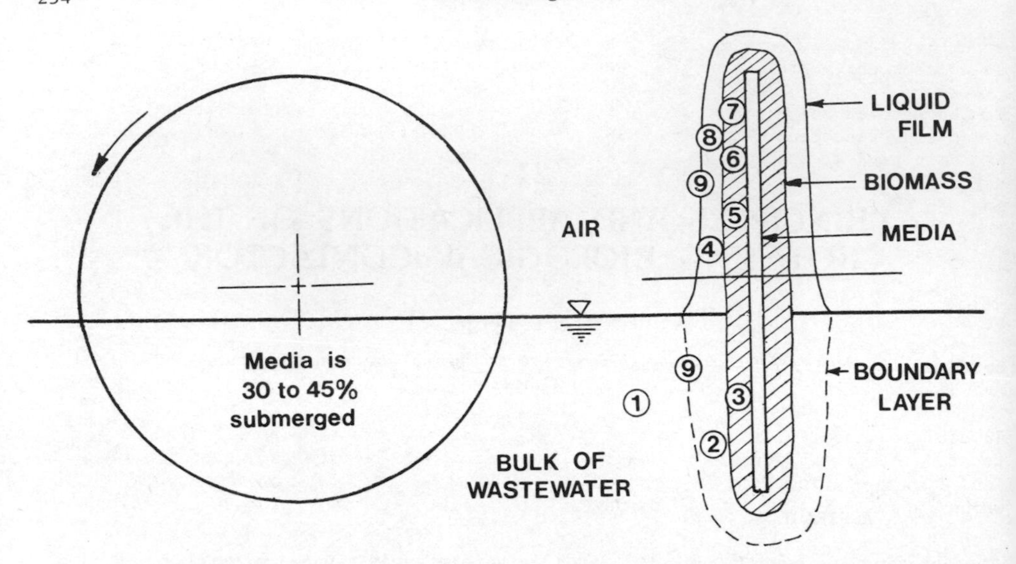

Fig. 1. Basic rotating biological contactor (RBC) principles with possible limiting mechanisms: 1. bulk fluid mixing; 2. substrate diffusion in boundary layer to biomass; 3. substrate diffusion in biomass layer to reactive site; 4. solution of oxygen and diffusion through liquid film layer; 5. diffusion of oxygen through biomass to reactive site; 6. microbial reaction rates; 7. diffusion of metabolic by-products outward in biomass layer; 8. transport of metabolic by-products to bulk liquid and the atmosphere (9). (after Friedman and co-workers,1979).

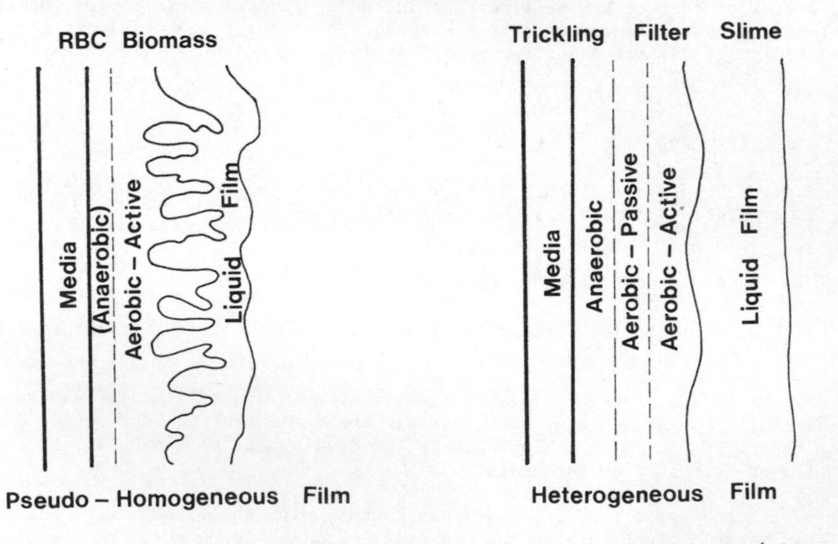

Fig.2. Comparison of RBC and trickling filter fixed-film characteristics (after Grieves,1972; Antonie,1976)

strate) and oxygen stimulates the aerobic biomass which removes nutrients, e.g. soluble organic carbon, for cell metabolism and growth.

The biomass on the media may be 1 to 4 mm thick, which would be the equivalent of a mixed liquor suspended solids (MLSS) concentration of 10,000 to 20,000 mg/L if it was dispersed growth (Antonie, 1976). As a result, the required detention times are much shorter than an activated sludge system for the same degree of treatment, e.g. 60 to 90 minutes versus 6 hours.

Compared to the smooth, gelatinous biological slime of trickling filters (percolating filters) the RBC biomass usually has a "shaggy" appearance. The combination of hydraulic drag in the bulk of the wastewater and gravity while in the atmosphere tends to build the microscopic, filamentous bacteria into braided, elongated macroscopic filaments and hence, a shaggy appearance. As a result, the RBC biomass has more than just the nominal surface area available and can be more deeply aerated and "active" when compared to a trickling filter slime.

In addition to providing a means of controlling aeration and wastewater contact, rotation of the RBC provides positive shearing of excess biomass due to the turbulent flow conditions around the biomass. This also helps to provide a biomass which is primarily active and aerobic by preventing anaerobic biomass conditions.

History

The term rotating biological contactor is a common generic term used to describe the wastewater treatment process which is also known as the rotating biological disc (or disk), rotating biological reactor, rotating biological surface and rotating biological filter. Commercial names include AeroSurf, BioDisc, BioDrum, BioRotor, BioShaft, BioSpiral, BioSurf, ENVIRODISC, RBS, ROTORDISK and SurfAct, among others.

Doman (1929) documented one of the first uses of the RBC principle, involving an experimental unit comprised of 14, 35.6 (14 inch) diameter galvanized steel discs rotated at 0.5 to 1.0 RPM. During the same time period investigators in Germany experimented with rotating wooden surfaces which (unfortunately) were subject to deterioration (Dallaire, 1979). Both received little subsequent attention.

Research recommenced in Europe in the 1950's with work conducted by Hans Hartmann and Franz Pöpel at the Technical University of Stuttgart. Hartmann (1960) described the operation of two experimental immersion-drip-filter-disc plants ("tauchtropfkörpern") and outlined their performance characteristics and economics for use in wastewater treatment. Pöpel (1964) and Hartmann (1964) described design formulae based on BOD load, disc area and rotational speed.

In 1959 J.C. Stengelin, Tuttingen, West Germany began manufacture and marketing of 2 m and 3 m diameter expanded polystyrene disk RBC systems. By 1972 (Beak, 1973) there were almost 1000 installations in Europe treating domestic, industrial and mixed wastewaters with sizes ranging from a single residence up to a 10,000 person equivalent, with most serving less than 1000 people (Antonie, 1976; Dallaire,1979).

In the U.S. Allis-Chalmers began independent development of the RBC system in the mid-1960's (Antonie, 1976). After learning of the European work they became a Stengelin licencee under the trade name BIO-DISC. After some limited marketing success, they sold their RBC technology to the Autotrol Corporation in 1970 (Dallaire, 1979).

To this point, the state of the RBC art made use of the Stengelin-type expanded polystyrene disc media which, because of low specific area (52.5 m^2/m^3(16 ft^2/ft^3)) and relatively high capital cost using 3.05 m (10 ft.) diameter, 5.2 m (17 ft.) long, 1950 m^2(21,000 ft^2) shafts, RBC's were not very cost competitive with other treatment alternatives, e.g. activated sludge. As a result very few RBC's had been sold in the U.S. by 1972 (Autotrol, 1979).

By 1972 Autotrol had developed a corrugated polyethylene media which increased the specific area to 121.4 m^2/m^3(37 ft^2/ft^3). The standard unit ("shaft") size was als increased to 3.6 m (11.9 ft.) diameter, 7.6 m (25 ft.) long and 9295 m^2(100,000 ft^2 in order to improve RBC cost effectiveness (Antonie 1976; Dallaire, 1979).

In North America RBC's have gained increased acceptance and use since 1972, going from treating 2250 m^3/day (0.6 US MGD) in 1972 to more than 1.8 million m^3/day (480 US MGD) by early 1980, with more installations planned (Autotrol, 1979; Hormel, 1979; CMS, 1980a). As a further sign of the acceptability of the process several new firms have entered the market with their own RBC media and process designs.

Such healthy competition has led to further process developments such as higher density media (e.g. 13,950 m^2(150,000 ft^2) per 7.6 m (25 ft.) shaft) for use in nitrification applications (where the biomass is thinner and less likely to clog the media) and an air drive system which replaced the typical mechanical drive. The air drive also provides additional oxygen and turbulence to the system which responds by producing a thinner, more efficient, aerobic biomass which may then per mit the use of high density media for soluble biochemical oxygen demand (BOD) removal (Autotrol, 1978; McCann and Sullivan, 1980).

There are currently approximately 1000 RBC installations in Europe, 300 in the U.S. 200 in Canada, and 785 in Japan in addition to smaller numbers elsewhere in the world. (Autotrol, 1979; Hormel, 1979; CMS, 1980a; Hynek and Iemura, 1980). A recent three day research symposia (Seven Springs, Pennsylvania, Feb., 1980) devote completely to the RBC is a further sign that this "new" 20 to 50 year old process has "arrived" and is here to stay.

Process Operation

Being a secondary treatment method, RBC's should be preceded by some type of pretreatment to remove large objects, and floatable and settleable materials. In lar scale installations this would be accomplished by conventional primary treatment (grit removal and primary sedimentation) whereas smaller installations may use hydrasieves or septic tanks as pretreatment (Antonie, 1976; Autotrol, 1978). Foll ing primary treatment the wastewater flows into the RBC zone for secondary treatme

The RBC zone is usually physically divided into at least four stages, each of whic acts as a completely mixed fixed-film reactor and develops a specific microbiological ecosystem which is suited to the particular organic and dissolved oxygen conce trations in that stage. The initial stages, which receive the highest organic con centrations, develop a mixed culture of filamentous and non-filamentous heterotrop bacteria, primarily involved in the conversion of soluble organic carbon to carbon dioxide, water and new cell material. In subsequent stages the biomass type chang with relatively higher trophic level microorganisms (e.g.bacteria, protozoa, rotifers) developing in response to the lower organic carbon levels and higher dissolv oxygen levels present. Once the BOD is reduced to 30 mg/L (total) (15 mg/L (solub nitrifying bacteria can out compete heterotrophic bacteria and grow quickly enough to dominate the latter RBC stages, converting ammonia (NH_4^+) to nitrate (NO_3^-) (Antonie, 1976; Mueller and co-workers, 1980).

Denitrification may also be achieved by using RBC's which have been completely sub

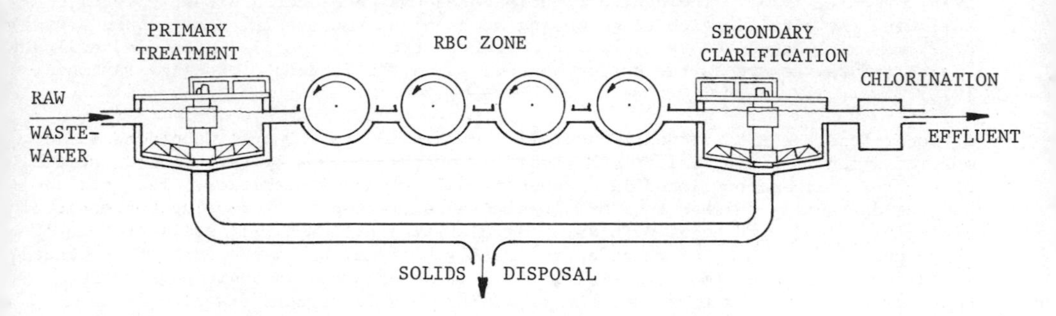

Fig. 3. Typical RBC wastewater treatment system

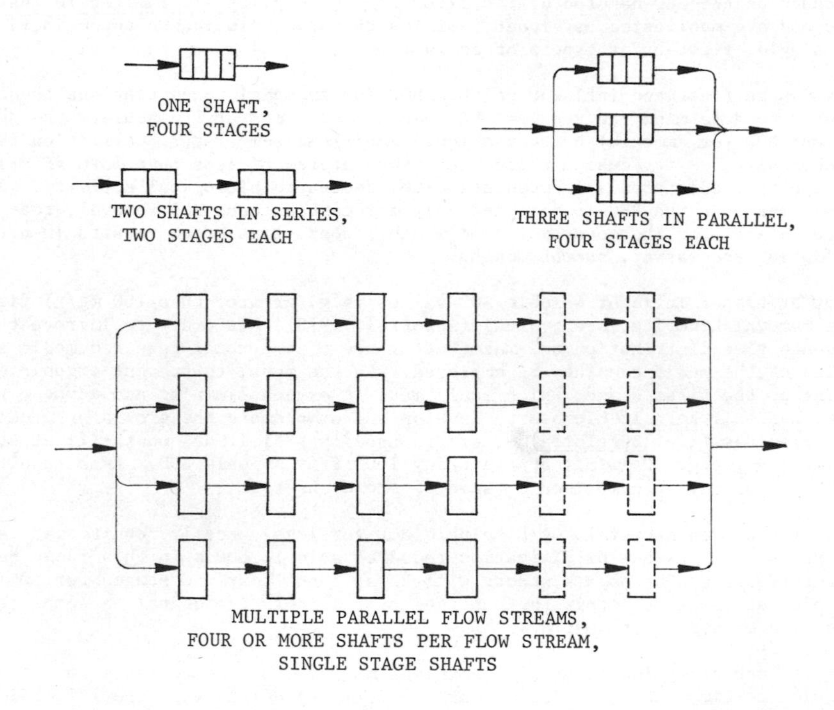

Fig. 4. RBC staging variations

merged and have developed a biofilm of the necessary bacteria (Davies and Pretorius, 1975; Antonie, 1976). The denitrification stage may be located either directly following the nitrification stage (using methanol as the supplemental carbon source) or it may be located before the aerobic stages, treating recycled nitrified effluent (using the raw sewage as the carbon source, as in the Bardenpho process) (Antonie, 1976).

Effluent from the RBC zone will generally contain 100 to 150 mg/L suspended solids, most of which is biological growth sheared from the biomass by turbulence. Removal of the solids is accomplished in a conventional secondary clarifier. The settled RBC solids usually thicken to 3 or 4% and can be treated further using conventional biological sludge treatment methods. The clarified effluent will still contain 10 to 20 mg/L (total) BOD, of which approximately half may be in the form of suspended solids that could be removed by filtration, if necessary. In addition, carbon adsorption can be used to remove the remaining soluble organic fraction if an extremely high quality effluent is required.

Process Design

Design of the RBC has been commonly accomplished by using empirical relationships between influent soluble BOD (or ammonia) concentrations, the hydraulic loading (m^3/m^2/day or gal/ft^2/day) and the desired effluent quality. These relationships are based on the RBC manufacturer's experience with pilot and full scale installations and are manifested as either families of curves similar to those in Fig. 5 or as a single, first-order type plot as in Fig. 6.

Using design flows and influent soluble BOD (or ammonia) concentrations these curves are used to determine the required RBC media area necessary to achieve the desired effluent BOD (or ammonia) concentration. Empirical temperature correction factors are commonly used for design wastewater temperatures of less than 13°C (55°F) to increase the media area in compensation for decreased biological activity. As a bonus, even when the RBC is designed only for carbonaceous BOD removal, the increased area for the governing cold weather conditions often results in nitrification during the warmer, summer months.

Except at higher influent soluble BOD values (e.g. greater than 100 mg/L) first order removal kinetics, are typical (Autotrol, 1978). In order to improve the residence time distribution and make better use of the total required media area, staging of the media can then be employed. Furthermore, to prevent organic overloading of the first stage, which could make it oxygen limiting and cause a very thick anaerobic-aerobic biomass to develop and/or promote the growth of undesirable microorganisms (e.g. *Beggiatoa*), there is usually a limit set on the first stage loading rate, e.g. 20 gm/m^2·day (4 lb/day/1000 ft^2) soluble BOD. As a result the first "stage" may be comprised of several media shafts.

Effluent requirements of 15 mg/L soluble BOD (or less) usually requires at least four RBC stages. Since nitrification requires soluble BOD's in this range before the autotrophs can propogate, there will be at least four RBC stages for BOD reduction plus at least one stage (but not necessarily only one shaft) to complete the conversion of ammonia to nitrate.

Although each stage functions as a complete mix reactor and there is no difference in residence time when the flow is parallel or perpendicular to the RBC media shaft most large scale municipal RBC plants are designed with the flow perpendicular to the shaft for easier shaft maintenance. Cast-in-place, wooden-baffled, rectangular concrete tanks are most often employed although hemi-hexagonal concrete tanks are also used. For wastewater strengths up to 300 mg/L BOD a tank volume to media area

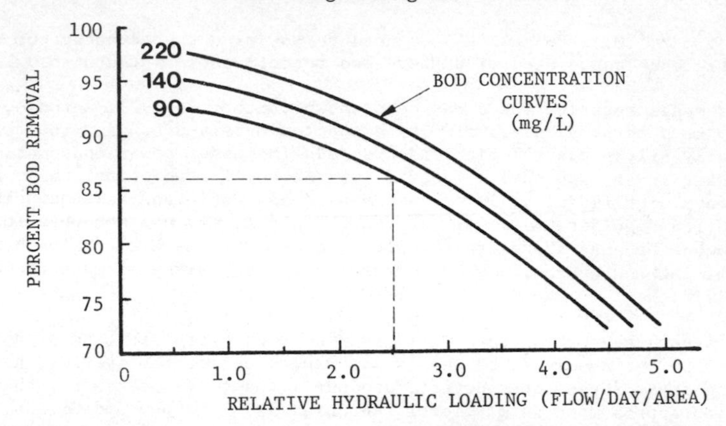

Fig. 5. Typical empirical RBC design curves for carbon removal

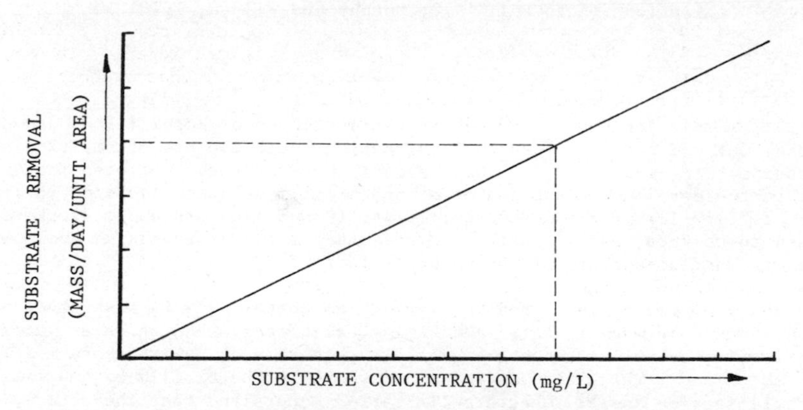

Fig. 6. A single line, first-order type RBC design curve

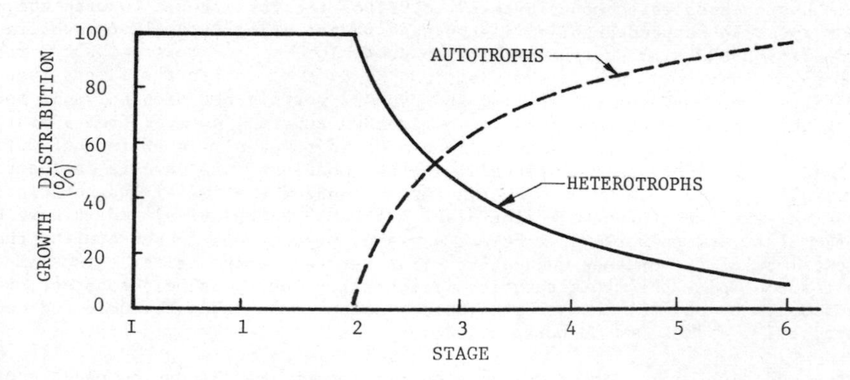

Fig. 7. Typical growth distribution for heterotrophic and autotrophic bacteria in an RBC (after Mueller and co-workers, 1980).

ratio of 0.005 m^3/m^2(0.12 US gal/ft^2) is used to maximize the advantage of the increased efficiency associated with increased retention times (Autotrol, 1978).

The full size media rotation speed has been shown (Autotrol, 1971; Antonie, 1976) to be optimum at a perpendicular velocity of approximately 0.3 m/sec (60 ft/min.) for mechanically driven plants: increased speeds increase power consumption considerably (power is an exponential function of the angular velocity) while removal does not improve significantly; lower speeds decrease efficiency without altering power consumption significantly. Air driven RBC plants may use somewhat lower rotational speeds because the extra turbulence caused by the air bubbles helps to provide the mixing and hydraulic shear necessary for a healthy biomass (McCann and Sullivan, 1980).

Although RBC's have been shown to be able to withstand hydraulic (and organic) sur loads because the biomass is fixed and is not washed out of the system, the efflue quality during these surges does deteriorate until the surge passes but then quick recovers to its approximate former level (Antonie, 1970; Murphy and Wilson, 1978). For this reason flow equalization tanks are recommended to maintain consistent effluent quality when the flow fluctuations are large, e.g. for small communities or package installations (Autotrol, 1978; Murphy and Wilson, 1978).

Mathematical Models

Mathematical models for RBC design were first developed by Pöpel (1964) and Hartmann (1964) but did not find frequent application, with the use of empirical relationships being more common. More recently however there has been increased interest in the development and use of mathematical models to simulate RBC kinetic (Grieves, 1972; Williamson and McCarty, 1976a, 1976b; Famularo and co-workers, 197 Filion and co-workers, 1979; Paolini and co-workers, 1979; Zeevalkink and co-worke 1979; Mueller and co-workers, 1980 and others).

Famularo and co-workers (1978) and Mueller and co-workers (1980) have developed models based on both mass transfer and biological kinetics and which are applicab to nitrification as well as soluble organic carbon removal. With these models it has been shown under which circumstances and where in the biofilm oxygen limitati diffusion limitation and/or substrate limitation occurs and that the active, aero layer is only 100 to 200 µm thick. The occurrence of denitrification within the lower, oxygen starved layers of biomass has also been indicated. Furthermore, th biofilm has been shown to contribute 89 to 96% of the total reaction with the mix liquour volatile suspended solids in the tank contributing 4 to 11% removal and t liquid film on the biomass, essentially no removal.

Models allow the designer to describe RBC process performance with only one set o temperature corrected coefficients for a given wastewater (derived from a minimal amount of pilot testing) and as such are useful in the analysis of further full scale and pilot plant studies. In addition the models can be used in design to quickly evaluate various process configurations and/or wastewater characteristics without the expense of further pilot plant studies. Mathematical models have bee used (Famularo and co-workers, 1978) to explain, predict and compensate for the apparent loss of performance when pilot plant derived design criteria is scaled u to full scale while the media rotation peripheral velocity is held constant, a problem noted by several researchers (Murphy and Wilson, 1978; Friedman and co-workers, 1979; Wilson and co-workers,1980).

With the addition of a cost calculating subroutine RBC models can be used to opti mize the design for maximum cost effectiveness.

Table 1 RBC Installation Summary: North America
(After Autotrol,1979;Hormel,1979;CMS,1980a)

Applications	No. of Installations	No. of Shafts	Total Flow	
			m^3/day	US MGD
Municipal	206	2328	1,970,000	519.9
Industrial	38	200	115,000	30.5
Package Plants*	138	151	10,600	2.8

* - not including approximately 120, 2.3 m^3/day (600 US GPD) units

Table 2 Municipal RBC Installation Summary: North America
(After Autotrol,1979;Hormel,1979;CMS,1980a)

Flow Range		No. of Plants	Percent of Total
m^3/day	US MGD		
< 3785	< 1	99	48.1
3785 to 18925	1 to 5	82	39.8
18925 to 37850	5 to 10	17	8.3
37850 to 75700	10 to 20	5**	2.4
> 75700	> 20	3***	1.4
		206	100.0

**- one of which is a 54500 m^3/day(14.4 US MGD)Nitrification plant;Guelph,Ont.
*** - 1. 204,400 m^3/day(54 US MGD), Alexandria, Virginia
2. 140,000 m^3/day(37 US MGD), Peoria,Illinois
3. 90,840 m^3/day(24 US MGD), Orlando,Florida

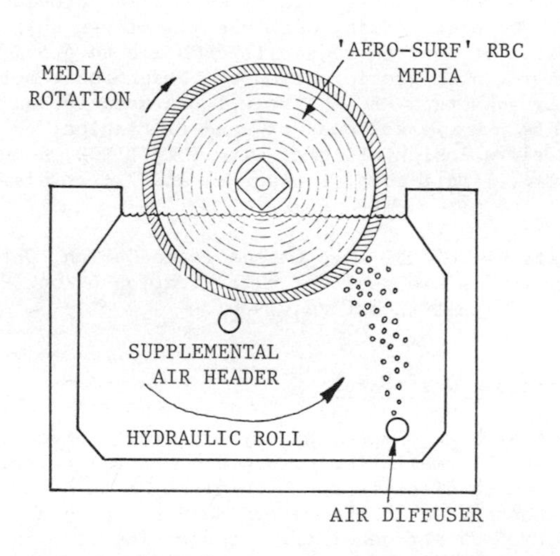

Fig. 8. The combined air-driven RBC ('Aero-Surf') and activated sludge system known as the 'Surfact' process (Autotrol,1978).

APPLICATIONS

Municipal Wastewater Treatment

As may be seen from Table 1 the most common application of the RBC is in the secondary treatment of municipal wastewaters. In this application the RBC may be used either (1) to upgrade an existing primary plant, (2) as part of the design of a new secondary wastewater treatment plant, or (3) as a means of upgrading an existing trickling filter or activated sludge secondary plant. In all cases it is possible to design the RBC for nitrification as well as the reduction of BOD.

Although well over 90% of these plants have design flows of 38,000 m^3/day (10 US MGD) or less there are plants designed to treat much higher flows, as may be seen in Table 2.

The Alexandria, Virginia installation is the largest in North America by a wide margin, using 56 standard density shafts to provide secondary treatment for a 204,000 m^3/day (54 US MGD) design flow. The next largest municipal installation is at Peoria, Illinois where 84 standard density shafts are used to provide nitrification of 140,000 m^3/day (37 US MGD) of activated sludge effluent.

A Philadelphia, Pennsylvania installation marked the first full-scale use of a combined activated sludge-RBC system called "Surfact" (Nelson and Guarino, 1977). In this process variation standard density air drive RBC shafts are partially submerg in the existing activated sludge aeration tanks and driven by a combination of diffused air and hydraulic roll. The resulting combination of fixed film and suspended culture reactor was able to achieve higher treatment efficiency at the same or lower energy expenditures (per unit BOD removed) (Cowee and Sullivan, 1980).

The most comprehensive use of RBC's in municipal wastewater treatment will be at t Orlando, Florida plant, due to come on stream in 1981 (Dallaire, 1979; Autotrol, 1979). This will be a regional treatment plant which will replace fifty small (le than 7500 m^3/day (2 US MGD)) package plants as well as provide tertiary wastewater treatment. Fifty-seven high density mechanically driven shafts and 114 high densi air driven shafts will be used to reduce the BOD and to provide nitrification (Aut trol, 1979). Following alum precipitation of phosphorous, methanol is added and forty-two completely submerged RBC shafts will be used to anaerobically denitrify the wastewater. This stage is followed by sand-filtration, chlorine contact and post-aeration to yield a design effluent with 5 mg/L BOD, 5 mg/L suspended solids, 3 mg/L total nitrogen, 1 mg/L total phosphorus and 7 mg/L dissolved oxygen (Dallai 1979).

The largest municipal use of RBC's in Canada is in Guelph, Ontario where 32 high density, air driven shafts are used for nitrification of 54,500 m^3/day (14.4 US MC of activated sludge effluent (Crawford, 1980).

Package Plants - Domestic Wastewater

Biological treatment package plants, of any type (e.g. extended aeration, RBC's) a usually applicable to (1) small flow situations and (2) relatively remote installa tions where the economics dictates that the plant be constructed at a central facility and then transported to the site. Because it is usually difficult to obtain and/or keep trained personnel the package plant should require a minimum o operation and maintenance. In addition, capital and operating costs should be lo so as to not strain a likely limited budget. The RBC has shown itself to fulfill all of these characteristics.

Table 3 RBC Package Plant Installation Summary: North America
(After Autotrol,1979;Hormel,1979;CMS,1980a)

Flow Range m^3/day	Flow Range US GPD	No. of Plants	Percent of Total
< 18.9	< 5000	32*	23.2
18.9 to 37.9	5000 to 10000	39	28.3
37.9 to 75.7	10000 to 20000	22	15.9
75.7 to 113.6	20000 to 30000	16	11.6
113.6 to 151.4	30000 to 40000	6	4.3
151.4 to 189.3	40000 to 50000	13	9.4
> 189.3	> 50000	10	7.3
		138	100.0

*- not including approximately 120, 2.3 m^3/day (600 US GPD) units

Table 4 Industrial RBC Installation Summary: North America
(After Autotrol,1979;Hormel,1979;CMS,1980a)

Flow Range m^3/day	Flow Range US MGD	No. of Plants	Percent of Total
< 379	< 0.1	14	36.8
379 to 1893	0.1 to 0.5	15	39.4
1893 to 3785	0.5 to 1.0	2	5.3
3785 to 7570	1.0 to 2.0	1	2.6
7570 to 11355	2.0 to 3.0	2	5.3
11355 to 15140	3.0 to 4.0	2	5.3
> 15140	> 4.0	2	5.3
		38	100.0

The tankage of a RBC package plant is often, but not necessarily, constructed in steel and usually includes a primary sedimentation-flow equalization tank, a RBC media zone, a secondary clarifier tank and a chlorine contact chamber, all in one compact unit. Variations on this generalization include the use of septic tanks as pretreatment and flow-equalization, hydrasieves as pretreatment, some on-site concrete tank construction, and alum addition for phosphorous removal. Effluent disposal may be made to either an existing watercourse, a holding pond or a tile-leaching bed depending on the regulatory requirements (Autotrol, 1978; CMS, 1980b).

As shown in Table 3 most package RBC plants treat less than 76 m^3/day (20,000 US GPD) but some treat as much as 1136 m^3/day (300,000 US GPD). Users include resorts, summer camps, campgrounds, airports, highway gas stations, shopping centres, small villages, industrial parks, subdivisions, apartment developments, trailer parks, work camps, oil drilling platforms, and individual homes (Autotrol, 1979; Hormel, 1979; CMS, 1980a).

Package RBC plants provide a viable wastewater treatment alternate when municipal services are not available and other options e.g. septic tanks, stabilization ponds, extended aeration package plants are not suitable because of the land area required, operation and maintenance requirements or lack portability.

Small villages and towns have used RBC's because of their low operation and maintenance requirements, low land area requirements and high quality effluent. Land developers have used RBC's for suburban subdivisions which would otherwise be undeveloped while awaiting connection to a municipal or regional wastewater treatment system. As an added bonus, the RBC package plant can be re-used elsewhere or sold when connection is finally made.

In other suburban land development situations small (2.25 m^3/day (600 US GPD)) RBC' and reduced size (80 to 90% smaller) leaching beds (for effluent disposal) have replaced septic tanks for individual household wastewater treatment (CMS, 1980b). This has proved cost effective because it permits a substantial increase in the housing density which would otherwise be restricted by the larger leaching beds required for the poorer quality septic tank effluent.

Because of the relative temporary nature of workcamps and offshore drilling platfor or artificial islands, RBC package plants have been designed, built and operated as semi-self sufficient units, housed in helicopter- or road-portable trailer-like units (Forgie and co-workers, 1974; Forgie, 1976; Heuchert, 1980). These units usua have their own heating system and small laboratory facilities, i.e. a testing bench in addition to housing the RBC package plant and necessary tankage.

In one case (CMS, 1980b) the tankage of the RBC package was designed to be self-supporting and was provided with skids so it could be slid from site to site if a flat-bed trailer was unavailable. In addition, this tank was insulated with urethane foam and fitted with a large foam-insulated fibreglass cover which provid enough head room for operator access to the plant proper.

Industrial Wastewater Treatment

The RBC process can be applied to the treatment of industrial wastewaters which contain biodegradeable organic materials. Examples of such applications include food processing plants (e.g. dairies, distilleries, wineries, malting plants, commercial bakeries, fish processing plants, meat and poultry processing plants, vege table and fruit processing plants) and non-food processing plants (e.g. tanneries, pulp and paper wastes, landfill leachate, textile wastes, coal mine drainage, refineries and petrochemical plants)(Autotrol, 1978; Olem and Unz, 1980; Hynek and

Iemura, 1980). Design flows for existing RBC installations have been primarily less than 1893 m^3/day (0.5 US MGD) (see Table 4), whereas BOD concentrations have ranged as high as 15,000 mg/L (Autotrol, 1978).

Because of its varied but special nature industrial wastewater may require special pretreatment and/or operations to ensure adequate treatment. Primary clarification may not be needed if the wastewater is already low in settleable solids. PH adjustment (to pH 6.5 to 8.5) may be required to maintain optimum growth conditions. Temperature may have to be reduced to below 32°C (90°F) to prevent undesired biological growth and/or weakening of the plastic media. Oil and grease concentrations should be reduced to at least 200 mg/L. Nitrogen (N) and/or phosphorus (P) may have to be added to overcome deficiencies and provide a BOD:N:P ratio of approximately 100:5:1 (Autotrol, 1978).

Some or all of these pretreatment requirements may be at least partially accomplished through the use of flow equalization tanks. Such tanks also provide for more uniform flows to the RBC zone and as a result, higher treatment efficiency.

Because the organic concentrations in industrial wastewaters may be high, the resulting biological kinetics may be in zero-order regime, rendering staging inappropriate, i.e. all the RBC area is in one stage (Autotrol, 1978).

Pilot plant testing may be necessary to provide accurate design information when there is insufficient experience with treating a specific industrial wastewater. Otherwise existing design criteria may be used.

Industrial RBC applications may be designed to provide complete treatment (including nitrification) before discharge to an existing water course or they may simply be used as a 'roughing' or 'pretreatment' before discharge to existing municipal sewers for further treatment.

RBC Cost Effectiveness

Because the RBC media shafts are essentially individual units the RBC process may not exhibit the economies of scale which may be associated with building activated sludge aeration tanks. For this reason the capital cost of an RBC system may be greater than that of an activated sludge or trickling filter system.

Unlike activated sludge or trickling filters the RBC does not require pumping to recycle sludge or effluent or compressors to provide air for oxygenation and mixing. The RBC process only requires that energy be expended in rotation of the media shaft through the wastewater.

Typically this requires a 5.6 kW (7.5 HP) motor per 7.6 m (25 ft.) shaft for mechanically driven RBC's. Air driven RBC's may require substantially less energy consumption, using a single compressor or blower to drive several shafts. Schirtzinger (1980) documented an air driven 6435 m^3/day (1.7 US MGD) RBC plant which used one 30 kW (40 HP) blower throttled to 18.6 kW (25 HP) to drive six shafts plus aerate a grit chamber and a sludge holding tank.

Because there is no process control parameter (e.g. MLVSS) to monitor and the amount of mechanical equipment is relatively small, the operation and maintenance requirements of RBC plants are much lower than activated sludge systems (Autotrol, 1978; Clark and co-workers, 1978).

When capital costs and operating costs are combined in either a life cycle or cost effectiveness analysis the energy savings of the RBC become highly advantageous when compared to other treatment alternatives. Lundberg and Pierce (1980) compared the RBC process (air and mechanical drives, standard and high density media) to the activated sludge process (air and pure oxygen types) over a 11,355 to 189,240 m^3/day (3 to 50 US MGD) range using present worth analysis: in all cases the RBC process was cheaper than the activated sludge process. The Orlando, Florida RBC plant cited earlier in the paper will have a capital cost 8 to 10% more than an oxygen activated sludge plant but will have a 20 year life cycle cost of 12 to 15% less (Dallaire, 1979).

SUMMARY AND CONCLUSIONS

Advantages

The rotating biological contactor (RBC) can consistently achieve high degrees (eg. greater than 90%) soluble organic carbon removal and/or low effluent soluble BOD concentrations (eg. less than 15 mg/L) without requiring effluent or sludge recycling. Furthermore, RBC operation does not require MLSS or sludge age control and is therefore relatively simple. Maintenance requirements are low, eg. 0.5 to 1.0 man-hour per shaft per week for larger installations (Autotrol,1978). Energy use is lowe than other conventional secondary treatment methods,eg. 2.0 kW/1000 m^3/day(10 HP/ US MGD) for a mechanically driven RBC versus 5.9 to 7.8 kW/1000 m^3/day(30 to 40 HP/ US MGD) for the activated sludge process (Dallaire,1979). Air driven RBC's offer further energy savings as well as increased process,ie. rotational speed, control.

Because 95% of the microorganisms in the RBC system are part of the fixed-film or biomass, there are no problems with either hydraulic washout or sludge bulking as in the activated sludge process. Furthermore, because the 'excess' biomass is continuously sheared off by turbulence there are fewer problems with the biomass build up into a thick trickling filter-like 'slime' which may become anaerobic or clog th media. RBC sloughed solids usually settle better than activated sludge solids and, therefore, the secondary clarifier can be made smaller and sludge thickening requir ments decreased relative to those required by an activated sludge system.

With RBC's nitrification occurs in conjunction with carbon removal and does not require separate aeration, sedimentaion and recycling or long sludge retention times as may be needed for nitrification by the activated sludge process. RBC's can easil be used to provide nitrification when required,eg. during the summer, because of a relatively short start-up time.

The RBC is flexible in its application to either new secondary or tertiary treatmen plants or in upgrading existing trickling filter or activated sludge secondary plan as well as in the treatment of domestic, municipal and biodegradeable industrial wastewaters. In some upgrading applications,eg. the Surfact process, the increased treatment may come with little or no extra energy expenditures.

Disadvantages

The RBC process is still relatively new and,as such, both engineers and regulatory agencies are somewhat unfamiliar with and, to some degree, skeptical of the process In at least one instance this has resulted in a regulatory agency setting a fixed a relatively conservative design criteria which does not necessarily reflect the RBC' actual capabilities (Murphy and Wilson,1978).

Organically overloaded RBC's may develop a thick growth which increases energy consumption and may cause anaerobic sloughing. Under the same operating conditions th presence of hydrogen sulphide will stimulate the growth of *Beggiatoa* which do not contribute to carbon removal and may inhibit the growth of carbon bacteria. This growth also forms a fine floc which is difficult to settle.

Mechanical RBC's do not permit easy process control,eg. increasing rotational speed to increase aeration and mixing, because variable speed gear controls or changing drive sprockets is not practical. Similarly, it is not practical to provide moveable inter-stage baffles for stage area process control.

Properly sized equalization tanks, which are required to maintain good treatment during hydraulic and/or organic load fluctuations, add to the cost of the RBC system. Similarly, covers or, in some cases, buildings are required to protect the RBC biomass from the extremes of the weather, and as such add cost but not necessarily any efficiency to the RBC treatment system.

There have been instances of structural failure of one or more components of the RBC system, with the most notable being the shaft and the media. Shafts have failed due to fatigue stresses in the stub-end welds or in the shaft proper as the result of cyclic or heavy loads,eg. changing section modulus during shaft rotation and/or heavy biomass growths. Media failures have been caused by ultra-violet degradation and embrittlement of unprotected,ie. coverless, white polyethylene RBC media.

Air driven RBC's may not be self-starting after a lengthly power failure if the biomass is thick and the unsubmerged growth is allowed to drain during the failure. To prevent such an occurrence it may be necessary to provide back-up power generation equipment and/or tank dewatering equipment.

The RBC may have a higher capital cost and land use requirement than the activated sludge process and therefore may not be suitable for some budget and physical constraints.

Conclusions

The rotating biological contactor (RBC) process is a viable secondary and/or tertiary wastewater treatment method which, because of its low operation and maintenance requirements and costs but high degree of treatment, should be considered for use for treating any biodegradable wastewater. With more and more plants successfully using the RBC process, there will be an increase in the acceptability and use of RBC's in the future. Research will continue in both the theoretical modelling and practical application of the RBC process.

ACKNOWLEDGEMENT

The author would like to express his gratitude to the Autotrol Corp.(BioSystems Division); CMS Equipment Ltd.,Mississauga,Ont.; PETWA Canada Ltd.,Calgary,Alta. and Napier-Reid Ltd.,Markham,Ont. for their assistance in providing the current RBC installation information and some of the slides used in the presentation.

REFERENCES

Antonie,R.L.(1970). Response of the Bio-Disc process to fluctuating wastewater flows. Proc. of the 25th Purdue Industrial Waste Conf.,Purdue University,Lafayette,Ind.

Antonie,R.L.(1974). Nitrification of activated sludge effluent:BioSurf process,Part 2. Water and Sewage Works, 121, No.12, 54-55.

Antonie,R.L.(1976). Fixed Biological Surfaces-Wastewater Treatment,CRC Press Inc., West Palm Beach, Florida.

Autotrol (1971). Application of the Rotating Disc Process to Municipal Wastewater Treatment, US Goverment WPC Research Series, Project No.17050 DAM,Nov.1971.

Autotrol (1978). Autotrol Wastewater Treatment Systems-Design Manual, Autotrol Corp. Bio-Systems Division, Milwaukee, Wisc.

Autotrol (1979). BioSurf Process Installation Summary: US and Canada, Autotrol Corp. Bio-Systems Division, Milwaukee, Wisc.

Beak, T.W., Consultants Ltd. (1973). An Evaluation of European Experience with the Rotating Biological Contactor, Canadian Environmental Protection Service Report No. EPS 4-WP-73-4 ,Oct.1973
Clark,J.H.,E.M.Moseng and A.Takashi (1978). Performance of a rotating biological contactor under varying wastewater flow. J.Wat.Pollut.Control Fed.,50, 896-911.
CMS (1980a). ROTORDISK Installation List, CMS Equipment Ltd.,Mississauga,Ont.,March.
CMS (1980b). Personal Communication Re:ROTORDISK installations, CMS Equipment Ltd., Mississauga,Ont., March.
Cowee,J.D. and R.A.Sullivan (1980). Surfact:current developments and process applications. Presented at the 1st Nat'l Symp. on RBC Technology,Seven Springs,Penn.,F
Crawford, P.M. (1980). Use of rotating biological contactors for nitrification at th City of Guelph water pollution control plant, Guelph, Ont., Canada. Presented at the 1st National Symposium on RBC Technology, Seven Springs, Pennsylvania, Feb.
Dallaire, G. (1979). US's largest rotating biological contactor plant to slash energ use 30%. ASCE Civil Engineering, 49,No.1, 70-73.
Davies,T.R. and W.A.Pretorius (1975). Denitrification with a bacterial disc unit. Water Research, 9, 459-463.
Doman,J. (1929). Results of operation of an experimental contact filter with partial submerged rotating plates. Sewage Works Journal, 1 ,No.5, 555-560 .
Famularo,J., J.A.Mueller and T.Mulligan (1978). Application of mass transfer to rot biological contactors. J.Wat.Pollut.Control Fed., 50, 653-671.
Forgie,D.,V.Christensen,K.Heuchert and T.Chong (1974). Workcamp wastewater treatmen rotating biological contactor and physical-chemical treatment. Proc. of the 26th Western Canada Water and Sewage Conference, Calgary, Alta. Sept. 1974.
Forgie, D.J.L. (1976). Evaluation of an Autotrol BioSurf Rotating Biological Contac Packaged Wastewater Treatment Plant. A report prepared for the Fed. Activities Environmental Branch, Environment Canada, Oct. 1976.
Filion,M.P.,K.L.Murphy and J.P.Stephenson (1979). Performance of a rotating biologi contactor under transient loading conditions.J.Wat.Poll.Control Fed.,51,1925-1933.
Friedman,A.A,L.E.Robbins and R.C.Woods (1979). Effect of disk rotational speed on rotating biological contactor efficiency. J.Wat.Pollut.Control Fed., 51, 2678-269
Grieves,C.G. (1972). Dynamic and Steady-State Models for the Rotating Biological Dis Reactor. Ph.D. Thesis, Clemson University, Clemson, S.Carolina.
Hartmann,H. (1960). Development and operation of dipping filter plants, Gas-u-Wasse 101, 281-285.
Hartmann,H. (1964). The use of dipping contact filters in biological treatment of s Verbandsbericht Nr.71/2 des Verbandes Schweizerischer Abwasserfachleute.
Heuchert,K. (1980). Personal communication RE: BioSurf installations, PETWA Canada Calgary, Alta. May, 1980.
Hormel, Geo. A. & Company (1979). A New Generation of Rotating Biological Surface Systems for Wastewater Treatment. Promotional Brochure, November,1979.
Hynek,R.J. and H.Iemura (1980). Nitrogen and phosphorus removal with rotating biolo gical contactors. Presented at the 1st Nat'l Symp. on RBC Technology,Seven Sprin
Lundberg,L.A. and J.L.Pierce (1980). Comparative cost-effectiveness analysis of rot biological contactor and activated sludge processes for carbon oxidation. Preser at the 1st Nat'l Symp. on RBC Technology, Seven Springs, PA.
McCann,K.J. and R.A.Sullivan (1980). Aerated RBC's - what are the benefits?. Preser at the 1st Nat'l Symp. on RBC Technology, Seven Springs,PA.
Mueller,J.A.,P.Paquin and J.Famularo (1980). Nitrification in rotating biological c tactors. J.Wat.Pollut.Control Fed., 52, 688-710.
Murphy,K.L. and R.W.Wilson (1978). Pilot Plant Studies of Rotating Biological Conta Treating Municipal Wastewater. A report prepared for the Central Mortgage and Ho Corporation (Ottawa,Ont.) by International Environmental Consultants Ltd.,Toron
Nelson,M.D. and C.F.Guarino (1977). New 'Philadelphia Story' being written by poll control division. Water and Wastes Engineering, 14,No.9, 22-30.
Olem,H. and R.F.Unz (1980). Rotating-disc biological treatment of acid mine draina J.Wat.Pollut.Control Fed., 52, 257-269.

Paolini, A.E., E. Sebastiani and G. Variali (1979). Development of mathematical models for treatment of an industrial wastewater by means of biological rotating disc reactors. Water Research, 13, 751-761.

Pöpel, Von F. (1964). Aufbau, Abbauleistung und Bemessung von Tauchtropfkörpern. Schweizerische Zeitschrift Für Hydrologie, 26, 394-407.

Schirtzinger,M.M. (1980). First USA air drive RBC units operational experience and performance,Indian Creek Wastewater Treatment Plant. Presented at the 1st Nat'l Symp. on RBC Technology, Seven Springs, Pennsylvania, Feb.

Williamson,K. and P.L. McCarty (1976a). A model of substrate utilization by bacterial films. J.Wat.Pollut.Control Fed., 48, 9-24.

Williamson,K. and P.L. McCarty (1976b). Verification studies of the biofilm model for bacterial substrate utilization. J.Wat.Pollut.Control Fed., 48, 281-296.

Wilson,R.W., K.L.Murphy and J.P.Stephenson (1980). Scale-up in rotating biological contactor design. J.Wat.Pollut.Control Fed., 52, 610-621.

Zeevalkink,J.A., P. Kelderman, D.C. Visser and C. Boelhouwer (1979). Physical mass transfer in a rotating disc gas-liquid contactor. Water Research, 13, 913-919.

22

RATIONAL DESIGN OF AERATED LAGOONS — A CASE STUDY

K. Subba Narasiah* and M. Larue**

**Dept. of Civil Engineering, University of Sherbrooke, Sherbrooke, Quebec, Canada*
***Service de Protection de l'Environnement, Quebec, Canada*

ABSTRACT

The present paper sums up results of a nine month long study conducted on an existing aerated lagoon serving a small municipality in the Eastern Townships, Quebec. The purpose of this study was to determine some of the kinetic coefficients of biodegradation for a particular type of wastewater on site, under actual conditions of flow and weather variations. For a rational design of any biological treatment process, knowledge of the various biokinetic constants and their variation with temperature is essential. The values of these constants reported in current literature are inadequate, as they are, quite often, based on laboratory-scale studies and not on on-site experience.

INTRODUCTION

Aerated lagoons can be essentially classified into two groups:

(a) Aerobic lagoon in which the biological solids are maintained in suspension by proper mixing of the lagoon contents. Complete-mix or quasi-complete-mix conditions are a prerequisite of this type of operation. (Figures 1.1 and 1.2).

(b) Aerobic-anaerobic lagoon in which the mixing power-input is such that all solids cannot be held in suspension but is just enough to maintain a uniform distribution of oxygen in the lagoon. Consequently, a portion of decomposable suspended solids settle to the lagoon bottom where they undergo anaerobic decomposition (Figure 1.3).

DESIGN FACTORS

Actually, the aerated lagoons are designed using a few empirical equations taking into account the following factors:

- overall BOD_5 removal anticipated
- degree of mixing in terms of power-input per unit volume of water
- effect of temperature variation on BOD removal.

Two different approaches are used at present in the application of mathematical models to aerated lagoon design. The first model given below assumes that the

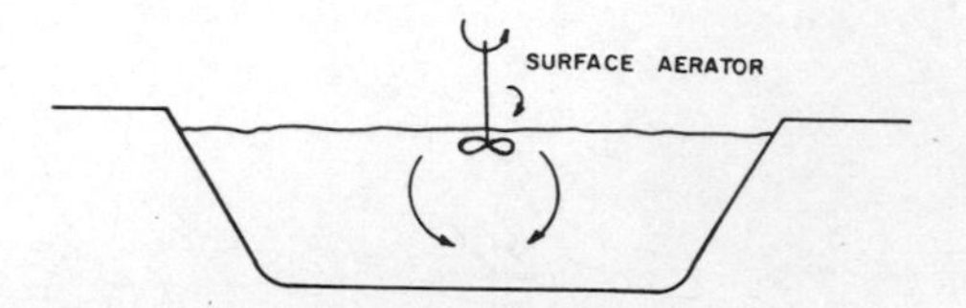

FIGURE 1.1 AEROBIC LAGOON (Complete mix)

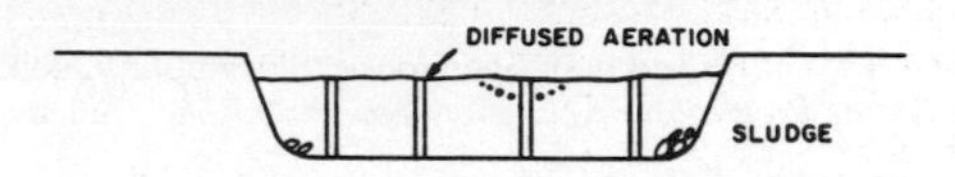

FIGURE 1.2 AERATED OXIDATION POND

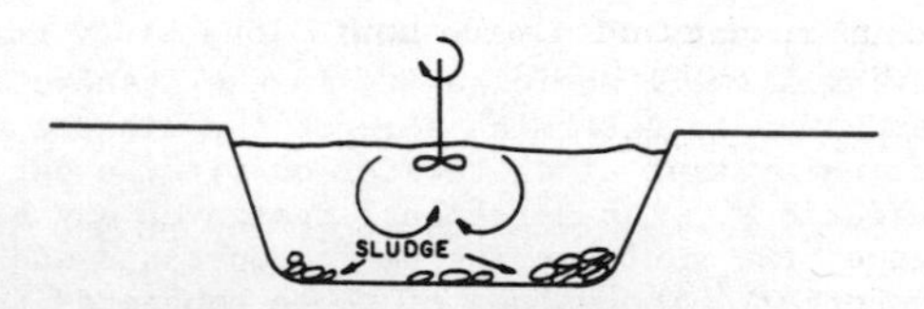

FIGURE 1.3 AEROBIC - ANAEROBIC LAGOON

Fig. 1. Classification of aerated lagoons

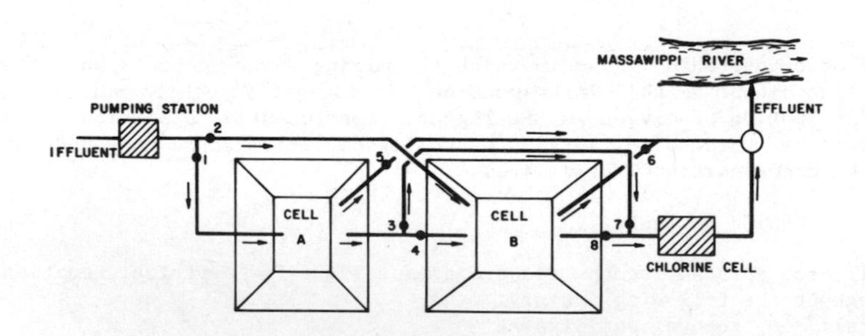

FIG. 2 _ WASTE WATER TREATMENT PLANT, NORTH HATLEY, QUEBEC.

BOD removal rate follows a first-order reaction (Metcalf and Eddy, Inc., 1979)

$$\frac{S_e}{S_a} = \frac{1}{1 + K \cdot t} \qquad (1)$$

Where S_e and S_a are the BOD_5 of effluent and influent wastes, mg/L,

K = BOD removal rate constant (day^{-1})
t = retention time of wastewater (days).

A modification of equation (1) was suggested by Eckenfelder (1967) that takes into consideration the suspended solids concentration in the lagoon. The modified form of equation (1) is the quasi-second-order equation as given below:

$$S_e = \frac{S_a}{1 + k \cdot X \cdot t} \qquad (2)$$

Where S_e = BOD soluble in the effluent (mg/L)
S_a = BOD total in the influent (mg/L)
k = BOD removal rate constant (mg/L·day)
X = average MLVSS concentration in lagoon (mg/L)
t = retention time of wastewater (days)

k is called the specific removal rate and $K = k \cdot X$ is called the overall removal rate constant. Thus, equation (2) can be rewritten as:

$$S_e = \frac{S_a}{1 + K \cdot t} \qquad (3)$$

The second model is based on the reaction kinetics of an activated sludge process without sludge recycle. The equation is given by Metcalf and Eddy, Inc. (1979) as

$$S_e = \frac{K_s\ (1 + k_d \cdot t)}{t\ (Yk - k_d) - 1} \qquad (4)$$

Where S_e = substrate concentration in the effluent (mg BOD_5 soluble/L)
K_s = half-velocity constant (mg/L)
k_d = coefficient of microorganism decay (day^{-1})
t = cellular retention time (days)
Y = growth-yield coefficient (mg-VSS/mg BOD_5 removed)
k = substrate utilization rate (day^{-1}) (base 10)

In order to predict the temperature effect on the reaction rate, Benedict and Carlson (1974) suggested the following relationship resulting from the van't Hoff-Arrhenius relationship and the Streeter-Phelps equation.

The van't Hoff equation is given by:

$$\frac{K_{T_2}}{K_{T_1}} = e^{(\mu/T_1T_2)\,(T_2-T_1)} \tag{5}$$

Where K_{T_2} and K_{T_1} are the reaction rate constants at temperatures T_1 and T_2, and

μ = temperature characteristic.

By assuming T_1T_2 to be constant,equation (5) can be rewritten as the Streeter-Phelps equation which is given by:

$$\frac{K_{T_2}}{K_{T_1}} = \theta^{(T_2 - T_1)} = \theta^{\Delta T} \tag{6}$$

Where Θ = temperature coefficient,
ΔT = temperature differential = $(T_2 - T_1)$

The values of the various constants appearing in equations (1) through (6) vary with the type of waste. Bartsch and others (1971), for example, give a value of k = 0.10-0.30 for domestic wastewaters. As for the value of Θ, Eckenfelder and Englande (1974) give an average value of 1.076 for domestic wastewater. On the other hand, Rich and White (1976) report several values of these constants based on bench-scale tests in the laboratory using synthetic wastewater. It is widely acknowledged that the best values one can obtain are those based on studies of existing facilities treating a particular waste. Hence, the present study to evaluate some of these constants was undertaken using an actual treatment plant.

EXPERIMENTS ON SITE

An existing wastewater treatment plant situated in North Hatley, in the province of Québec, was chosen for these studies. The plant, consisting of a sewage pumping station, two aerated lagoons and a chlorine contact chamber, has been in operation since 1973 and serves a population of about 1500. Figure 2 shows a schematic of the plant.

The plant has been designed for a flow of 570 m^3/d (150,000 gpd US). There are two aerated lagoons with mechanical aeration by floating aquajet aerators. There is a 7.5 HP aerator in the first lagoon and two 5 HP aerators in the second lagoon. According to the consultant, both these lagoons were originally designed as complete-mix tanks. The retention time of the wastewater is nine days. A chlorine cell with plug flow conditions is located at the end of the second lagoon. After 20 minutes of contact the effluent passes through a 90° V-notch and discharges into the Massawippi River. In order to evaluate the biological performance of the lagoon with changing detention times, the following steps were taken:

The existing wooden V-notch was removed from its guides in the chlorine contact chamber. In its place, a metallic 90° V-notch made up of 6 segments each 6" high was installed into the guides. Thus, by adding or removing one or more of these segments, it was possible to control the depth of the wastewater from about 2½ to 4 m (7½ to 12 ft) in both the lagoons, and consequently to vary the retention time from about 7½ days to 13 days.

The normal operation of the plant was not disturbed except for brief periods of 15 to 20 minutes whenever the wastewater levels were changed. Consequently the cables retaining the aerators had to be tightened or loosened to keep the aerators taut. During winter operation lagoon B is closed by removing the two aerators. A thick cover of ice up to 0.50 m (1½ ft) is formed and this cell acts more or less like a sedimentation tank during this period extending from December through April. The flow through the plant was measured both at the pumping station and at the V-notch. Whenever the depth of the lagoons was altered, a period of at least 6 hours was given so that stable flow conditions were reestablished. This was observed as level fluctuations in the V-notch rounded off gradually.

In all, six series of experiments were conducted, three in summer and three others in winter, each series consisting of the following steps:

- Note level of wastewater in lagoon,
- Take wastewater samples at pumping station, in the lagoon and at the V-notch,
- Change wastewater level i.e. the period of retention by adding or removing the segments,
- After stable conditions were established, take wastewater samples in the lagoon and effluent.

The following measurements were taken during each series.

- Temperature of air and wastewater,
- BOD_5 total and soluble of influent and effluent,
- SS and VSS of influent and effluent.

During test-periods the chlorine dosing equipment was not used so that effluent sampling was not affected by desinfection. In order to insure proper disinfection of the effluent, the chlorine was added after all sampling before the effluent was discharged into the river stream.

ANALYSIS AND DISCUSSION

Evaluation of Kinetic Constants

Equation (2) can be rearranged to give the value of K, the overall BOD removal rate constant.

That is

$$\frac{S_a - S_e}{X \cdot t} = k \cdot S_e \quad \text{or} \quad \frac{S_a - S_e}{t} = K \cdot S_e$$

where

$$K = k \cdot X$$

A plot of $\frac{S_a - S_e}{t}$ vs. S_e gave a linear relationship (Figures 3 and 4), the slope of which is the value of K. Then (U.S. EPA, 1971)

$$k = \frac{K \text{ mg/L BOD removed}}{X \text{ mg/L VSS} \cdot \text{retention time}}$$

Next, applying equations commonly used in studying biokinetics (Rich and White,1976) we can represent equation (4) in the following linearized form for graphical purposes:

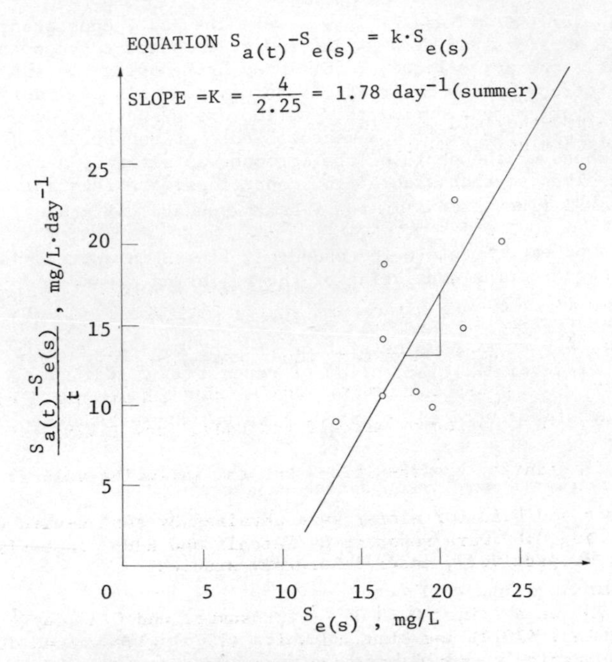

FIG. 3: EXPERIMENTAL EVALUATION OF KINETIC CONSTANT K

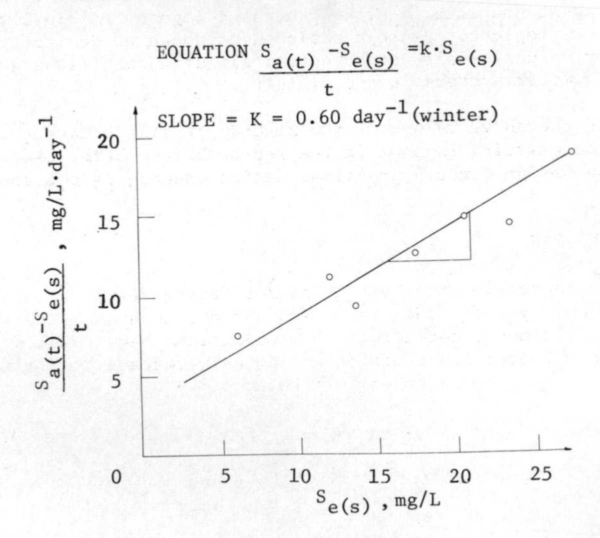

FIG. 4: EXPERIMENTAL EVALUATION OF KINETIC CONSTANT K

$$\frac{1}{t} = Y \frac{(S_a - S_e)}{X \cdot t} - k_d \tag{7}$$

The values of the growth-yield coefficient Y and the microorganism decay coefficient k_d can be determined by plotting

$$\frac{1}{t} \text{ versus } \frac{S_a - S_e}{X \cdot t} \quad \text{(See Figures 5 and 6).}$$

We obtained a value of $K = 1.78\ day^{-1}$ for summer and $0.60\ day^{-1}$ for winter conditions. Reported values in literature are as follows:

$K = 8.00\ day^{-1}$ under laboratory conditions (U.S. EPA, 1971),
$K = 5.00\ day^{-1}$ under laboratory conditions (Rich and White, 1976). A value of
$K = 3.10\ day^{-1}$ is reported by Metcalf and Eddy, Inc. (1979)

Similarly the coefficient of growth-yield Y has the following values:

Y = 0.68 for summer and 0.13 for winter were obtained by our on-site experiments where as Y = 0.50 and 0.60 were reported by Metcalf and Eddy, Inc. (1979) and Rich and White (1976), respectively, under laboratory studies.

As for values of k_d, we obtained $0.13\ day^{-1}$ for summer and $0.22\ day^{-1}$ for winter. Metcalf and Eddy, Inc. (1979) and Rich and White (1976) give values of $0.05\ day^{-1}$ and $0.2\ day^{-1}$, respectively, for laboratory experiments.

It may be emphasized here that values of kinetic constants for lower temperatures are few and far between in current literature. This often leads to arbitrary choice of design constants when designing aerated lagoons or similar treatment processes for colder regions. Hence, a rational approach to designing would be to determine their values on-site under actual climatic conditions and applying these results to similar systems in the vicinity.

It is obvious that the above values of the kinetic constants have to be confirmed by similar tests on existing lagoons in the region before being used extensively for aerated lagoon design. Hence, a practical design example is not attempted at this point.

CONCLUSIONS

On-site experience is perhaps the best guide for future design of aerated lagoons or similar biological systems. Studies conducted during summer and winter months on existing aerated lagoon under actual flow conditions have enabled the evaluation of some of the kinetic constants of biological treatment. Similar tests could be conducted to determine other coefficients.

ACKNOWLEDGEMENTS

The above study was made possible by a grant from the National Research Council on Science and Engineering, Ottawa, Canada.

The cooperation of the authorities of the Municipality of North Hatley, Québec, in realizing the above study is gratefully acknowledged.

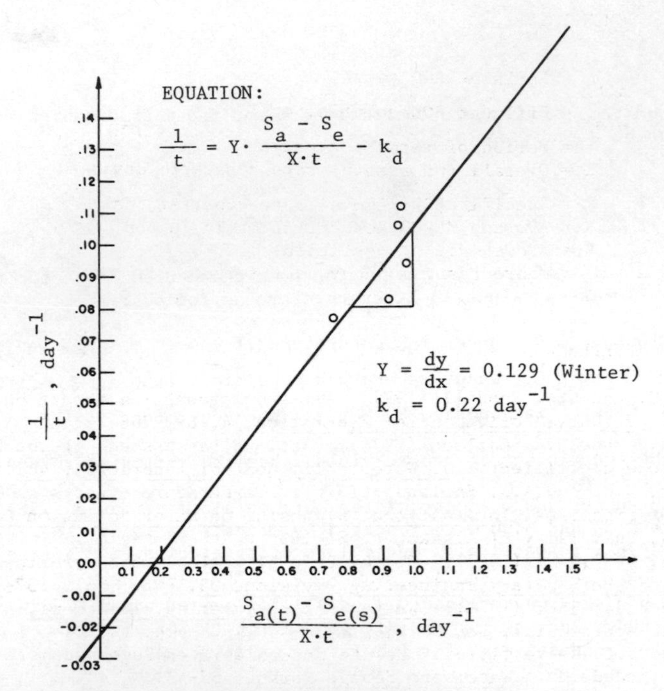

FIG. 5 :EXPERIMENTAL EVALUATION OF KINETIC CONSTANTS Y AND k_d.

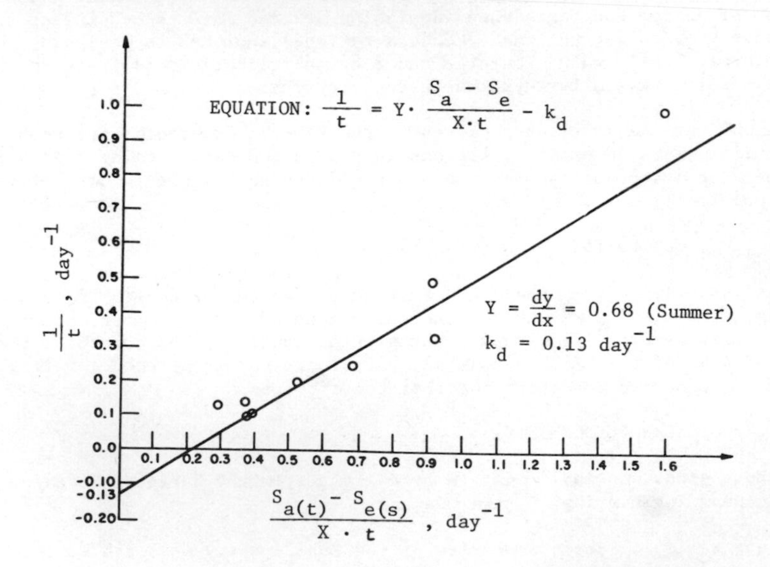

FIG. 6 : EXPERIMENTAL EVALUATION OF KINETIC CONSTANTS Y AND k_d.

NOMENCLATURE

$S_{a(t)}$ = Influent BOD total, mg/L
$S_{e(s)}$ = Effluent BOD soluble, mg/L
t = Period of retention, days
K = Overall BOD removal rate constant, day^{-1}
k = Specific BOD removal rate constant, day^{-1}
X = Average MLVSS concentration in lagoon, mg/L
Y = Growth-yield coefficient
k_d = Microorganism decay coefficient, day^{-1}

REFERENCES

Bartsch, E.H. and C.W. Randall (1971. Aerated lagoons - a report on the state of the art. J. Water Poll. Control Federation, 4, 699-708.
Benedict, A.H. and D.A. Carlson (1974). Rational assessment of the Streeter-Phelps temperature coefficient. J. Water Poll. Control Federation, 46, 1792-1799.
Eckenfelder, W.W. and A.J. Englande (1974). Temperature effects on biological waste treatment processes, International Symposium on Water Pollution Control in Cold Climates, University of Alaska, p.180-189.
Eckenfelder, W.W., Jr. (1967). Comparative biological waste treatment design Journal of the Sanitary Engineering Division, 93, (No. SA6), 157-170.
Metcalf and Eddy, Inc. (1979). Wastewater Engineering, Treatment, Disposal, Reuse. 2nd Edit. McGraw Hill, New York, p. 438-440, 523-534.
Rich, L. and S.C. White (1976). How to design aerated lagoons systems to meet 1977 effluent standards. Water and Sewage Works, 82-83.
United States Environmental Protection Agency (1971). Design Guides for Biological Wastewater Treatment Processes. Research report 11010 ESQ, p. 119-144.

23

ROLE OF MIXING IN BIOCHEMICAL STABILIZATION OF WASTEWATER ORGANICS

M. A. Aziz

Department of Civil Engineering, University of Singapore, Kent Ridge, Singapore

ABSTRACT

This paper represents an attempt to show experimentally the role of mixing (velocity) in the biochemical stabilization of wastewater organics. A number of experiments were conducted both in static (standard BOD bottle) and dynamic (simulated treatment plant) systems using domestic wastewaters from a local source in order to study the effect of mixing on the biochemical stabilization of the organics present. From the results of the laboratory investigation, it was observed that there is a significant difference in the biochemical stabilization of organics done under static and dynamic experimental conditions. The mixing (liquid flow velocity) is found to play a very significant role in the process of organics stabilization. The paper also critically discusses some new facts which were observed during the laboratory investigation.

KEYWORDS

Biochemical oxygen demand; biochemical stabilization; dissolved oxygen; domestic wastewater; dynamic system; effluent quality; mixing; static system; wastewater organics.

INTRODUCTION

Research in the field of wastewater treatment technology has been directed mainly towards the improvement in wastewater effluent quality as a prerequisite for the protection of the aquatic environment. The resultant technology has progressed to a stage where it is now possible to design wastewater treatment systems capable of producing any desired effluent quality.

In order to evaluate the degree of treatment of wastewater in a treatment plant and the degree of pollution in an aquatic environment, the standard BOD test is used as the most important single criterion. However, the stabilization of wastewater organics in a standard BOD bottle is a static process, whereas in a wastewater treatment plant it is a dynamic one where the mixing (velocity) plays a vital role. The organics present in the wastewater of a treatment plant usually are not stabilized in the same manner as in the laboratory BOD bottle. There is a significant difference between the conditions controlling the rates of biochemical stabilization of organics in a static system and a dynamic system. The difference between these two rates of biochemical stabilization of wastewater organics generally can be attributed to the difference in the biochemical and physical characteristics of each environment. In the standard BOD bottle test, the effect of velocity or mixing upon the matabolic activity of microorganisms has not been precisely defined and there is no certainty that the rate of biochemical stabilization of wastewater in a BOD bottle is valid for a treatment plant. The merit of applying the data of biochemical stabilization of wastewater organics obtained in such a non-dynamic system (BOD bottle) directly to a dynamic system (treatment plant or stream) is definitely questionable (Azia, 1970; Nejedly, 1966). Under laboratory conditions, all the organic matter is subjected to incubation simultaneously under artificial nutrient conditions and for the same duration of time; under treatment plant or stream conditions, the longitudinal or turbulent mixing produces a metabolic gradient between organics and the microorganisms. The longitudinal or axial mixing brings into mutual contact the differential portions of organics in different stages of biochemical change with microorganisms in different stages of their development. As a result, the rate of biochemical stabilization of wastewater organics and the total biochemical oxygen utilization are expected to be higher in a treatment plant or in a stream than those in the laboratory static system.

Phelps (1944) first proposed the first order monomolecular reaction for the BOD satisfaction of organic matter present in domestic wastewaters and organically-polluted river waters. Teh accepted formulation for BOD reaction is as follows:

$$y_t = L(1 - 10^{-k_1 t}) \tag{1}$$

in which y_t is the biochemical oxygen demand (mg/L) at anytime t (days), L is the ultimate biochemical oxygen demand (mg/L)and k_1 is the biochemical oxidation reaction rate constant (day^{-1}).

The smooth curve expressed by equation (1) does not occur in practice. Some investigators (Aziz, 1970; Hartman and Wilderer, 1969; Nejedly, 1966; Simpson,1965) have objected to the use of the first order monomolecular reaction (Eq. 1) as the measure of the biochemical activity. The mathematical solution to the BOD exertion of wastewaters depends upon bacterial metabolism. The pattern of the BOD satisfaction curves is mainly dictated by bacteria and not by simple mathematics (Aziz, 1970; Gannon, 1963; Nejedly, 1966). If conditions change so that the bacteria change the rate of metabolism, the application of the common mathematical model (Eq. 1) to determine BOD values is not justified.

Theriault (1927) revised equation (1) and Fair, Moore and Thomas (1941) proposed a modification to it by introducing some additional parameters. Work done by Anderson (1964) has shown that the BOD curves have four or five distinct phases even when dealing with the biochemical oxidation of pure substrates like glucose. Some authors (Revelle, Lynn and Rivera, 1965; Simpson, 1965; Woodward, 1953) proposed second order reaction equations for wastewater BOD satisfaction, but still kinetic modelling remains a subject of further investigation and research. Therefore, it is thought that there is a real need to study in the laboratory a controlled dynamic system which will, to a great extent, simulate the conditions prevailing in a wastewater treatment plant. In view of the above fact, both physical and mathematical models were developed to study the nature of the stabilization of wastewater organics in a dynamic system and to compare the same with that of the conventional bottle BOD test. Of course, the studies of BOD satisfaction of wastewaters have been carried out before under dynamic conditions by some researchers (Nejedly, 1966; Rand, 1966; Revelle, Lynn and Rivera, 1965) but the effect of mixing on the biochemical stabilization of wastewater organics, the relationship between bacteria and the rate of biochemical oxygen uptake and also the relationship between the bottle BOD and the dynamic BOD were not precisely established.

THEORETICAL CONSIDERATIONS

The mathematical representation for the course of biochemical stabilization of wastewater organics requires basically knowledge of two main factors: (a) the total oxygen demand of the organics, and (b) the rate at which this demand is satisfied. The biochemical stabilization of wastewater organics is, of course, a function of the type of bacteria present, nature and type of organics, pH, temperature, and liquid flow velocity (mixing).

The quantity of oxygen utilized by bacteria in stabilizing organics in a wastewater treatment plant during a short time interval may be taken as the algebraic sum of the observed decrease in the dissolved oxygen (DO) concentration plus the quantity of oxygen introduced into the wastewater by the process of atmospheric reaeration and is represented by:

$$R_b = \pm\Delta y_d + \Delta y_c \qquad (2)$$

where R_b = rate of biochemical oxygen uptake (mg/L·h)

Δy_d = dissolved oxygen decrease (mg/L·h)

and Δy_c = oxygen introduced into the wastewater by atmospheric reaeration process (mg/L·h)

The change in DO concentration in the wastewater is equal to the difference in DO concentration measured at the beginning and at the end of the short time interval. If an increase in DO concentration is found to occur during the short time interval, Δy_d is taken to be negative.

The process of biochemical stabilization of organics utilizes first the DO initially present in the wastewater. The primary replacement of DO in a wastewater treatment plant occurs through atmospheric reaeration which is the penetration of atmospheric oxygen into the wastewater through the exposed surface. As postulated by Adeney and Barker (1919), the rate of atmospheric oxygen transfer to water or wastewater under constant conditions of temperature and turbulence (liquid flow velocity) is directly proportional to the saturation deficit. This is represented by:

$$\frac{dy}{dt} = -k_2(Y_s - y) \qquad (3)$$

where $\frac{dy}{dt}$ = rate of atmospheric oxygen transfer into water or wastewater (mg/L·h)

y = DO concentration (mg/L) at time t (hours)

Y_s = saturation DO concentration (mg/L)

k_2 = atmospheric reaeration rate constant (h^{-1}).

It is apparent that the maximum rate of atmospheric oxygen transfer into water or wastewater will occur when the oxygen concentration in the air-water interface is zero. The rate of transfer is also a function of liquid flow velocity, liquid depth and liquid characteristics.

The variation of DO content of water or wastewater with time is given by the integration of equation (3). If the DO concentrations are y_1 and y_2 at times t_1 and t_2, respectively, integration results in the following:

$$\ln[(Y_s - y_1)/(Y_s - y_2)] = -k_2(t_2 - t_1) \qquad (4)$$

Putting $Y_s - y_1 = d_i$, $Y_s - y_2 = d_t$, and $t_2 - t_1 = t$, equation (4) becomes

$$\ln(d_i/d_t) = -k_2 t \tag{5}$$

where d_i and d_t represent the initial and final DO deficits (mg/L), respectively, after a time interval t (hours).

If the true DO saturation value (Y_s) for the specific water or wastewater is known, the plot of d_t versus t (on semilog paper) will result in a straight line, the slope of which yields the atmospheric reaeration rate constant, k_2 (Aziz, 1970; Rand, 1966

For a short time interval, the quantity of atmospheric oxygen introduced into water or wastewater can be determined by the following relationships:

$$\frac{\Delta y_c}{\Delta t} = k_2\ d_{av} \tag{6}$$

or

$$\Delta y_c = k_2(d_{av})(\Delta t) \tag{7}$$

where d_{av} is the average DO deficit $= \frac{(d_1 + d_2)}{2}$ (mg/L)

Once the values of Δy_d and Δy_c are known, equation (2) can be used to compute the biochemical oxygen utilization rate R_b of wastewater organics for a number of short time intervals. The biochemical oxygen uptake curve for a specific wastewater can be obtained by plotting the cumulative biochemical oxygen uptake against time for a 5-day period. The average rate of biochemical oxygen uptake can then be computed from the slope of the plotted cumulative oxygen uptake curve.

EXPERIMENTAL

Apparatus

The experimental apparatus developed is basically a cylindrical tank having two concentric walls fixed at the base on a common horizontal circular plate (Fig. 1). The inner cylinder is closed at the top by a circular plate which is centrally connected to a vertical metallic rod attached to a shaft driven by a motor. The inner diameter of the outer cylinder and the outer diameter of the inner cylinder are 450 mm and 150 mm, respectively, and thereby the channel width is 150 mm, the total depth of channel being 300 mm. The apparatus is placed in a waterbath in which the temperature is controlled by a thermostat and a cooling coil. This arrangement allows each experiment to be conducted at desired temperature, liquid depth, and liquid flow velocity. When the apparatus is rotated with a desired depth of wastewater at a desired speed, the water in the channel after a short period of time attains a velocity profile closely representing the dynamic conditions prevailing i a wastewater treatment plant. Since the experimental channel is very small, any velocity measuring device such as pitot tube or an ottmeter will disturb the equilibrium of the liquid flow in the channel. Therefore, a simple method as shown in Fig. 2 is adopted to test the steady state conditions of the liquid flow in the channel and measure the average linear velocity of liquid flow knowing the angular velocity. The device shown in Fig. 2 is first placed at the bottom of the channel which is then filled up with wastewater to a desired level. The apparatus is then switched on to rotate at a desired speed and allowed to equilibrate approximately for 20 minutes. At the beginning, the nylon threads attached to each of the floats (Fig. 2) are inclined. The sides and the bottom of the channel provide a constant shear force on the enclosed prism of wastewater. This constant application of shear force causes the wastewater to acquire speed gradually to that of the channel itself, and the thread of each float gradually approaches vertical position. After about 20 minutes when equilibrium is reached, the wastewater in the channel maintains a uniform flow condition and the float threads at different depths become vertical. This phenomenon verifies that the wastewater in the channel flows at the same speed as that of the apparatus itself.

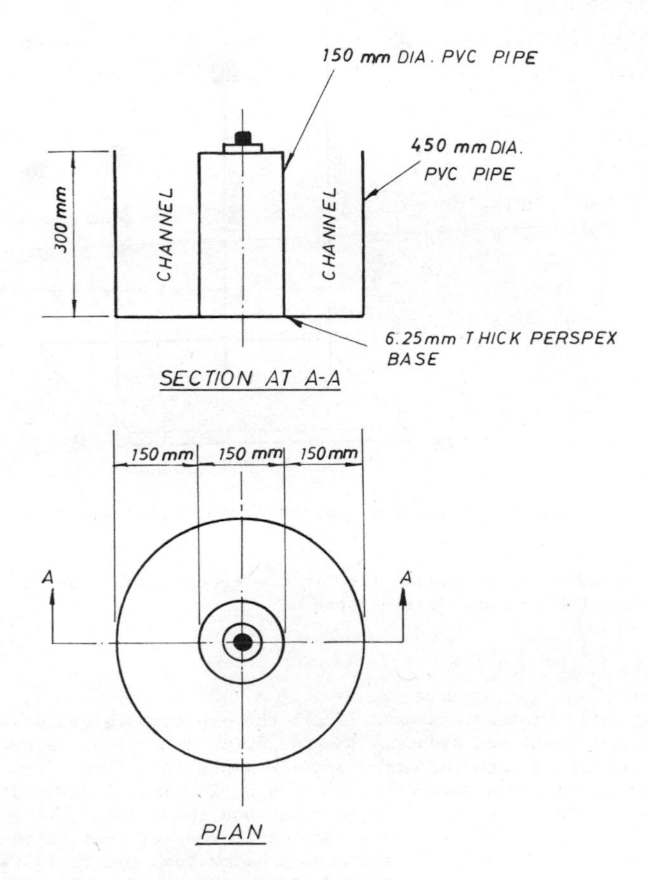

Fig. 1. Experimental channel

The average linear fluid-flow velocity in the channel is then calculated as follows:

$$v = \omega r = 2\pi rn/60 \tag{8}$$

where v = linear liquid-flow velocity (m/s)
ω = angular liquid-flow velocity (s^{-1})
r = mean radius of the channel (m)
n = revolutions per minute

Putting r = 0.15 m, the expression for the mean linear velocity of liquid flow becomes

$$v = \frac{2\pi \times 0.15 \times n}{60} = 0.0157\ n\ (m/s) \tag{9}$$

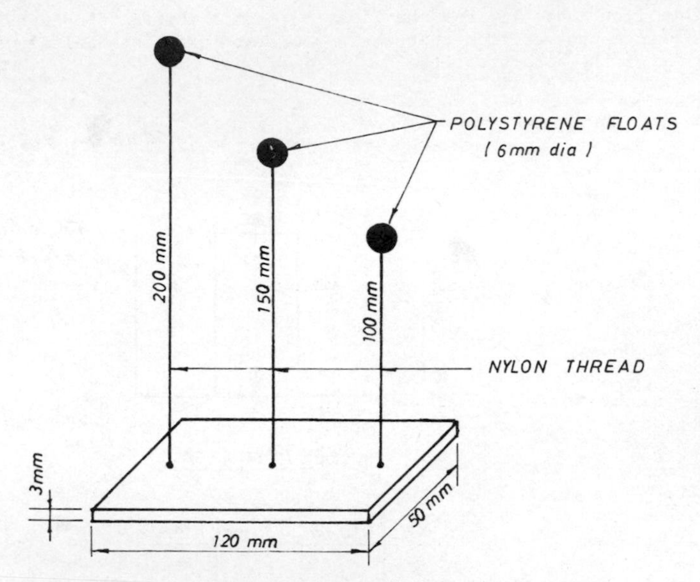

Fig. 2. Device for testing steady state liquid flow conditions

The mean velocity of liquid flow in the channel can, therefore, be computed using equation (9) for any desired speed.

Methods

The domestic wastewaters used in all the experimental runs were collected from loca wastewater treatment plants. The following general procedure was followed in all the experiments reported herein. The channel was first filled with filtered and thoroughly mixed wastewater to a depth of 250 mm. A desired speed was set up by adjusting the pulley between the motor and the shaft. The apparatus was then switc on. The electric cooling coil and the thermostat unit in the waterbath were starte simultaneously to bring down the wastewater temperature in the bath to 20°C. The water level of the bath and that of the channel were kept same during each experime The whole system was then allowed to equilibrate for about 10 minutes. The tempera ture of the wastewater in the channel was maintained at 20 $\pm$ 0.2°C. When steady state conditions were attained, the calibrated DO electrode of a DO meter was immer sed slowly and carefully creating minimum possible disturbance to the hydrodynamics of the wastewater flow in the channel. The DO electrode was kept fixed at the cent of the channel at mid-depth by attaching it to a metallic rod suspended over the channel. Since each experiment was conducted for 5 days, the DO meter was connecte to an automatic recorder. After fixing the DO electrode in the channel, the waste-water was brought to the desired DO saturation level using diffused air. The syste was then allowed to equilibrate for another 5 minutes and the DO concentration was marked on the recorder in order to establish the initial DO level in the wastewater of the channel. The apparatus was then left running for 5 days and the DO profile was recorded automatically. At hourly intervals, 10 mL wastewater samples were collected from the channel in screw-cap sterilized bottles for standard plate counts.

Upon completion of an experiment, mercuric chloride (a biological inhibitor) was added to the wastewater to give a final concentration of 20 mg/L in order to inhibit further DO uptake by bacteria. This was done to study the atmospheric reaeration rate in the wastewater. A period of two hours was allowed for the complete kill of bacteria (checked by plate counts) present in the wastewater and then the wastewater was deoxygenated by bubbling nitrogen through it. When there was a slight but significant amount of DO present (7 to 10%), the nitrogen supply was stopped. The zero time for the atmospheric reaeration was established at that time when the DO in the wastewater began a steady increase and the apparatus was left running for 2 days. After completion of each complete run, the channel was washed thoroughly by boiling water to remove the traces of mercuric chloride, followed by a thorough rinse of the sides and bed of the channel with ethanol. Laboratory analyses for BOD, suspended solids, pH and bacterial counts of each sample of wastewater were done as per standard methods (APHA, AWWA and WPCF, 1978).

RESULTS

Typical results of two experiments are presented in Figs. 3 and 4. The atmospheric reaeration rate constant, k_2 was determined under varying liquid-flow velocity conditions keeping the depth of liquid constant as shown in Fig. 5. From each experimental run, the DO concentration in percentage saturation at hourly intervals was taken from the graphic recorder sheets and the computation of various parameters involved in Eq. (2) was made using a computer programme in order to determine the rate of biochemical oxygen uptake. From the experimental results, the influence of the liquid-flow velocity (mixing) on the average rate of biochemical oxygen uptake is shown in Fig. 6. It can be seen from Fig. 6 that the liquid-flow velocity plays a significant role in the biochemical stabilization of wastewater organics. The optimum velocity was found to be around 0.48 m/sec. Hence, all further experiments were conducted at a constant velocity of 0.471 m/sec.

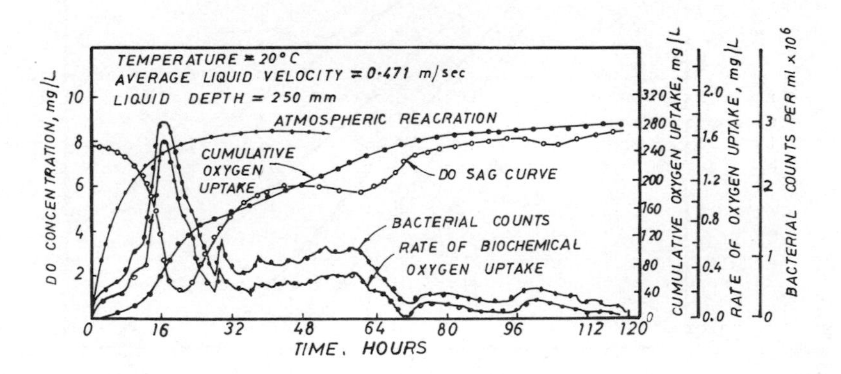

Fig. 3. Dissolved oxygen sag curve and related atmospheric reaeration, biochemical oxygen uptakes, and bacterial counts

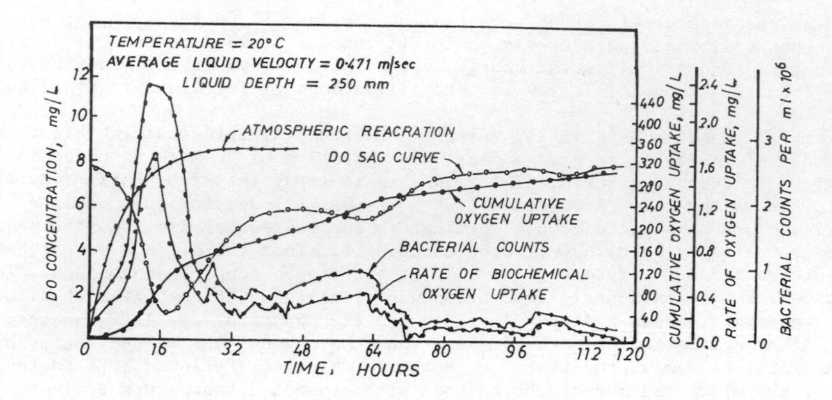

Fig. 4. Dissolved oxygen sag curve and related atmospheric reaeration, biochemical oxygen uptakes, and bacterial counts

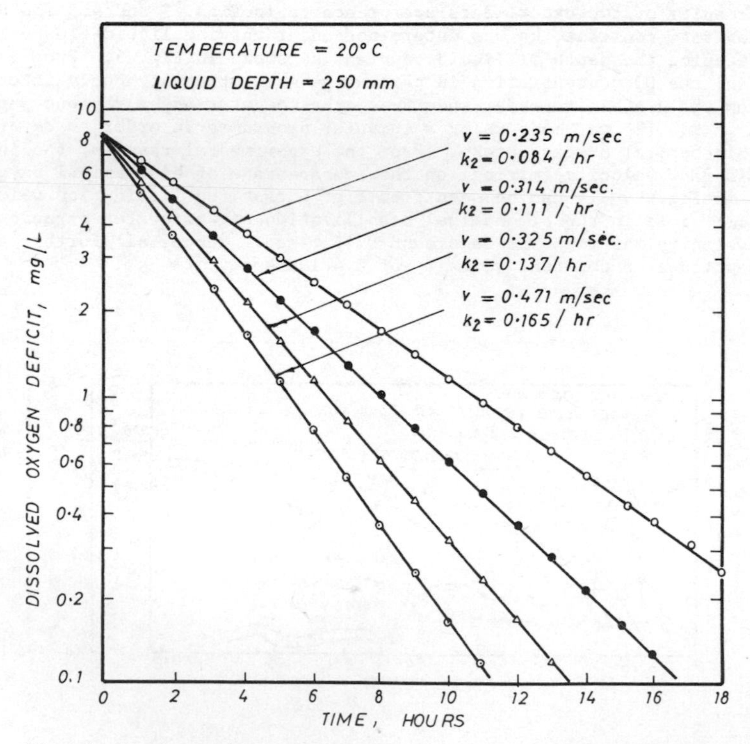

Fig. 5. Atmospheric reaeration rate constants under different liquid flow conditions

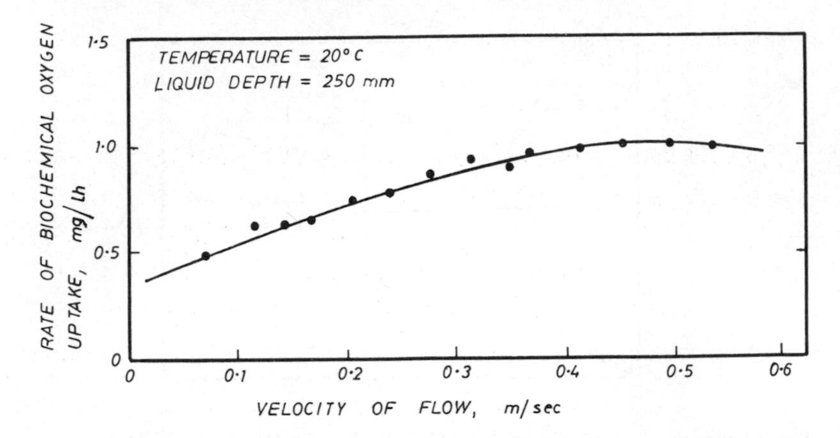

Fig. 6. Influence of liquid flow velocity (mixing) on the average rate of biochemical oxygen uptake

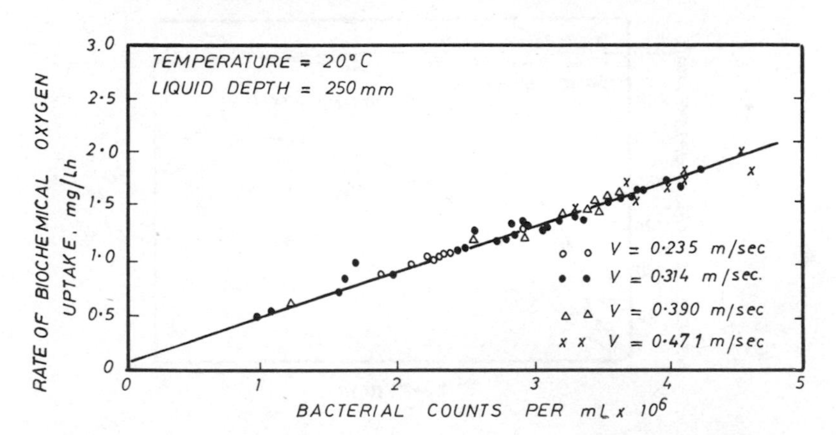

Fig. 7. Relationship between bacterial counts and rate of biochemical oxygen uptake

The relationship between bacterial counts and the average rate of biochemical oxygen uptake is shown in Fig. 7. The relationship is found to be linear under varying wastewater flow conditions. Figs. 8 and 9 show the variation of biochemical oxygen uptakes between static and dynamic systems for the two samples of wastewater reported in the Figs. 3 and 4, respectively. It is observed that in both the samples, there is a significant difference in the rate of biochemical oxygen uptake at various times and in the total oxygen uptake for the 5-day period between the static and dynamic systems. Multiple sags are observed in the total biochemical oxygen uptake in the dynamic system whereas in case of static system, there is virtually no indication of such sags.

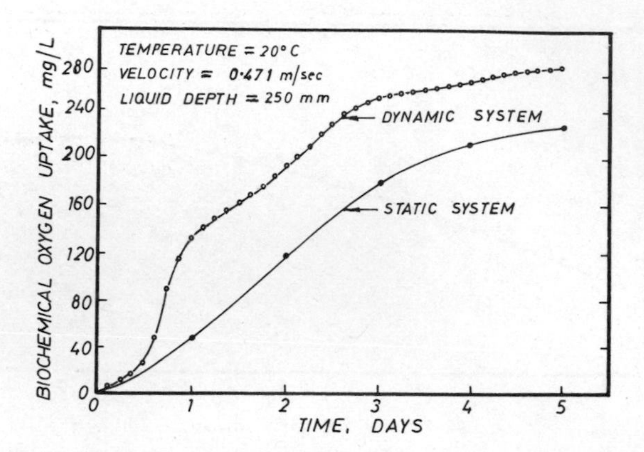

Fig. 8. Comparison of biochemical oxygen uptakes between static and dynamic systems

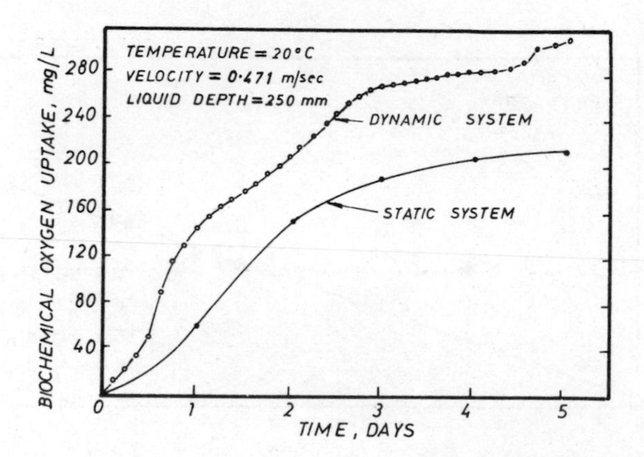

Fig. 9. Comparison of biochemical oxygen uptakes between static and dynamic systems

The relationship between the biochemical oxygen uptakes of domestic wastewater and organically-polluted river water in both static and dynamic systems is shown in Fig. 10, which shows that the 5-day biochemical oxygen uptakes in the dynamic syste are up to about 35% higher than those in the static system (bottle BOD test). Ther fore, mixing plays a vital role in stabilizing the organics present in wastewaters and also in organically polluted river waters.

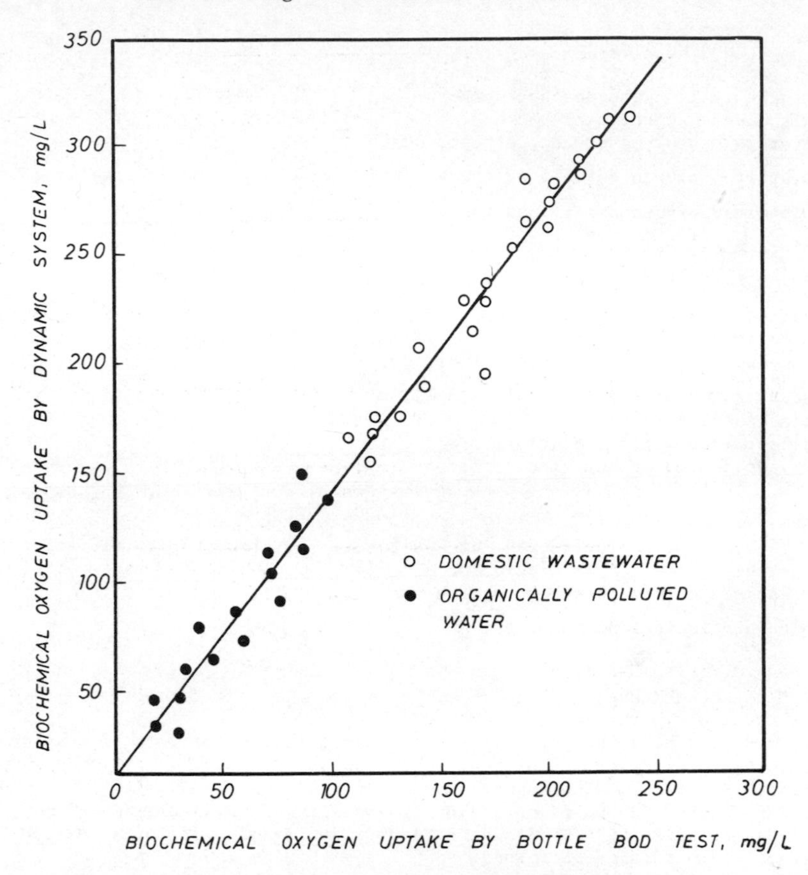

Fig. 10. Relationship between biochemical oxygen uptakes by static bottle BOD test and by dynamic system

CONCLUSIONS

From the experimental findings reported herein, the following concluding statements are made:

1) Bottle BOD test does not simulate well actual field conditions and underestimates the actual BOD values.

2) The experimental apparatus developed to determine the biochemical oxygen uptake better simulates actual field conditions, leading to the determination of more representative BOD values.

3) The new approach for determining the biochemical oxygen demand (BOD) of wastewaters is a radical departure from existing known techniques. However, the standard bottle BOD test remains an important tool to assess the degree of wastewater treatment needed and to evaluate the wastewater effluent quality.

NOMENCLATURE

BOD Biochemical oxygen demand, mg/L
DO Dissolved oxygen, mg/L
d_{av} Average dissolved oxygen deficit, mg/L
d_1 Dissolved oxygen deficit at the beginning of short time interval, mg/L
d_2 Dissolved oxygen deficit at the end of short time interval, mg/L
k_1 Biochemical oxidation reaction rate constant, day^{-1}
k_2 Atmospheric reaeration rate constant, h^{-1}
L Ultimate oxygen demand, mg/L
n Rotational speed, rpm
R_b Rate of biochemical oxygen uptake, mg/L·h
r Mean radius of experimental channel, m
t Time interval, day
v Linear liquid-flow velocity, m/s
y_t Oxygen demand at any time, mg/L

Δt Short time interval
Δy_c Dissolved oxygen introduced into wastewater by atmospheric reaeration process, mg/L·h
Δy_d Dissolved oxygen decrease, mg/L·h
ω Angular liquid-flow velocity, s^{-1}

REFERENCES

Adeney, W.E. and H.G. Becker (1919). The determination of rate of solution of atmospheric nitrogen and oxygen. Phil. Mag., 317-337.
Anderson, G.K. (1964). The kinetics of biological reactions using large volume respirometers. Ph.D. Dissertation, University of Newcastle-upon-Tyne (UK).
APHA, AWWA and WPCF (1978). Standard Methods for Examination of Water and Wastewater. 13th Edition, New York.
Aziz, M.A. (1970). Prediction of river water quality for polluted rivers. Ph.D. Dissertation, University of Strathclyde, Glasgow, 57-66.
Fair, G.M., E.W. Moore and H.A. Thomas (1941). The natural purification of rive muds and polluted sediments. J. Sew. Wks., 13, 270-307.
Gannon, J.J. (1963). River BOD abnormalities. USPHS Report No. WP-187, Universit of Michigan, 270-278.
Hartman, L. and P. Wilderer (1969). Physical and biochemical aspects of BOD kinet Proc. 4th Int. Conf. Wat. Poll. (Prague), 241-249.
Nejedly, A. (1966). An explanation of the difference between the rate of BOD progression under laboratory and stream conditions. Proc. 3rd Int. Conf. Wat. Poll. (Munich), 1, 23-36.
Phelps, E.B. (1944). Stream Sanitation. John Wiley & Sons, Inc., New York.
Rand, M.C. (1966). Laboratory studies of sewage effects on atmospheric reaeratio J. Sew. Ind. Wast., 31, 1187-1199.
Revelle, C.S., W.R. Lynn and M.A. Rivera (1964). Bio-oxidation kinetics and a second order equation describing the BOD reaction. J. Wat. Poll. Cont. Fed., 34, 1678-1686.
Simpson, J.R. (1965). The biological oxidation and synthesis of organic matter. J. Inst. Sew. Purif., 2, 171-179.
Theriult, E.J. (1927). The oxygen demand of polluted waters. PH Bulletin, No. 1 USPHS, 141-143.
Woodward, R.L. (1953). Deoxygenation of sewage - a discussion. J. Sew. Ind. Was 25, 918-925.

24

THE DESIGN AND OPERATION OF A 1400 L ANAEROBIC DIGESTER UTILIZING SWINE MANURE

J. Pos, R. H. te Boekhorst, B. Walczak and V. Pavlicik

School of Engineering, University of Guelph, Guelph, Ontario, Canada

ABSTRACT

Anaerobic digestion is discussed as an integral component of commericial livestock waste management and energy conservation operations. The design and operation of a 1400L "plug flow" anaerobic digester, without or with intermittent mixing, is described for fresh swine manure processing. In general, no major differences were found between the intermittently-mixed and unmixed operations, except for short-term stability of gas production which was improved significantly by intermittent mixing. Less difference between the two mixing cases was observed with respect to time-average biogas production rates. Results for pH, volatile fatty acid, bicarbonate alkalinity, total nitrogen, organic nitrogen and ammonia also are given.

KEYWORDS

Anaerobic digestion; swine manure; methane; biogas.

INTRODUCTION

The anaerobic digestion of organic wastes to produce a gaseous fuel can be an integral component of a combined waste management and energy conservation system for commercial livestock operations. The application of this natural process may also assist in the abatement of environmental and social problems by reducing the amount of odors from conventional manure handling systems and providing an alternative source of energy from a renewable natural resource.

In the past, farmers have utilized the ageless process of photo-synthetic conversion of solar radiation for the production of feed for animals. The feed is then converted to meat, eggs and dairy products, and the crop residues and animal manures are returned to the land as nutrients for crop production and as a soil conditioner. The possibility of utilizing the stored solar energy in organic wastes by anaerobic digestion to produce a useful fuel, prior to land disposal, has made this alternative a promising proposition, particularly with the rising cost of petroleum fuels. However, even though environmentally a more acceptable effluent for land disposal can be produced, the digestion process and reactor design are complex biological and engineering systems that require basic knowledge and efficient management for successful operation.

The biological process traditionally has been defined as a two-phase digestion system where volatile acids are produced in the first phase, are converted in the second phase by methanogenic organisms to methane, carbon dioxide and trace amounts of other gases. Analysis of the effluent has shown that, in addition to a reduction in odor, primarily only the organic matter has been reduced and, with minor exceptions, the nutrient content remained virtually unchanged. The effluent can be disposed on the land with minimum risk of environmental pollution.

ANAEROBIC DIGESTION

Two-phase Concept

As mentioned earlier, anaerobic digestion has been described in simplified terms as a two-phase process as illustrated in Figure 1. Phase I involves the breakdown

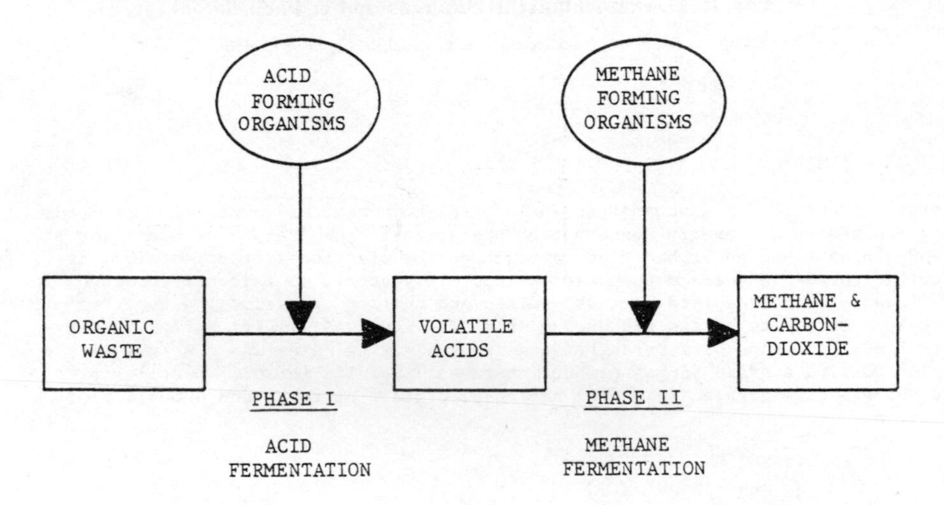

Figure 1. Simplified process for methane production.

of complex organic material into volatile fatty acids by acetogenic bacteria. The acids were then believed to be converted in Phase II into CH_4 and CO_2 by methanoge bacteria (McCarty, 1965; Bryant and Davis, 1975). A number of single stage (Jewel and others, 1976; Goodrich and others, 1978; Fischer and colleagues, 1979), and tw stage (Persson and others, 1976) farm-scale digesters were designed and put into operation. However, it soon became apparent that a physical separation of the two phases was not successful; furthermore, more recent work (Bryant and others, 1967; Bryant and Davis, 1977) suggests that methanogenic bacteria are unable to cataboli the longer chained fatty acids, indicating that an intermediate phase is present, therefore yielding the two-phase concept unsatisfactory.

Three-phase Concept

A simplification of Bryants' (1979) perceived three-phase microbial process is illustrated in Figure 2. The first phase is essentially the same as previously stated, i.e. fermentative bacteria hydrolyze polysaccharides to produce long

>2 carbon) and short (<2 carbon) chain organic acids, alcohols, H_2 and CO_2. Proteins and lipids also have a similar fate. In the second phase, the H_2- producing acetogenic bacteria produce acetate and H_2 from the longer chained fatty acids and alcohols of the first phase. In the third phase, the methanogenic bacteria utilize the H_2, CO_2 and acetate from the first two phases, to produce CH_4 and CO_2.

Although the three phases are shown separated for illustrative purposes, the symbiotic nature of the organisms must be emphasized, as each group depends on the other two for its efficient metabolism. For example, the breakdown of organic material by the fermentative bacteria, and the production of H_2 by the H_2-producing bacteria, are dependent on the H_2-utilizing methanogenic bacteria maintaining the partial pressure of H_2 at an extremely low level. Therefore, the degradation of

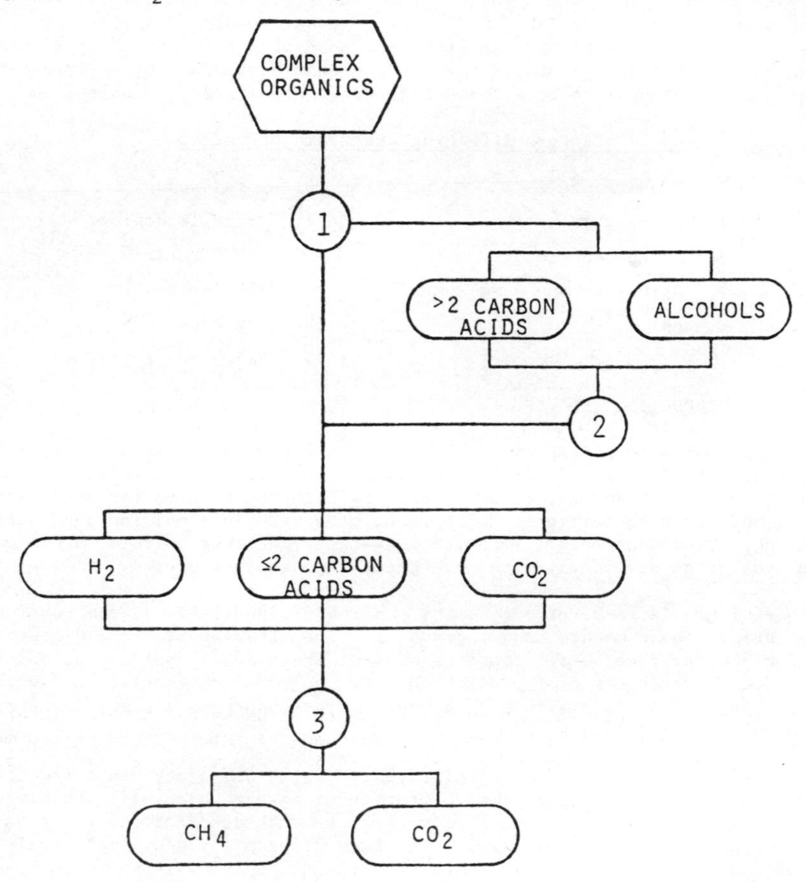

Figure 2. Three-phase concept for methane production.

organic material is directly related to CH_4 production and vice versa. Lawrence (1971) reported that H_2 pressures of 10^{-2} atm are toxic to the H_2-producing bacteria and will retard the fermentation of acetate to CH_4. Kaspar and Wuhrmann (1978) found that the steady state value for H_2 gas was less than 10^{-4}atm. This symbiosis would suggest that a single stage reactor, in which the organisms are in a mutually beneficial environment, would prove to be the most efficient.

Fermentative bacteria are most prolific, resulting in a rapid breakdown of proteins and carbohydrates; on the other hand, bacteria fermenting fatty acids are very fastidious in their environmental requirements and exhibit a relatively slow growth rate. Hence the rate limiting reactions must be related to the breakdown of fatty acids or methane fermentation, which is related to the efficient utilization of H_2 by the methanogenic bacteria. Therefore, in order to optimize gas production with respect to time, the design of an efficient anaerobic digestion system must cater to the metabolic requirements of the methanogenic bacteria. Other important enviro mental and operational factors to be considered are listed in Table 1.

TABLE 1. Factors Affecting Anaerobic Digestion

1. Temperature	6. Total solids
2. Loading rate	7. Volatile solids
3. Retention time	8. C/M/P ratio
4. Agitation and scum removal	9. Alkalinity
5. pH	10. Toxicity
	11. Salts concentration

PRODUCTION PROCESS

An anaerobic digester designed to utilize manure from a livestock operation for the production of a gaseous fuel can best be integrated into a liquid manure handling system. Normally, liquid manure is either spread on the land with tank wagons, applied with irrigation equipment, or stored in a holding pond before fielc spreading. Good management is required to ensure healthy animals, avoid environmental pollution and comply with government regulatory agencies.

If a useful gas is to be produced, the anaerobic digester is simply coupled into the existing liquid manure handling system. The effluent from the digester, altho relatively odour-free, still contains most of the original nutrients, and to obtai the maximum benefit for crop production must be spread on the land and worked into the soil as quickly as possible. The only major change is a significant reduction of the volatile solids.

Ideally, animal manure should be collected, treated and placed into the digester as soon as possible to reduce the absorption of oxygen which will inhibit the grow of methanogenic bacteria and also to utilize as much as practical the sensible hea in the manure before it cools down. The flow diagram for a typical single-stage anaerobic digestion system, with partial effluent recycle, is shown in Figure 3. Crop residues may also be a source of organic matter for anaerobic digestion.

When associated with a livestock operation, the gas production facility is general integrated into the manure handling system and usually follows close behind the collection of manure in or near the animal building. A slurry tank becomes part o the manure collecting facility to receive the daily production with sufficient capacity to contain at least two days of production. Any required pre-treatment such as temperature adjustment, solids dilution and pH buffering can be accommodat at this time. A portion of the effluent, or digested sludge, may be recycled as seed material or to dilute the raw material which may have a high solids content.

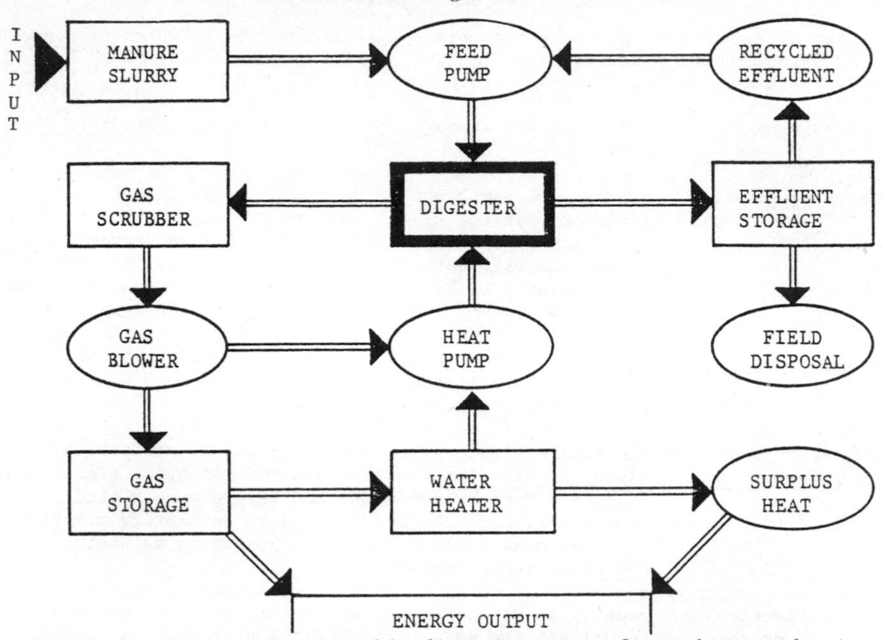

Figure 3. Single-stage anaerobic digestion system for methane production.

A schematic diagram of a simple gas production facility, coupled to a livestock operation, is illustrated in Figure 4.

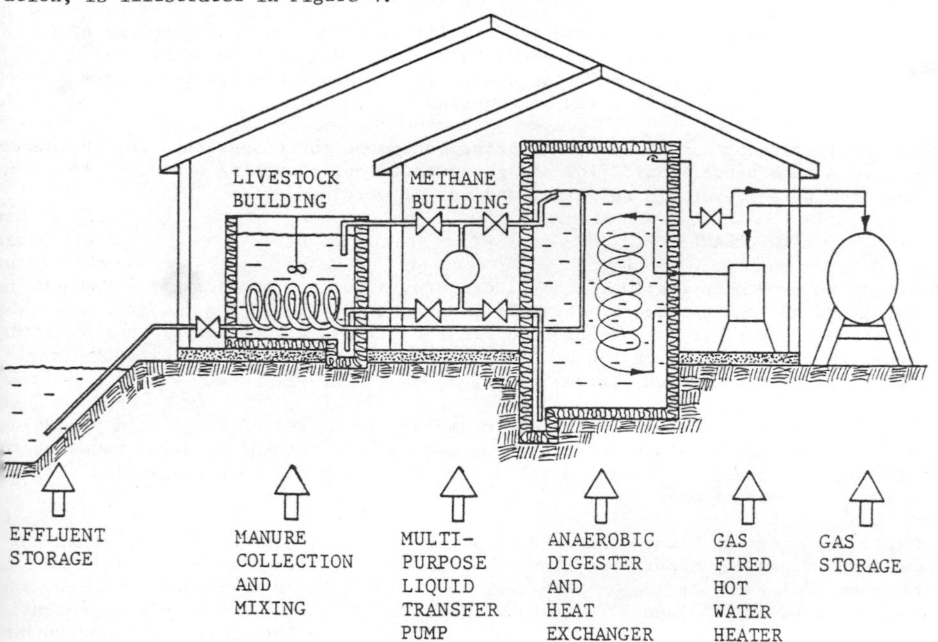

Figure 4. Schematic layout of a full-scale anaerobic digestion system coupled to a livestock production building.

The essential components for an on-farm biogas production plant, which is connected into the animal manure handling system at the point where the manure has been collected from the animals, are summarized as follows:

1. Slurry tank - for temporary storage of manure after collection to permit pre-treatment.

2. Transfer pump - one or more pumps may be required to (a) transfer influent from the slurry tank to the digester; (b) transfer effluent from the digester to storage; if the hot effluent is pumped through a heat exchanger in the slurry tank, the influent can be preheated before being pumped into the digester; (c) transfer effluent from the digester to the slurry tank for dilution and (d) to recirculate the contents of the digester to prevent scum formation and provide a uniform distribution of organisms in the digester.

3. Digester - the size will depend on the daily manure production, amount of dilution required and the retention time. Digesters must be gas tight and may be constructed from steel, concrete and fiberglass reinforced plastics; they may also include a heat exchanger to maintain operating temperatures.

4. Gas collection and gas storage system - usually about 20 percent of the digester volume is used to collect the gas. Short term gas storage is required for flexibility in gas utilization. A gas-fired hot water heater may be used to provide hot water for heating purposes and to maintain digester temperature.

5. Effluent storage- since there is very little reduction in the volume of manure processed through the digester (possibly 5-7 percent), the usual liquid manure storage facilities will still be required.

Other considerations may include monitoring equipment for temperature and pH contro flow meters to measure liquid flow and gas production, scrubbing equipment to remov carbon dioxide and hydrogen sulphide, and safety warning devices.

PILOT PLANT

Following an extensive review of the literature, a methane producing pilot plant was designed and constructed at the Arkell Research Station. The digester was mold ed to shape using fiberglass reinforced plastic; it was of horizontal configuration and oval in cross-section having an effective volume of 1400 L. Details of the digester are illustrated in Figure 5.

The pilot plant was put into operation in January 1979, and operated as a plug-flow system using an immersed hot water heat exchanger to maintain operating temperature Gas production, temperature, pH, total solids (TS), ash and volatile solids (VS) were monitored during operation.

Conventional start-up procedures were employed using digester sewage sludge as the inoculent. Fresh pig manure was blended with heated effluent in a 2:1 ratio to provide an influent with 10 to 15 percent TS of which 70 to 80 percent was VS. Th equivalent loading rate was 1.75 kg VS/m^3 digester volume. The calculated retenti time was 80 days as recycled effluent was utilized for dilution purposes and as a supplemental heat source. The digester was loaded every third and fourth day. Th

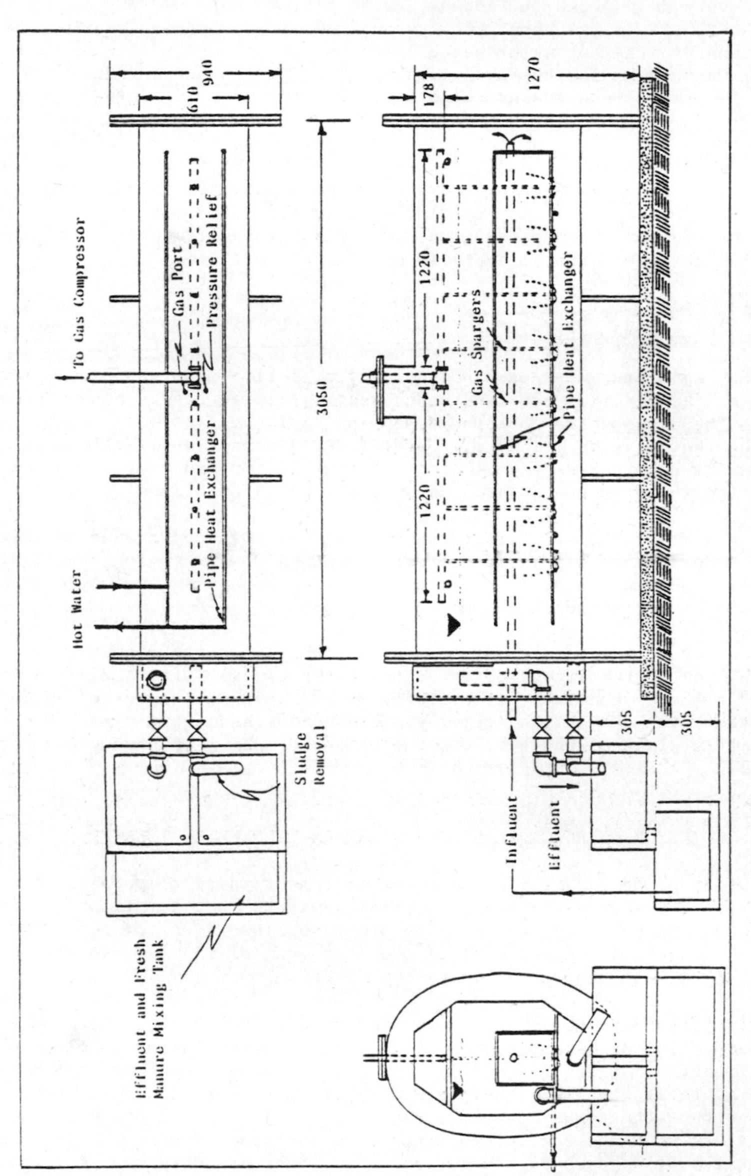

Figure 5. Pilot plant anaerobic digester. Dimensions given in mm.

cyclic variation in gas production was attributed to the intermittent loading schedule. Peak gas production occurred within 24 hours after each loading. Under stable conditions, the average gas yield was 1167 L/day, which is equivalent to 205 L/pig-day (7.23 ft^3/pig-day).

During three very cold weeks in February, with minimum outside temperatures of -20°C the average daily yield was about 700 L of biogas with a heating valve of 24.3 MJ/m^3 the energy input to the hot water heater was 25 L of propane with a heating value of 91.3 MJ/m^3. On this basis, the equivalent of 13.4 percent of the energy produced wa required to maintain the digester temperature during the cold winter period.

Suddenly, between the 124th and 127th days of operation, scum production accelerated plugging the gas lines and releasing the safety valve. Pipe lines were cleared and normal operation resumed. Within 7 days, scum and foam again plugged the system and operation was suspended. Since very little information had been published concernin the problem of scum and foam production, a laboratory scale model of the pilot plant was built (Pavlicik and others, 1979) to examine the problem.

SCUM REMOVAL EXPERIMENT

The laboratory apparatus consisted essentially of a 21.5 L horizontal cylinder, 200m in diameter as the digester and two gas collecting/storage tanks each with a floatir steel dome. The digester was built with an internal overflow weir at the gas discharge end of the cylinder for scum removal. The digester was inclined from the horizontal and a return line from the gas storage was provided at the opposite end from gas collection with the inlet located below the liquid level.

An electric heating cable, with rheostat control, was wrapped around the digester t control the temperature. The weight of the floating domes provided the equivalent of 120 mm water column pressure. With the addition of a 6 kg removable weight, the pressure could be increased to 180 mm water column.

The procedure for scum removal was to simultaneously load the digester and reintroduce the gas, under increased pressure, to generate a wave action to fragment the scum and allow it to flake-off over the weir. This method worked satisfactoril so long as the thickness of the scum layer did not exceed about 5 mm. Water spargi was also used in place of the gas, but the effect of dilution produced a reduction in gas yield.

OPERATION OF PILOT PLANT AS AN INTERMITTENT-MIX DIGESTER

The 1400 L pilot plant was modified and new equipment added to permit operation as conventional, intermittent-mix reactor. Mixing was to be accomplished by pneumatic gas sparging of the substrate using a low pressure blower. A compressor was considered, but the potential corrosion from the hydrogen sulfide and the high energy requirements ruled out that possibility.

During the intial start-up, it soon became evident that a positive pressure could r be maintained in the digester because of gas leaks in the blower impeller housing a around the immpeller shaft. An attempt was made to seal the entire motor and blowe assembly in a single container, but this also proved unsatisfactory. However, it w soon apparant from the temperature observations that a minimum of four minutes of g sparging per day provided sufficient mixing for uniform temperature distribution a daily dispersion of organisms. Therefore, the blower assembly was isolated from t gas collection and transport system by shut-off valves and connected into the syst only when needed (4 minutes/24 hours).

PILOT PLANT RESULTS

In addition to monitoring the same parameters as before, it was decided to include as well volatile acid concentration, bicarbonate alkalinity and ammonia concentration. Organic overloading will bring about an increase in volatile acid concentration and a decrease in bicarbonate alkalinity and, ultimately, pH. This in turn leads to an increase in unionized volatile acids, further inhibiting the methanogenic bacteria and allowing greater accumulation of volatile acids and further lowering of pH. This cyclic process will eventually lead to total collapse of the process. To avoid the problem, it is first necessary to employ proper start-up procedures and then maintain a high bicarbonate alkalinity to provide ample buffering capacity. This is not usually a problem with animal manures having a high C/N ratio, although gas yields can be increased by increasing C/N to 16/1 (pig manure is normally 10/1) by adding more carbon in the form of rice hulls, ground corncobs and wood waste.

The results for this experiment are summarized in Figures 6 to 14 and cover a period of 46 days from April 25 to June 10, 1980. "No mixing" of digester contents was employed during the first 13 days and the last 13 days. Mixing was provided during the 20 days from May 8 to May 28, 1980.

In general, there does not appear to be any major difference in total performance between operating with mixing or without mixing. Percent dry matter (Figure 6), ash (Figure 7) and volatile solids (Figure 8) for influent and effluent is fairly constant within normal expectations of variability of manure supply and quality of manure. However, mixing does have a significant effect on uniformity of biogas production over the short term as shown in Figure 9. The low point on the 25th day occurred at a time when the gas sparging tubes malfunctioned and mixing was delayed for two days. Otherwise, it would appear that even with minimum mixing of 4 minutes per day, daily production did not vary more than 7%/day as compared to a total range of more than 18 percent for the no mixing periods. Nevertheless, daily gas production, as an average over the "no mixing", "mixing" and "no mixing" periods, was 27.2 , 27.2 and 28.9 ft^3/day, respectively.

A comparison of the influent and effluent in Figures 10 and 11 shows an average pH of 6.8 for influent and an average pH of 8.15 for effluent with a cumulative increase of 1.24% for effluent from day 1 to day 43, which increase parallels the increase in volatile acid concentration over the same period.

A review of the nitrogen analysis, as shown in Figures 12 and 13, indicates an average reduction from 8×10^3 ppm in the influent to 7×10^3 ppm in the effluent. This may not be a true indication of the losses as a portion of these losses would include those incurred in removing, transporting and storing the samples for analysis. Further research in losses during analysis procedures will be required. A comparison of the two figures 12 and 13 shows a decrease of 58% in organic N and an increase of 92% in ammonia.

The biogas produced was analyzed for CH_4 and CO_2 content by gas chromotography (1830 mm x 6.4 mm stainless steel column; Chromosorb 102 on 80-100 mesh packing). For the plug flow digestor operation with intermittent mixing, the maximum biogas CH_4 content detected during the 46-day period was 54.1% CH_4 (concurrently with 40.5% CO_2 and 5.4% other gases). This methane content is somewhat less than values customarily given in the literature. Overall, the time-average methane content of the biogas produced was about 51%.

CONCLUSION

From our experience over the past two years, both the 20L laboratory reactor and the 1400L pilot plant have performed as good facilities to be used singly or in combination to monitor the effects and analyze the results of changing the various parameters on anaerobic digestion of animal manures.

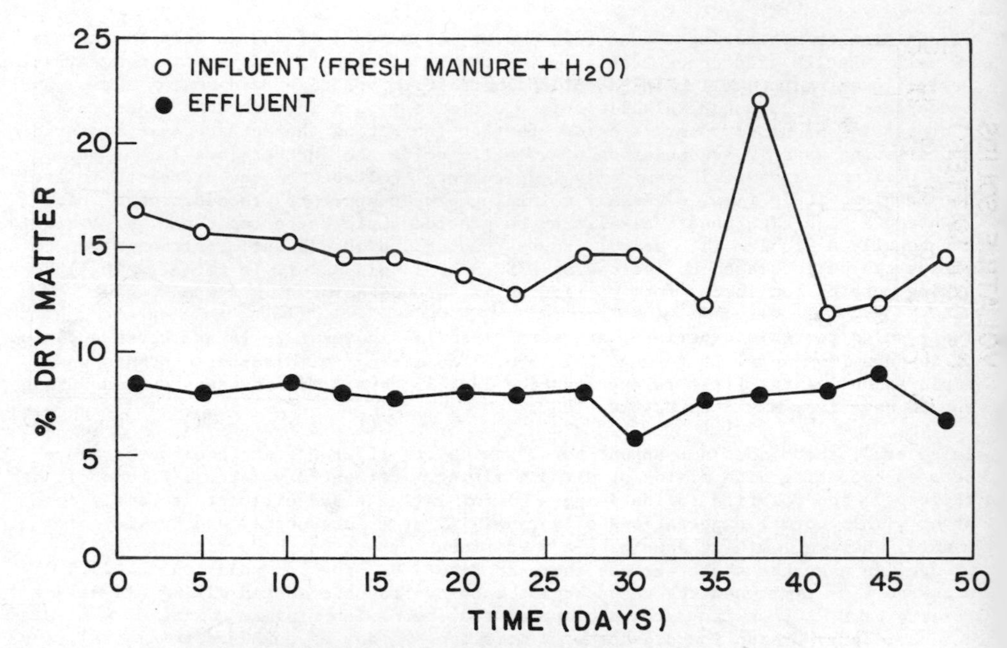

Figure 6. Dry matter variation during the operating period (80-04-25 to 80-06-10)

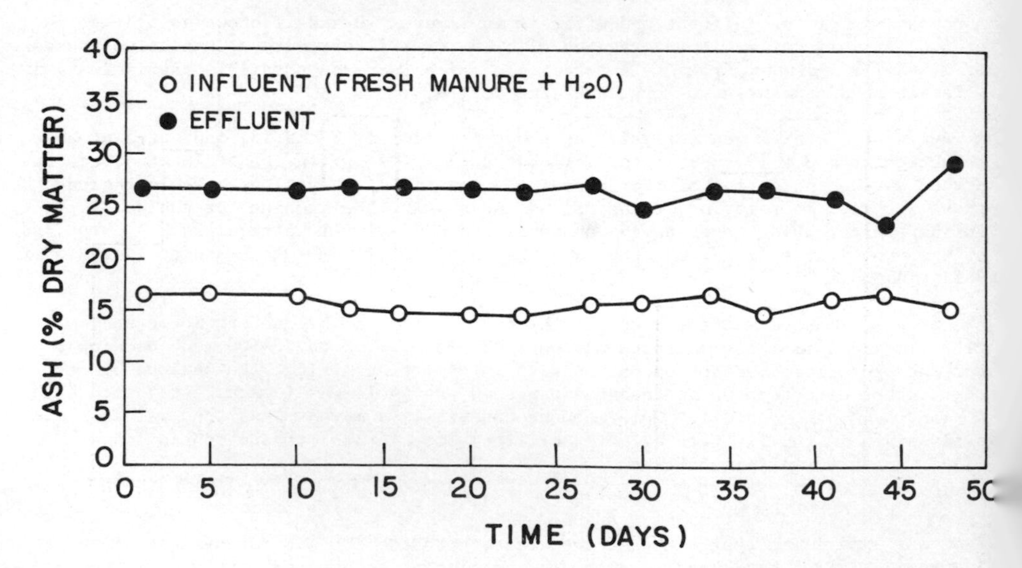

Figure 7. Ash as percent of dry matter

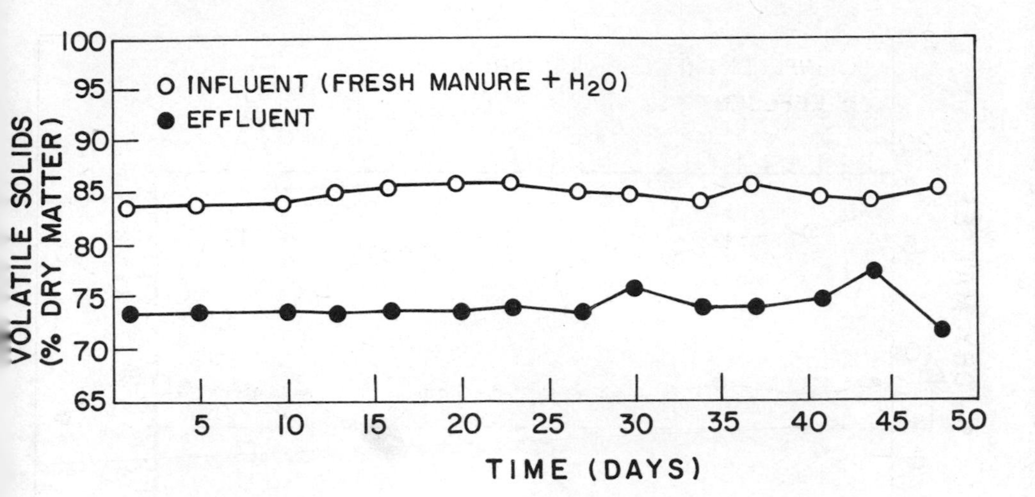

Figure 8. Volatile solids as percent dry matter.

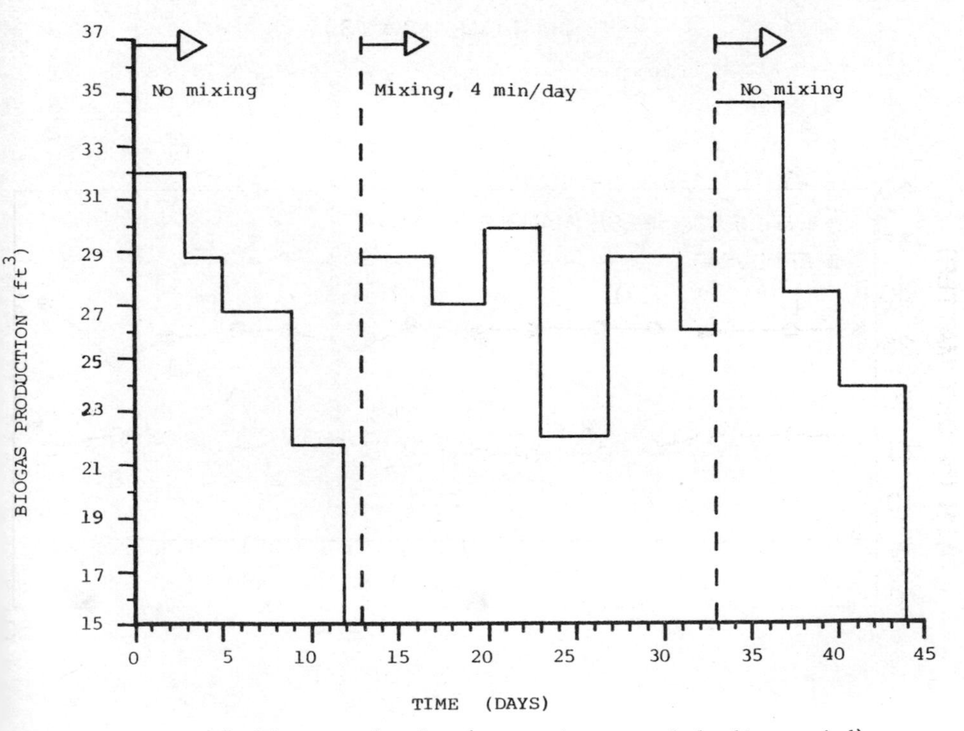

Figure 9. Biogas production (averaged over each loading period).

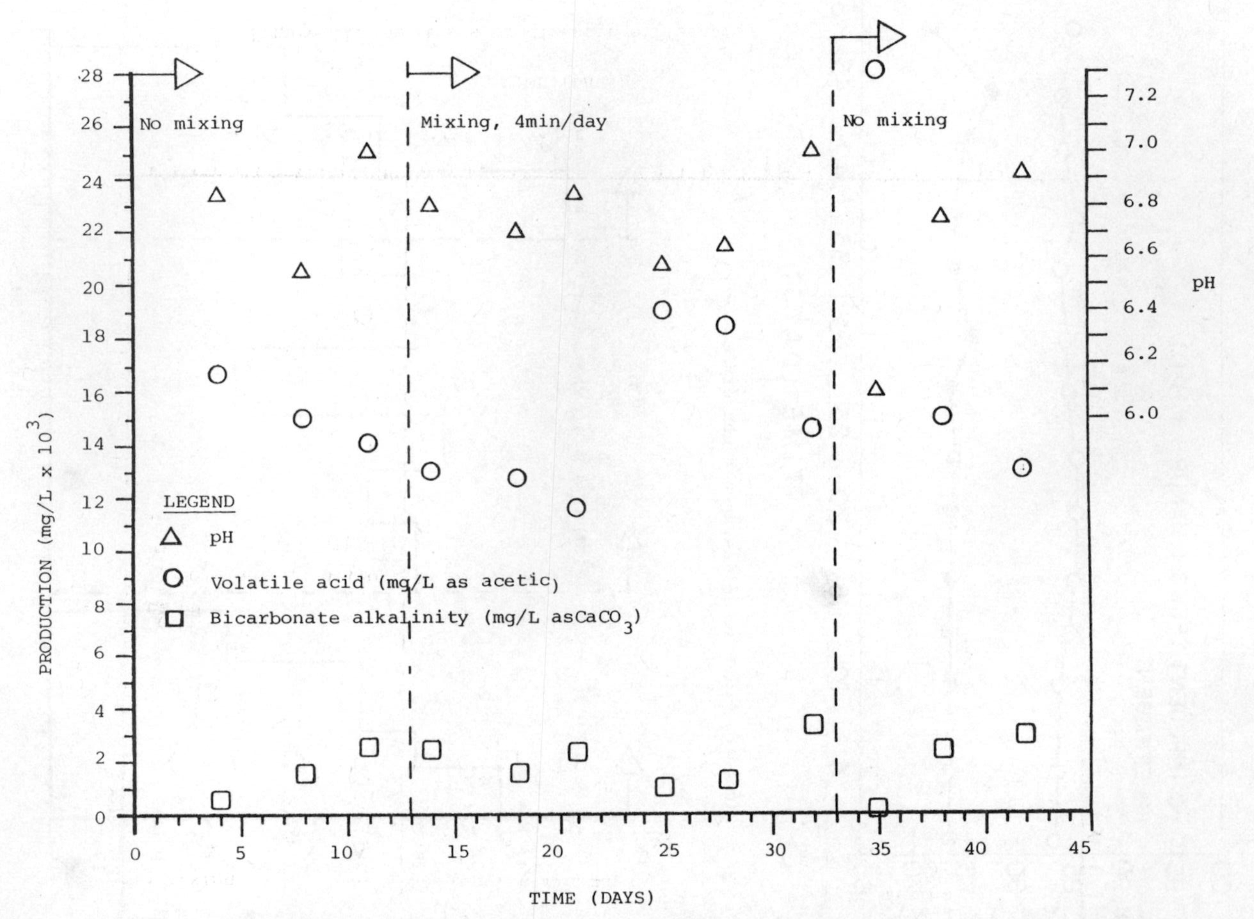

Figure 10. Analyses of influent (fresh pig manure + water)

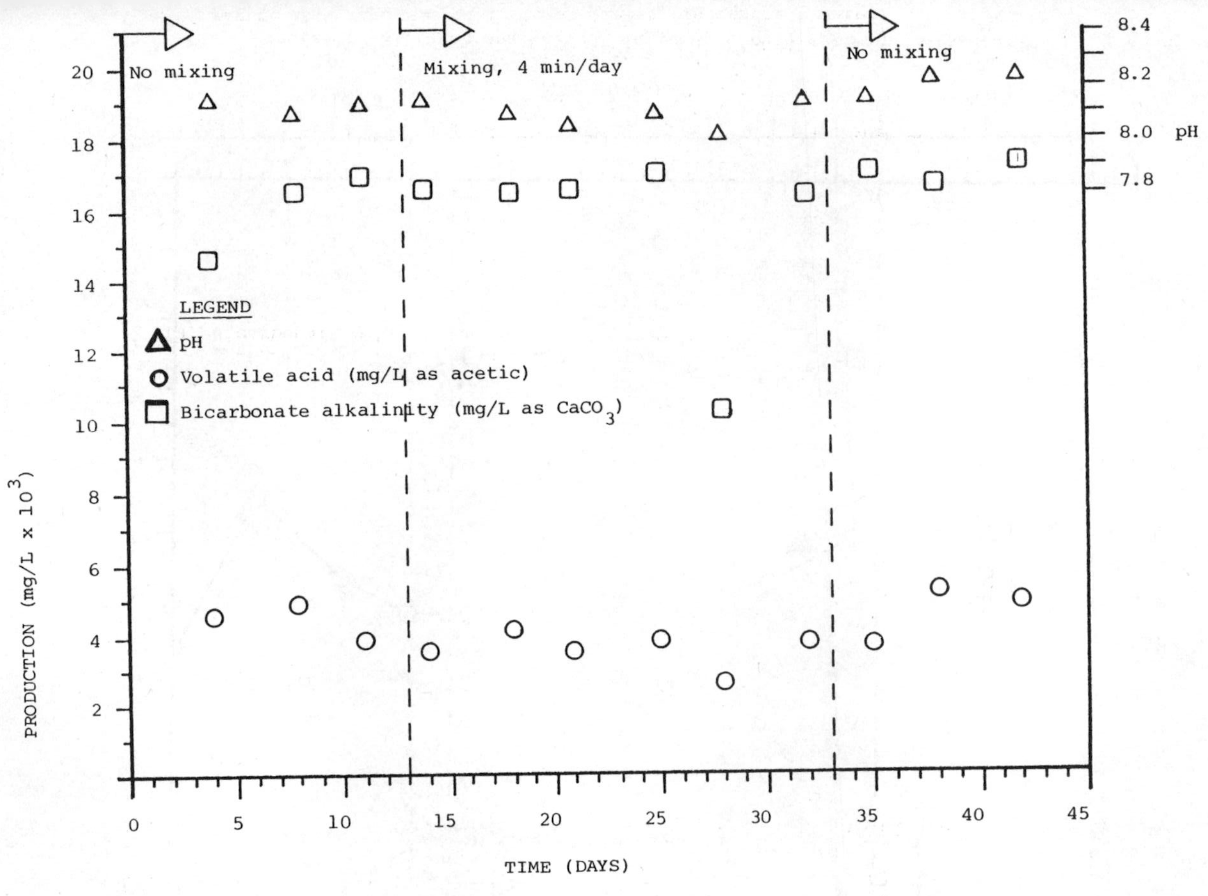

Figure 11. Analyses of digester effluent.

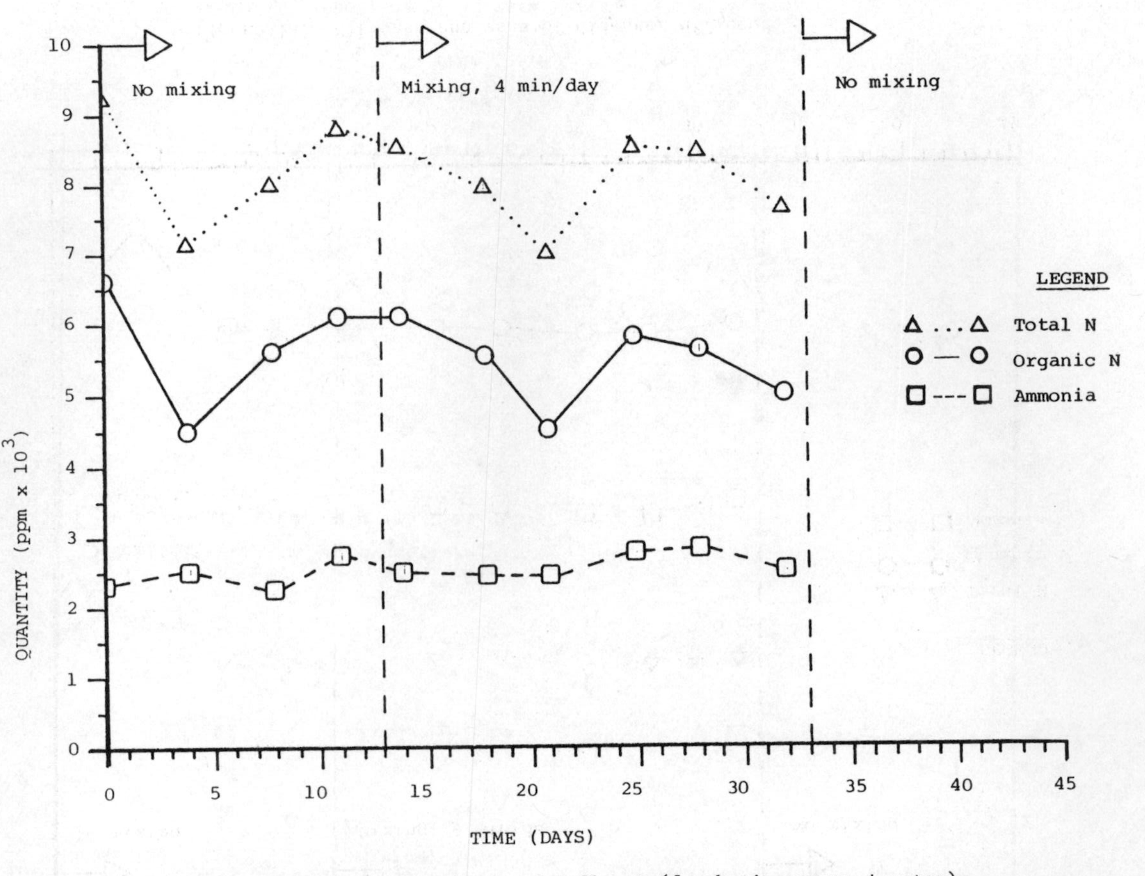

Figure 12. Nitrogen analyses of influent (fresh pig manure + water)

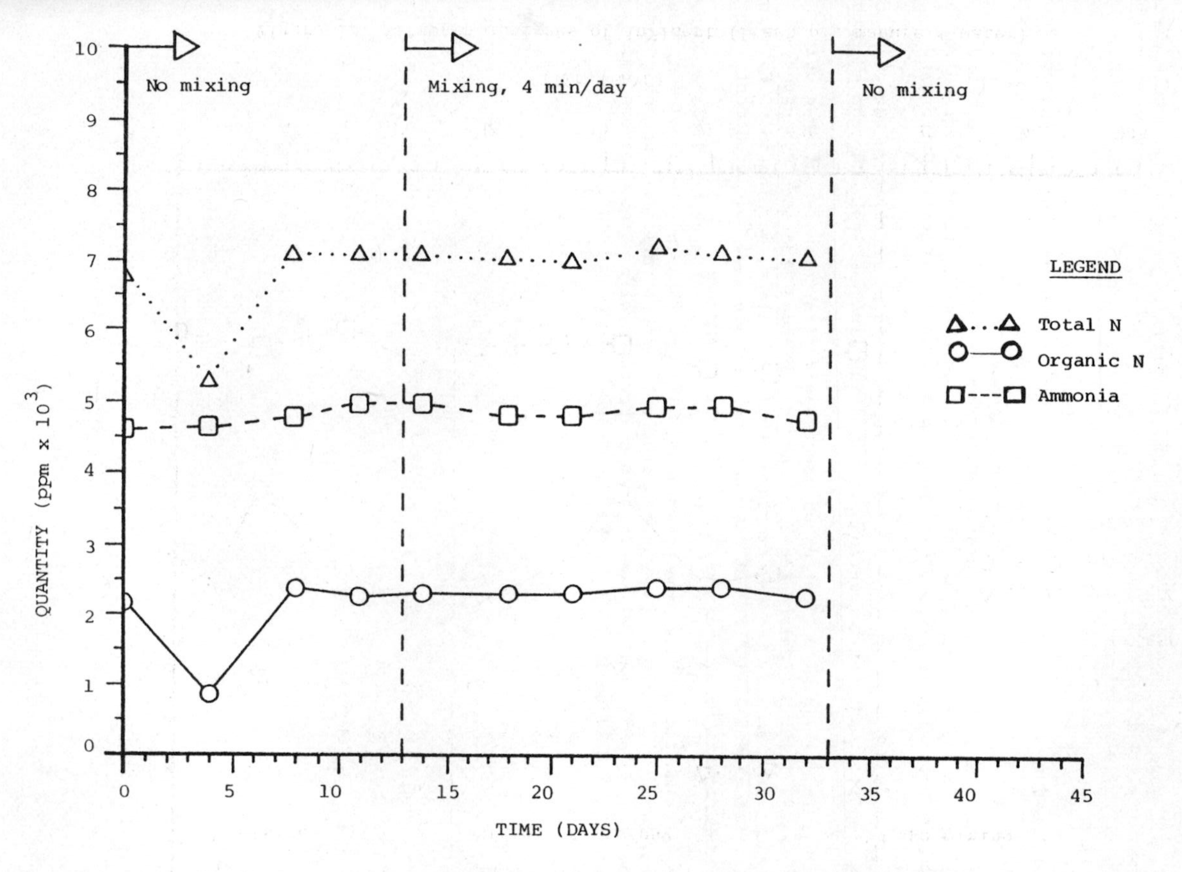

Figure 13. Nitrogen analyses of digester effluent

REFERENCES

Bryant, M.P., E.A. Wolin, M.J. Wolin and R.S. Wolfe (1967). *Methanabacillus omelianskii*, a symbiotic association of two species of bacteria. Archiv Mikrobi 59, 20-31.

Bryant, M.P. and C.L. Davis (1975). Thermophilic methanogenesis from cattle wastes University of Illinois, Urbana Illinois.

Bryant, M.P. (1979). Microbial methane production - theoretical aspects. Universit of Illinois, Urbana, Illinois.

Fischer, J.R., F. Meador, D.M. Sievers, C.D. Fulhage and E.L. Iannotti (1977). Design and operation of a farm-size anaerobic digester for swine. ASAE 77-4052.

Goodrich, P.R., et al., 1978. Utilization of energy from methane generation. Fina report submitted to the Minnesota Energy Agency.

Jewell, W.J., H.R. Davis, R.W. Guest, D.R. Price, W.W. Gunkel, H.R. Capener, G.L. Casler and P.J. Van Soest (1976). Anaerobic fermentation of agricultural wastes - potential for improvement and implementation. Cornell University, Ithaca, N.Y.

Kasper, H.F. and K. Wuhrmann (1978). Kinetic parameters and relative turnovers of some important catabolic reactions in digesting sludge. Appl. Environ. Microbio 36, 1-7.

Lawrence, A.W. (1971). Application of process kinetics to design of anaerobic processes. In: F.G. Pohland (Ed.). Anaerobic Biology Treatment Processes, Advances in Chemistry Series, American Chemical Society, Washington, D.C. p.163-

McCarty, P.L. (1965). Kinetics of waste assimilation in anaerobic treatment. In Proceedings of the Society for Industrial Microbiology.

Pavlicik, V., J. Pos and B. Walczak (1979). A laboratory anaerobic digester for biogas production. Extension paper. School of Engineering, University of Guel

Persson, S. et al. (1977). Experiences from operating a full size anaerobic diges Paper No. 77-4053. Annual Meeting, ASAE, North Carolina State University, Rawleigh, N.C. June 1977.

25

PERFORMANCE OF AN ANAEROBIC-AEROBIC TREATMENT SYSTEM FOR PALM OIL SLUDGE

K. K. Chin* and K. K. Wong**

**Civil Engineering Department, University of Singapore, Singapore*
***Environmental Science Department, University Pertanian, Malaysia*

ABSTRACT

In this study, an anaerobic-aerobic treatment train using 20-day solids retention time for anaerobic digestion followed by 30 day aerated lagoon was able to reduce the BOD content of palm oil sludge from 25,000 mg/L to 50 mg/L. Biological treatment alone was not able to remove the colour of the sludge. Chemical treatment with lime following the anaerobic-aerobic treatment train brought the colour level from 5,000 Hazen units to around 540. Biokinetic coefficients were evaluated for the biological treatment systems.

KEYWORDS

Anaerobic-aerobic treatment; palm oil; wastewater; mesophilic; nutrient balance; biokinetic coefficients; discharge standards.

INTRODUCTION

The Southeast Asian nations are major producers of palm oil with Malaysia alone producing some 2 million metric tons per annum. The conventional crude oil production techniques generate an average of 0.6 metric ton of palm oil sludge for every ton fresh fruit processed. This sludge contains a high concentration of biodegradable organic compounds, suspended solids, colours, and other pollutants which cause damages on their discharge into receiving waters (Table 1). Current Malaysian standards for effluents discharged to watercourses involve a progressive reduction in pollutional discharges over a four-year period (Table 2). When discharging into catchment reserved for potable water supply, effluent BOD is limited to 50 mg/L.

Biological wastewater treatment processes are well suited for treating palm oil wastewater. Work on anaerobic digestion of palm oil sludge at both the mesophilic and thermophilic ranges showed a high degree of organics removal (Kirat and Ng, 1968; Seow, 1977; Sinnappa, 1978; Southworth, 1979; Chin and Tan, 1979a,b; Wong and co-workers, 1980). Activated sludge treatment of the raw sludge removed more than 98% of the BOD and COD at less than 15 day solids retention time (Chin, 1978).

As BOD:N:P ratio for the palm oil wastewater averages around 100:3:0.5, an anaerobic process is reasonable for use as the first stage of wastewater treatment as it

TABLE 1 Characteristics of Palm Oil Waste

Parameter *		Range	Mean
pH		3.5-5.2	4.0
BOD_5	mg/L	20,000-35,000	25,000
COD	mg/L	30,000-60,000	45,000
NH_4-N	mg/L	26-84	45
Organic-N	mg/L	500-800	600
Nitrate-N	mg/L	20-60	30
Total Solids	mg/L	30,000-40,000	35,000
Ash	mg/L	4,000-5,000	4,500
Oil and Grease	mg/L	4,800-12,000	7,500
Total Phosphate	mg/L	68-250	140
Chloride	mg/L	480-1,200	960
Magnesium	mg/L	120-380	200
Calcium	mg/L	40-180	80
Sodium	mg/L	6-50	30
Potassium	mg/L	780-1,600	1,400

* all in mg/L, except pH

TABLE 2 Malaysian Palm Oil Mill Effluent Standards

Parameter	Standards to be implemented at time indicated			
	1 July 1978 to 30 June 1979	1 July 1979 to 30 June 1980	1 July 1980 to 30 June 1981	1 July 1981 to 30 June 1982
BOD_3 @ 30°C, mg/L	5,000	2,000	1,000	500
COD, mg/L	10,000	4,000	2,000	1,000
Total Solids, mg/L	4,000	2,500	2,000	1,500
Total Suspended Solids, mg/L	1,200	800	600	400
Oil and Grease, mg/L	150	100	75	50
Ammonia Nitrogen, mg/L	25	15	15	10
Total Nitrogen, mg/L	200	100	75	50
pH	5-9	5-9	5-9	5-9
Temperature, °C	45	45	45	45

improves the nutrient balance when much of the organic carbon is biologically converted to methane and carbon dioxide. The BOD:N:P ratio would thus be made more favourable for subsequent aerobic treatment.

This paper summarizes the results of an anaerobic-aerobic treatment train for palm oil sludge.

MATERIAL AND METHOD

The anaerobic digestion runs were carried out in 20-litre PVC digesters. Mixing was accomplished by recycling the digester's content with a pump. The anaerobic digestion units were started by slurrying the palm oil sludge with sewage sludge-acclimated culture from domestic sewage treatment plant. The progress of the digestion runs was monitored by daily measurements of gas production and composition. Parameters measured at regular intervals were alkalinity, volatile acids, BOD, COD, volatile solids, pH, nitrogen, and phosphate in accordance with the A.W.W.A.-A.P.H.A. standard methods (1975). Both the settled and non-settled effluents from the anaerobic systems were used as feed to the aerobic systems. Settled effluent had an average holding time of eight hours. The aerobic systems were carried out in 2-litre self-contained reactors fashioned after a similar continuous feed unit developed by Ludzack (1960). Each unit was fed continuously with the effluents from the anaerobic units. The characteristics of the composite effluent from the anaerobic digesters are given in Table 3. Environmental conditions such as pH, dissolved oxygen, temperature, and agitation were held relatively constant.

Temperature and pH were maintained respectively at $30^{\circ}C \pm 1$ and 7.8. Dissolved oxygen was kept at a level of at least 1 mg/L.

For both the anaerobic and aerobic treatment units, specific values of the solids retention time (θ_c) were estimated by $X/(\Delta x/\Delta t)$, where X is volatile solids in the reactor and $(\Delta x/\Delta t)$ represents the total volatile solids wasted per day. The effluents were analyzed daily and the excess amount of biological solids accumulated in the system was withdrawn to keep the mixed liquor suspended solid concentrations on relatively uniform level. To insure that steady state conditions prevailed each unit was operated for a time period equal to at least three times the solids retention time before operational parameters were measured and used in the system analysis.

RESULTS AND DISCUSSION

Previous work on anaerobic digestion showed that effluent BOD, COD, and volatile acids decreased as the solids retention time increased. BOD and COD reduction were higher than 50% for units run on seven-day solids retention time. The growth yield coefficient (Y), the saturation constant K_s, the microbial decay coefficient (k_d), the maximum rate of substrate utilization (k), and the minimum biological solids retention time (θ_c^m) for systems with cell recycle were determined respectively to be 0.052 gVSS/gBOD degraded, 270 mgBOD/L, 0.02 day^{-1}, 2.26 gBOD/gVSS-day, and 10.3 days (Chin and Tan, 1971a). For the once through system, Y, K_s, k_d, k, and θ_c^m were respectively 0.23, 227, 0.036, 0.44, and 15.3 (Chin and Tan, 1979b). In this study, anaerobic units running on 20-day solids retention time were chosen as the first stage treatment process based on effluent quality and favourable BOD:N:P ratio of 100:6.97:1.35 for aerobic treatment. Steady state parameters and the quality of the settled effluent for this unit are given in Table 3. Settling alone removed 5% BOD, 25% COD, 64% total suspended solids and 46.8% total solids from the

TABLE 3 Steady State Parameters of Anaerobic Unit

		Non-settled	8-hour settled
BOD	mg/L	4,720	1,650
COD	mg/L	8,800	6,600
TSS	mg/L	6,700	2,400
TS	mg/L	8,900	4,720
Alkalinity	mg/L	2,850	2,850
Phosphate	mg/L	64	52
Nitrogen	mg/L	320	280
pH		7.1	7.5
Temp.	^oC	30	30
Colour	Hazen unit	5,000	5,000
BOD:N:P		100:6.97:1.35	100:16.97:3.15
Gas production L/g BOD utilized		0.68	-

effluent. After settling the BOD:N:P ratio was 100:16.97:3.15. Colour in the effluent was around 5,000 Hazen units.

Both settled and non-settled effluents were used as feed for the aerobic treatment systems. The steady state treatment parameters using the settled effluent are given in Table 4 and those using the non-settled effluent are given in Table 5.

TABLE 4 Steady State Parameters Using Settled Anaerobic Effluent

θ_c	days	5	10	15	20	25	30
BOD	mg/L	780	320	215	132	92	50
COD	mg/L	2,370	1,840	1,390	1,180	1,075	568
MLVSS	mg/L	430	380	320	283	236	233
TKN	mg/L	182	160	94	78	63.6	32.3
Total P	mg/L	42	29	23	16.2	10.2	9.4
pH		7.8	7.8	7.8	7.8	7.8	7.9
Colour	Hazen unit	5,000	5,000	5,000	5,000	5,000	5,000

The anaerobic-aerobic treatment train using 20-day solids retention time for anaerobic treatment and 25-day solids retention time for aerobic treatment produce an effluent of 297 mg/L BOD and 907 mg/L COD which satisfied the fourth generation Malaysian effluent standards. For treatment train where settled effluent from anaerobic units was fed to the aerobic units, the effluent BOD concentration of th 30-day aerobic unit was 50 mg/L which satisfied the Malaysia effluent standards fo

TABLE 5 Steady State Parameters Using Non-Settled Anaerobic Effluent

θ_c	days	5	10	15	20	25	30
BOD	mg/L	1,309	828	702	562	297	246
COD	mg/L	2,268	1,443	1,277	1,214	907	510
MLVSS	mg/L	1,636	1,220	891	749	676	568
TKN	mg/L	198	174	150	122	91	35
Total P	mg/L	46	32	24	21	14	10.3
pH		7.8	7.8	7.8	7.8	7.8	7.8
Colour	Hazen unit	5,000	5,000	5,000	5,000	5,000	5,000

discharge into protected catchment. The COD concentration of 568 mg/L was however higher than the allowable discharge standard of 100 mg/L. Both the anaerobic and aerobic treatment systems were unable to remove colour in the effluent which was maintained at a level of 5,000 Hazen units. Lime dosage of up to 800 mg/L $Ca(OH)_2$ resulted in a reduction of 89.2% of the colour to a level of 540 Hazen unit. After lime treatment, COD was reduced to 64 mg/L. Much of these COD were difficult to degrade biologically. The sludge settled well with around 2% solids after biological treatment. A SVI of around 100 was observed for all aerobic treatment units.

Applying materials balances to the aerobic microbial system, the following basic biokinetic model for completely mixed systems were developed as follows (Lawrence and McCarty, 1970):

$$S = \frac{K_s[1 + k_d\theta_c]}{\theta_c[YK - k_d] - 1} \tag{1}$$

and

$$X = \frac{Y[S_o - S]}{1 + k_d\theta_c} \tag{2}$$

where S_o and S are respectively the influent and effluent substrate concentrations. By plotting $\frac{S_o - S}{X\,\theta_c}$ against $\frac{1}{\theta_c}$, Y and k_d were determined. Similarly, K_s and k were evaluated through a plot of $\frac{\theta_c}{1 + \theta_c k_d}$ against $\frac{1}{S}$. The biokinetic coefficients evaluated are given in Table 6.

Biokinetic coefficients evaluated were not significantly different from those obtained for raw palm oil sludge, skim milk, and yeast. The settled effluent from anaerobic digestion units contained a higher proportion of COD that was difficult to degrade biologically. Although the BOD:N:P ratio was more favourably for aerobic growth system, θ_c^m evaluated was higher than systems fed with non settled anaerobic effluent and raw palm oil sludge.

TABLE 6 Biokinetic Coefficients for Aerobic Systems

Feed	Y gVSS/g	k_d day^{-1}	K g/gVSS-day	K_s mg/L	θ_c^m day	Coeff. Basis	References
Raw Palm Oil Sludge	0.8	0.2	0.43	442	7.0	BOD_5	Chin 1978
Anaerobically, Treated Palm Oil Sludge Eff.							
(a) Settled	0.89	0.19	0.35	276	13	BOD_5	This study
	0.17	0.12	2.10	7,350	20	COD	This study
(b) Non-Settled	0.9	0.20	0.44	220	5.6	BOD_5	This study
	0.45	0.18	1.11	1,038	3.7	COD	This study
Skim Milk	0.48	0.045	5.1	100	0.42	BOD_5	Gram,1956
Yeast	0.877	0.079	0.91	680		BOD_5	Wu and Kao, 1976
Yeast	0.944	0.102	1.23	554		BOD_5	Wu and Kao, 1976

CONCLUSION

The anaerobic-aerobic treatment system for palm oil sludge using 20-day solids retention time for anaerobic digestion followed by aerated lagoons produced an effluent quality satisfying the Malaysian effluent discharge standards. The system, however, was not effective in colour removal. Treatment with 800 mg/L $C_a(OH)_2$ brought the colour level of the biologically treated effluent down from 5,000 Hazen units to 540 and the COD concentration level to 64 mg/L. The biokinetic coefficients evaluated for the aerobic units were not significantly different from those obtained for raw palm oil sludge without prior treatment by anaerobic digestions.

REFERENCES

American Water Work Association, American Public Health Association (1975). Standards Methods for the Examination of Water and Wastewater 14th Ed., Washington, D.C.

Chin, K.K. (1978). Palm oil waste treatment by aerobic process. Proceedings of the International Conference on Water Pollution Control in Developing Countries. Bangkok, Thailand. pp. 505-511, Feb.

Chin, K.K. and G.T. Tan (1979a). Anaerobic treatment of palm oil sludge. Proceedings of the Third Turkish-German Environmental Engineering Symposium. Istanbul Turkey, July.

Chin, K.K. and G.T. Tan (1979b). An anaerobic treatment system for palm oil mill effluent. Proceedings Intern'l Conf. on Agricultural Engineering in National

Development. Universiti Pertanian Malaysia, Sedang, Malaysia, Sept.
Gram, A.L. (1956). Reaction Kinectics of Aerobic Biological Processes, I.E.R. Series 90, Report No. 2 Sanitary Engineering Laboratory, University of California, Berkeley, May.
Kirat, S. and S.H. Ng (1968). Treatment and disposal of palm oil effluent. Malaysian Agric. J., 46, 316-323.
Lawrence, A.W. and P.L. McCarty (1970). A unified basis for biological treatment, design and operation. J. Sanitary Engineering Div., American Soc. Civil Eng., 96 (SA3), 757-778.
Ludzack, F.J. (1960). Laboratory model activated sludge unit, Jour. Water Pollution Control Federation, 32, 605.
Seow, C.M. (1977). The role of chemistry in anaerobic digestion of palm oil effluent. Proc. Symposium on Improving the Quality of Life in Malaysia: The Role of Chemistry, Institute of Chemistry, Malaysia, Kuala Lumpur, Malaysia, March.
Sinnappa, S. (1978). Treatment studies of palm oil mill waste effluent. Proceedings, International Conference on Water Pollution Control in Developing Countries. Bangkok, Thailand, pp. 525-537, February.
Southworth, A. (1979). Palm oil factory effluent treatment by anaerobic digestion in lagoons. Proc. 35th Ind. Waste Conference, Purdue University, Lafayette, Indiana, U.S.A., Ann Arbor Science Publishers, Michigan.
Wong, K.K., Omar b. Muhammad Zain and Ramdzani b. Abdullah (1980). Jour. of the Institution of Engineers, Malaysia Petaling Jaya, Malaysia (In Press).
Wu, Y.C. and C.F. Kao (1976). Activated sludge treatment of yeast industry wastewater. Jour. Water Pollution Control Federation, 48, 2609-2618.

26

POTATO PROCESSING WASTEWATER TREATMENT USING ANAEROBIC FERMENTER-FILTER SYSTEMS

T. Viraraghavan*, R. C. Landine*, A. A. Cocci*, G. J. Brown* and K. C. Lin**

**ADI Limited, Consulting Engineers, Fredericton, N. B., Canada*
***Department of Civil Engineering, University of New Brunswick, Fredericton, N. B., Canada*

ABSTRACT

A recent evaluation conducted by the authors demonstrated the economic advantage of an anaerobic system over other alternatives for treating potato processing wastewater. The analysis also showed that an anaerobic fermenter-filter system would be the best choice. A system of this type has already been adopted in the treatment of potato processing wastewater in England under our advice and similar systems are being developed now for installations in Canada and USA. Commencing in January 1978, laboratory model studies were conducted on the use of an anaerobic fermenter and an anaerobic filter (upflow modes for both) in series (as separate reactors), horizontal anaerobic filters, and a unified reactor integrating the anaerobic filter at the exit end of the fermenter. The studies generally indicate excellent BOD and COD removals (greater than 90%) through the combined system.

This paper presents a brief review of anaerobic treatment for food processing wastewaters, a summary evaluation of alternative treatment systems for potato processing wastewater, summary results of research studies and essential design information on the system located in England.

KEYWORDS

Potato processing wastewater treatment; anaerobic fermenter; anaerobic lagoon; anaerobic filter; horizontal anaerobic filter.

INTRODUCTION

Stabilization of many food processing wastewaters by aerobic processes is rather difficult, energy intensive and costly, essentially due to their variable nature and high strength (BOD and SS). Such wastewaters are, however, quite amenable to anaerobic treatment.

The low rate of microbial production in anaerobic treatment results in minimal sludge disposal problems and low nutrient requirements. Anaerobic processes exert low power demands, have an ability to lie dormant for considerable time periods without causing problems, and have the potential for energy recovery through the utilization of biogas produced. Some disadvantages include concern over odours, low temperature and start-up.

REVIEW OF ANAEROBIC SYSTEMS

The anaerobic process most commonly applied to food processing wastewater is the anaerobic lagoon. Anaerobic lagoon treatment has met with both successes and failures in various applications. Lagoon failures can often be traced to any one or combination of these factors: improper startup procedures, unfavourable operating conditions, and poor design. Table 1 provides data on some operating lagoon systems treating food processing wastewaters.

TABLE 1 Anaerobic Lagoon Systems

Type of waste	Location	Scale	Operating temperature, °C	Loading rate kg BOD/m^3.d	Removal effic., %
Tomato processing	Indiana	Full	--	0.174	73.6
Potato processing	S. Dakota	Full	21	0.728	73.7
Potato processing	Idaho	Pilot	16-20	0.075-0.373	Minimal
Municipal plus potato processing	N. Dakota	Full	4-20	0.08 -0.16	30

Two other anaerobic processes (the anaerobic contact process and the anaerobic filter) have also been applied in the treatment of food processing wastewaters. The anaerobic contact process is suitable for wastewaters high in both soluble and suspended organics. The major constraint in the use of anaerobic contact process is often poor solids separation, due to the presence of evolved gases. Full scale application of this process is quite limited in the food processing industry. Table 2 includes data from bench scale, pilot plant and full scale applications.

TABLE 2 Anaerobic Contact Process Systems

Type of waste	Location	Scale	Operating temperature, °C	Loading rate kg/m^3.d	Removal effic., %
Potato processing	Idaho	Pilot	22	1.92	43 BOD
Rum distillery slops	Puerto Rico	Pilot	35	0.56 - 3.57	81-57 COD
Pear peeling	Ontario	Bench	35	5.00	91 COD
Bean blanching	Ontario	Bench	35	7.30	88 COD
Sugar beet processing	Sweden	Full	35	--	98 BOD
Meat packing	Minnesota	Full	32-34	2.40	91 BOD

The anaerobic filter is better suited for treatment of wastewaters high in soluble organics and low in suspended matter. Anaerobic filters have been used in a number of bench and pilot scale investigations involving food processing wastewaters. A full scale anaerobic filter was used to treat wheat starch wastes; recently a full scale filter installation has become operational in England treating potato processing wastewaters. Table 3 furnishes information on some of the anaerobic filters treating food processing wastewaters.

Additional information on these anaerobic systems have been provided by McKim and others (1979). A recent system developed for anaerobic treatment in the Netherland is the "Upflow Anaerobic Sludge Blanket" (UASB) process, described in detail by Lettinga (1979). With this process, a high solids retention time (SRT) is accomplishe

TABLE 3. Anaerobic Filter Systems

Type of waste	Location	Scale	Operating temperature, °C	Loading rate $kg/m^3.d$	Removal effic., %
Wheat starch processing	Washington	Full	32	3.79	56 BOD
Potato processing	Idaho	Pilot	19	0.93	85 BOD
Brewery	Kentucky	Bench	35	1.60	90 BOD
Tapioca starch	Thailand	Bench	26-30	4.00	92 COD
Molasses	W. Germany	Bench	35	8.40	85 COD

in an unpacked reactor by equipping the upper part of the reactor with a gas separator/settler device. Its performance has been demonstrated in the laboratory with sugar beet waste, cannery, methanolic and dairy wastes, and also in a 6 m^3 pilot plant and in 200 m^3 and 800 m^3 full scale plants with sugar beet wastes. High loading rates in the region of 15-30 kg $COD/m^3.d$ at 25-35°c were achieved in the pilot plant, with 95 percent removal of dissolved organic matter. In the 200 m^3 full scale plant at a CSM sugar factory in Holland, loadings of 15 kg $COD/m^3.d$ were maintained successfully.

A two-stage reactor of a new design was recently developed in Sweden for anaerobic treatment by Norrman and Frostell (1977). The first stage reactor is basically a holding tank used for hydrolysis of the substrate and the second stage is an anaerobic filter. Results of experiments using a synthetic wastewater showed greater than 95% COD removal through the entire reactor for a loading rate of 0.43-1.70 kg $COD/m^3.d$ calculated on the total reactor volume (hydrolysis reactor plus anaerobic filter).

EVALUATION OF ALTERNATIVE SYSTEMS

Eleven secondary treatment system alternatives were evaluated to choose the most economic scheme for treating potato processing wastewater with an estimated average flow of 8000 m^3/d and BOD of 27,000 kg/d. Treatment systems considered and their corresponding costs are presented in Table 4. The treatment system (Number 9 in Table 4) consisting of two anaerobic lagoons, in parallel, incorporating an anaerobic filter at the exit end of each lagoon, followed by an aerated lagoon and clarifier (existing) offered the most economical choice. The major advantages of the proposed anaerobic - aerobic system, design criteria and other special design features are available elsewhere (Cocci and others, 1980).

SUMMARY RESULTS OF RESEARCH

A bench-scale system consisting of an anaerobic fermenter and anaerobic filter, both in the upflow mode was operated at the University of New Brunswick, Department of Civil Engineering, on potato processing wastewater. Figure 1 shows a schematic diagram of the model.

Table 5 presents system loading rates, while Table 6 shows system performance. The organic loading rates on the anaerobic fermenter were in the 0.149 - 0.719 kg $BOD/m^3.d$ range; the overall removals of BOD and SS for the anaerobic fermenter-filter system were in the ranges of 97.1% - 98.4% and 88.8% - 97.3%, respectively.

TABLE 4 Estimated Costs for Alternative Treatment Systems

Systems	Capital Costs (Million Dollars)	Annual Costs: Energy	O&M+	Total*
		(thousand 1978 Dollars)		
1. Carbon absorption	7.00	320	2910	3830
2. Biotower	7.00	100	330	1250
3. Biodrum	7.00	-	-	-
4. Deep shaft	3.69	80	410	895
5. Activated biofilter	3.25	127	486	913
6. Aerated lagoon	1.77	276	372	605
7. Anaerobic filter + aerated lagoon	1.95	86	293	459
8. Anaerobic lagoons + aerated lagoon	1.93	97	246	500
9. Anaerobic lagoons and anaerobic filters + aerated lagoon	2.00	35	169	432
10. Anaerobic lagoons and anaerobic filters + biodrum	2.57	23	160	498
11. Anaerobic lagoons and anaerobic filters + biotower	2.21	13	160	450

+O & M costs include maintenance and sludge removal but do not include energy recovery from methane.
*Including debt charges for a 15 year write-off at 10% interest.

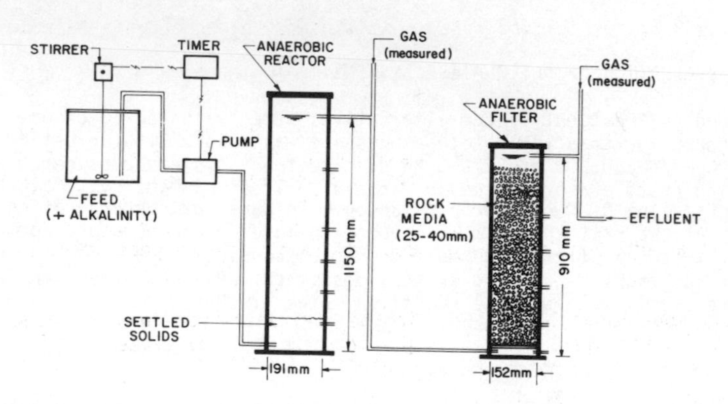

Fig. 1. Schematic diagram of "Brown Model".

The system performance was excellent and the effluent quality was found to be fair consistent. Detailed information on the performance of the experimental system has been reported by Lin and Brown (1980).

During 793 days of operation, 4170 g of VSS were fed to the fermenter and the VSS accumulation in the fermenter was 315 g. Therefore, 92.4% of the VSS input was an aerobically decomposed. Further solids degradation information has been given by

TABLE 5 System Loading Rates

Loading rate number	BOD loading, fermenter	$kg/m^3.d$ filter*	SS loading, fermenter	$kg/m^3.d$ filter *
1	0.149	0.067	0.128	0.015
2	0.290	0.069	0.273	0.021
3	0.719	0.080	0.156	0.095

*based on total filter volume (media + liquid)
$kg/m^3.d \times 62.43 = lb/1000\ ft^3.d$

TABLE 6 System Performance

Loading rate number	Sample	BOD mg/L	SS mg/L
1	lagoon influent*	2080	1790
(day 58 to 152)	lagoon effluent	467	107
	fermenter effluent	60	107
2	lagoon influent*	2030	1910
(day 153 to 341)	lagoon effluent	241	74
	fermenter effluent	54	51
3	lagoon influent*	5030	1090
(day 341 to 518)	lagoon effluent	277	332
	fermenter effluent	78	122

*fermenter influent is primary clarifier influent for loading rates 1 and 2; for loading rate 3, it includes blancher water also.

Brown and others (1980).

At the conclusion of loading rate number three (Table 6), a facultative aerated lagoon was added to the bench scale system. The aerated lagoon was fed anaerobic filter effluent and had a detention time of 2-1/2 days. The lagoon achieved 66% BOD removal at an average loading rate of 0.018kg $BOD/m^3.d$, over a four month period, resulting in an effluent of 15 mg/L BOD and 33 mg/L SS respectively.

A laboratory model anaerobic filter in the horizontal mode was operated using potato processing wastewater. Figure 2 shows the horizontal filter model.

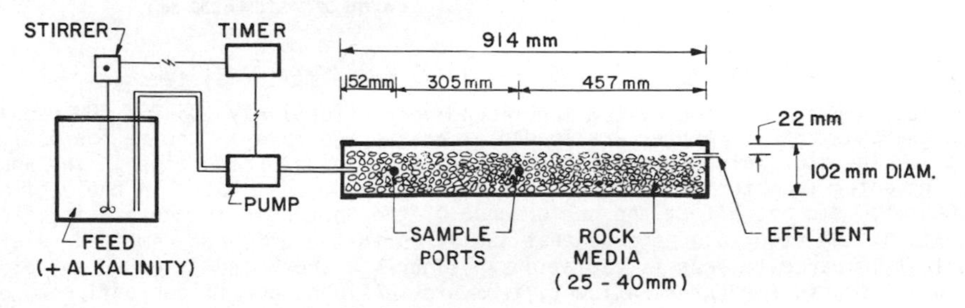

Fig. 2. Horizontal filter.

Table 7 presents a summary of the performance of the horizontal filter.

TABLE 7 Performance of Horizontal Anaerobic Filter

Description	Characteristics BOD, mg/L Range	Average	COD, mg/L Range	Average	SS, mg/L Range	Average
Influent	770-1320	1090	1310-2310	1890	155-635	341
Effluent	68-140	98	122-240	165	40-60	50
Efficiency of removal (%)		91		91		85

NOTE: 1. Averages are time weighted
2. Average BOD loading on filter - 0.78 kg/m^3 d
3. Average COD loading on filter - 1.35 kg/m^3 d
4. Average SS loading on filter - 0.24 kg/m^3 d
5. Lime dosage of 500 mg/L added to the feed
6. Average liquid retention time = 0.60 day

The filter achieved BOD and COD removals of 91% at loading rates of 0.78 kg BOD/m^3.d and 1.35 kg COD/m^3d. Details are presented elswehere (Landine and others, 1980).

An experimental study of potato processing wastewater treatment using a unified fermenter-filter system was carried out over a period of approximately one year. Figure 3 shows the model of the unified anaerobic fermenter - filter system.

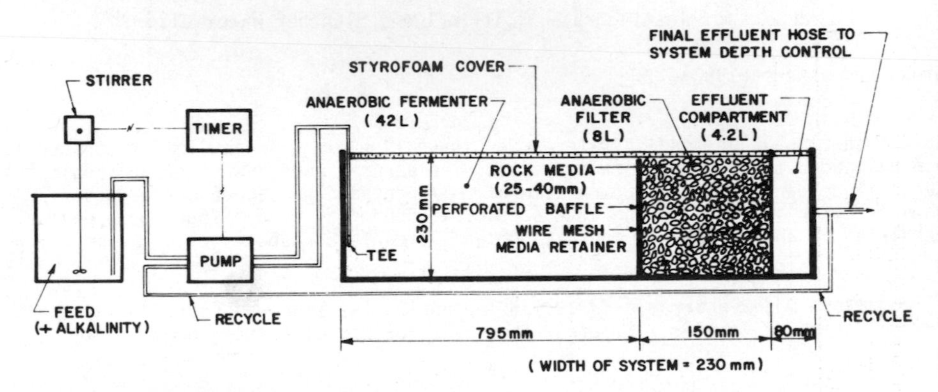

Fig. 3. Unified anaerobic fermenter - filter system

The study showed that the system functioned very effectively (85-95% COD removal) when the anaerobic fermenter was loaded at an average rate in the region of 1 kg CO m^3.d. The study also revealed that decreasing the detention time in the anaerobic fermenter from ten to four days, while keeping loading rates in the region of 1 kg COD/m^3.d did not affect the performance of the anaerobic fermenter significantly Results of the study also showed that SS levels in the effluent remained relatively constant compared to feed SS variations. Figures 4 and 5 show the COD and SS level in the anaerobic fermenter influent, fermenter effluent and filter effluent, respectively.

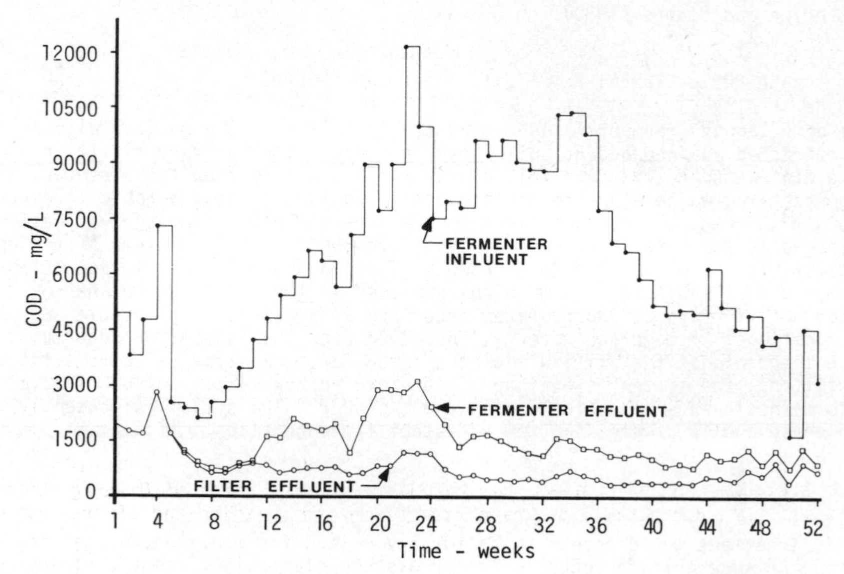

Fig. 4. COD of anaerobic fermenter influent (●), fermenter effluent (□) and filter effluent (○)

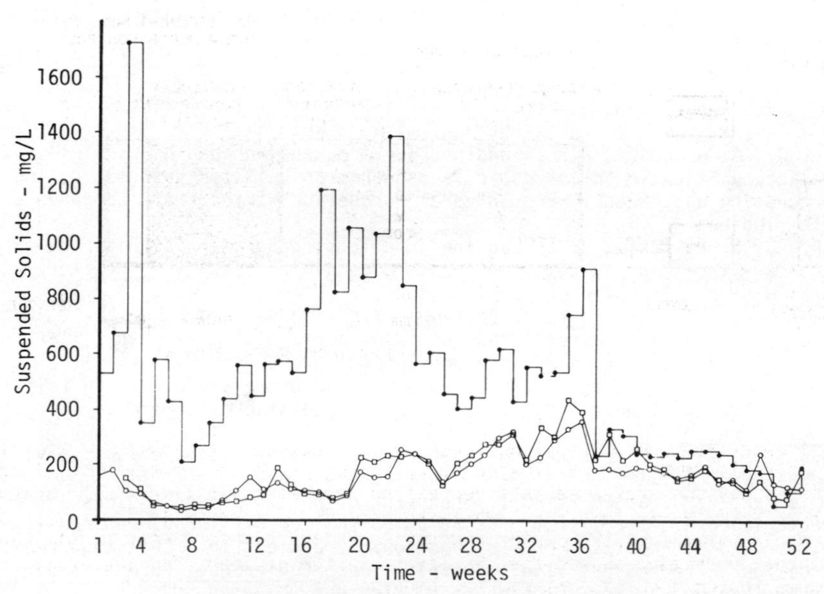

Fig. 5. SS in anaerobic fermenter influent (●), fermenter effluent (□) and filter effluent (○)

A detailed account of the unified system and its performance is available in a paper by Landine and others (1980).

ANAEROBIC FERMENTER - FILTER SYSTEM IN ENGLAND

An anaerobic lagoon (fermenter) and an anaerobic filter in the horizontal mode have been constructed as independent units between the existing primary clarifier and existing biotowers, to reduce significantly the organic load on the biotowers in a potato processing wastewater treatment plant in England. The anaerobic lagoon is of 22 000 m^3 liquid volume, equivalent to ten days detention time and was designed to be loaded at the rate of approximately 0.5 kg BOD/m^3.d. The volume of the anaerobic filter following the lagoon is 2600 m^3, with a design hydraulic retention tim of 12 h. Rock media (75-125 mm size) was used in the filter. The anaerobic fermenter - filter system incorporates necessary arrangements for recycle and alkalinity additions. A membrane cover is installed over the lagoon for odour control and to collect biogas for utilization at a later date. A straw mat covers the anaerobic filter. The system is designed to achieve approximately 75% BOD removal in order to maintain an acceptable load on the biotower. The system is presently in startup phase and is progressing towards stability conditions.

SUMMARY

Anaerobic treatment would appear to be ideally suited for many food processing wastewaters. An anaerobic fermenter - filter system offers significant cost savings and process advantages for the treatment of potato processing wastewater. Laboratory studies have shown that such a system performs exceedingly well, removing approximately 90% of the organic load at loadings rates of 1 kg COD/m^3.d. In view o the simplicity of construction and operation along with low energy costs and prospects of energy recovery, these systems offer significant advantages in compariso to other systems.

REFERENCES

Brown, G. J., A. A. Cocci, R. C. Landine, T. Viraraghavan and K. C. Lin (1980). Sludge accumulation in an anaerobic lagoon-anaerobic filter system treating potato processing wastewater. Presented at Purdue Industrial Waste Conference, Lafayette, Indiana.

Cocci, A. A., M. P. McKim, R. C. Landine and T. Viraraghavan (1980). Evaluation of secondary treatment systems for treating potato processing wastewater. Agricultural Wastes. Accepted for publication.

Landine, R. C., G. J. Brown, A. A. Cocci, T. Viraraghavan and K. C. Lin (1980). Potato processing wastewater treatment using horizontal anaerobic filters. Canadian Institute of Food Science and Technology Journal ,(in print).

Landine, R. C., A. A. Cocci, T. Viraraghavan, G. J. Brown and K. C. Lin (1980). Anaerobic fermentation/filtration of potato processing wastewater. Journal Wa Pollution Control Federation,(in print).

Lettinga, G. (1979). Direct Anaerobic treatment handles wastes effectively. Ind trial Wastes, 25 (1).

Lin, K. C. and G. J. Brown (1980). Treatment of potato processing wastewater by anaerobic lagoon-filter system. Canadian Journal of Civil Engineering,7 (2), 373

Norrman, J. and B. Frostell (1977). Anaerobic treatment in a two-stage reactor a new design. Proceedings 32nd Industrial Waste Conference, Purdue University, Ann Arbor Science Publishers, Ann Arbor, Michigan, USA.

McKim, M. P., A. A. Cocci, T. Viraraghavan and R. C. Landine (1979). Combined a obic-anaerobic treatment produces low cost, quality effluent. Industrial Waste 25 (3), 53-55.

27

BACTERIAL DEGRADATION OF LEATHER WASTES

S. Datta and A. L. Chandra

Dept. of Microbiology, Bose Institute, Calcutta-700 009, India

ABSTRACT

In this investigation, the leather degrading capacity of a soil bacterium has been assessed and the optimum conditions for degradation were determined.

KEYWORDS

Leather degradation; nitrogen solubilized; amino nitrogen produced; enzyme assays; optimal requirements.

INTRODUCTION

Bacteria play an important role in the biodegradation of hides and skins, both in the raw and salted conditions (Rao, Nandy and Santappa, 1976, 1977; Thompson, Woods and Welton, 1972; Welton and Woods, 1973), but reports on the biodegradation of leather, manufactured from hides and skins, are rare. In this paper, bacterial degradation of leather wastes is being reported.

MATERIALS AND METHODS

Soil bacteria were screened for proteolytic activity, initially by the nutrient gelatin agar-$HgCl_2$ method (Williams and Cross, 1971) and, finally, by the 4% gelatin liquefaction method. The bacteria which liquefied gelatin stabs completely in 18 h were selected and tested for leather degrading capacities.

Inoculum

Bacteria were suspended in sterile distilled water from 18 h grown cultures on nutrient agar slants. A standard inoculum size of 10^7-10^8 cells was used to inoculate 500 mg leather.

Growth on Leather

Calf leather shavings were collected from tannery refuse. Vegetable and chrome tanned leathers were cut into small pieces and sterilized at 5 lb/in^2 for 10 min. A basal salts medium (Noval and Nickerson, 1959) containing mg/L:K_2HPO_4,1500; $MgSO_4.7H_2O$, 50; $CaCl_2$, 50; $FeSO_4.7H_2O$, 15 and $ZnSO_4.7H_2O$, 5 was sterilized at 15 lb/in^2 for 15 min in 500 mL flasks. 500 mg leather was suspended in 100 mL basal salts medium containing the standard inoculum. The flasks were shake incubated at 37°C for 72 h, centrifuged at 5000 rev/min for 40 min. Incubation for longer than 72 h did not increase the amount of degradation. The supernatants were used for analyses and enzyme assays.

Analyses

Amino nitrogen was estimated by the ninhydrin colorimetric method (Yemm and Cocking 1955) and expressed as glycine equivalents. Total nitrogen in the substrates and supernatants was estimated with Nessler's reagent (Vogel, 1961). The undegraded substrates and the degraded substrate sediments were digested with concentrated H_2SO_4 according to standard procedure, prior to treatment with Nessler's reagent.

Enzyme Assays

Proteolytic activity was measured as an increase in absorbancy in the supernatant at 280 nm with the supernatant of uninoculated leather as the control. Proteolytic activity was also measured by the modified Anson's (1938) method with casein. Five mL of the supernatant was added to 10 mL of 5% casein and digested at 45°C for 30 min. Five mL of the digested mixture was treated with 10 mL of 0.3 N trichloroacet acid (TCA), warmed and filtered. To 5 mL filtrate, 10 mL 0.5 N NaOH and 3 mL twice diluted Folin phenol reagent were added, shaken and the colour intensity was read with a Klett-Summerson colorimeter using filter No. 66. A control set with casein was used to which TCA was added before adding the enzyme. Enzyme activity was expressed tentatively as mg tyrosine.

Optimal Requirements

Effects of inorganic salts (K_2HPO_4, $MgSO_4.7H_2O$, $CaCl_2$, $FeSO_4.7H_2O$ and $ZnSO_4.7H_2O$ added separately/in combinations at the basal salts medium concentrations), 1% sugars (glucose, fructose, sucrose, mannose and galactose), 1% inorganic nitrogen sources (NH_4NO_3, $NaNO_3$, $NaHPO_4$ and NH_4Cl), 1% amino acids (glycine, alanine, lysin and tyrosine), pH (5-9), temperature (28-50°C), incubation periods (24-96 h) were studied. Estimations were made by the release of amino nitrogen in the supernatan

RESULTS AND DISCUSSION

Out of the 300 soil bacteria tested, one bacterium designated as PB_4 could degrade both vegetable and chrome tanned calf leather in 72 h. The results are presented in Table 1.

TABLE 1 Degradation of Leather by Bacterium PB_4[a]

Type of calf-leather	Solubilised nitrogen(%)	Glycine equivalent[b]	Proteolytic Activity A_{280}[c]	tyrosine[d]
Vegetable tanned	22.3	8.0	0.55	1.6
Chrome-tanned	20.5	6.5	0.237	2.0

[a]Average of 3 digestions

[b]Amino nitrogen, mg per 500 mg substrate

[c]Readings taken at a dilution of 1:100

[d]mg of tyrosine per 500 mg substrate

The bacterium appeared to be a *Bacillus* species. The effects of optimum requirements are expressed as mg glycine equivalents per 500 mg leather. Incorporation of glucose, fructose, mannose and galactose into the basal salts medium could not increase degradation of chrome tanned leather from 6.5 to 9.5 glycine equivalents. None of the nitrogen sources could yield better results. Addition of 1.5 g/L K_2HPO_4 to plain distilled water was found to be as suitable as the basal salts medium; in the phosphate solution, glycine equivalents were 9.2 and 6.9 in vegetable and chrome tanned leathers, while in the basal salts medium, 8.0 and 6.5 glycine equivalents were formed in vegetable and chrome tanned leathers, respectively. When incubated at 28°C, the amounts of glycine equivalents produced from vegetable and chrome tanned calf leather were 8.2 and 6.4, while at 37°C the amounts were 8.0 and 6.5. So degradation was optimum at a range of 28-37°C. Degradation of both types of leather occurred at higher temperature; at 50°C, glycine equivalents were 7.4 and 6.0 in vegetable and chrome tanned leather, respectively. pH was found to be optimum at 7-8 for both types; at pH 7 and 8, glycine equivalents were 8.0 and 8.2 in vegetable tanned leather and in chrome tanned leather it was 6.7 at both pH 7 and 8.

CONCLUSIONS

Considerable interest has developed in composting as a means of solid waste disposal. While this interest has centred mainly around the cellulosic wastes, this work has shown that a similar interest could also develop around the nitrogenous wastes like leather. The leather wastes could be degraded by bacteria into utilizable nitrogenous products. The bacterium isolated by us appeared to be ideal for this purpose, since it could degrade both vegetable and chrome tanned leathers in an inexpensive K_2HPO_4 solution and could withstand a temperature of 50°C.

ACKNOWLEDGEMENT

The authors are grateful to the Director and to the Head of the Microbiology Dept. for laboratory facilities.

REFERENCES

Anson, M.L. (1938). The estimation of pepsin, trypsin, papain and cathepsin with haemoglobin. J. Gen. Physiol., 22, 79-89.

Noval, J.J. and W.J. Nickerson (1959). Decomposition of native keratin by Streptomyces fradiae. J. Bacteriol., 77, 251-263.

Rao, R.S., S.C. Nandy and M. Santappa (1976). Hydrolytic action of some anaerobic strains of the genus Clostridium on raw skin. Leather Science, 23, 263-271.

Rao, R.S., S.C. Nandy and M. Santappa (1977). Skin hydrolysis by two newly isolated facultatively anaerobic organisms. Leather Science, 24, 1-7.

Thompson, J.A., D.R. Woods and R.L. Welton (1972). Collagenolytic activity of aerobic halophiles from hides. J. Gen. Microbiol., 70, 315-319.

Vogel, A.I. (1961). A Text-book of Quantitative Inorganic Analysis Including Elementary Instrumental Analysis. Longmans & Co., London, 3rd ed. pp. 783-784.

Welton, R.L. and D.R. Woods (1973). Halotolerant collagenolytic activity of Achromobacter iophagus. J. Gen. Microbiol., 75, 191-196.

Williams, S.T. and T. Cross (1971). Actinomycetes. In C. Booth (Ed.), Methods in Microbiology, Vol. 4. Academic Press, London. pp. 295-334.

Yemm, E.M. and E.C. Cocking (1955). Determination of amino acids by ninhydrin. Analyst, 80, 209-213.

SECTION VI

PHYSICAL CHEMICAL PROCESSES FOR POLLUTION CONTROL

28

REMOVAL OF RADIUM FROM URANIUM MINING EFFLUENTS — CURRENT STATUS AND NEW DIRECTIONS

P. Huck

Faculty of Engineering, Univ. of Regina, Regina, Canada

ABSTRACT

The effluents from Canadian uranium mining and milling operations are treated to remove radium-226. Historically, treatment has been accomplished by the addition of barium chloride to form a finely divided $(Ba,Ra)SO_4$ precipitate which then settles in large ponds or lakes. Recently, a considerable amount of effort has been directed towards developing effluent treatment systems capable of meeting proposed stricter effluent standards and providing easier recovery of sludge for disposal. This paper describes recent Canadian research and development work in this area. Improvements to conventional pond systems are discussed, as is the development of mechanical systems incorporating coprecipitation and solids separation. Also reported are investigations of methods such as ion exchange which provide an alternative to systems based on barium chloride coprecipitation. Radium removal during mill processing is discussed as a potential long term alternative to effluent treatment, and areas requiring further research are identified.

KEYWORDS

Radium removal; uranium mining effluents; treatment ponds; coprecipitation; flocculation; filtration; ion exchange; barium chloride; barium-radium sulphate; tailings decontamination.

INTRODUCTION

The mining and milling of uranium ores produces large quantities of waste solids or tailings in slurry form which are impounded at mine sites. The tailings contain all the radioactive isotopes from the uranium, actinium and thorium decay series. During the life of a mine, large volumes of effluent are discharged from the tailings basins. Of the various dissolved species contained in the effluents, the one of particular concern for environmental protection is radium-226.

Various jurisdictions have established objectives or requirements for radium levels in mining effluents or receiving waters (Table 1). Although current effluent requirements are based on dissolved radium-226, it has been recognized that where dissolved radium activities are measured, significant quantities of suspended

TABLE 1 Regulatory Levels - ^{226}Ra

Existing			
Federal	(E)*	10pCi/L	(3μm filter)
Ontario	(E)	3pCi/L	(1.2μm filter)
Saskatchewan	(RW)	3pCi/L	(Total)
Proposed			
AECB	(E)	10pCi/L	(Total)

*E-Effluent; RW-Receiving water

radium may be present. Consequently, the Atomic Energy Control Board has proposed an effluent target activity of 10 pCi/L total radium-226 (Averill and co-workers, 1978).

Currently, all Canadian uranium mining and milling operations provide effluent treatment for radium removal. Treatment consists of the addition of barium chloride to form a finely divided $(Ba,Ra)SO_4$ precipitate which then settles in large ponds or lakes. Within the last several years, a considerable amount of effort has been directed towards developing improved treatment systems. This is in response to both the proposed stricter effluent standards and the concerns of regulatory authorities regarding the feasibility of sludge recovery and disposal from existing systems.

This paper describes recent Canadian research and development on improved processes for removing radium from uranium mining effluents. To provide background, conventional treatment systems are described briefly first. Also considered are alternative uranium milling processes which could greatly reduce the discharge of radionuclides to the tailings areas, and perhaps obviate the need for treatment of tailings decants.

This paper is not intended to be an exhaustive cataloguing of Canadian research on radium removal. Rather, it describes work which is representative of the major directions in which investigations are proceeding.

CONVENTIONAL TREATMENT

Moffett (1979) has described the treatment system in use at Rio Algom's Quirke mill in Elliot Lake, Ontario. The system, which consists of two ponds in series, is shown in Fig. 1. The effluent from the tailings basin is discharged through a spillway and 9 mg/L of barium chloride is added just prior to the waterfall into the first settling pond. This pond provides about 10 hours' detention time before discharging through a baffled weir into the second pond. At the weir, a further 2 mg/L of barium chloride is added. After a detention time of 30-35 hours in the second pond, the effluent is discharged over a rock dam to a shallow natural basin and thence to the Serpent River. The rock dam is Rio Algom's final point of control, where effluent quality is measured for regulatory purposes.

Typical radium activity levels through the pond system are shown in Table 2. Although 98.5 percent of the dissolved radium was converted to the particulate form in the first pond, only 70 percent of this total radium had settled out. The final effluent contained total radium-226 levels of 120 pCi/L. At the actual point of discharge to the Serpent River, dissolved radium levels were less than the Ontario dissolved criterion of 3 pCi/L.

A system of natural lakes is used for tailings disposal and effluent treatment by

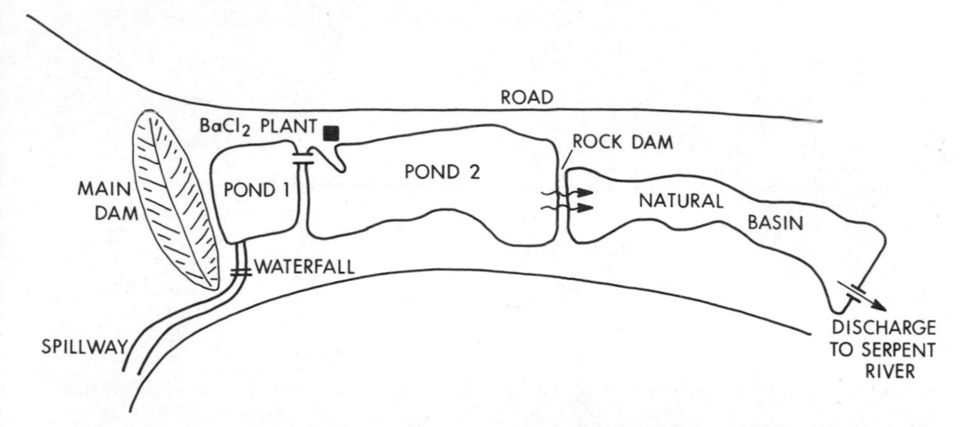

Fig. 1. Schematic diagram of Rio Algom's Quirke effluent treatment system (after Moffett, 1979).

TABLE 2 Radium-226 Levels in Quirke Treatment System (pCi/L)*

	Untreated Effluent	Prior to 2nd $BaCl_2$ Addition	Final Effluent	Discharge To River
Total	820	250	120	-
Dissolved	800	12	5	<3

*After Moffett (1979).

Eldorado Nuclear Ltd. at Uranium City in northern Saskatchewan (Ruggles and Rowley, 1978). As shown in Fig. 2, tailings are discharged to Fookes Lake and the decant overflows to Marie Lake. Some radium removal is achieved in Marie Lake through natural sedimentation and possibly adsorption. The discharge from Marie Lake is treated with barium chloride and flows into Meadow Lake. This is an artificial lake created by constructing a dam which is now Eldorado's final point of control. Ruggles and Rowley reported a barium chloride dose of 27 mg $BaCl_2$/L, although this could vary throughout the year. The mine water is treated separately with barium chloride and then discharged into Meadow Lake. Although radium levels at the final point of control were reported to be below the federal regulatory limit of 10 pCi/L dissolved, it was noted that dilution played a role in the achievement of these levels.

PROCESS DEVELOPMENTS

Pond Systems

Rio Algom Ltd. constructed a pilot plant as part of a research program to produce an effluent with a radium-226 activity of less than 10 pCi/L total and 3 pCi/L dissolved (Moffett and co-workers, 1979). This pilot plant allowed process variables to be altered and assessed in a systematic manner.

The pilot plant, shown schematically in Fig. 3, had a capacity of 2-8 L/s (25-100 Igpm). It consisted of an agitated precipitation reactor where barium chloride was

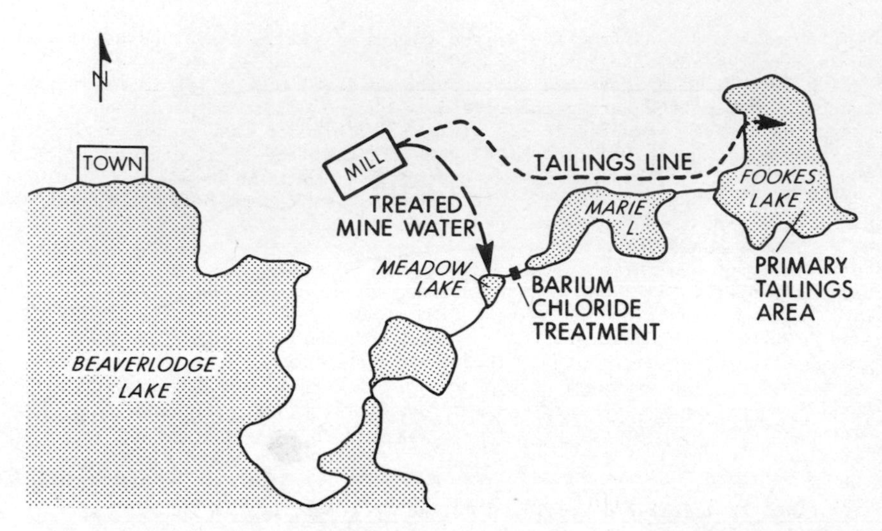

Fig. 2. General arrangement of tailings management and effluent system at Eldorado's Uranium City operation (after Kilborn, 1979).

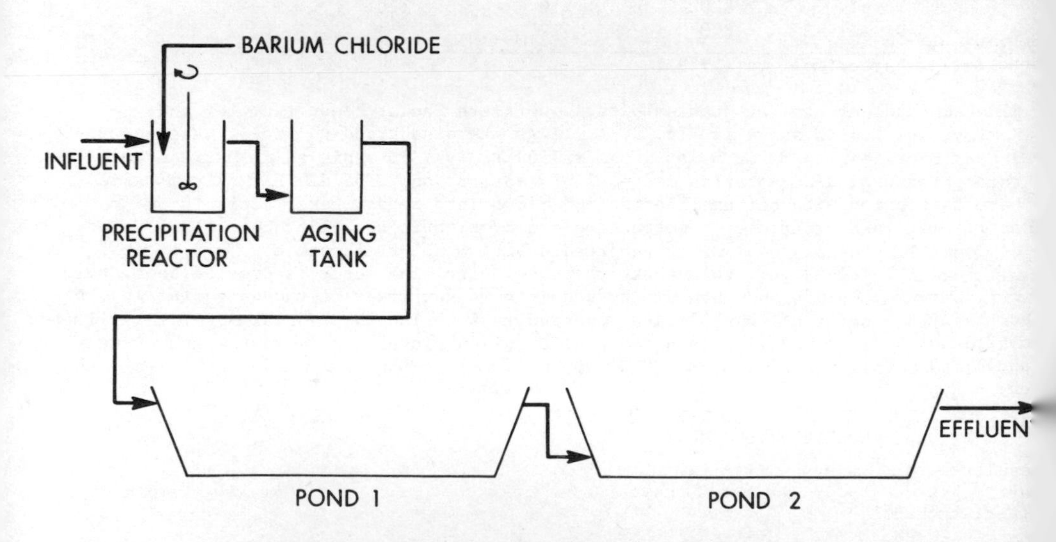

Fig. 3. Schematic diagram of pilot scale treatment pond system at Rio Algom Ltd. (after Moffett and co-workers, 1979).

added, a three-chamber aging tank, and two series connected settling ponds. Nominal retention time for each settling pond at 8 L/s (100 Igpm) was 60 hours. After baffles were installed in Pond 1 to reduce short circuiting, tracer studies showed actual residence time to be about 64 percent of theoretical.

The plant influent was decant from the Quirke tailings area, containing radium-226 levels ranging from approximately 600 to 1000 pCi/L. A baseline run in September 1978 using 15 mg/L $BaCl_2$ gave a dissolved radium level of 6 pCi/L in the final pond effluent. Doubling the barium chloride dose had no effect on performance. When the same tests were conducted in the winter, performance was poorer, with approximately 20 pCi/L dissolved radium being reported in the final effluent. Increases in reaction time obtained by decreasing the flowrate through the plant reduced the effluent ^{226}Ra level to 13 pCi/L under winter conditions. An increase in barium chloride dose to 90 mg/L produced a further reduction to 5 pCi/L.

In an effort to improve performance, an analysis of laboratory, pilot plant and historical data from this mill was conducted. This showed that the effluent radium-226 level was a function of influent radium activity level, barium chloride dose, and pH. Subsequent laboratory tests showed a dramatic increase in the residual soluble radium level as pH was increased above 9.5. This was noteworthy because the pH of the pilot plant feed was normally at or above 9.5 from December through April. An attempt to demonstrate the beneficial effect of pH control in the pilot plant was not successful. It was postulated that perhaps the dissolved radium-226 activity level being achieved was too low to show the effect of pH control, or that an unidentified change had occurred in the tailings basin characteristics prior to the pilot plant test.

It was concluded by Moffett and co-workers that an effluent containing consistently less than 10 pCi/L dissolved radium-226 could likely be produced. The need to provide pH control or an opportunity for solids build-up in the precipitation reactor was not established definitely.

Schmidt and Moffett (1979) have reported total radium results for the same pilot pond system. As Table 3 indicates, effluent total radium levels from the ponds ranged from approximately 50 to 400 pCi/L. As would be anticipated, the best

TABLE 3 Effluent Characteristics from Pilot Plant Employing Settling Ponds*

Date (1978-1979)	Barium Chloride Dose (mg/L)	Retention Time (Days)	Influent ^{226}Ra (pCi/L) Total	Effluent ^{226}Ra (pCi/L) Total	Effluent ^{226}Ra (pCi/L) Dissolved
ep. 8-Sep.20	15	2.5	700	299	7
ov.28-Jan. 8	15	2.5	1000	420	25
an.11-Feb. 3	30	2.5	1000	216	30
eb. 4-Feb.19	90	10	1000	55	9

After Schmidt and Moffett (1979).

esults were obtained with the highest barium chloride dose and the longest ettling time. However, the 90 mg/L barium chloride dose would likely be impracically high for full scale operation. Work with this pilot plant has been ontinuing and it is possible that better results have been obtained in subsequent ests.

ew design features have been incorporated into several recently constructed full cale effluent treatment systems. For example, Madawaska Mines at Bancroft, ntario selected a baffled concrete tank instead of a pond to allow settling f the (Ba,Ra)SO_4 precipitate.

Mechanical Systems

Mechanically based treatment systems represent the first major departure from treatment ponds. Work on a mechanically based system was initiated as a result of recommendations made by the Radioactivity Sub-group of a Federal Mining Task Force established under the auspices of Environment Canada. With respect to radium, the sub-group recommended that research be undertaken to achieve an improve understanding of the barium-radium coprecipitation process, to examine the effects of flocculants on the settling of barium-radium sulphate, and to investigate more effective means of radium removal including the development of mechanical treatment systems as an alternative to ponds (Averill and co-workers, 1978). In response to these recommendations, the Wastewater Technology Centre of Environment Canada initiated a bench scale study in 1976. Wilkinson and Cohen (1977) reported initial attempts to improve settling rates to permit the use of mechanical clarifiers for sedimentation. They reported large increases in settling rates using either ferric chloride or alum as coagulants. In batch bench scale tests, filtered radium levels of 3 pCi/L and total radium levels of 10 pCi/L were achieved.

On the basis of this feasibility study, a bench and pilot scale process development and demonstration project was formulated, and was jointly funded by government and industry. The objectives were 10 pCi/L and 3 pCi/L for total and filtered ^{226}R respectively.

Results of the initial bench scale work were reported by Huck and co-workers (1979) The unit processes examined were precipitation, coagulation, flocculation and sedimentation. The principal findings from batch tests were as follows:

1. The barium-radium coprecipitation reaction proceeded rapidly initially and was apparently first order. Radium-226 activity levels of approximately 7 to 10 pCi/L were achieved in this rapid phase, which required approximately 15 minutes. Beyond this point, the reaction proceeded extremely slowly.
2. Flocculation of the precipitate with either alum or ferric chloride gave equivalent total radium levels after settling. Ferric chloride was selected on the basis of its ability to form flocs faster.
3. For rapid mix, the optimum dimensionless energy input (Gt) was found to lie within the range 5 000 to 15 000. There was considerable flexibility in rapid mix time, with times between three seconds and two minutes giving similar acceptable results at the same overall energy input. For flocculation, a time of 20 minutes and an overall energy input (dimensionless Gt product) in the range 20 000 to 30 000 appeared to be optimum.

A bench scale continuous-flow process was operated to confirm the results of the batch experiments. The apparatus (Fig. 4) consisted of two (later three) precipitation reactors in series, a rapid mix unit, two flocculators in series and a clarifier. The only available clarifier was significantly oversized and had a surface loading rate of only 0.93 m^3/m^2.d(19 Igpd/ft^2).

The continuous flow process produced an effluent containing total ^{226}Ra activity levels below 10 pCi/L although at a relatively low clarifier overflow rate. The filterable radium results from continuous flow operation confirmed the predictions made from batch kinetic data but the objective of 3 pCi/L was not met. Because of timing constraints, pilot scale work was initiated immediately with this proces design. Later in the program, bench studies were used to complement pilot scale experiments in investigating alternative process designs.

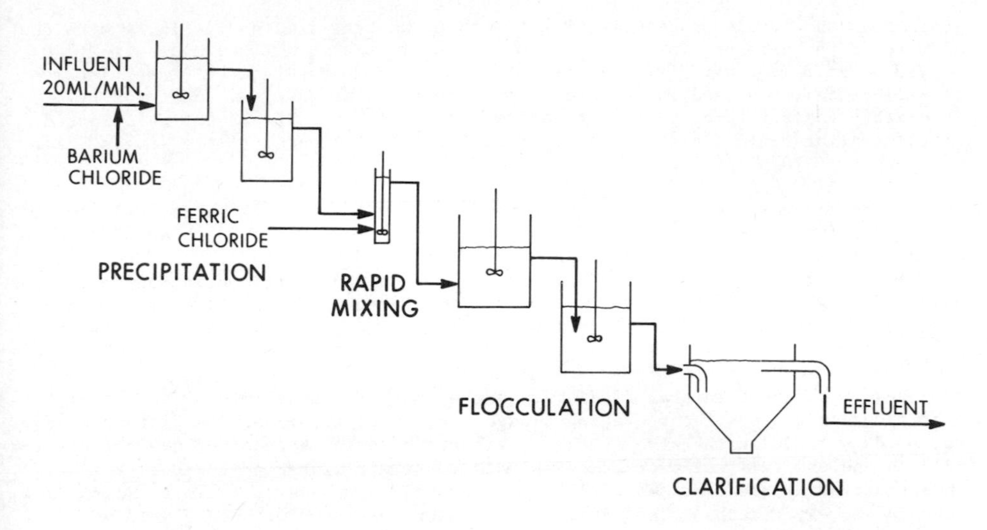

Fig. 4. Schematic diagram of bench scale continuous-flow apparatus (after Huck and co-workers, 1979).

Pilot scale testing was initiated in July 1978 at the Quirke Mine of Rio Algom Ltd. in Elliot Lake, Ontario and is continuing at present. The pilot plant is located below the main tailings dam and obtains tailings decant prior to the existing treatment. Chemical characteristics of the decant are given in Table 4.

TABLE 4 Quirke Tailings Basin Effluent Characteristics*

Parameter	Mean Value
Total Radium-226 (pCi/L)	720
Dissolved Radium-226 (pCi/L)	640
pH	$\simeq$9
Sulphate (mg/L)	1150
Alkalinity (mg/L as $CaCO_3$	125
Suspended Solids (mg/L)	$\simeq$2
Nitrate Nitrogen (mg/L)	70
Ammonia Nitrogen (mg/L)	21

*Batch samples received at the Wastewater Technology Centre of Environment Canada, Burlington, Ontario, January 1978 to March 1979. After Averill and co-workers (1980a).

The pilot plant consists of an enclosed 14 m (45 ft) highway trailer and an 11 m (36 ft) flatbed trailer which contain process tanks, chemical feeding equipment and a circular clarifier. Additional, separate process units include a second circular clarifier, an inclined plate clarifier and a 20 cm diameter gravity filter. The nominal capacity of the pilot plant is 23 L/min (5 Igpm). The process capabilities of the pilot plant include precipitation, coagulation, flocculation, sedimentation, sludge recycle, granular media filtration and automatic pH control.

Initial pilot plant experiments were designed to test correspondence between pilot scale and bench scale processes. The flowsheet was the same as in Fig. 4 with the addition of a third precipitation reactor. The initial clarifier hydraulic load of 7.3 $m^3/m^2.d$ (150 $Igpd/ft^2$) was selected as a compromise between conventional clarification rates used in wastewater treatment and those rates indicated by laboratory tests to produce target radium-226 residuals. It was not anticipated that the pilot scale clarifier would produce total effluent activity levels below the target of 10 pCi/L.

Averill and co-workers (1978) reported results from the initial pilot scale work. Mean total ^{226}Ra activity levels in the range of approximately 50 to 60 pCi/L were achieved at a clarifier hydraulic load of 7.3 $m^3/m^2.d$ (150 $Igpd/ft^2$). The addition of sludge recycle and a polymer to the coagulation/flocculation operation gave improved total ^{226}Ra activity levels. On the basis of preliminary screening experiments, granular media filtration appeared to be a feasible alternative to clarification.

As a result of these initial findings, further work was performed on the clarification process incorporating sludge recycle with polymer. A plate clarifier was used to minimize the size of the clarification unit, because of the anticipated need to enclose full scale treatment plants to permit winter operation. Further investigation of granular media filtration was also conducted, as were additional bench scale experiments. The studies undertaken consisted of both development work for process optimization and a demonstration phase to determine process variability and reliability. The two processes under study, clarification and filtration, are shown schematically in Fig. 5. The precipitation step is common to both.

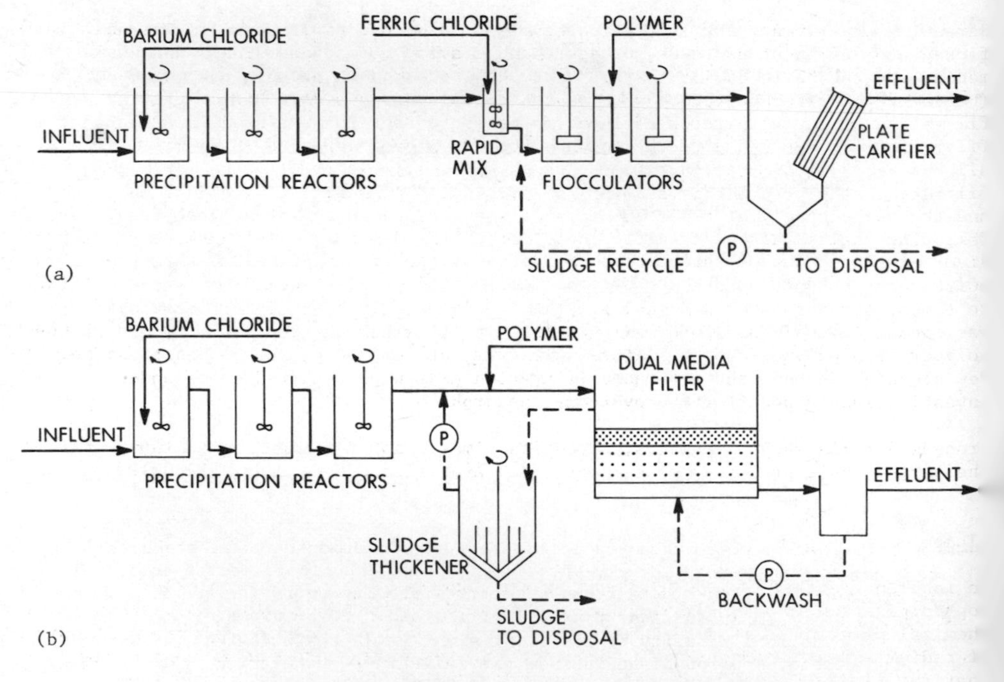

Fig. 5. Schematic diagrams of pilot scale treatment processes: (a) clarification, (b) filtration (after Averill and colleagues, 1980a).

The results of the further work were reported by Averill and colleagues (1980a). The precipitation step did not consistently produce an effluent meeting the dissolved radium-226 target of 3 pCi/L. Precipitation seemed to be influenced more by seasonal effects than by operational factors over the range within which these were varied. A small but significant deterioration in performance was observed during summer but a cause and effect relationship was not established. The performance of the precipitation operation was considered to be provisionally acceptable in view of the additional dissolved radium removal achieved by downstream process units.

For the clarification process incorporating sludge recycle, the suspended solids concentration in the flocculators was shown to be important. Using the plate clarifier at a hydraulic loading of 29 $m^3/m^2.d$ (600 $Igpd/ft^2$), the program goal of 10 pCi/L total radium-226 was achieved at a flocculator suspended solids content of 3,000 mg/L but not at 1500 mg/L. Hydraulic loads were eventually increased to 490 $m^3/m^2.d$ (10,000 $Igpd/ft^2$) while still meeting the target of 10 pCi/L. Eventual deterioration of the sludge did occur, however, and was tentatively attributed to excessive polymer age. The very high hydraulic loadings could not be reproduced after a "cold start" of the process in which the sludge was synthesized by massive chemical additions. It was suspected that increased water viscosity in the winter may have adversely affected performance. During a brief test under winter conditions, a hydraulic loading of 150 $m^3/m^2.d$ (3,000 $Igpd/ft^2$) on the plate clarifier appeared to be successful in meeting the 10 pCi/L total ^{226}Ra target. This loading rate was considered to be equivalent to a conventional clarifier loading of approximately 29 $m^3/m^2.d$ (600 $Igpd/ft^2$), a value which is within the range of normal wastewater treatment practice.

Screening experiments for filtration examined both coprecipitator effluent and flocculator effluent as feeds, pressure and gravity modes, and single and dual media. On the basis of jar tests, ferric chloride and an anionic polymer (Percol 727, Allied Colloids) were tested as filtration aids. Direct gravity filtration of coprecipitator effluent was selected for further study. A chemical filtration aid applied directly to the filter influent appeared to be necessary at all but very low filtration rates. Subsequent testing showed that a maximum filtration rate of 6.5 $L/m^2.s$ (8.0 $Igpm/ft^2$) could be achieved with dual media and the anionic polymer aid. The polymer was applied intermittently at a low dose following an initial bed conditioning. During the demonstration phase, a 20 cm diameter filter was operated. The effluent met the total radium-226 target of 10 pCi/L and the unit generally performed well. The process seemed to be insensitive to temperature or other seasonal effects. Some deterioration in effluent quality was caused by coal-sand intermixing. This intermixing was apparently due to either polymer or a polymer-gypsum mixture which was not removed by normal bed cleaning. New cleaning methods and alternate filtration aids were reported to be under investigation in an effort to overcome the problem.

Process details such as dosages, detention times, flow rates and power inputs for the various unit processes were reported by Averill and co-workers (1980b).

Sludge

An inherent feature of all treatment systems incorporating barium chloride coprecipitation is that they produce a sludge which must be disposed of. From a chemical point of view, this sludge is primarily barium sulphate with trace amounts of radium and other impurities. However, the radium is sufficiently concentrated that the sludge is considered as a long lived medium level waste by the Atomic Energy Control Board (Ontario Environmental Assessment Board, 1979). Since AECB requires that precipitation ponds be designed to facilitate retrieval of the settled precipitate (Coady and Henry, 1978), it would seem that use of the ponds for ultimate disposal of the sludge will not be permitted. Sites must therefore be found for sludge from ponds and from any mechanical systems that are constructed Possible options range from simply returning the sludge to the tailings areas to placing it underground, after suitable processing.

If the environmental consequences were shown to be acceptably small, the option of returning the sludge to the tailings area would be attractive, since it would be the simplest and cheapest method. Historically, AECB has been opposed to this method (Coady and Henry, 1978). However, this policy is being re-evaluated, as evidenced by AECB's recent funding of a study to examine the environmental consequences of sludge placement in tailings areas (Supply and Services Canada, 197 The results of this study have not yet been released.

Regulatory policy regarding sludge disposal has an important impact on the merits of various effluent treatment systems. If disposal is complicated and expensive, alternative treatment technologies which do not produce a sludge become more attractive. Of the barium chloride systems, mechanical systems offer the advantag of producing an easily retrievable, consistent sludge relatively free of impurities. Pond systems would produce a sludge containing more impurities. Existing unlined ponds or lakes would yield a sludge mixed with organic material. Techniques for retrieval of sludge from ponds and lakes have not been demonstrated

If placement of sludge in tailings areas is permitted, the production of sludge ceases to be a significant liability. Since treatment systems are typically located close to tailings areas, handling would be greatly simplified. The sludge could probably be pumped to the tailings area without any processing. Although mechanical systems would offer easier retrieval of sludge, the relative merits of ponds and mechanical systems would be influenced by the desired mode of sludge placement. Mechanical systems would produce sludge continuously, facilitating dispersal of the sludge throughout the tailings area. The sludge from pond systems presumably would be recovered when the ponds were decommissioned In this case, placement of all sludge in one location within the tailings area would be easier.

Once the issue of sludge disposal has been resolved, the relative merits of variou effluent treatment systems can be better resolved.

NEW APPROACHES

Alternative Treatment Technologies

Techniques other than barium chloride precipitation have been used to remove radium-226 from solution in certain applications. Typically, these alternative technologies involve advanced treatment operations such as ion exchange, reverse osmosis or solvent extraction. In general, these unit operations are considerabl more expensive than the more conventional waste treatment processes of precipitat sedimentation and even sand filtration. Thus their consideration for mining effluents likely could be justified only if the more conventional processes could not meet regulatory requirements or if sludge disposal was sufficiently expensive to offset the higher costs of these other processes. All of these alternative processes produce a concentrated stream requiring further treatment and disposal. Also, in the case of reverse osmosis the feed stream must be free of suspended solids.

In some cases these alternative techniques have been used to treat streams quite different from mining effluents, both in volume and in radium content. Many of t references relate to potable water treatment in certain geographical areas where background radium levels exceed drinking water criteria. In such instances, initial radium levels typically are well below 50 pCi/L - i.e. they are very low compared to most tailings basin decants. Also, radium removal in potable water treatment is normally only one step in a process train. Great care must therefore be taken in assessing the applicability to mining effluents of results obtained for radium removal in potable water treatment.

Trace Metal Data Institute (1978) has reviewed the use of ion exchange and reverse osmosis for both potable water treatment and mining effluents. They have noted the use of manganese impregnated fibres to sorb radium. Bland (1980) has also investigated manganese impregnated fibres for radium removal. Although excellent results were obtained in small scale tests, he cautioned that the process may not be economically viable.

Itzkovitch and Ritcey (1979) concluded that ion exchange showed more promise than reverse osmosis or solvent extraction for removing radium-226 from uranium mining and milling effluents. Eldorado Nuclear Ltd. is currently undertaking bench scale screening tests of ion exchange for removal of dissolved ^{226}Ra, ^{230}Th and ^{210}Pb (Itzkovitch, 1980). The work is jointly funded by Eldorado and CANMET. Five resins have been tested on both carbonate and acid leach tailings effluents obtained prior to barium chloride treatment. Both batch and column tests were performed, the latter in the upflow mode with the bed slightly fluidized. Stripping of the resin was not investigated, since it was assumed that the spent resin would be disposed of. Although results appeared encouraging, the economic feasibility has not yet been determined.

One of the few reported studies using ion exchange for uranium mining effluents was performed at Oak Ridge National Laboratory in the 1960's (Arnold and Crouse, 1965). These workers conducted bench scale tests on simulated lime-neutralized acid mill waste. A number of exchangers were screened in batch tests. The three materials selected for column testing were barytes, Decalso (a synthetic zeolite) and clinoptilolite. In all cases, the size range tested was 20 x 50 mesh. The columns were 0.8 cm in diamater, the bed volume was 8.0 mL and the solution flow rate was equivalent to 1.0 L/m^2.s (1.5 US gpm/ft^2).

As shown in Table 5, the number of bed volumes treated prior to breakthrough ranged

TABLE 5 Bench Scale Ion Exchange Results*

Exchange Material	No. Bed Volumes to Breakthrough	Resin Loading: pCi/L Wet Settled Exchanger	Resin Loading: pCi/g Dry Exchanger**
Barytes	3000	6.0×10^6	3000
Decalso	2600	6.0×10^6	9000
Clinoptilolite	480	1.1×10^6	1400

*Data from Arnold and Crouse (1965);
**Calculated from authors' data.

rom 3000 for barytes down to less than 500 for clinoptilolite. The resin loadings eported on a wet-settled basis have been converted to loadings per gram of dry esin using the authors' data on the weight of exchanger per bed volume. The ighest loading, approximately 9,000 pCi/g, was obtained for the synthetic zeolite. he authors noted that in a process application, higher loadings could be obtained by operating columns in series.

The effects of solution flow rate and exchanger particle size were evaluated using Decalso. As would be expected, the efficiency of radium adsorption decreased with increase in solution flow rate and exchanger particle size. Elution tests showed that the radium was stripped efficiently from clinoptilolite but not from Decalso or barytes.

Although the feasibility of treating uranium mining effluents using these ion exchange materials cannot be assessed from these laboratory results, a preliminary indication of practicability can be obtained. An assumed tailings decant flow of 0.3 m^3/s (4000 Igpm) containing 1000 pCi/L is representative of several Canadian operations. Assuming 99 percent removal of radium-226, calculated resin requirements on a once-through basis are 20 tonnes per day for clinoptilolite, 10 tonnes per day for barytes and 3 tonnes per day for Decalso. A process with these high resin requirements would not likely be economically feasible. While ion exchange cannot be dismissed on the basis of these test results, it is obvious that before it could be used in most existing Canadian mills, either viable elution schemes or order of magnitude improvements in factors such as resin loadings would be required.

Alternative Ore Processing Methods

An alternative long term solution to treating tailings decants would be to prevent the radium from reaching the tailings area by extracting it in the mill and disposing of it separately. Such approaches are being examined, although their primary objective is to create tailings sufficiently low in radionuclides that they will not present special management problems or constitute a long term hazard. If this goal is achieved, decant treatment might be eliminated.

These approaches involve changes to the milling process itself and as such are potential solutions over the medium or long term. A detailed discussion of milling processes is outside the scope of this paper. Most Canadian mills use sulphuric acid to extract the uranium from the ore, while one mill uses a carbonate leach. The important aspect for tailings management is that only small amounts of the othe radionuclides are extracted along with the uranium. Thus the radium-226 and the other radionuclides remain with the tailings. A possible medium term solution involves the addition of a radium removal step in the milling process. This could conceivably be retrofitted to existing mills and incorporated into new designs. The long term solution involves completely different uranium extraction processes and as such would only be applicable to new mills. An example of each type of process is presented. Additional investigations are reviewed by Itzkovitch and Ritcey (1979).

Raicevic (1979) has described a flotation approach for the treatment of Elliot Lake uranium tailings. The Elliot Lake ore contains pyrite, which reports to the tailings and causes environmental problems through oxidation and the resultant production of sulphuric acid. The objectives of the investigation were to develop a process to remove pyrite and radioisotopes from the tailings as concentrates.

The initial laboratory work was performed on old tailings from an abandoned pond. The best results, obtained when the pyrite and radioactive materials were floated separately, showed a recovery of more than 95 percent of the sulphide sulphur (mainly pyrite) and 80 percent of the radium. The pyrite and isotopes concentrat combined comprised approximately 25 percent of the weight of the original tailings The radium in the tailings was reduced from 290 to 50-60 pCi/g. Similar results were obtained with fresh tailings in continuous flow tests.

The technology was demonstrated at pilot scale at the Denison mill in Elliot Lake. The pilot plant work was carried out using the same flotation reagents used in the previous tests except that "mill repulping solution" was used instead of fresh water for the repulping of the tailings prior to flotation. The "mill repulping solution" was found to contain dissolved radium-226 levels ranging between approximately 600 and 1200 pCi/L.

The pilot-plant operation rejected 98% of the pyrite, 54% of the uranium, 65% of the radium and 75% of the thorium in combined pyrite and isotopes concentrates. These two concentrates together comprised approximately 30% by weight of the mill tailings. The decontaminated tailings assayed 123-151 pCi/g radium-226. The pilot plant results were therefore similar to those obtained in earlier testing, with the exception of a higher radium content in the decontaminated tailings. It was felt that this difference was due to the high radium content in the Denison "repulping solution".

Although no economic evaluation of the process was conducted, the decontaminated tailings appeared to be suitable for mine backfill. Since backfill typically uses about 50 percent of plant tailings, this would eliminate the surface storage of about 50% of the current uranium tailings. Backfilling would also reduce the number of pillars in the mines, increasing the recovery of uranium. The pyrite concentrate produced would be suitable for sulphuric acid production after further processing.

Methods for the disposal of radium and thorium from the concentrates produced in this process were reported to be under study.

Skeaff (1979) has noted that the development of new technology to remove the radionuclides such as radium which form sparingly soluble sulphates or sulphate coprecipitates will necessitate a fundamental departure from the conventional sulphuric acid uranium extraction process. Regardless of whether the sulphate ion originates from acid addition or pyrite oxidation, its presence in solution is an impediment to radium-226 recovery. Skeaff reported on a bench scale investigation of high-temperature chlorination of uranium ore. Testwork was directed toward obtaining high extractions of uranium, thorium and radium-226, as well as iron, sulphur and the rare earths. The method consisted of chlorinating samples of an Elliot Lake uranium ore at elevated temperatures and repulping the resulting calcine in dilute HCl. The best conditions yielded extractions of uranium, iron and sulphur (all as chlorides) greater than 95 percent. Residues were obtained which contained approximately 20 pCi/g radium-226 and virtually no pyrite. Further investigation is required to determine whether an economically viable process can be developed.

Although alternative uranium milling processes may ultimately eliminate or reduce the need for tailings decant treatment, their implementation is many years away and by no means assured. Much work remains to be done. Itzkovitch and Ritcey (1979) noted that although the alternative processes are in the early stages of development, they appear to be complex relative to existing technology; the corrosive nature of the proposed leachants would increase mill capital costs substantially; and almost total leachant recycle would be necessary to make operating costs viable.

SUMMARY

n Canada, conventional treatment for removing radium-226 from uranium mining ffluents involves barium chloride addition to precipitate the radium, followed by ettling of the precipitate in large ponds or lakes. These treatment systems were esigned to meet dissolved radium criteria. However, regulatory authorities are now considering effluent standards which would include suspended radium. An example is the Atomic Energy Control Board's proposed requirement of 10 pCi/L total radium-226. Existing systems are generally incapable of meeting such a standard and they do not facilitate the recovery of the precipitate for disposal.

Recently, a considerable amount of work has been undertaken in Canada to develop improved effluent treatment systems for uranium mines and mills. Pilot scale studies are being conducted in an effort to improve the performance of pond systems. Although early results from these investigations have indicated that good dissolved radium removals can be obtained consistently, a limit of 10 pCi/L total radium has not been met.

A mechanical treatment system has also been developed and is being tested at pilot scale. The process utilizes barium chloride addition to achieve $(Ba,Ra)SO_4$ coprecipitation in a series of stirred tank reactors. Two solid-liquid separation options are under investigation: granular media filtration and coagulation/flocculation/clarification incorporating sludge recycle and polymer addition. Both processes have produced effluent total radium levels below 10 pCi/L, and are currently being demonstrated to establish their reliability. Some operating difficulties remain to be resolved, particularly with clarification.

Alternative treatment technologies not involving barium-radium sulphate, coprecipitation have been investigated for radium removal, but generally for applications other than mining effluents. Ion exchange appears to be the most attractive of these alternative techniques and is currently being tested on Canadian mining effluents. On the basis of earlier laboratory work with simulated effluents, resin requirements for ion exchange would appear to be prohibitively high for most existing Canadian operations.

The relative merits of various processes can be assessed more completely when regulatory requirements regarding $(Ba,Ra)SO_4$ sludge disposal have been finalized and the cost implications determined.

A potential long term alternative to effluent treatment is the extraction of radium in the mill, followed by separate disposal. Although this approach is under active study, investigations are at an early stage.

Further work is required in several areas relating to radium removal from uranium mining effluents. A more fundamental understanding of factors affecting barium-radium sulphate coprecipitation is needed so that consistent dissolved radium removals can be obtained using that process. The technical and economic feasibilit of ion exchange and other alternative processes should be determined. Research into radium extraction processes and alternative mill flowsheets should be continued.

REFERENCES

Arnold, W. D. and D. J. Crouse (1965). Radium removal from uranium mill effluents with inorganic ion exchangers. Ind. Eng. Chem. Process Design Develop., 4, 333-337.

Averill, D. W., D. Moffett, V. Lakshmanan and R. T. Webber (Members of the Technical Working Group)(1978). Interim Report No. 1. Joint Government-Industry Program for the Removal of Radium-226 from Uranium Mining Effluents. Wastewater Technology Centre, Environment Canada, Burlington, Ontario.

Averill, D. W., G. H. Kassakhian, D. Moffett and R. T. Webber (1980a). Developmen of radium-226 removal processes for uranium mill effluents. Proceedings First International Conference on Uranium Mine Waste Disposal, Vancouver, B.C.

Averill, D. W., G. H. Kassakhian, D. Moffett and R. T. Webber (Members of the Technical Working Group) (1980b). Interim Report No. 2. Joint Government-Industry Program for the Removal of Radium-226 from Uranium Mining Effluents. Wastewater Technology Centre, Environment Canada, Burlington, Ontario.

Bland, C. J. (1980). Personal communication.

Coady, J. R. and L. C. Henry (1978). Regulatory principles, criteria and guidelines for site selection, design, construction and operation of uranium tailings retention systems. Proceedings Seminar on Management, Stabilization and Environmental Impact of Uranium Mill Tailings, OECD Nuclear Energy Agency, Albuquerque, New Mexico.

Huck, P. M., P. Wilkinson and D. W. Averill (1979). Process development for the removal of radium-226 from uranium mining and milling effluents. Water Pollution Research in Canada, 14, 19-33.

Itzkovitch, I. J. (1980). Personal communication.

Itzkovitch, I. J. and G. M. Ritcey (1979). Removal of radionuclides from process streams - a review. CANMET Report 79-21. Canada Centre for Mineral and Energy Technology, Energy Mines and Resources Canada, Ottawa, Ontario.

Kilborn, Ltd. (1979). An Assessment of the Long Term Suitability of Present and Proposed Methods for the Management of Uranium Mill Tailings. Study for the Atomic Energy Control Board, Ottawa, Ontario.

Moffett, D. (1979). Characterization and disposal of radioactive effluents from uranium mining. CIM Bulletin, 72, June, 152-156.

Moffett, D., E. Barnes and J. N. Hilton (1979). Removal of radium from tailings basin effluents. Presented at Canadian Uranium Producers' Metallurgical Committee Meeting, Elliot Lake, Ontario.

Ontario Environmental Assessment Board (1979). The Expansion of the Uranium Mines in the Elliot Lake Area. Final Report.

Raicevic, D. (1979). Decontamination of Elliot Lake uranium tailings. CIM Bulletin, 72, August, 109-115.

Ruggles, R. G., and W. J. Rowley (1978). A Study of Water Pollution in the Vicinity of the Eldorado Nuclear Limited Beaverlodge Operation 1976 and 1977. Surveillance Report EPS 5-NW-78-10, Environmental Protection Service, Environment Canada.

Schmidt, J. W. and D. Moffett (1979). Overview of Canadian environmental research in the uranium mining industry. Presented at B. C. Water and Waste Assocation and FACE Uranium Mining and the Environment Seminar, Vancouver, British Columbia.

Skeaff, J. M. (1979). Chlorination of uranium ore for extraction of uranium, thorium and radium and for pyrite removal. CIM Bulletin, 72, August, 120-125.

Supply and Services Canada (1979). Research and Development Bulletin, No. 76, p.35.

Trace Metal Data Institute (1978). The Current Status of Radium Removal Technology. Bulletin 607, El Paso, Texas.

Wilkinson, P. and D. B. Cohen (1977). The optimization of filtered and unfiltered ^{226}Ra removal from uranium mining effluents, status report (1976-1977). Proceedings ^{226}Ra Workshop, Canadian Uranium Producers' Metallurgical Committee, ERP/MSL 80-14 (TR), Energy, Mines and Resources Canada, pp. 10-1 to 10-20.

29

HYPOCHLORITE WASTE MANAGEMENT IN CHLOR-ALKALI INDUSTRY

S. D. Gokhale and L. L. Frank

Divisional Technical Center, Diamond Shamrock Corporation, P.O. Box 191, Painesville, Ohio, USA

ABSTRACT

In chlorine-alkali producing plants, sparger tanks filled with sodium hydroxide solution are kept in readiness to absorb any and all chlorine gas from the electrolytic cells or gas compression room during outages and shutdown of parts of the system and during emergencies. The hypochlorite produced as a result of absorption of chlorine in caustic solution has become a pollution liability.

In our plants, the hypochlorite is decomposed catalytically to sodium chloride and oxygen. After filtering, the sodium chloride solution can be optionally recycled or disposed of in the effluent. Advantages of this method over other methods and previous work on the subject are briefly reviewed. Two different catalysts and three different procedures developed at our laboratory for decomposing the hypochlorite are described. Formulae derived for computing the weight of catalyst to be added are given. Some of the facilities for handling hypochlorite solution at our plants are illustrated.

KEYWORDS

Sodium hypochlorite; hypochlorite; decomposition of hypochlorite; catalysts for decomposing hypochlorite; disposal of waste hypochlorite; synergistic effect of Ni and Fe on hypochlorite decomposition.

INTRODUCTION

Chlor-alkali plants generate most of the nation's chlorine production and supply all of the alkali (NaOH as well as KOH) demands. The process is the electrolytic decomposition of alkali chloride solutions. Chlor-alkali plants always have sparging tanks filled with 18-20% NaOH solution kept ready to absorb chlorine gas during plant malfunctions, operational breakdowns, or some emergency situations. When pure or mixed chlorine gas cannot be pumped, compressed, liquefied, or stored for a period of time, it is diverted to these emergency scrubbers, as they are called in some plants. The last fraction of uncondensed gases in the first stage of the liquefaction of chlorine is a mixture of chlorine and more volatile gases like hydrogen, nitrogen, oxygen, and carbon dioxide and is called "off-gas". When it is combined with other impure chlorine sources such as from decompression of chlorine cylinders and tank cars from the clients, it becomes "snift-gas". A second elaborate stage recovers chlorine from this mixture which contains 35-55% chlorine. The snift-gas recovery constitutes about 4-10% of the total chlorine production. All the snift-gas is bubbled through the emergency scrubbers whenever the second stage system develops problems. Only during some emergencies and only for a short while is all of the chlorine gas produced sparged through the scrubbers. The chlorine absorbed in the alkaline solution gets converted to oxidizing compounds. At low and moderate temperatures up to 120°F the reaction is:

$$Cl_2 + 2NaOH_{aq} \rightarrow NaOCl_{aq} + NaCl_{aq} + H_2O \quad (1)$$

while at high temperature the reactions are:

$$Cl_2 + 2NaOH_{aq} \rightarrow NaOCl_{aq} + NaCl_{aq} + H_2O \quad (2)$$

$$3NaOCl_{aq} \rightarrow NaClO_{3aq} + 2NaCl_{aq} \quad (3)$$

The chlorine absorption tanks are used on "ready on call" basis, intermittently, and at very irregular rates. Between two absorption periods, the hypochlorite can and does undergo a slow decomposition due to impurities present in the liquors and exposure to sunlight.

$$NaOCl_{aq} \rightarrow NaCl_{aq} + \tfrac{1}{2}O_2\uparrow \quad (4)$$

When scrubbers are used intermittently, the concentration of hypochlorite varies widely. Therefore, the hypochlorite produced cannot be made to order and delivered regularly, or sold as a bleach to a customer. When the NaOH strength is depleted enough, the hypochlorite was, until recently, sewered via an impounding basin as the chlorine gas is switched from one scrubber to the second scrubber. The emptied scrubber is immediately refilled with a NaOH solution before the second scrubber becomes depleted. Thus, there is no interruption in coverage of the emergency.

Effluents containing hypochlorite have, because of their highly oxidizing nature, become environmentally objectionable and a pollution liability. The amount of waste hypochlorite depends on the individual plant size, type of compression pumps, snift-gas recovery system, operational difficulties, power outages and market logistics. The total production of chlorine in the United States for 1979 was 12,271,519 T[1] and in the rest of the world it was about 15.5×10^6 T[2]. Snift-gas recovery systems represent, as indicated above, about 4-10% of total production. These figures give some indirect idea about the quantities involved in waste hypochlorite.

An industrial and an environmentally acceptable method of disposing of the hypochlorite is needed. This method in addition has to be economical and suitable for the plant in question. Most of the pollution abatement methods resort to either neutralization, decomposition, recovering, and/or recycling the objectionable chemical species. Conversion or destruction of the hypochlorite ion OCl^- prior to its disposal would be a logical step to follow. There are several methods, as shown in Table 1, one could choose from, each one having some advantages and some disadvantages.

---

[1] Woodard, K. E. Jr. and E. J. Rudd (1980). Report of the Electrolytic Industries J. Electrochem. Soc. 127, 251C

[2] Silver, M. M. (1980). Hybrid Chlor-Alkali Plant Evaluation. Spring Meeting (St. Louis) The Electrochem. Soc.

TABLE 1 Comparison of Methods for Conversion or Destruction of Hypochlorite

Method	Principle	Advantages	Disadvantages
1. Thermal Decomposition	$3NaOCL \xrightarrow{\Delta H} NaClO_3 + 3NaCl$	Simple; no chemicals to buy/add.	Too slow, energy intensive, formation of chlorates.
2. Acidification	$NaOCL + HCl \rightarrow NaCl + HOCl$ $HOCl + HCl \rightarrow Cl_2 + H_2O$ Side reaction at pH < 9 $NaOCl \rightarrow NaCl + 1/2\ O_2$	Some of the chlorine is recovered. Brine can be recycled.	Needs many tanks for storage, needs special reactor, danger of uncontrolled reaction. Some oxygen is evolved. Needs dilute HCl.
3. Overchlorination (Auto Acidification)	$Cl_2 + H_2O \rightarrow HOCl + HCl$ $NaOCl + 2HCl \rightarrow NaCl + HOCl$ $HOCl + HCl \rightarrow Cl_2 + H_2O$ Also at pH < 9 $NaOCl \rightarrow NaCl + 1/2\ O_2$	No chemicals added. No alkali wasted. Brine can be recycled.	Releases chlorine with oxygen, needs special reactor (yet to be developed). Needs more tanks for absorption.
4. Chemical Treatment	$NaOCl + Na_2SO_3 \rightarrow Na_2SO_4 + NaCl$ $NaOCl + NaHS \rightarrow NaCl +$ Sulfur Products $NaOCl + H_2O_2 \rightarrow NaCl + H_2O + O_2$	Stoichiometric. Very clean with H_2O_2 and the effluent brine can be recycled.	Expensive-to-use chemicals. Additional hazards and unpleasant odors. Formation of sulfates and sulfur. Low pH (7 to 9) adjustment required for the efficient use of H_2O_2.
5. Catalytic Decomposition	$NaOCl \xrightarrow{Catalyst} NaCl + 1/2\ O_2$	Fast but controllable reaction rate. Filtered brine can be recycled. Catalyst can also be recycled (with some fresh addition).	Need of separating the catalyst after the decomposition, before disposing of the liquor.

After conducting some exploratory experiments and calculations, it appeared that catalytic decomposition was the most practical and safest method of disposing of the waste hypochlorite in a chlor-alkali plant.

PREVIOUS WORK

Catalyzing the decomposition of hypochlorite by transition elements and their compounds has been known for some time and has been studied extensively (cf. CA references at the end). The very first study was made by W. Kind who reported in 1922 that Cu, Cu soaps, Fe and Fe rust catalyze the oxidation of cellulose by hypochlorite solution. The following year, O. R. Howell (1923) reported on his investigations on catalytic decomposition of sodium hypochlorite by "cobalt peroxide". He postulated that the reaction proceeded in the following stepwise manner:

$$Co{:}O{:}O + NaOCl \longrightarrow Co{:}O(OCl)ONa \rightarrow CoO + O_2 + NaCl \quad (5)$$

$$2CoO + O_2 \longrightarrow 2Co{:}O{:}O \quad (6)$$

According to Howell, the reaction rate is directly proportional to the amount of the catalyst added and to the square root of the $[Na^+]$. Increased concentration of NaOH in the solution tends to form more stable Co:O(OH)ONa to the exclusion of Co:O(OCl)ONa and decreases the rate. He found the reaction rate increases by 2.37 times for every 10°C. The activation energy is 16.574 kcal/g mol and the overall reaction is of the first order. Many scientists have studied the catalytic decomposition since then. They added Fe, Co, Ni, Cu, Mn, Ir, Hg or rare earths as a metal, a salt or in an oxide form to act as a catalyst. At times, carriers such as sand, silicates and polymers were used as the catalyst supports. Some authors have mentioned CaO and MgO among other compounds as promoters when Cu is used as a catalyst.

Studies of binary, ternary or more complex systems of the above-mentioned catalysts were also made by a few investigators. Chirnoaga recognized the synergistic effect as early as 1926 when he noted that mixtures of Co and Ni peroxides are more active than either compound used by itself. Studies made with mixtures of Co, Ni, and Fe oxides by J. R. Lewis in 1928 led him to conclude that mixed oxides showed a marked "promoter" effect and that the reaction could be represented by a so-called zero-order equation. Twenty-one years later Perel'man published his study in 1948 on ternary catalyst mixtures. He studied Co-Ni-Cu, Co-Cu-Fe, Co-Ni-Fe, and Ni-Cu-Fe hydroxide systems to determine the connection between the compound diagram of a ternary system and its activity. Perel'man and Zvorykin (1949) studied Co-Mn-Cu and Cu-Mn-Ni hydroxide systems which decompose NaOCl solutions to illustrate that the catalytic activity varies linearly with the composition if there is no chemical interaction in the components in a binary mixture. Addition of the third and the fourth component does not appreciably increase the catalytic activity of the best binary composition. The additions only broaden the maxima and may improve the mechanical and chemical stability of the compound catalyst.

Patel and Mankad (1954) found cobalt sulfate slightly more effective than cobalt chloride in decomposing NaOCl and whereas $NiCl_2$ showed only slight effect, $NiSO_4$ showed no effect at all. Perel'man and Zvorykin published another paper in 1955, six years after their first paper, giving results of their investigation of activation energy of the catalytic decomposition of NaOCl. They used $Co(OH)_2$ and $Ni(OH)_2$ and found that the activity of the first catalyst to be much greater. The decomposition constant is higher by almost an order of magnitude. The activation energy for NaOCl solutions containing 4% active chlorine and a) 2%, b) 1% alkalinity was found to be: 21.110; 23.890 kcal / mol for uncatalyzed reactions, 14.070; 14.970 kcal / mol for Co catalyzed reactions; and 15.980; 16.860 kcal / mol for Ni catalyzed reactions. Lister (1956) studied the kinetics of uncatalyzed decomposition of NaOCl and the catalyzed reaction with Mn, Fe, Co, Ni, and Cu. He showed that in no case did the catalyst accelerate the decomposition of chlorate. Only the decomposition to Cl^- and O_2 was affected. Mn and Fe did not catalyze the reaction. Co and Ni catalyzed, with the rate being proportional to the amount added but nearly independent of the NaOCl ion concentration. He found that decomposition rates were affected by total ionic strength of the solution rather than by the individual concentration of NaOH and NaCl. Lister (1956) also discussed a possible mechanism of the catalytic decomposition and suggested:

$$2CoO + ClO^- \longrightarrow Co_2O_3 + Cl^- \tag{7}$$

$$Co_2O_3 + ClO^- \longrightarrow [Co_2O_3ClO^-] \tag{8}$$

$$[Co_2O_3ClO^-] \longrightarrow 2CoO + Cl^- + O_2 \tag{9}$$

Makarov and Shreibmann (1956) described their experiments for the removal of hypochlorite from a mixture of $Ca(OCl)_2$, $CaCl_2$, $Ca(OH)_2$, or NaOCl, NaCl and NaOH in effluents and removal of chlorine in off-gas using Co, Ni, Cu, Fe hydroxides and their mixtures. They found that mixtures of Cu, Fe and Ni added in the ratio of 1:1:3 and Cu and Fe in the ratio of 1:2 were the most effective catalysts; they were added as hydroxides. A total concentration of 0.01 g/L of Cu + Fe + Ni (1:1:3 mixture) catalyst decomposed 2.7 g/L hypochlorite at 60-80° in 30-60 minutes. They also reported on bubbling chlorine through hot solutions of milk of lime containing the catalysts. The Cu + Fe + Ni (1:1:3) mixture in the amount of 0.2 g/L milk of lime was most effective and temperatures of 60-80°C were recommended. They reported that recycling the catalyst again and again in their last series of the experiments did not yield good results. Iohdanov in 1962 found Cu added as a solution of $CuSO_4$ and 0-200 mg/L of CuO catalyzed the decomposition of NaOCl 15-60 g/L solution containing 4 g/L free NaOH. He theorized that CuO in the active complex is oxidized on its surface to CuO_2 which has significantly higher oxidizing potential than ClO^- in an alkaline media. Energy of activation was 16.000 kcal / mol.

Vasilev and Mikhailova (1964) reported on this work with oxide coated wires of Co, Ni and Cu used as catalysts to study the kinetics of decomposition of NaOCl. Relative catalytic action was in the same order. The rate of decomposition increased with temperature and concentration of NaOCl but decreased with increased concentration of free NaOH. Birykuova and Mekhova (1969) obtained a patent for removing chlorine from waste gases. Their process consists of treating the gases with alkali solutions followed by catalytic decomposition of the hypochlorite at high temperatures. The chlorate contents are decomposed by SO_2. Another USSR patent was issued to Denisoys and Agaltsov (1970) for "Catalyst for Decomposition of Hypochlorite Solution". They prepared the said catalyst by mixing compounds of Ni and/or Co and/or Cu with $Fe(OH)_3$, NaCl, and a solution of Plexiglass or vinyl chloride in acetone. The mixture was molded and subjected to leaching out operation at ≤80°C.

In 1971, Pichukov, Mirnov and Sergeyer were issued a USSR patent for complete decomposition and making NaOCl solutions harmless. According to this method, the solution of NaOCl is heated to 80°-100°C in the presence of a ternary catalyst (Co or Ni) + Fe + Cu and passed through a layer of a friable material like quartz sand containing 1-5 g/L Ni or Co hydroxide. Kurlyandskaya and co-workers obtained a USSR patent in 1972 for making a catalyst for decomposition of hypochlorite solutions from Fe, Ni and Cu hydroxides precipitated from solutions of appropriate salts containing Al salts and/or silicates, with subsequent coagulation of the resulting gel, drying and pulverization.

Kinosz of Alcoa received a patent in 1976 for an antipollution method to decompose hypochlorites formed when solutions of alkali or alkaline earth metal hydroxides are used for scrubbing chlorine containing gases. Co, Ni, Cu or Ca catalysts in concentrations 9-1000 ppm, according to this patent, decompose hypochlorite formed when the catalyst is added as salt, fused metal, or metal powder and the mixture is reacted in a baffled tank at 80° for 0.5-1 hour. Kinosz (1976) stated that there was no benefit in using the catalyst with a zeolite carrier. He determined that the rate of decomposition (r) of NaOCl catalysed by Co catalyst is given by

$$\log r = 0.034T + 1.3703 \log Z + 0.16699\,pH - 4.6162 \qquad (10)$$

$$r = (\log N_o - \log N_f)/t \qquad (11)$$

where T = temperature, °C
Z = concentration of Co catalyst (expressed as element), ppm
N_o, N_f = initial and final concentration of NaOCl, g/L
t = residence time, min.

Most recently, Caldwell and Fuchs (1978) of Dow Chemical patented a method and presented a paper on "Catalytic Destruction of Hypochlorite in Chlorine Plant Effluent". They found that cobalt oxide spinel (Co_3O_4) can be used for that purpose. The catalyst is deposited on aluminum silicate, alumina, chemical stoneware, and titanium sponge. Incidently, the same catalyst was found by Caldwell and Hazelrigg (1976) to be an electrocatalyst for making stable anodes for chlorine. Caldwell found that the catalytic decomposition was a first order reaction and that, in practice, a catalyst contact time of 5-6 half-lifes decomposes 97-98% of the NaOCl. The half-life is given by the equation:

$$t_{1/2} = 2.155\,(10^{0.232pH - 0.0312T}) \qquad (12)$$

where $t_{1/2}$ = half-life, min.
T = °C

Catalyst bed volume is then given by:

$$B.V.\ (cu.ft.) = 0.987Ft_{1/2}\log(C_{in}/C_{out}) \qquad (13)$$

where F = flow rate, gallon/min.
C = concentration of NaOCl, g/L

EXPERIMENTAL

The data in the literature regarding the relative effectiveness of the various catalysts were conflicting and did not agree with the results of our preliminary experiments with respect to optimal conditions. We at the Electro Chemicals Division of Diamond Shamrock Corporation began independent investigations with an objective to develop the quickest and most economical methods of destroying hypochlorite for subsequent environmentally acceptable disposal. The method would have to be fast enough to enable, in case of a continued emergency, decomposition and safe disposal of the contents of one absorption tank/scrubber and to refill it with NaOH solution before the solution in the other absorption tank/scrubber became depleted.

Cobalt Catalyst

We found that cobalt was by far the most powerful among the single element catalysts in the metallic form or as compounds capable of catalyzing the decomposition of the hypochlorite and that adding it as a 5-10% solution of cobalt chloride was the most effective and economical way. The heat of decomposition helps to increase the rate of decomposition. The decomposition is affected by pH in an unexpected manner as shown by the following experiment: the same amount of cobalt chloride solution was added to a series of NaOCl solutions adjusted to different levels of pH's. The decomposition rate of these solutions was followed by iodometric titrations at known times. The results obtained are given in Table

TABLE 2 Effect of pH on the Cobalt Catalyst Decomposing Hypochlorite

Time (min)	NaOCl (mole/liter)							
	pH 9.0	pH 9.5	pH 10.0	pH 10.5	pH 11.0	pH 11.5	pH 12.0	pH 13.0
0	2.04	2.27	2.53	2.36	2.41	2.41	2.49	2.16
2		2.08						
5		1.98						
10		1.71						
15	0.033	1.43	1.98	2.09	2.02	2.04		
30	0.023	0.72	1.46	1.80	1.85	1.47	1.64	
45		0.29		1.64	1.58	0.95		
60		0.09	0.185	1.45	1.28	0.52	0.48	1.45
90				1.11	0.86	0.08		
120			0.005	0.86	0.59	0.005	0.005	0.90

A plot was made of percentage decomposition after one hour versus initial pH and given in Fig. 1. The decomposition rate, as it can be noted, has a minimum betwe pH of 10.5 and 11. The decomposition is very slow above pH 13. There may be twc

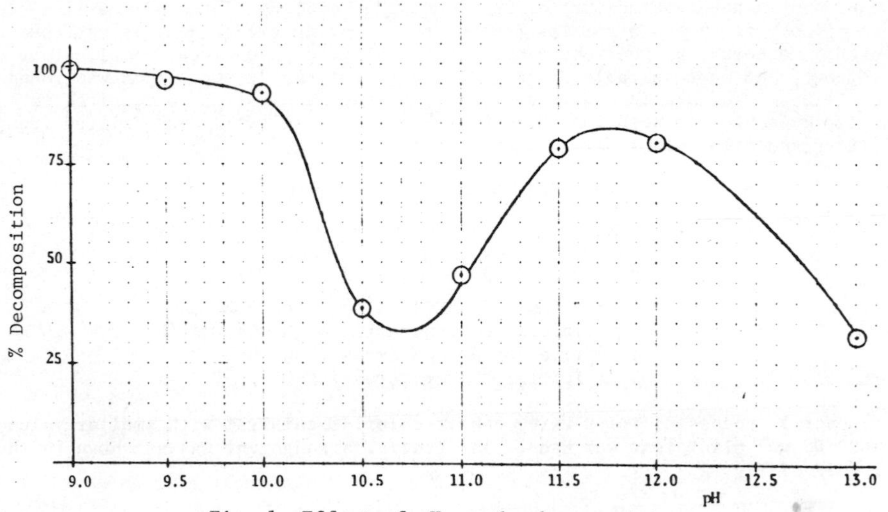

Fig. 1. Effect of pH on the decomposition of sodium hypochlorite in the presence of cobalt catalyst

different mechanisms of decomposition in the two pH ranges resulting in such an observation. This, incidently, coincides with the fact that the proportion of concentration of HOCl begins to increase at pH <10.5-11 in a concentrated brine solution containing OCl^- ions. It was also noted that the decomposition of hypochlorite was accompanied by increasing amounts of chlorine at pH ≤ 9. There seems to be not only a different mechanism but also an overlapping of two types of reactions below pH 11; one reaction forming oxygen according to:

$$2OCl^- \rightarrow O_2 + 2Cl^- \tag{14}$$

and another reaction forming chlorine:

$$OCl^- + Cl^- + 2H^+ \rightarrow Cl_2 + H_2O \tag{15}$$

Ayres and Booth (1955) have studied the decomposition of hypochlorite at pH = 8, 9, and 11 with iridium as the catalyst. They found that if the initial pH was 10, the pH does not change very much with the decomposition. If the initial pH was 9, it will decrease and then increase with time. If the initial pH was 8, it decreased sharply. They also found that some chlorine was formed simultaneously with the decrease in the pH.

There is also the possibility that the composition and activity of the catalyst produced in situ in the form of insoluble hydroxide complex may also depend upon the pH. The differing compositions of the complex may in turn have some effect on the rate and products of the reaction.

Although it was observed that the amount of catalyst required for the decomposition of the hypochlorite at pH ≤ 9 is very small, it was decided not to operate the future industrial decomposer units at that pH. First, adjusting the pH of the hypochlorite solution to 9 or any pH below 11 by chlorination and/or acidification is difficult from the operational point of view because the rate of change of pH with chlorination is very steep in that range. Second, the reaction is too rapid at pH ≤ 9 and is accelerated further by the heat generated. Thus, the reaction can go out of control unless a special reactor for removing excess heat is designed and used. Moreover, as mentioned above, some chlorine is evolved, complicating the matter. The most suitable pH range to carry out the decomposition was found to be 11-12. A formula for the amount (W_{80}) of the catalyst as Co required to decompose hypochlorite solution at 80°F with moderate agitation was derived from laboratory and pilot plant data:

$$W_{80} = 0.000792\ C_0 V/t,\ \text{lb-hr/mol} \tag{16}$$

and

$$W_T = 5.85\ W_{80} e^{-0.022\ T} \tag{17}$$

where V = volume of NaOCl in solution , US gal
C_0 = initial concentration of NaOCl, mol/L
t = time for 99% decomposition, hr
T = initial temperature of NaOCl, °F

After several successful tests using cobalt-chloride catalyst with smaller volumes, a 10,000 US gal plant test was made. The general arrangement was as shown in the upper half of Fig. 2.

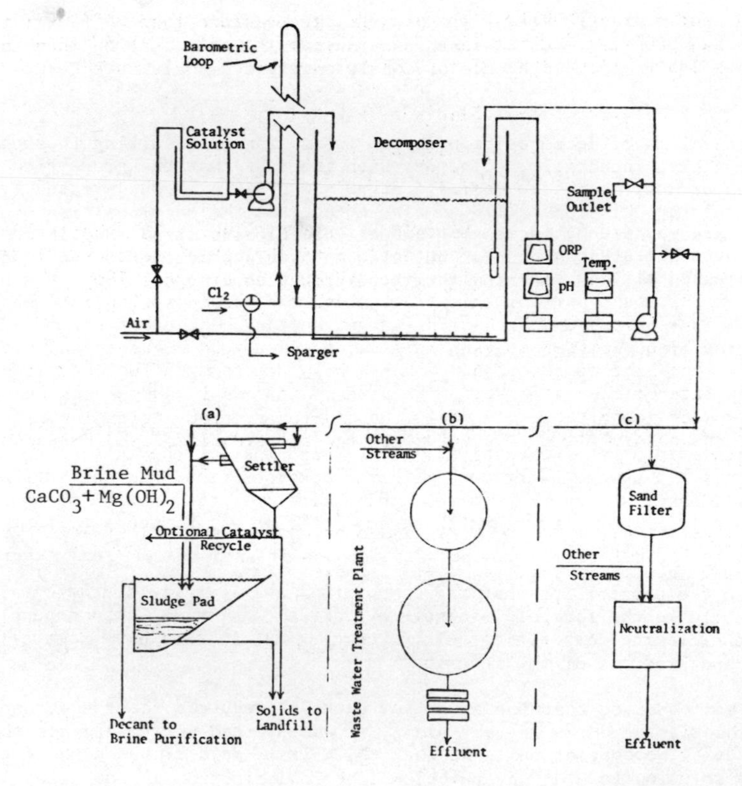

Fig. 2. Decomposition of sodium hypochlorite with cobalt salt and disposal of the resultant solution

The pH of NaOCl solution was adjusted by bubbling through the snift-gas (35-55% Cl_2) once the concentration of free NaOH was depleted to 3.9% by absorption of chlorine. The contents of the tank were monitored closely with a pH meter and also by titrating hypochlorite samples. When the snift-gas was finally stopped, the initial analysis of the contents of the tank just prior to the catalyst addition was:

NaOCl	=	72.24 g/L = 0.97g mol/L
NaOH	=	2.0 g/L
pH	=	11.6
Temp.	=	84°F

The 2 g/L free NaOH remaining in the 10,000US gal solution is equivalent to about 167 lb of anhydrous NaOH. This means that before overchlorination could occur, it would have required about another five minutes of snift-gas flow if all of the off-gas (~50% Cl_2) from a 250 T Cl_2/day plant were bubbled through the decomposer. The amount of catalyst required to decompose 99% of hypochlorite in the solution in an hour was computed to be:

$$W = 0.97 \times \frac{10,000}{1} \times 0.000792$$
$$= 7.68 \text{ lb. Co} = 31 \text{ lb. } CoCl_3 \cdot 6H_2O$$
$$\simeq 77 \text{ ppm}$$

The weight was not corrected for slightly higher temperature than 80°F. If the decomposition had been carried out immediately after the chlorination when the temperature was 105°F, the weight of the catalyst required would have been:

$$W_{105} = 5.85 \times 31.0 \times e^{-0.022 \times 105}$$
$$= 18 \text{ lb. } CoCl_3 \cdot 6H_2O$$
$$= 0.56 \, (W_{80})$$

The catalyst salt was dissolved in about 50 US gal of water by air agitation and added to the hypochlorite tank. The contents of the hypochlorite tank were agitated by compressed air and also recirculated through an external loop by a pump. Agitation is an important part of this process and the decomposition rate depends upon it. It is the agitation which brings about better contact between the catalyst and the hypochlorite, quicker release of the oxygen bubbles, and even distribution of the heat of reaction. It also helps to form a fine hydroxide complex precipitate of the catalyst, to keep the particles in suspension and separated. Too slow or no agitation is dangerous and too vigorous agitation is wasteful. It was observed that the volume of the solution increased by about 20% due to the minute oxygen bubbles dispersed in the solution during the decomposition.

Since the catalyst in the above test was added by a pump, the addition was gradual. There was a time lag before all of the catalyst was in and the reaction started in full force. The analytical results of decomposition in the test as depicted in Fig. 3 are within the set goal of 99% decomposition in one hour. The decomposed solution was pumped through a sand filter as shown in Fig. 2c and analyzed for cobalt. The analysis indicated cobalt concentration in the filter outlet was .1 ppm.

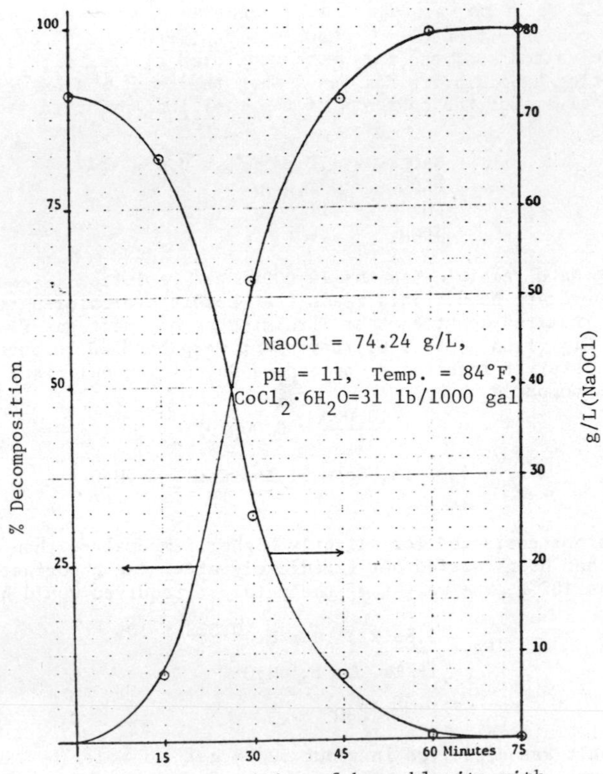

Fig. 3. Decomposition of hypochlorite with cobalt chloride catalyst

Alternatively, the decomposed solution can be sent to the waste water treatment area for the whole plant as shown in Fig. 2b, or to the sludge pads where sludges (mainly $CaCO_3$ and $Mg(OH)_2$) from the brine purification are collected as shown in Fig. 2a. Optionally, a settler can be interposed before the sludge pad depending upon flowrate and pad size. Most of the catalyst metal hydroxide complex settles out with the sludges and is removed to a reserved landfill area. If a settler is used, a thick slurry of the catalyst can be withdrawn from the bottom. The slurr can be either recycled with some makeup or sent to landfill. The clear decant liquor is transferred to either a waste water treatment stream or to the plant brine system. The decomposed decant liquor, which essentially is almost a saturat sodium chloride solution, is purified with the rest of the brine, filtered, and electrolyzed. The recycle of the brine is better explained by the flow diagram i Fig. 4.

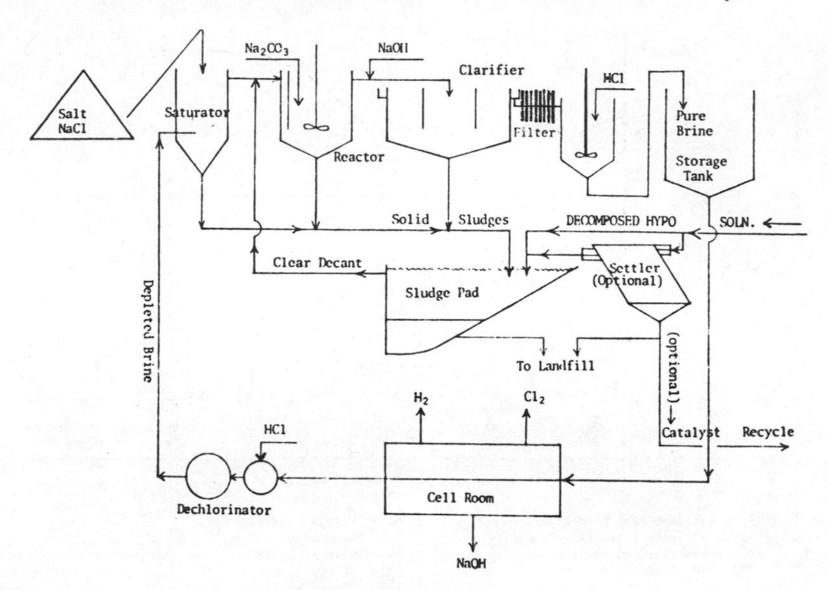

Fig. 4. Recycle of hypochlorite in the brine system after its decomposition

Plant tests have resulted in the following recommended procedure:

1) Chlorine is allowed to flow into one of the two scrubbers filled with 18% NaOH until the free caustic is depleted to about 2%. The depletion of caustic is monitored by an ORP (oxidation reduction potential) probe. (Starting with NaOH concentrations >18% leads to a super-saturated brine in the decomposer at the end, resulting in deposition of sodium chloride crystals which are harmful to valves and pumps.)

2) The contents of the scrubber are pumped to a decomposition tank while chlorine is switched to the other scrubber. The size of the decomposition tank should be 20% larger than that of the scrubber to allow for the expansion in the volume of the solution due to the minute oxygen bubbles.

3) In the decomposition tank, the alkali content is allowed to deplete further by allowing a part or all of the chlorine-containing gases to bubble through it. The pH is monitored by (if possible) two pH probes. A dual system is recommended to ensure that the monitoring system is functioning properly. Such a system is schematically shown in Fig. 5.

4) When the pH reaches 12.0, an alarm is activated and the automatic chlorine valve shown as AV in Fig. 5, is closed to the decomposition tank and all of the chlorine is now bubbled through the other emergency scrubber. This is schematically shown in Fig. 5.

5) Agitation is started on the hypochlorite tank before the addition of the catalyst and is continued until decomposition is complete. Agitation is important and should be done properly, either by compressed air or better still with a mechanical agitator.

6) An amount of dilute 5-10% cobalt chloride solution calculated using eqs. (16) or (17) is added to the decomposition tank by means of a metering pump.

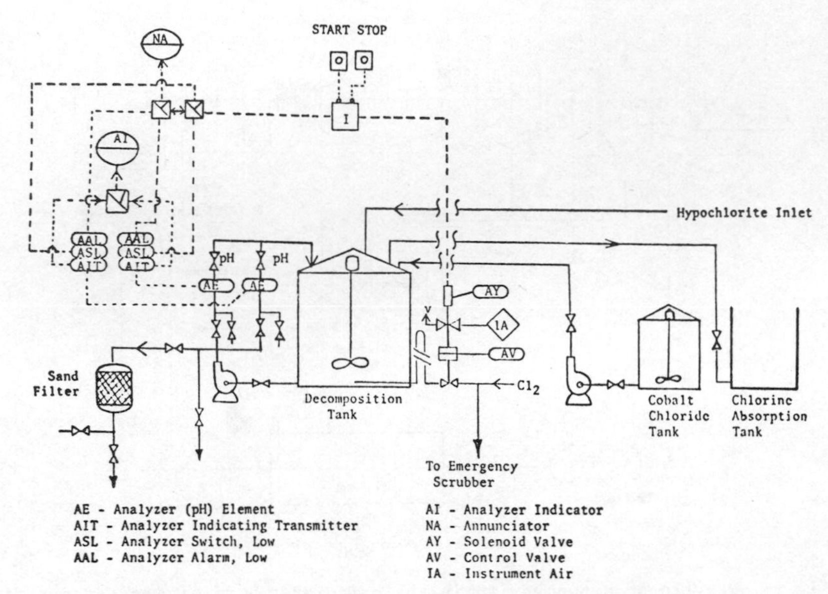

Fig. 5. Hypochlorite decomposition at controlled pH 11-12

7) The solution, mostly containing sodium chloride and a few grams per liter of sodium chlorate, is decanted or filtered to separate the metal hydroxide complex.

8) The catalyst is reused or added to other solid waste and landfilled.

9) The clear brine, depending upon the individual plant, is added to the plant effluent or to the plant brine system for further treatment. In the unlikely event of a malfunction of both pH monitors, resulting in overchlorination, a chloride absorption tank filled with 18% NaOH is available to absorb any chlorine evolved. With judicious use of piping and valve arrangement one could use the scrubber in operation for this purpose.

10) The chlorine valve (A) in Fig. 5 could be controlled manually by pressing the on-off switch (I) if such need arises.

Binary Ni-Fe Catalyst

Some of the chlorine plants may not want to spend large amounts on capital equipment and duplicate pH control systems if they are used only occasionally. It was necessary to find a catalyst which would be effective even at a pH >12 where the rate of change of pH with chlorine absorption is not that rapid. This will also eliminate fine and close control of the final pH of the hypochlorite before decomposition to between 11 and 12. In our search for this kind of a catalyst we found that a Ni-Fe binary catalyst was much more effective than the cobalt at pH ≥12.

Effectiveness of single element catalysts, binary, and ternary mixtures are compared with cobalt chloride in Table 3. In each test, similar amounts of catalysts were used. The test consisted of preparing an NaOCl solution, adjusting the pH, adding the catalyst, stirring, and measuring the undecomposed NaOCl concentration periodically to determine the rate of decomposition.

TABLE 3 Comparison of Catalytic Activity

pH = 12 Temp. 80°F.

Catalyst	Initial NaOCl Concn. g/L	Catalyst(s) Concn. x 10^3 mol/L	% Decomposition 15 min.	30 min.	45 min.	60 min.
Co	167.7	3.09	22.7	52.4	84.4	99.3
Ni	167.7	3.09	0.5	4.1	4.2	8.5
Fe	146.5	3.26	0	0	---	0.8
Cu*	118.9	4.18	11.7	13.1	16.0	17.5
La	167.7	1.24	0.9	1.1	1.1	1.1
Mn	146.5	3.07	0	0	---	1.3
Co-Ni	159.3	1.55 + 1.55	8.1	13.4	23.3	32.3
Co-Fe	144.4	1.55 + 1.63	29.3	75.3	90.1	98.7
Co-Mn	144.4	1.55 + 1.03	4.4	10.2	13.0	17.8
Cu-Fe*	118.9	3.34 + 0.84	6.8	14.8	23.0	25.4
Ni-Fe	173.0	1.55 + 1.63	96.6	99.6	99.6	99.6
Co-Mn-Ni	144.4	1.03 + 1.02 + 1.03	2.9	5.0	9.4	15.1
Co-Ni-Fe	144.0	1.03 + 1.03 + 1.09	97.2	99.6	99.6	99.6

*pH = 13

The results in Table 3 indicate that a Co-Ni-Fe ternary combination appears to be the best, but considering the cost of cobalt and the complication of adding a third chemical, the slight advantage over the Ni-Fe combination is not significant.

The ratio of Ni-Fe was varied to determine if an optimum ratio exists. The data collected is shown in Fig. 6.

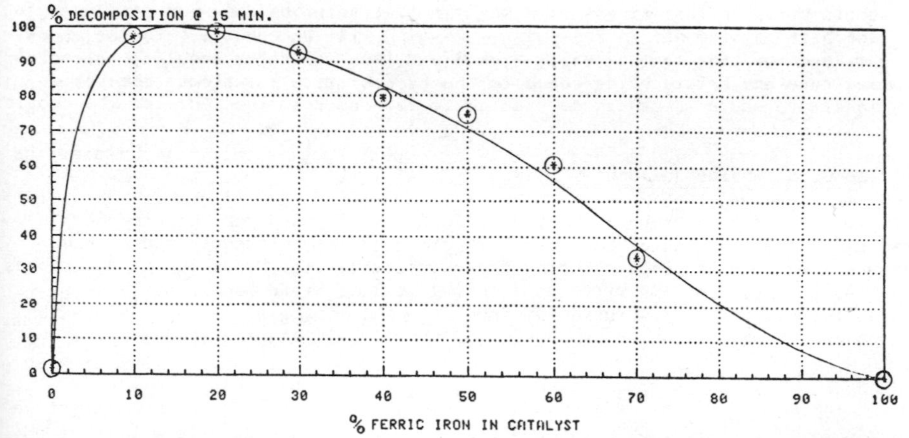

Fig. 6. Decomposition of hypochlorite as a f[(Ni+Fe) catalyst composition]

The results indicate that the higher the Ni, the better the decomposition, up to about 85% Ni. However, considering the fact that the nickel salts cost as much as [illegible] times more than the corresponding iron salts, it was decided to use a 50:50 mixture. A formula for the weight of the Ni + Fe catalyst required was obtained on that basis. Our laboratory experiments and plant test showed that the weight of catalyst required as (0.5 Ni + 0.5 Fe) is given by:

At pH 12.5 $$W_{80}^{12.5} = 0.000277VC_o/t, \text{ lb-hr/mol} \quad (18)$$

At pH 13.0 $$W_{80}^{13} = 0.000742VC_o/t, \text{ lb-hr/mol} \quad (19)$$

and

$$W_T = 5.85\ W_{80}e^{-0.022T} \quad (20)$$

where V = volume of NaOCl solution, U.S. gal
C_0 = initial concentration of NaOCl, mol/L
t = time for 99% decomposition, hr
T = initial temperature of NaOCl, °F

The above formulae are obviously applicable at specific and narrow ranges of pH. The formula below can be used for relating the required weight of the (0.5 Ni + 0.5Fe) catalyst to not only the initial concentration and volume of the hypochlorite, but also the NaOH concentration. Equation (21) applies to the range of 0.4% $\leq$NaOH $\leq$10%. This will make the use of a monitoring pH meter redundant, although we do not advise its elimination.

$$W_{80} = VC_oF/t \quad (21)$$

$$\text{if } F = e^{-8.2745 + 1.2944 [\ln (NaOH)] - 0.0686 [\ln (NaOH)]^2} \quad (22)$$

$$W_T = 5.85\ W_{80}e^{-0.022T} \quad (20)$$

where (NaOH) = excess NaOH (wt.% solution)

When the hypochlorite is decomposed at a higher pH, the resultant solution will need neutralization and pH adjustment. If that solution is, however, recycled, care should be taken that excess of alkali in that solution does not get added to the plant brine system all in one pulse. High pH will be conducive to the undesirable carryover of iron to the cell. Nickel and iron chlorides should be dissolved and mixed together before being added to the hypochlorite solution. The general operational procedure is otherwise similar to the one described earlier for cobalt.

Simultaneous Scrubbing and Decomposition

In many plants, depending upon the type of compressors used for chlorine and the number of outages they have, the scrubbers are used intermittently and have long intervals when they are idle. During these intervals, the hypochlorite decomposes slowly due to light and impurities. If catalyst like Ni:Fe binary combination is added to the starting 18% solution of NaOH, the chlorine gas will be absorbed on one hand and the hypochlorite produced thereby will simultaneously decompose on the other.

$$Cl_2 + 2NaOH_{aq} \rightarrow NaOCl_{aq} + NaCl_{aq} + H_2O \quad (23$$

$$NaOCl_{aq} \xrightarrow{\text{Catalyst}} NaCl_{aq} + \tfrac{1}{2}\ O_2 \quad (24$$

$$Cl_2 + 2NaOH_{aq} \rightarrow 2NaCl_{aq} + \tfrac{1}{2}\ O_2 + H_2O \quad (25$$

The second reaction is slower of the two. The rate of the decomposition will mainly depend upon the temperature, the amount of catalyst, the physical characteristics of the scrubber, the concentrations of NaOH and NaOCl. From experimental data it was found that:

$$r = \frac{\text{mol NaOCl decomposed}}{\text{liter - hr.}}$$

$$= 0.59288e^{0.059T}(NaOCl)^{1.21}(CAT)^{1.49}\ (NaOH)^{-0.35} \quad (26)$$

where (NaOCl) = NaOCl concentration, mol/L
(CAT) = catalyst concentration, g[Ni + Fe]/L
(NaOH) = NaOH concentration, wt.%
T = temperature, °F

A dynamic computer simulation program was developed using the above rate equation to model the operation of an emergency chlorine neutralizer with or without decomposition. The following input is required in order to run the program.

Input Parameters:

NOPT1 - Sodium Hypochlorite Decomposition?
0 - No
1 - Yes
NOPT2 - Print a Table of Results versus Time
0 - Print a Maximum of 100 Iterations
2 - No Table Printed
NOPT3 - Print Plots of Parameters versus Time
0 - No Graphs Plotted
1 - NaOCl
2 - NaOCl and NaOH
3 - NaOCl and NaCl
4 - NaOCl, NaOH and NaCl
5-8 - As 1-4 With Temperature
Height - Inside Height of Scrubber, ft.
UAO - Overall Heat Transfer Coefficient, BTU/ft^2-hr-°F
ODIA - Outside Diameter of Scrubber, ft.
Vol - Initial Volume of Caustic, US gal
TS - Temperature of Surroundings, °F
TCL2 - Temperature of Chlorine Feed, °F
FLOCL2 - Feedrate of Chlorine, lb/hr.
CCL2 - Concentration of Chlorine in Feed, wt. fraction
CNAOH - Initial NaOH Concentration, wt. fraction
CFNAOH - Final NaOH Concentration, wt. fraction
TNAOH - Initial Temperature of Strong Caustic, °F
CCAT - Catalyst Concentration, lb. (Ni + Fe)/1000 US gal (0 for NOPTI = 2)
Time - Total Time for Simulation, hours
N - Number of Iterations

Output available from the program included a listing of the important input data, a table of concentrations, temperature, volume and density for each iteration and plots of concentrations (NaOCl), (NaCl), and (NaOH) and temperature versus time. The program was used to determine the minimum catalyst concentration required to insure that sodium hypochlorite remained at a safe level and was completely decomposed when scrubbing was stopped.

Results from the simulation of the operation of a scrubber at one of our plants during a snift recovery outage are given in Figs. 7 and 8. Parameters used in the simulation are given in Table 4. Fig. 7 shows the concentration profile for sodium hypochlorite formed without catalyst and no decomposition and Fig. 8 is the profile with the catalyst and simultaneous decomposition.

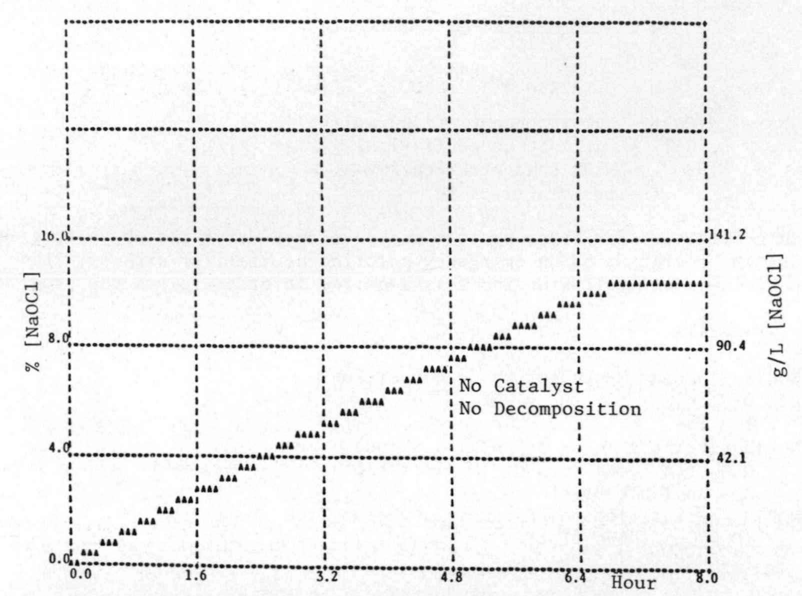

Fig. 7. Computer simulation - simultaneous scrubbing and decomposition

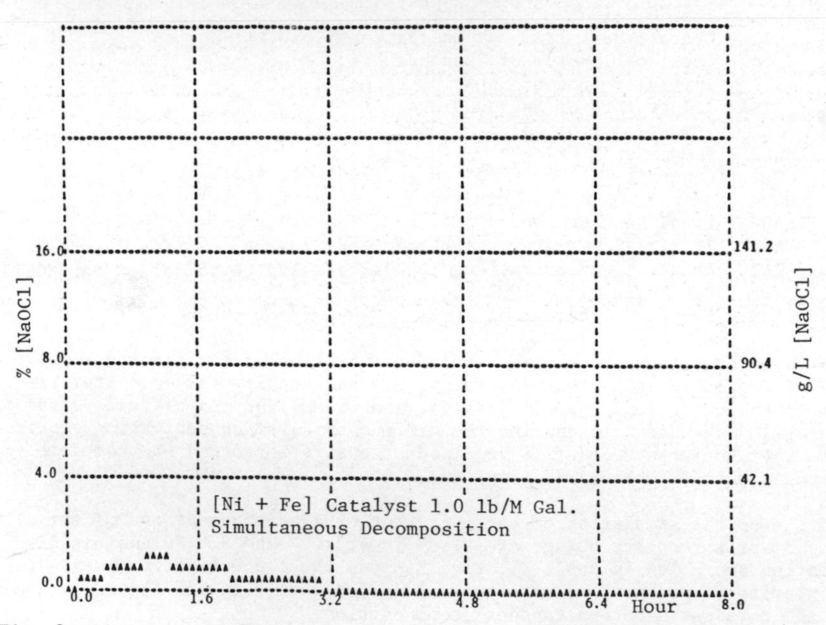

Fig. 8. Computer simultation - simultaneous scrubbing and decomposition

TABLE 4 Simulation Parameters

Volume	30,000 US gal
Caustic Strength	
Initial	18%
Final	5%
Initial Caustic Temperature	80°F
Chlorine Stream Parameters	
Total Flow	9026 lb /hr
Percent Chlorine	53%
Temperature	10°F
Temperature of Surroundings	80°F

The catalyst concentration was 1.0 lb/1000 US gal scrubber solution. The peak sodium hypochlorite concentration with no decomposition is 10.5% versus 1.0% (12 g/L) with decomposition. Figure 9 illustrates the computed relationship between peak sodium hypochlorite concentration and catalyst concentration for initial temperature of 60°F and 80°F under above stated conditions.

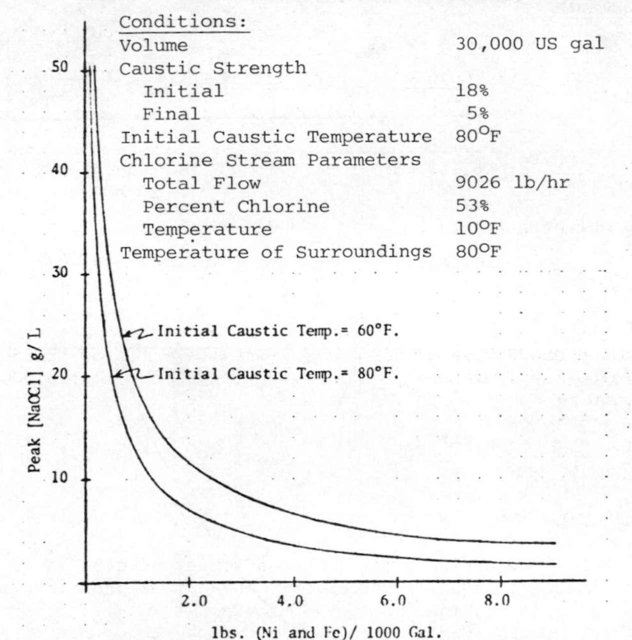

Fig. 9. Peak sodium hypochlorite vs. catalyst concentration

It should be noted that the reaction rate used in this program was based on the laboratory data obtained from a mechanically well-agitated reaction. Since 60°F represents the worst case at the plant and no mechanical agitators are used in the scrubbers, it was recommended that a catalyst concentration of 4.0 lbs/1000 US gal be maintained in the scrubbers. This concentration should result in only a low level of hypochlorite formation during a snift outage. Analytic followup of two runs made at two different plants are depicted by graphs in Figs. 10 and 11. The

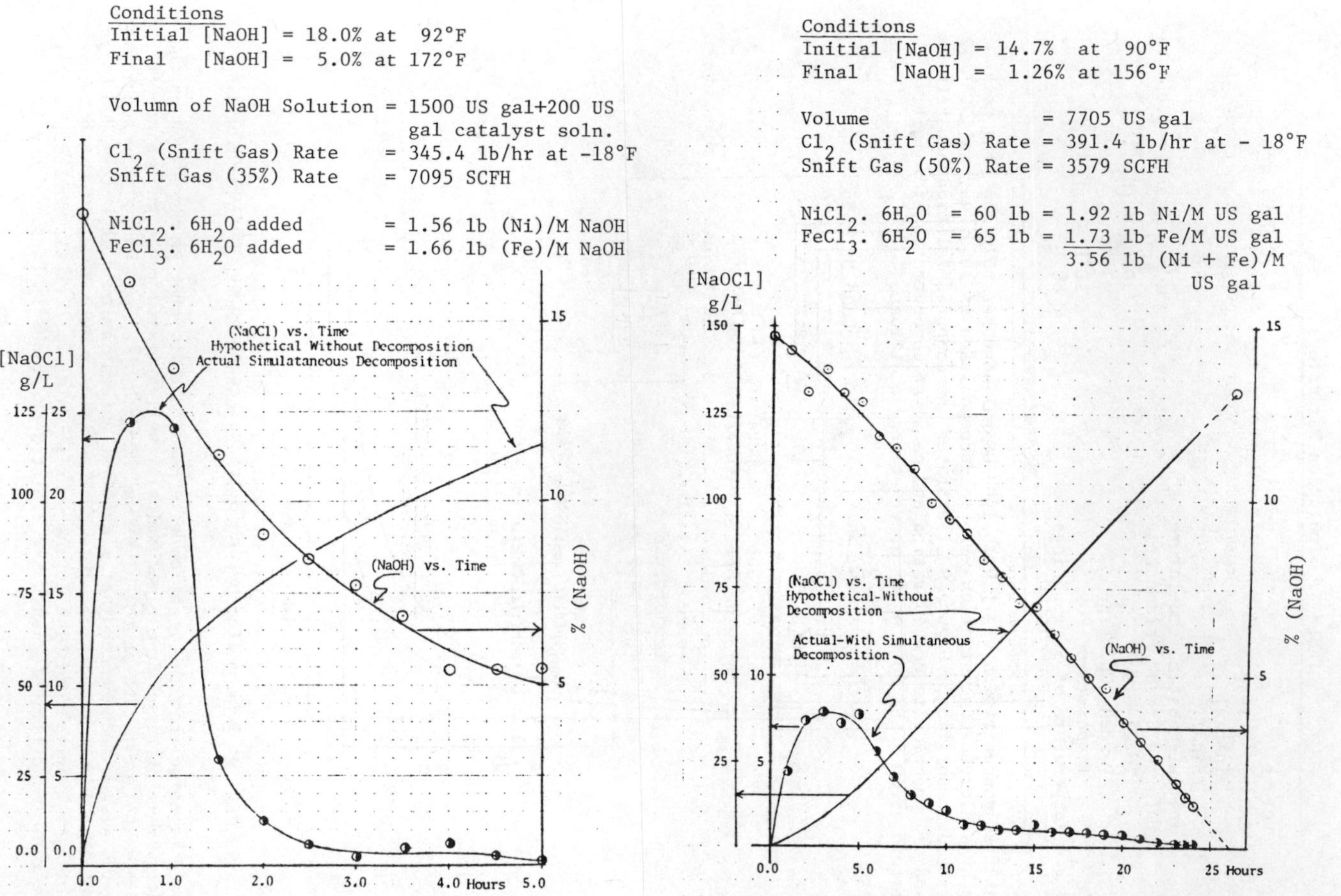

Fig. 10. Simultaneous scrubbing and decomposition plant test.

Fig. 11. Simultaneous scrubbing and decomposition plant test.

peak heights were 25 g/L and 15.5 g/L respectively with 3.33 lb of the combined catalyst as Ni and Fe per 1000 US gallons of the solution.

Our laboratory experiments and plant tests proved that the simultaneous absorption and decomposition can be accomplished without the problem of release of chlorine gas or violent eruptions.

Since both absorption of chlorine and decomposition of the resultant hypochlorite can be done simultaneously, a separate tank for the later decomposition is no longer required. The schematic flow diagram of the process is given in Fig. 12.

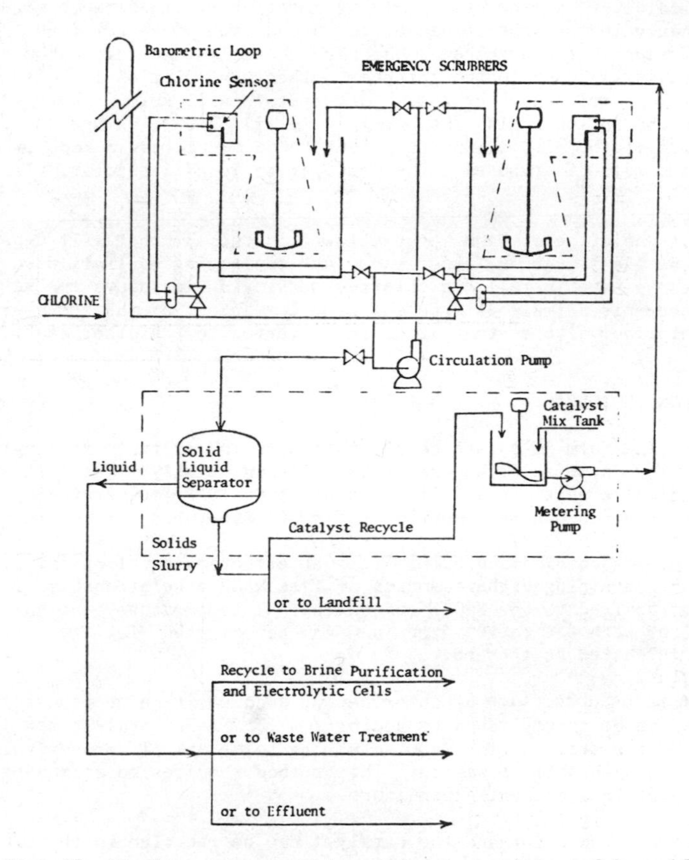

Fig. 12. Simultaneous scrubbing and decomposition of hypochlorite

The additional equipment besides the scrubber tanks (which should exist in any case) needed for this process, as shown in that diagram, is a catalyst mix tank with a metering pump and a pH or chlorine monitoring device. Allowance should be made for the 10-15% expansion of the volume of the solution due to dispersed oxygen gas bubbles during decomposition. The decomposed solution is, as in the previous case, filtered by itself or combined with other streams for waste water treatment before sending out of the plant. Alternatively, the solution can be decanted by itself in a settler and/or in the existing sludge pad/tank as mentioned earlier and then combined at a regulated rate with the brine system of the main plant. In the brine system, the trace residues of the catalyst will coprecipitate with $CaCO_3$ + $Mg(OH)_2$ precipitate and the sodium chloride portion will be electrolyzed to produce chlorine and alkali. On the other hand, if a settler is used for separation, the thick slurry collected from its bottom with some make-up amount of catalyst from the mix tank can be re-used for the next batch. The make-up amount should compensate for the physical loss and the loss in the catalytic activity. Care should be taken that the hydroxide complex does not dry out if it is to be recycled.

Nickel-iron catalyst is made by dissolving together their salts to give the predetermined equivalent weight of nickel as Ni and iron as Fe per 1000 US gal of 18% NaOH solution in the scrubber. The binary catalyst solution should be 5-10% dilute before it is added to the scrubber. When the solution is added to the scrubber, a brown gelatinous hydroxide precipitate will form. Shortly after the chlorine gas absorption begins the precipitate will slowly change to a black, grainier precipitate. It is important that these particles be kept in maximum suspension and well distributed. The precipitate loses its activity if separated and allowed to dry.

The extent of depletion of NaOH to be allowed in the scrubber will depend upon the urgency of the situation, experience, and monitoring controls used. In one of our plants, we have installed a chlorine sensing device above the scrubber. The moment there is a trace of free chlorine coming out of the scrubber, it stops the chlorine valve so that it can be diverted to the other tank.

CONCLUSIONS

1. Cobalt in the form of dilute cobalt salt solution is the best single element catalyst for decomposing hypochlorite solution quickly. The most suitable and practical conditions are 11-12 pH and 100-110°F temperature. Weight of catalyst required can be computed (Eq. 17). Some, but not too much, NaOH is wasted.
2. A 50:50 binary combination of Ni-Fe is an effective catalyst for hypochlorite solutions containing higher amounts of free NaOH. Relationship of weight of catalyst required to the NaOCl concentration, temperature, and volume at pH 12.5, 13 or with 0.4 to 10% free NaOH can be computed (Eq. 20). Some amount of NaOH is wasted at the end.
3. A simultaneous absorption of chlorine and decomposition of resultant hypochlorite can be accomplished by adding 1-4 lb. Ni-Fe catalyst per 1000 US gallons of the starting 18% NaOH scrubbing solution. If properly carried out, very little NaOH will be wasted. This method requires more amounts of catalyst but less capital expenditure.
4. The decomposed solution and the catalyst can be recycled to the electrolytic cell after separating the catalyst.
5. The above methods have been tested and proven in plant operations.

REFERENCES

Ayres, G. H. and M. H. Booth (1955). J. Am. Chem. Soc., 77, 825-7, & 828-33.
Biryukova, L. V. and E. V. Mekhova (1969). All Union Sci. Res. and Design Inst. of the Al, Mg, and Electrode Ind., U.S.S.R. 237,724 (Cl. C. 02 C), 12 Feb., 1969, Appl. 03 Jan., 1967.
Caldwell, D. L. and R. J. Fuchs (1978). (Dow Chem. Co.) U.S. 4,073,873 (Cl 423,499, Cl 01 D 3/04), Feb. 14, 1978, Appl. 29 Mar. 1976. Also "Catalytic Destruction of Hypochlorite in Chlorine Plant Effluent" presented at the Spring (1978) Meeting (St. Louis), The Electrochem. Soc.
Caldwell, D. L. and M. J. Hazelrigg (1976). Cobalt Oxide-Based Chlorine Cell Anodes, Spring (1978) Meeting, The Electrochem. Soc.
Chirnoga,E. (1926). J. Chem. Soc.,1693-1703.
Denisoya, A. I. and A. M. Agaltsov (1970). U.S.S.R. 285,903 (Cl.B 01j) 10 Nov. 1970. Appl. 04 Mar. 1969.
Howell, O. R. (1923). Proc. Roy. Soc. (London), 104 A, 134-52.
Iohdanor,Khr. (1962). Khim. Ind. (Sofia), 34, (1), 7-9.
Kind, W. (1922). Textilber., 3, 131-4.
Kurlyandskaya, I. I., A. A. Furman, S. A. Rogozhina, V. L. Kucher, A. N. Antonova and B. G. Robovskii (1972). U.S.S.R. 348,223 (Cl.B. 01j), 23 Aug. 1972, Appl. 30 Jun 1970.
Kinosz, D. L. (1976). (Al. Co. of Am.) Anti-pollution method. U.S. 3,965,249, (Cl. 423,497; C 01 B 11/06); 22 Jun 1976, Appl. 18 Sep. 1972.
Lewis, J. R. (1928). J. Phys. Chem., 32,213-34.
Lister, M. W. (1956). Can. J. Chem., 34,465-78.
Makarov, S. Z. and S. S. Shraibman. (1956). Khim. Prem., 202-8.
Patel, M. C. and B. N. Mankad. (1954). J. Indian Chem. Soc., Ind. and News Ed., 17,236-7.
Perel'man, F. M. (1948). Problemi Kinetiki Kataliza, Akad. Nank. S.S.S.R., Inst. Fiz. Khim, 5, Metody Izucheniya Katalizatora, 145-64.
Perel'man, F. M. and A. Ya Zvorykin (1949). Problemy Kinetiki i Kataliza G. Geidrgen. Kataliz, Akad. Nauk S.S.S.R., 203-5.
Perel'man, F. M. and A. Ya Zvorykin (1955). Zhur. Fis. Khim., 29,980-2.
Pichukov, A. P., A. M. Mironov and V. V. Sergeer (1971). All Union Sci. Res. and Design Inst. of Al, Mg, and Electrod. Ind., U.S.S.R. 311,267 (Cl. C01b), 19 Aug. 1971, Appl. 22 May 1968.
Vasilev, Khr. and M. Mikhailova (1964). Godishnik Khim. Tehnnol. Inst., 10 (2), 25-32.
CA 14:2530[1]; 16:2510[9]; 17:3636[9]; 19:2157[7]; 20:3331[8], 3375[3]; 22:1522[1]; 23:1341[6], 2298[8]; 2636[9], 5279[4]; 25:4773[3]; 26:260[1], 362[5], 2390[1]; 28:402[2], 6652[9]; 29:2312[3]; 30:6270[7], 7977[9]; 31:2942[4], 4458[7]; 32:362[4], 1553[6], 4415[8]; 33:1605[7]; 35:7806[9]; 36:2467[8], 4751[8]; 37:25[7]; 38:3561[5]; 39:2023[7]; 43:6515g; 45:5008f; 46:4930d; 47:6100g; 48:13383g; 49:1415c; 8682b, 1585c; 50:11152g; 51:637e, 827a; 52:3487a, 5186e, 6999f; 53:9878e; 57:2902c; 65:2460c; 66:60352; 75:10750g, 10857x, 10879f, 142409n; 76:37878n; 77:169291w, 129015u.

30
THE INFLUENCE OF POLYELECTROLYTES IN PROMOTING COAGULATION OF DYES FROM WASTE WATER

A. Darmawan*, N. M. Surdia and A. Sahilima***

**Institute of Textile Technology, Bandung, Indonesia*
***Chem. Dept., Institute of Technology Bandung, Bandung, Indonesia*

ABSTRACT

Dyes commonly used in polyester-cotton dyeing are a combination of dispersed- and vat-dyes. A treatment for waste waters from such industrial plants has been developed, i.e. by neutralization, aeration, coagulation, flocculation, filtration and adsorption using activated carbon.

The coagulating agents investigated were iron sulphate and aluminium sulphate combined with some cationic, anionic, and nonionic polyelectrolytes. The optimum conditions for coagulation found in this work were with 40 ppm iron sulphate combined with 0.2 ppm nonionic polyelectrolyte and a flocking time of 15 minutes.

A possible mechanism for the interaction of dye, metal salt and polymer aid has been proposed.

KEYWORDS

Coagulation; dispersed dyes; vat dyes; flocculation; iron sulphate; aluminium sulphate; polyelectrolytes; waste water treatment.

INTRODUCTION

A new textile dyeing and finishing plant mainly processing blended polyester-cotton fabrics with dispersed- and vat-dyes, is now under construction in Indonesia. The effluent treatment plant should have a constant capacity of 25 m^3/h with a possibility to be enlarged to 50 m^3/h temporarily. Half of the treated effluent has to be recycled.

To meet these objectives the biological/physical treatments that have been selected are the following major processes: equalization, neutralization, preliminary settling, aeration, secondary settling, filtration and adsorption.

For designing the system and mainly for the clari-flocculators, where the capacity can be doubled whenever required, the use of polyelectrolytes as coagulant aid has been investigated.

THEORETICAL BACKGROUND

Effluent from Polyester-Cotton Dyeing and Finishing Plants

Dyeing processes using dispersed- and vat-dyes produce effluents which are darkly colored, alkaline and containing organic matter (Darmawan, 1978; Nemerow, 1971). Although the organic matter can be removed by biological means, it is normally difficult to remove the residual dyes. Some of the chemical structures of dispersed- and vat-dyes are as follows:

$O_2N-C_6H_4-N{=}N-C_6H_4-N(C_2H_5)(C_2H_4OH)$

Azo Scarlet (dispersed dye)

Indigo (vat dye)

Dispersed dyes are chemically stable, nonionic, crystalline and have functional groups like $-NH_2$, -NHR, and -OH (Darmawan, 1980). They are finely dispersed in aqueous phase with a crystal size of 0.5 to 5 μm, and have a relatively low solubility.

During the dyeing process, the nonsoluble vat dyes are converted to the soluble leu co compounds by reduction with hydrosulphite and caustic soda. After the dyeing pro cess, they are oxidized by air or H_2O_2 back to the nonsoluble state.
The leuco compounds have the ability to form hydrogen bridges and tend to aggregate They are more easily destabilized by the addition of coagulants and polyelectrolytes than are the dispersed dyes.

Since caustic soda, soaps, surface active agents and sometimes a large quantity of sodium chloride are used in the process, the effluent characteristics will be highly alkaline with a high content of total disssolved matter.

Neutralization

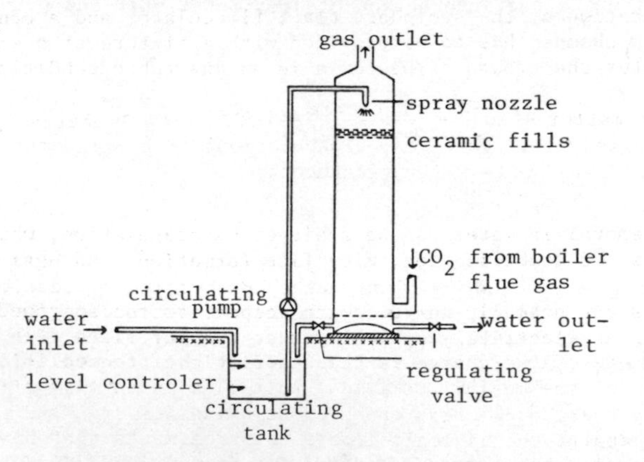

Fig. 1. Neutralization by CO_2 flue gas

The high pH, which often exceeds 11, can be lowered by flowing CO_2 gas obtained fro boiler waste gas in the neutralizing tower (Fig. 1). The process is more economica than neutralizing with mineral acid. One kg of light fuel oil will produce approximately 1.5kg of CO_2 gas.

Aeration

In the activated sludge process (Fig. 2), purification is effected by a wide range of aerobic bacteria, whose growth is encouraged by the addition of oxygen to the aeration tank (Siddeley), as illustrated by the following overall reaction:

$$\text{Organic mátter} + O_2 \xrightarrow[\text{bacteria}]{\text{aerobic}} CO_2 + H_2O + \text{energy}$$

The vital part of this process is the aeration device, which must be able to transfer and circulate the oxygen from the compressed air through the liquid to the micro-organisms and keep the sludge in suspension.

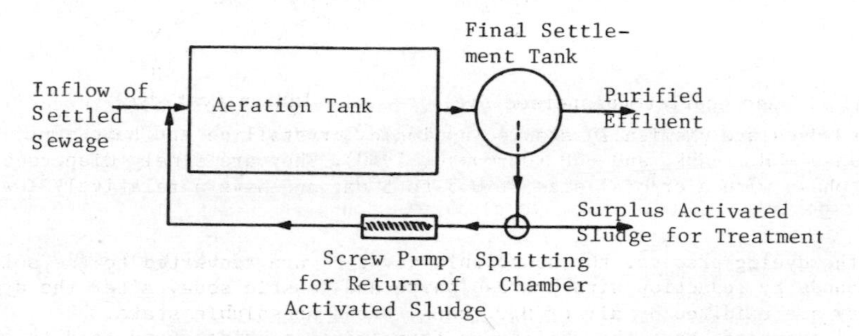

Fig. 2. Main components of activated sludge system

Textile effluent may contain ammonia, and according to Culp, Wesner and Culp (1978) the removal reaction is as follows:

$$NH_4^+ + 2\,O_2 + 2\,HCO_3^- \xrightarrow[\text{bacteria}]{\text{Nitrosomonas}} NO_3^- + 2\,H_2CO_3 + H_2O$$

To provide denitrification and inhibition of filamentous micro-organisms, which will give good operation of the secondary clari-flocculator and a continously good effluent, an anoxic chamber has to be provided with a mixture of a small quantity of air. This promotes the growth of bacteria in an anaerobic condition :

$$\text{organic matter} + NO_3^- \xrightarrow[\text{(anaerobic)}]{\text{bacteria}} CO_2 + N_2 + \text{energy}$$

Coagulation

Colloidal solids removal in water can be achieved by coagulation, which consists of the following steps : destabilization, microfloc formation, and agglomeration.

Colloidal particles are normally charged with respect to the surrounding medium due to the formation of an electrical double layer at the interface (Schroeder, 1977). The sign and magnitude of the charge is characteristic of the colloidal material and the composition of the medium. Colloidal particles in textile effluents are normally negatively charged, whereas the hydrous oxides of iron and aluminium are usually positively charged.

Coagulants are inorganic, metal hydroxide-forming substances, such as salts of aluminium and iron. These trivalent salts will dissociate in water; the hydrated metal ions in aqueous solution will reduce the repulsive forces between colloids by compressing the diffuse double layer surrounding individual particles. The particles will be destabilized and the interaction is increasingly governed by attractive van der Waals forces.

The hydrolysis of hydrated iron is as follows:

$$\left[Fe(H_2O)_6\right]^{+++} + H_2O \rightleftharpoons \left[Fe(H_2O)_5(OH)\right]^{++} + H_3O^+$$

$$\left[Fe(H_2O)_5(OH)\right]^{++} + H_2O \rightleftharpoons \left[Fe(H_2O)_4(OH)_2\right]^{+} + H_3O^+$$

The ferric hydroxo complexes have a pronounced tendency to polymerize towards a di- meric ion, in which the two iron ions are bound by hydrous bridges :

$$(H_2O)_4Fe \langle {}^{O(H)}_{O(H)} \rangle Fe(H_2O)_4$$

This dimer can undergo further hydrolytic reactions and form higher hydroxide com- plexes with the formation of more hydrous bridges (Fair and colleagues, 1968).

Aluminium behaves in much the same way as Fe(III), but the stepwise hydrolysis can convert positive Al into negative aluminate ion :

$$\left[Al_6(H_2O)_6\right]^{3+} \xrightarrow{OH^-} \left[Al(H_2O)_5(OH)\right]^{2+} \xrightarrow{OH^-} \left[Al(H_2O)_4(OH)_2\right]^{+}$$

$$\xrightarrow{OH^-} \left[Al_6(OH)_{15}\right]^{3+}_{aq} \text{ or } \left[Al_8(OH)_{20}\right]^{4+}_{aq} \xrightarrow{OH^-} Al(OH)_3(H_2O)_{3(s)} \xrightarrow{OH^-} \left[Al(OH)_4\right]^{-}$$

So the effect of these multivalent metal ions on coagulation is not brought about by the ions themselves, but by their hydrolysis products.

Flocculation

Modern flocculants which are now widely used are linear or long branched, long- chained polymers or polyelectrolytes, based on acrylamide, ethylene-amine or ethy- lene-imine with a degree of polymerization ranging from 50,000 to 150,000 for cat- ionic, and 200,000 for nonionic and anionic types (Reuter, 1978; Stockhausen). According to the number and type of group that will dissociate in aqueous solution nonionic, anionic and cationic polymers has to be distinguished. The following che- mical structure shows an anionic polyelectrolyte, where Y may comprise 0-40% of the total and X + Y may reach 200,000 units.

$$\left[\left[\cdots \underset{H}{\overset{H}{C}} - \underset{\underset{NH_2}{C=O}}{\overset{H}{C}} \cdots \right]_X \left[\cdots \underset{H}{\overset{H}{C}} - \underset{\underset{O^{(-)}\ Na^+}{C=O}}{\overset{H}{C}} \cdots \right]_Y \right]_n$$

Flocculation refers to aggregation of the particles. Once the colloids have been destabilized, collisions occur and the particles form small microflocs. Continued mixing enables the microflocs to form macroflocs (Fig. 3).

The long-chained polymers provide bridging between particles by attaching themselves to the absorbant surfaces of colloids, building larger flocculated masses.

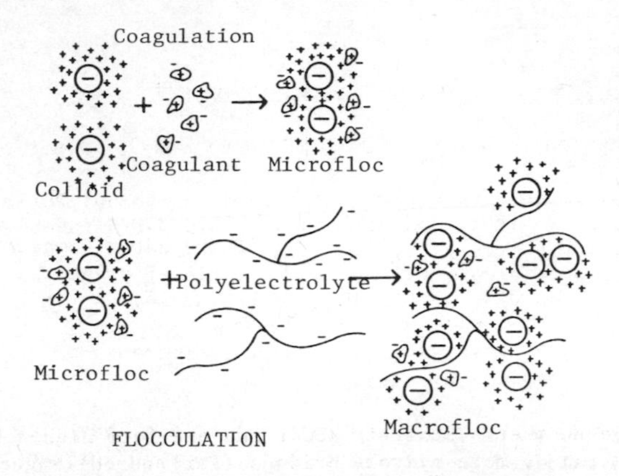

Fig. 3. Agglomeration resulting from coagulation with metal salt and polymer aid

Filtration and Adsorption

In a textile effluent treatment plant where the treated effluent is to be recycled, a vertical, pressurized, high rate, in-depth, multi-media filter with a flow rate between 3-7 m/h is suitable. The underdrain, normally fitted with plastic nozzles, supports the lowest bed having the smallest particles and highest density. The upper bed consists of plastic pellets (Parker, 1978), which have the biggest size and lowest density. Due to the different sizes and densities, the filtering media will be always positioned back to the former location, so that the upper media will remove coarser impurities, while the middle and lower parts serve to remove the medium sized and finest particles. This type of filter is compact, has better filtering capabilities, lower headloss and longer backwashing intervals as compared to the normal single media filters.

Adsorption by granular activated carbon is meant to reduce mainly dissolved organic impurities and residual dyes. According to Lurgi (1976), the characteristic feature of activated carbon is its specific pore structure and hence its large internal surface area, which is about 600-1600m^2/g. Depending on the diameter and pore distribution, the pores are classified into three orders: macropores of over 200 Å, which are of minor importance for actual adsorption processes, transitional and micropores of under 200 Å at which the actual adsorption takes place (Lurgi, 1976).

EFFLUENT TREATMENT DESIGN

Design Bases

Although the average B O D and C O D will be around 200 mg/L BOD_5 and 600 mg/L COD, the following design bases have been used (Darmawan, 1979).

Effluent flow : normal operation, 25 m^3/h; temporary operation, 50 m^3/h.

pH	: 5 - 12
B O D	: 500 mg/L
C O D	: 1500 mg/L
NH_3(N)	: 40 mg/L
PO_4(P)	: 24 mg/L
Detergent	: biodegradable
Alpha	: 0.8
Beta	: 0.9

Alpha factor is the ratio of oxygen transfer into waste water compared with the transfer into clean water, and beta factor is the ratio of oxygen saturation value of a waste water compared to that of clean water (Siddeley).

Estimated final effluent : less than 20 mg/L BOD_5, less than 5 mg/ L NH_3(N); pH = 6.5 - 8.5

If the effluent output has to be increased to 50 m^3/h, BOD_5 may rise to 50-100 mg/L and ammonia may not be oxidized to nitrate at all. About 50% of the treated effluen will be disposed for irrigation with an acceptable pH between 6.5-8.5 and a BOD_5 be tween 60-100 mg/L at the moment, although in the future the maximum allowable BOD_5 might be reduced. Since the rest of the effluent has to be recycled, treatment for lowest yield of BOD_5 would be recommended to reduce the load of the activated carbon.

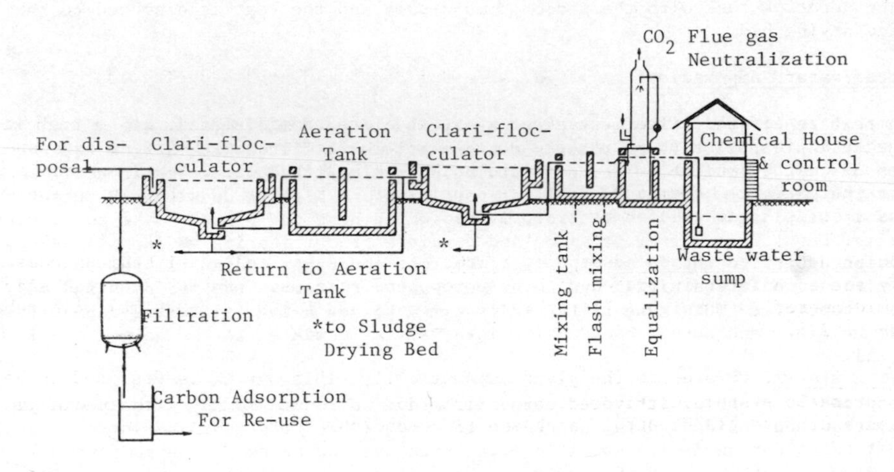

Fig. 4. Diagram of textile effluent treatment plant

Construction Design

Waste water sump of concrete, located underground, size 5x5x2 m height, capacity 50 m^3 with sump screen and two sewerage pumps of 25 m^3/h at 6 m H_2O

Equalization tank of concrete with a surface aeration blower and mixer, size 7x5x height, capacity 100 m^3.

Flash mixing tank of concrete with a high speed mixer and iron sulphate or alum sing installation; size 1.4 x 1.4 x 2 m height, capacity 3.5 m^3

Neutralization tank of concrete with medium speed mixer, automatic pH controller and neutralization installation for the dosing of calcium hydroxide slurry by a mono-pump and hydrochloric or sulphuric acid by a PVC control valve . Polyelectrolyte will be dosed with a diaphragm metering pump. Size 2.5 x 2.5 x 2 m height, capacity 12.5 m^3.

Neutralization with boiler fluegas to replace the mineral acid neutralization system will be applied in the near future. CO_2 from the boiler waste gas is blown into the neutralizing tower and comes into contact with falling water droplets. pH can be controlled by adjusting the flow rate of the circulation pump. Size 1.5x1.5x3.75 x height, made of fiber glass with stainless steel reinforcement and ceramic fills.

Preliminary clari-flocculator of concrete, 8 m diameter, 2 m side-wall and 7.5^o floor slope with rotating half bridge scraper fabricated of mild steel blades and driving gear.

Aeration tank of concrete, size 20.5 x 5 x 4 m depth, divided into 4 aeration chamber of 5 x 5 m and one anoxic chamber of 0.5 x 5 m. This tank will be provided with a fine bubble dome diffuser where 750 m^3/h of air at 5 m H_2O will be blown by the high pressure blower.

Secondary clari-flocculator with the same construction as the preliminary clari-flocculator involves the removal of biological solids that has been formed in the activated sludge process. To maintain proper MLSS, a part of the sludge is returned to the aeration tank with the use of a mono-pump and the rest is disposed to the sludge drying bed.

Treated water sump made of concrete, size 9 x 3 x 5 m depth, capacity 135 m^3.

Automatic rapid sand filter, quantity 2 units, size 2.5m diameter, 2.4 m high with an under drain having polypropylene nozzles with 0.2 mm openings. Lowest bed consists of quartz sand (0.4-0.5 mm), middle bed of anthracite coal (0.8-0.9 mm) and upper bed of polypropylene pellets (1.2 mm). Two pumps of 25 m^3/h at 30 m H_2O which functions also as a backwash pumps are provided.

Granular activated carbon adsorption filter of mild steel coated with bituminous epoxy and an underdrain fitted with polypropylene nozzles. Quantity 2 units, size 2.5 m diameter, 3 m height. Filter material is Calgon F 100 (0.8-0.9 mm) with a bed depth of 2 m.

Control system, to operate the plant automatically. This system is designed to be an all-automatic system, with water level controller, pH controller, control valves, pressure differential control, backwash timer and pump motor actuators.

EXPERIMENTAL PROCEDURE AND RESULTS

Dyeing

Blended 65% polyester and 35% cotton cloth was dyed with dispersed- and vat dyes. The dispersed dyes used were Palanil Brilliant Red BEL, Resolin Blue FBL, and Palanil Gold Yellow GG, while the vat dyes used were Indanthrene Red FBB Coll, Indanthrene Blue RS Coll, and Indanthrene Gold Orange Coll. Biologically degradable Teepol CH- 53 and Optimol MNS were used as wetting agent and as carrier, respectively. Waste water from the dyeing processes, totally three different batches, was collected in plastic drums of about 200 L for each type of color.

Water for the dyeing process was analyzed before and after being used (Table 1).

TABLE 1 Analysis of Water

Characteristics	Before dyeing	After dyeing		
		Red	Blue	Yellow
pH	6	11.6	11.6	11.5
Turbidity (mg SiO_2/L)	2	30	29	30
C O D (mg/L)	–	565	539	520
B O D (mg/L)	–	180	170	165
Color	Colorless	dark red	dark blue	dark yellow
Dissolved matter (mg NaCl/L)	150	3920	2400	3980

Waste Water Treatment

Neutralization. As can be seen from Table 1, waste water has a pH of 11; therefore, it has to be neutralized. Neutralization has been done on laboratory scale using concentrated H_2SO_4 until a pH of 8.5 is obtained. The amount of concentrated H_2SO_4 used was 1.100 kg/m^3 waste water. If CO_2 gas is used for neutralization, about 0.700 kg/m^3 waste water is required.

Aeration. To lower the B O D , the neutralized water is then aerated with compresse air using a diffusion system, i.e. air is blown through porous carborundum speciall made for this purpose, with the formation of activated sludge.

Coagulation and flocculation have been carried out with variations in concentratic of coagulants and flocculants. The coagulants used were iron sulphate and aluminium sulphate, while the polyelectrolytes examined were Praestol reagents made by Stock-hausen, Germany, of the following types :

444-K)	cationic	2935/74)	anionic	2810/74)	nonionic
423-K)		2900/74)		3000/74)	

Evaluation of results was done using a spectrophotometer (Hach Model DR-EL/2), BOD apparatus (Hach Model 2173), COD reactor (Hach Model 16500-10) and a temperature regulator (Hach Model 2597 Incutrol/2). The results are shown on Fig. 5 and Table

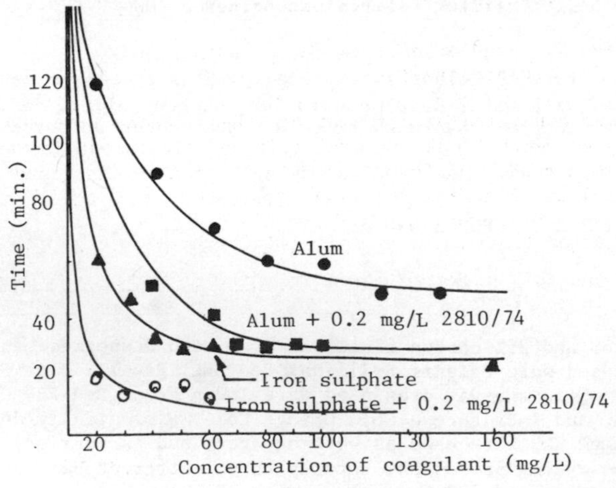

Fig. 5. Time vs. concentration of coagulant

Flocking time as function of concentration and the size as well as the quality of flock were being used to determine the optimum condition of coagulation.

Optimum concentration for Fe(III) sulphate is 40 mg/L with a flocking time of 30-35 minutes. Optimum concentration for Al(III) sulphate is 80 mg/L with a flocking time of 60 minutes.

TABLE 2 Flocking Time as Function of Polyelectrolyte Type in Combination with Metal Salts

Praestol Code No.	Type	Flocking time (min.)	
		Fe salt + polymer	Al salt + polymer
444-K	cationic	25	50
423-K	cationic	25	50
2935/74	anionic	25	50
2900/74	anionic	25	50
2810/74	nonionic	15	30
3000/74	nonionic	20	40

These results were obtained from red dyed waste water, but the same results were obtained for the other two types of waste water. A combination of a nonionic polyelectrolyte with Fe(III) sulphate seems to be the best one.

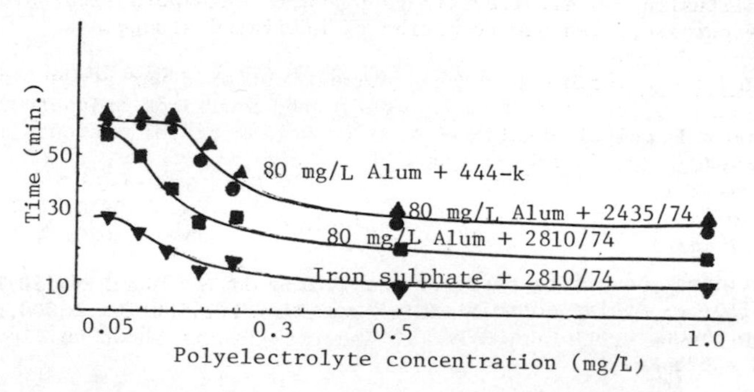

Fig. 6. Coagulation time vs. polyelectrolyte concentration

To obtain the optimum condition of coagulation, the flocking time was examined as function of varying concentrations of metal salt and flocculant in waste water (Fig. 6). The optimum condition found was 30 mg/L iron sulphate and 0.2 mg/L of a nonionic polyelectrolyte (Praestol 2810/74).

The water obtained after treatment was analyzed (Table 3).

TABLE 3 Analysis of Water after Treatment using Iron Sulphate, Nonionic Praestol 2810/74, and Filtration through Activated C.

Characteristics	Red dyed	Blue dyed	Yellow dyed
pH	8.3	8.2	8.3
Turbidity (mg SiO_2/L)	7	8	7
C O D (mg/L)	143	152	147
B O D (mg/L)	52	58	48
Color (Pt - Co)	70	80	80
Dissolved matter (mg NaCl/L)	5384	3900	5550

These results should be compared with the results given in Table 1.

COSTS OF CHEMICALS FOR TREATMENT

The costs of chemicals for neutralizing the waste water with CO_2 obtained from flue gas are much lower as compared to that with H_2SO_4 (Darmawan, 1980).

Although there are not much differences in the chemical costs when treating the waste water with or without the addition of polyelectrolyte, the coagulation time is shorter (Table 4).

TABLE 4. Chemical Costs for Treatment of Waste Water Without and With the Addition of Polyelectrolyte

Coagulant	Neutralization with H_2SO_4 (1.100kg/m^3)	Neutralization with CO_2 from light fuel oil (0.700kg/m^3)	Neutralization with CO_2 from Flue gas	Coagulation time
Iron sulphate (40 g/m^3)	Rp.244.40/m^3	Rp. 22.70/m^3	Rp.4.04/m^3	30 minutes
Iron sulphate (30 g/m^3) +Praestol 2810/74 (0.2 mg/L)	Rp.244.50/m^3	Rp. 23.16/m^3	Rp.4.50/m^3	15 minutes

The costs mentioned above were based on the prices of H_2SO_4 at Rp.250/L; iron sulphate at Rp. 110/kg; polyelectrolytes (Praestol 2810/74) at Rp.6,000/kg and light fuel oil at Rp. 40/kg.
1.0 US Dollar = 625 Rp.

CONCLUSIONS

1. Waste water from cloth dyeing using dispersed- and vat-dyes has a high COD, BOD, pH, high suspended matter and conductivity. This is due to the significant quantity of chemicals used during the dyeing process.

2. After neutralization, aeration, coagulation, filtration and adsorption, the COD and BOD values will be reduced about 70%.

3. Neutralization has been carried out with H_2SO_4,using 1.100 kg/m^3 waste water.However, CO_2 gas would be more economical, because it is cheaper, the amount needed is only 0.700 kg/m^3, and it can be obtained as flue gas from boilers.

4. Iron sulphate is a better coagulant than aluminium sulphate. Optimum conditions were found with 30-40 mg/L iron sulphate in combination with a nonionic polyelectrolyte with a concentration of 0.2 mg/L and with 15 minutes flocculation time.

5. The fact that aluminium sulphate is less effective in the coagulation process is probably due to the higher hydration possibility of Al, so that its metal hydroxide has a higher water content, but at the same time a lower density, resulting in very small settling velocities. Also the possibility that Al can reverse its charge with time is not favorable for coagulation.

6. The use of polyelectrolyte in combination with iron sulphate will reduce the coagulation time from 30 minutes to 15 minutes. This means that, by the addition of polyelectrolyte, the capacity of the clari-flocculators can be doubled.

7. By passing the treated waste water into a column of activated carbon, the COD and BOD values will be lowered, also the color and turbidity. This is due to the structure of activated carbon, which is porous and which has a high absorbility. Activated carbon can still be regenerated by heating and oxidation, or by chemical methods (Chesemisinoff and Ellerbush, 1978).

REFERENCES

Chesemisinoff, P.N. and F. Ellerbush (1978). Carbon Adsorption Handbook, Ann Arbor Science, Michigan.
Culp, R.L., G.M. Wesner and G.L. Culp (1978). Handbook of Advanced Wastewater Treatment, Van Norstrand Reinhold Company, New York.
Darmawan, A. (1978). Waste Water Treatment of Textile Industries by Flotation, Upgrading in Industrial Waste Water Treatment, Institute of Technology Surabaya (ITS).
Darmawan, A. (1979). Waste Water Treatment of Textile Industries for FIVE STAR INDONESIA LTD. Number D/L-20a/79.
Darmawan, A. (1980). The Biological and Physical Treatment of Waste Water Obtained from the Dyeing of Blended Polyester Cotton Fabric With Dispersed and Vat Dyes, Institute of Textile Technology, Bandung.
Fair, G.M., J.C. Geyer and D.A. Okun (1968). Water and Waste Water Engineering, Vol. 2, John Wiley & Sons, New York.
Hammer, M.J. (1975). Water and Waste Water Technology, John Wiley & Sons, New York.
Lurgi, Express Informations T.1188/3.76.
Nemerow, N.L. (1971). Liquid Waste of Industry: Theories, Practices and Treatment, Addison-Wesley Publ., Reading, Mass.
Parker, H.W. (1978). Waste Water Systems Engineering, Prentice Hall of India, New Delhi.
Reuter, J. (1978). Improvements in the Purification of Municipal, Industrial Sewage Waters with Coagulation and Flocculating Agents, International Congress for Water Treatment, Bangkok, Feb. 21-25.
Sandmann, H. and U. Morgeki (1977). Waste Water Treatment by Modern Methods, International Textile Bulletin 3/77, p. 414-420 I.T.S.-Zurich.
Schroeder, E.D. (1977). Water and Waste Water Treatment, McGraw-Hill Inc., New York.
Siddeley, H., Water Engineering Ltd., England, Publication No. FBDD. 676.
Stockhausen, Praestol Information No. 1.00.

31
TREATABILITY STUDY OF WASTEWATER GENERATED FROM PULP AND PAPER INDUSTRY USING RICE STRAW AS RAW MATERIAL

H. Mitwally, O. El-Sebaie, M. Akel and M. Attia

High Institute of Public Health, Alexandria University, Alexandria, Egypt

ABSTRACT

The pollution of streams by industrial wastewater in Egypt started with the beginning of the second half of this century. The discharge of industrial wastewater of the different companies into the Abu-Kier drain increases its organic load and the problem is intensified due to the accumulation of other chemical pollutants, including toxicants and carcinogens, together with the physical pollutants which interferes with the colour, odour and taste of the drain water. All these factors affect public health, as a considerable number of farmers use this drain water for irrigation, bathing and sometimes drinking.

Sedimentation is considered as the first stage of treatment to remove the settleable solids and floating matter and, accordingly, it reduces the suspended solids content. Studies of the coagulation process on the final wastewater of the Rakta Company indicated that lime was the best and most economic coagulant. The application of aluminium sulfate and ferric chloride for the coagulation of final wastewater resulted also in a good removal of COD, volatile solids and colour, but these coagulants are much more expensive than lime itself, in addition to producing a more voluminous sludge. Optimum pH was found to be 4.0 for ferric chloride and 5.4 for aluminium sulfate.

Oxidation of the final wastewaters using calcium hypochlorite was tried and found to be uneconomical; also, the percentage removal of COD and volatile solids were smaller than those given by the aforementioned coagulants. Segregation of the black liquor for separate treatment by acidification reduces the strength and difficulty of treating the final wastewater. A slow settling, voluminous sludge was obtained after the acidification process. High-purity lignin could be isolated from the sludge and prepared for use as a tanning agent.

INTRODUCTION

Industrial waste is roughly comparable, in its nature, to municipal sewage and other sanitary and domestic wastes. With the rise of new industries and the expansion of older ones, the problem of handling industrial wastes tends to grow and impose direct dangers to the safety of receiving water bodies.

Widespread and serious water pollution problems have occurred in the developed countries over the period of their industrialization and have more recently occurred in many of the developing countries. The situation started to deteriorate in Egypt after the beginning of the present local industrial revolution. Urgent measures should be taken to control water pollution, to prevent further deterioration of water quality and to improve the situation wherever possible.

The Rakta Pulp and Paper Company, the biggest pulp and paper mill in Egypt, is considered to be one of the most dangerous and serious pollution sources in Alexandria Governorate since it discharges about 2000 - 3000 m^3/hr. The plant uses local rice straw and imported pulp as raw materials and applies the soda process. The importance of this work could be attributed to the fact that rice straw is very rarely used as a raw material in the paper industry. It includes all aspects of paper manufacture which discharge typical wastewaters containing large quantities of suspended and dissolved materials of both organic and inorganic nature, in addition to a considerable amount of greasy substances which leak from the fuel tanks.

Suspended solids consist of silts and foreign matter originating in the straw preparation stage, coatings and fillers from the paper making processes, and fiber particles from both pulping and papermaking. Dissolved organics in the waste originate almost entirely from the raw straw. Lignin imparts a dark brown colour to process wastewater from the chemical pulping stage.

Rakta Pulp and Paper Company discharges its wastewater without pretreatment into the Abu-Kier Drain. The drain flows through cultivated areas to the Tabia pumping station which is located on the drain terminal at the south west corner of Abu-Kier Bay (map of locations available on request from the authors).

Abu-Kier Bay is considered as an extension of Alexandria Sea Cost from the eastern side. The pollution in the bay is caused by the disposal of wastewaters pumped from Abu-Kier Drain. The wastewater in the drain consists of the irrigation drainage from the agricultural areas of the neighbouring Governorate and the different industrial wastes from the Kafr El-dawar and Tabia districts.

MATERIALS AND METHODS

Wastewater treatment was investigated using the total effluent as well as black liquor and white water, as the treatment of wastewater originating from certain parts of the process could represent a means of reducing the pollutional load of the final waste.

Plain sedimentation, chemical coagulation and oxidation were the three different treatment processes applied to the total effluent, while oxidation and acidification were used for the black liquor treatment. Plain sedimentation was the only treatment process applied on white water.

Plain Sedimentation. Six graduated cylinders were used for plain sedimentation. Each cylinder was filled with one litre of wastewater which then was allowed to settle for different periods (0.5, 1, 1.5, ..., 6 hr etc). The resulting wastewater was characterized by collecting the supernatant liquid from each cylinder.

Chemical Coagulation. The coagulants used in this treatment were:
(i) Aluminium sulfate, $Al_2(SO_4)_3.18\ H_2O$: solutions were prepared by dissolving an exact amount of the coagulant in one litre of bidistilled water so that 1 mL contained 40 mg of the salt. The pH of the samples were adjusted to 5.4 using HCl 1:1 in order to reach the optimum pH zone for coagulation with alum.
(ii) Ferric chloride, $FeCl_3$: solutions were prepared following the same technique

except that one drop of concentrated hydrochloric acid was added to avoid hydrolysis. The pH of the samples were adjusted to 4.0 using HCl 1:1 to obtain the optimum pH zone for coagulation with this salt.
(iii) Lime, $Ca(OH)_2$: Lime was added to increase the pH to various levels. The pH values ranged from the pH of the initial waste to about pH 12.5. The technique of lime addition differed from that used for ferric chloride and aluminium sulfate; the appropriate amount of lime (1, 1.5, 2, ..., grams) was weighed and added to the wastewater samples and the pH was recorded after the process of coagulation was completed.

The chemical coagulation-flocculation jar test was carried out to find the optimum chemical dosages and conditions to improve the quality of the different wastewater samples for each coagulant.

Oxidation:

Aeration: in this process, compressed air was applied to a 10 liter volume of wastewater samples under test in big plastic tanks for a period of two hours; then, samples from the tanks were allowed to settle in graduated cylinders for different settling periods. Characterization of the waste following aeration and sedimentation was carried out on samples of the supernatant liquid in each cylinder.

Calcium Hypochlorite: a stock solution of calcium hypochlorite was prepared so that 1 mL of this suspension contained 50 mg of calcium hypochlorite. The available chlorine produced from $Ca(OCl)_2$ was determined using the sodium thiosulphate standard procedure. The optimal dosage of calcium hypochlorite and the characteristics of the treated wastewater were determined following the same procedure as for the coagulants.

Acidification: In this treatment, different volumes of HCl 1:1 were added to one liter of the sample and mixed using a magnetic stirrer, for 2 minutes, then the sample was left for about 2 hours to settle. The supernatant was taken and its characteristics evaluated.

The physico-chemical analyses were carried out according to the Standard Methods For the Examination of Water and Wastewater (A.P.H.A., A.W.W.A. and W.P.C.F., 1971).

RESULTS AND DISCUSSIONS

Treatment of the Total Effluent

Plain Sedimentation. Plain sedimentation of the combined wastewaters from the papermill could remove 42-57% COD, 37-56% volatile solids and 60-84% suspended solids with a settling period ranging from 2 to 3 hours. The effect of plain sedimentation on the mean percentage removal of COD volatile solids and suspended solids after various settling periods for six combined wastewater samples are presented in Table 1 and Fig. 1. Subrahmanyam and Mohanrao (1973) have previously reported that the sedimentation of pulp and papermill wastewaters in a clarifier reduces the suspended solids by as much as 70%, which was confirmed by our work.

Coagulation.

Aluminium sulfate, $Al_2(SO_4)_3.18\ H_2O$: The addition of aluminium sulfate in the range of 300-400 mg/L was found to reduce COD, colour and volatile solids by 62-88%, 46-75% and 60-88%, respectively, without pH adjustment. With this treatment, the final pH values of the treated wastewater have been decreased to 6.1-6.7. It was found that a dosage of aluminium sulfate higher than 400 mg/L brought only a further slight percentage reduction in the different tested parameters.

TABLE 1. The Effect of Plain Sedimentation on the Percentage Removal of COD, Volatile Solids and Suspended Solids after various Settling Periods for the Total Effluent.

SETTLING TIME (HRS)	PERCENTAGE REMOVAL					
	COD		VOLATILE SOLIDS		SUSPENDED SOLIDS	
	MEAN	RANGE	MEAN	RANGE	MEAN	RANGE
0.5	33.8	32 - 36	31.5	29 - 38	40.8	25 - 67
1	48.9	34 - 57	43.3	35 - 54	59.3	50 - 68
1.5	50.8	41 - 57	44.0	35 - 54	62.7	55 - 68
2	50.9	42 - 57	45.0	37 - 54	68.6	60 - 75
3	51.1	42 - 57	45.6	38 - 56	72.4	62 - 84
4	52.0	46 - 57	46.2	39 - 56	73.5	62 - 84
5	53.6	46 - 59	46.5	39 - 56	74.4	63 - 85
6	53.9	46 - 60	46.7	39 - 57	75.2	63 - 85

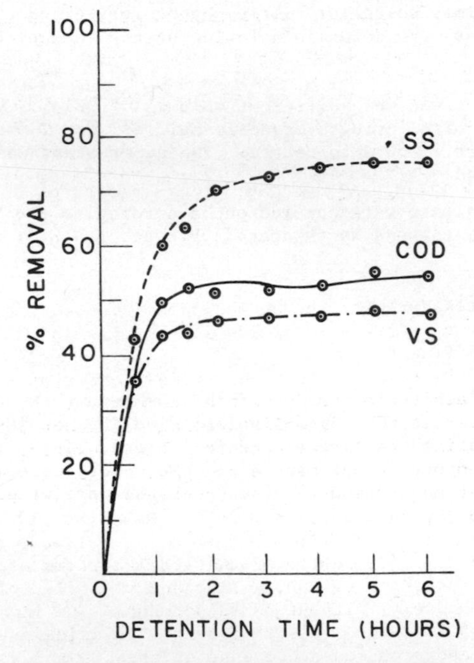

Fig. 1. The effect of plain sedimentation on the mean percentage removal of COD, volatile solids and suspended solids after various detention periods for final wastewater samples.

When the pH of the coagulation stage was adjusted to 5.4, the optimal dosage required was decreased to the range of 200 - 300 mg/L so as to reach 52-79% reduction of COD, 47-98.5% reduction of colour, and 45-77% reduction of volatile solids. The final pH of the treated wastewater was in the range of 4.8 to 5.4.

The effect of aluminium sulfate addition on the percentage removal of COD, colour and volatile solids with and without pH adjustment of the wastewater samples is presented in Table 2.

TABLE 2. The Effect of Aluminium Sulfate Addition on the Percentage Removal of COD, Colour and Volatile Solids with and without pH Adjustment for the Total Effluent.

DOSE (mg/L)	PERCENTAGE REMOVAL (Without pH Adjustment)						PERCENTAGE REMOVAL (After pH Adjustment to 5.4)					
	COD		COLOUR		VOLATILE SOLIDS		COD		COLOUR		VOLATILE SOLIDS	
	Mean	Range	Mean	Range	Mean	Range	Mean	Range	Mean	Range	Mean	Range
150	47.3	31-75	36.9	23-55	45.9	30-73	48.8	38-60	32.0	24-41	46.8	35-59
200	59.1	42-87	47.9	31-67	57.5	40-86	63.9	45-77	78.5	47-98	61.1	41-74
300	73.6	62-87	64.9	46-84	72.1	60-86	69.3	60-79	9.14	83-99	66.9	54-77
400	80.8	72-88	92.2	88-95	79.2	72-88	70.5	60-81	95.7	94-99	68.8	57-79
500	81.9	74-90	95.2	91-98	80.6	74-90	71.6	64-82	96.0	95-99	70.1	61-80
600	83.1	76-91	96.7	93-99	82.0	76-91	71.6	64-83	96.5	95-99	70.3	61-80
700	83.2	77-91	97.0	93-99.5	82.4	77-91	71.5	64-82	96.4	95-99	70.4	62-80
800	83.6	77-92	97.1	93-99.5	83.2	77-92	71.3	65-81	96.2	95-99	70.5	62-79
1000	85.1	80-92	97.1	94-99.5	85.1	80-92	72.5	65-82	96.2	95-99	70.8	62-79

Coagulation of the wastewater with the optimal aluminium sulfate dosage (200-300 mg/L) produced a volume of settleable solids in the range of 135-140 mL/L.

Smith and Christman (1969) reported that an optimal dosage of 150 mg/L aluminium sulfate is required to achieve 89% reduction of colour for the treatment of a hardwood kraftmill wastewater at pH 5.3, while a dosage of about 300 mg/L at the same pH is required for treating the softwood kraftmill wastewater. On the other hand, Varbanov, Papazov and Gogoev (1969) have found that the best results were obtained with 100-200 ppm alum at pH 5.0-6.5 for the treatment of various types of paper and board wastewaters by coagulation.

Ferric chloride, $FeCl_3$: It was found that the addition of 400 mg/L ferric chloride resulted in a reduction of COD, colour and volatile solids of 68-94%, 45-95% and 68-93%, respectively, without pH adjustment. The final pH of the treated waste was decreased to 5.2. Higher doses of ferric chloride gave a brown colour to the supernatant liquid and produced a decrease of the COD and volatile solids removal. This may be due to the excess residual iron and the phenomenon of charge reversal on the colloidal particles, caused by the increasing dose of coagulant; accordingly, the supernatant contained more organic matter, while lower doses of ferric chloride had a poor flocculation effect. When the pH of the wastewaters have been adjusted to 4.0, the required dosage of ferric chloride was reduced to 100-160 mg/L to give 55-77%, 70-94% and 50-77% reduction of COD, colour and volatile solids,respectively, as presented in Table 3. The final pH of the treated wastewaters was in the range of 3.75 to 3.9.

TABLE 3. The Effect of Ferric Chloride on the Percentage Removal of COD, Colour and Volatile Solids with and without pH Adjustment for the Total Effluent.

DOSE (mg/L)	PERCENTAGE REMOVAL (Without pH Adjustment)						PERCENTAGE REMOVAL (After pH Adjustment to 4.0)					
	COD		COLOUR		VOLATILE SOLIDS		COD		COLOUR		VOLATILE SOLIDS	
	Mean	Range	Mean	Range	Mean	Range	Mean	Range	Mean	Range	Mean	Range
80	46.2	24-89	6.1	3-12	45.9	25-88	54.0	46-65	71.9	61-78	49.5	36-64
120	53.6	35-90	10.3	4-21	54.0	36-89	64.0	55-71	81.6	70-94	58.5	44-70
160	61.9	47-91	14.3	5-30	62.4	48-90	67.9	62-77	87.0	78-94	63.4	52-77
200	69.5	58-91	18.0	6-38	70.9	61-90	70.3	67-76	89.2	87-92	65.7	60-76
400	80.2	68-94	75.9	45-95	80.4	68-93	69.4	66-75	84.3	80-87	63.6	57-73
600	81.9	81-83	97.4	96-99	81.5	81-83	68.5	65-74	78.4	57-86	61.7	56-71
800	79.0	75-81	63.7	33-96	76.7	74-80	67.4	64-73	75.5	50-86	60.0	54-70
1000	64.7	61-68	32.4	5-83	64.7	63-66	66.4	61-72	73.0	46-85	57.8	50-69

Coagulation of the total effluent with the optimal dose of ferric chloride (100-160 mg/L) produced a settleable solids volume between 150-155 mL/L, which was slightly higher than that obtained in the alum coagulation process. Smith and Christman (1969) have reported that ferric chloride coagulation of the softwood kraftmill wastewater required a dose of 286 mg/L to obtain 87% overall colour removal. They have also reported that the pH for optimum colour removal by ferric chloride was 3.9, which is nearly the same pH obtained in this study (4.0).

Lime, $Ca(OH)_2$: It was found that a dosage in the range of 1500-2000 mg/L resulted in removal of COD, colour and volatile solids of 54-80%, 51-98.8% and 44-82%,respectively, from the wastewater. The final pH of the treated effluent was highly alkaline, reaching 10.8-11.8.

The effect of lime addition on the percentage removal of COD, colour and volatile solids on the total effluent samples is presented in Table 4. A compact settleable solids volume of 110 mL/L was obtained with this coagulation process at an optimal dosage of lime of 1500-2000 mg/L.

Lime coagulation could be followed by recarbonation by CO_2 to precipitate the calcium ion (Ca^{++}) as calcium carbonate, resulting in a dense floc: the recarbonation lead to an improvement of the decolouration effect on the treated wastewater.

Gould (1974) used a high dosage of lime (2000 mg/L) to treat the highly coloured effluent from the caustic extract stage of a kraft bleach plant and obtained a 90% reduction in colour and approximately 45% removal of BOD. Spruill (1974) used a 1000 mg/L lime addition to achieve 80 to 90% colour reduction in a kraftmill effluent. These results are concordant with these reported in this paper although the raw material is different.

TABLE 4. The Effect of Lime Addition on the Percentage Removal of COD, Colour, and Volatile Solids for the Total Effluent.

DOSE (mg/L)	PERCENTAGE REMOVAL					
	COD		COLOUR		VOLATILE SOLIDS	
	MEAN	RANGE	MEAN	RANGE	MEAN	RANGE
1000	53.4	42 - 72	7.5	5 - 11	48.3	32 - 71
1500	62.9	54 - 76	65.0	51 - 99	58.1	44 - 76
2000	67.6	59 - 80	85.4	77 - 99	64.3	55 - 82
2500	68.3	59 - 81	86.6	76 - 99	65.5	57 - 82
3000	68.9	60 - 81	87.9	78 - 99	67.3	58 - 83
3500	70.1	61 - 81	89.3	81 - 98	69.3	60 - 84
4000	70.6	62 - 81	89.95	84 - 98	70.0	61 - 85

Oxidation.

Aeration: The aeration of the total mill effluent for two hours followed by sedimentation did not improve the wastewater characteristics compared to plain sedimentation alone. The percentage removal of COD, volatile solids and suspended solids were 41.5%, 28.8% and 50.3%, respectively, after a one hour settling period, which was found to be the optimal settling time.

Calcium Hypochlorite $Ca(OCl)_2$: It was found that an optimal dosage of calcium hypochlorite of 2000 mg/L produced 46-74%, 80-99%, and 55-76% reduction of COD, colour, and volatile solids, respectively. (The stock suspension contained 16.3 mg/mL available chlorine). The final pH of the treated wastewaters was in the range of 10-11.3.

The effect of calcium hypochlorite addition on the percentage removal of COD, colour and volatile solids for the samples of mill effluent shown in Table 5. It was also found that most of the colour reduction took place during the first two hours of contact; further reduction in colour was small after this period as shown in Fig. 2. The volume of settleable solids resulting from the hypochlorite treatment using the optimal dosage of calcium hypochlorite (2000 mg/L) was found to be 110 mL/L.

Treatment of the Black Liquor

Acidification. It was found that when 8 mL of HCl 1:1 was added to 1 liter of the black liquor, the percentage removal of COD and colour was 48.6% and 92.7%, respectively; a large volume of settleable solids was also produced, which decreased with time from 860 mL/L after the first settling hour to 710 mL/L after a 3 hour sedimentation period. As indicated by the final pH, the treated effluent was very acidic.

Saleh and co-workers (1978) concluded that high-purity lignin could be isolated from alkaline black liquor of rice straw by acid treatment, and could be converted into nitro and amino-lignin for use as tanning agents for pelts.

TABLE 5. The Effect of Calcium Hypochlorite Addition on the Percentage Removal of COD, Colour and Volatile Solids for the Total Effluent.

DOSE (mg/L)	PERCENTAGE REMOVAL					
	COD		COLOUR		VOLATILE SOLIDS	
	MEAN	RANGE	MEAN	RANGE	MEAN	RANGE
1000	45.6	22 - 67	--	--	40.6	11 - 69
1500	55.4	45 - 72	41.9	3 - 99.0	50.97	32 - 74
2000	61.4	46 - 74	78.6	5 - 99.9	56.9	35 - 76
2500	64.1	55 - 75	93.8	81 - 99.96	59.0	40 - 77
3000	64.9	57 - 75	96.1	91 - 99.6	60.5	44 - 77
3500	65.9	59 - 76	97.5	96 - 99.6	60.8	45 - 77

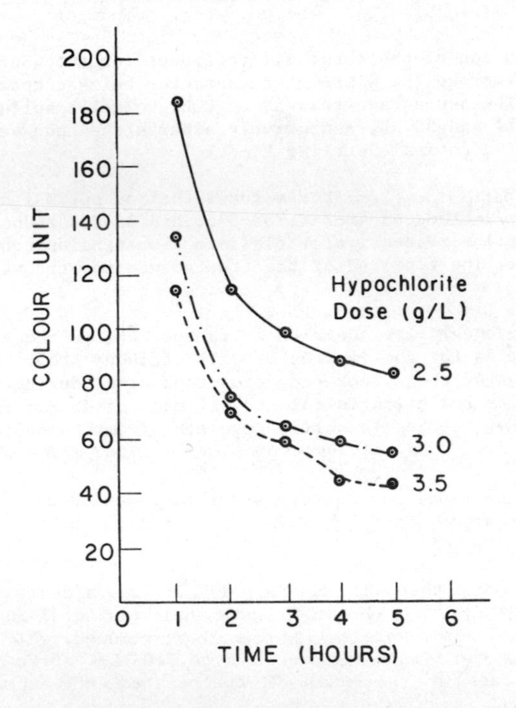

Fig. 2. The effect of time on the colour percentage removal of final wastewater when calcium hypochlorite was used as oxidizing agent.

Aeration. When black liquor was aerated for two hours and allowed to settle subsequently, only a very slight reduction in COD (about 8.1%) was observed, which is significantly lower than that obtained with other treatments.

Plain Sedimentation Treatment of the White Water

It was observed that sedimentation for a period of one hour removed from the white water 88-92%, 89-92% and 83-95% of the COD, volatile solids and suspended solids, respectively. The clarified white water should be recycled to reduce the volume and the pollution load of the total mill effluent. Treating the white water by sedimentation increases the possibility of applying the closed white water system as described by Brecht and co-workers (1975) and Hommel (1976) for the reduction of the excess white water discharge.

CONCLUSIONS

A good reduction of the suspended solids content of the wastewater was achieved by the plain sedimentation process when treating the final effluent of the Rakta Pulp and Paper Company.

The use of aluminium sulfate, ferric chloride, and lime coagulants for the final wastewater treatment revealed that lime was the best and most economic coagulant used. Lime coagulation could be followed by recarbonation with CO_2 to precipitate the calcium ion which remains in solution and improves the percentage removal of colour. It should be mentioned here that the silica removal may facilitate the lime recovery from the sludge, a process which adds to the economy of the treatment of lime.

The percentage removal of COD, colour and volatile solids of the final wastewater with aluminium sulfate coagulant is competitive with lime treatment. The pH of the effluent was low enough to permit direct polishing treatment on granular activated carbon to be carried out. But the application of aluminium sulfate was more expensive than the lime itself, and the recovery of aluminium sulfate from the sludge is very difficult. Also a voluminous incompact settleable solids resulted, a property which favours the use of lime more than alum.

The application of ferric chloride coagulant is not recommended due to its relatively higher cost than alum and, consequently, lime. Also it is not readily handled and applied; further-more, it is very corrosive, and it gave a voluminous settleable solids.

Aeration treatment of the final wastewater is not recommended because it gave no more improvement than that obtained by plain sedimentation; also, it is much more expensive.

The use of calcium hypochlorite is not preferable because it is expensive. On the other hand, residual chlorine may combine with phenol constituents producing "chlorophenol compounds" which remain in the effluent contributing to an additional polutional load to the receiving stream.

The acid treatment of the black liquor gave a good reduction in colour and COD. The high-purity lignin could be isolated from the sludge and converted to nitro- and amino-lignin which could be used as a tanning agent.

The plain sedimentation process was found to be a very successful one for the treatment of white water. Clarified white water should be reused as much as possible to reduce the volume of the final waste, and the fresh water consumed.

It is recommended that the chemical used for digestion process (soda) should be recovered from the black liquor as much as possible to be reused again. This will require further research to remove the high silica content from the black liquor to facilitate the alkali recovery process.

Further research studies are recommended to cover the following points:
(i) The determination of Abu-Kier drain flow, self purification process, and its maximum ability to assimilate some wastewater components for the maximum benefit of this natural process.
(ii) Study on the synergistic effect of the combined action of coagulants (e.g. ferric chloride with lime) for treatment of this waste and possibly assisted by synthetic polyelectrolyte coagulant aids.
(iii) The application of activated carbon for the treatment of the wastewater after applying lime coagulation, settling, and neutralization (to prevent calcium carbonate build up on the carbon columns).

Finally the pollution loads of effluents discharged into Abu-Kier drain from Rakta Pulp and Paper Company can be reduced to the required levels by conscientious application of established in-plant process controls and water recycle measures together with well designed and properly operated external treatment facilities.

REFERENCES

A.P.H.A., A.W.W.A. AND W.P.C.F. (1971). Standard Methods for the Examination of Water and Wastewater. 13th Edition.

Brecht, W. and co-workers (1975). Closed white water systems in paper mills using waste paper. Chem. Abs., 81, 2, 6059V (1974), J. Water Poll. Control Fed., 47, 6.

Gould, M. (1974). Colour removal from kraft mill effluent by an improved lime process. Tappi, 56, 79, (1973); J. Water Poll. Control Fed., 46, 6.

Hommel, K. (1976). Closed systems at Rhumspringe. J. Water Poll. Control Fed., 48, 6.

Saleh, M.S.E. and co-workers (1978). Pollution of environment by wastes of pulp and paper industry and its treatment for recovery of its valuable chemicals. Chemical Engineering Department, Alexandria University, Egypt.

Smith, S.E. and R.F. Christman (1969). Coagulation of pulping wastes for the removal of colour. J. Water Poll. Control Fed., 41, 222, Part (1).

Spruill, E.L. (1974). Colour removal and sludge recovery from total mill effluent Tappi, 56, 98, (1973); J. Water Poll. Control Fed., 46, 6.

Subrahmanyam, P.V.R. and G.J. Mohanrao (1973). Pulp and paper mill waste treatment. In Indian Association for Water Pollution Control, 10th Anniversary Commemoration Volume, p. 120-136.

Varbanov, D., I. Papazov and S. Bogoev (1969). Treatment of the wastewaters from the Petko Napetov paper factory. Water Pollution Abstracts, 42, 8, 1656.

32

ON THE DEPOSITION OF AEROSOLS A REVIEW AND RECENT THEORETICAL AND EXPERIMENTAL RESULTS

***P. L. Douglas, F. A. L. Dullien and D. R. Spink**

Department of Chemical Engineering, University of Waterloo, Waterloo, Ontario, Canada

*(*Present address: Dept. of Chem. Eng., Queen's Univ., Kingston, Ontario, Canada)*

ABSTRACT

An introduction to the literature on the deposition of aerosols on surfaces under various flow conditions is given; for example, on a cylinder in potential flow and on duct walls under turbulent flow conditions.

A theoretical approach is then outlined which explains experimentally observed deposition rates of an aerosol on a cylinder in turbulent cross flow. The parameter which best characterizes the system is the turbulent Peclet number ($Pe_t = U_0\delta/\varepsilon_0$).

KEYWORDS

Aerosol deposition; cylindrical collector; turbulent crossflow; continuity equation.

INTRODUCTION

The purpose of this paper is to update a paper presented at the last IWTU Conference by Dullien and Spink (1978) entitled"The Waterloo Scrubber Part I: The Aerodynamic Capture of Particles". In that paper,preliminary experimental results on the deposition of an aerosol on a cylinder in turbulent crossflow were presented. The preliminary results indicated that turbulent flow enhanced the collection efficiency to levels substantially higher than were predicted using existing predictive models.

In this paper, a continuation of the work on the deposition of aerosols on a cylinder in turbulent crossflow is presented, beginning with an introduction to the literature on aerosol deposition on various surfaces under different hydrodynamic conditions.

There are two broad classifications under which the deposition of aerosols may be discussed. The first is the deposition of aerosols on surfaces placed normal to the direction of the bulk flow. For example, consider the flow of an aerosol laden gas stream passed a cylinder, sphere, or ribbon immersed in the flow. These surfaces are often called nonstreamline surfaces, in hydrodynamic terms. The second classification is the deposition on streamlined surfaces such as the walls of a duct containing the flowing gas.

A brief review of some of the pertinent literature is presented here. It should be emphasized that this is not intended as comprehensive literature survey, but an introduction to the literature, terminology, and concepts with which some may be unfamiliar. If further information is required, the reader may, as a starting point, consult one or more of the references cited in this paper.

LITERATURE

Nonstreamlined Surfaces

For the flow against a nonstreamlined object,such as a cylinder,Strauss (1975) gives three basic collection mechanisms. They are: inertial impaction, interception and Brownian diffusion.

Since, in this paper, we are ultimately concerned with the deposition, under turbulent flow conditions, of an aerosol on a cylinder,only the deposition on cylinders under flow conditions which may approximate turbulent flow will be considered.

Inertial impaction. The following general assumptions characterize the development of the inertial impaction model.

(a) streamline flow.

(b) aerosol particles are considered to be point masses,except for the calculation of the drag forces.

(c) potential flow models are assumed,or experimental streamlines are used.

Consider the case where particulate-laden air is flowing past a cylinder, referred to as the collector. The aerosol particles follow the gas streamlines as they approach the collector; then at a point, depending on their inertia, a particle may leave the diverging streamlines and subsequently impact on the collector or it may miss it. This phenomena is illustrated by Fig. 1.

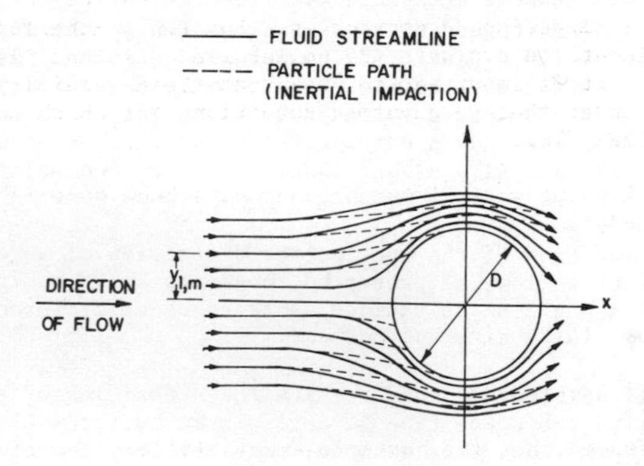

Fig. 1 Particle trajectory around a cylinder (due to inertial impaction)

It has been shown that the characteristic parameter which best defines the system is the inertial impaction parameter, Ψ.

$$\text{where} \quad \Psi = C\rho_p d^2 U_o / 18\mu D \qquad (1)$$

Figure 2 is a plot of the collection efficiency of a cylinder versus the inertial impaction parameter for potential flow.

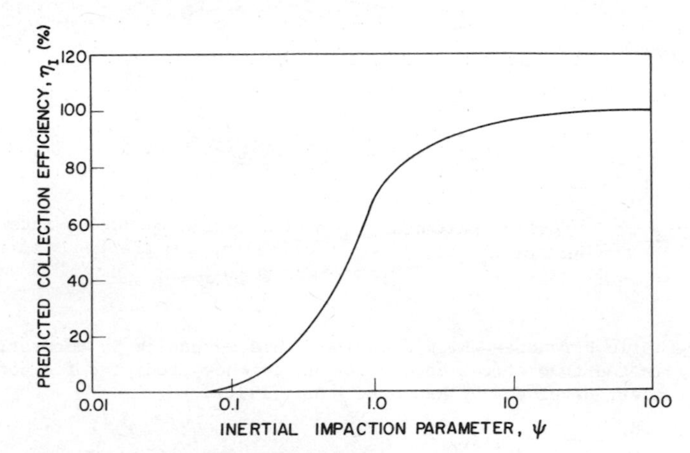

Fig. 2 Theoretical aerosol collection efficiency by cylinders - inertial impaction in potential flow

This theoretical curve has been predicted by others with only slight variations; Sell (1931), Albrecht (1931), and Langmuir and Blodgett (1944). A number of experimental studies have been conducted to determine the feasibility of the various theoretical predictions. The most extensive predictions were made by Wong and Johnstone (1953),who measured the collection efficiency of cylinders in potential flow. Their results are in good agreement with the theoretical predictions for values of the inertial impaction parameter less than 2.5. Ranz and Wong (1952) also performed experiments to evaluate the collection efficiency by inertial impaction on cylinders. It is important to note that these carefully designed experiments were performed under the hydrodynamic conditions for which the models were designed,i.e., potential flow.

Interception. The following general assumptions have been made in the development of the interception model.

(a) streamline flow.

(b) the aerosol has a finite size but no mass.

(c) potential flow is assumed.

In this case,it is assumed that the particle always follows the fluid streamlines. If the streamline on which the particle is travelling comes closer than d/2 to the collector,then the particle will be captured by interception. Figure 3 illustrates this process.

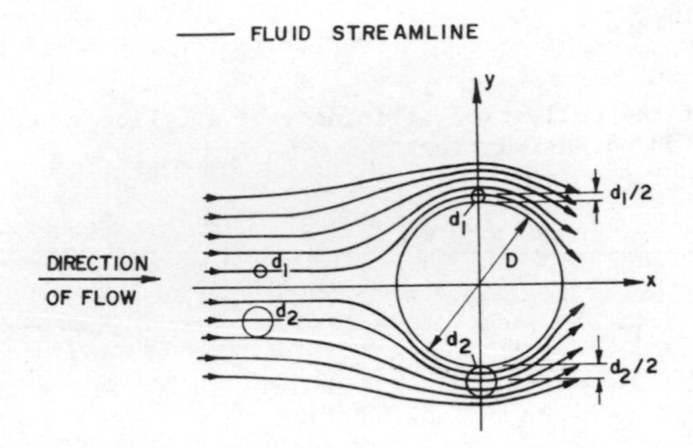

Fig. 3 Particle trajectory around a cylinder (due to interception)

The characteristic parameter which describes this mechanism is the ratio d/D. Equation (2) is the theoretical collection efficiency predicted for potential flow around a cylinder, developed by Ranz and Wong (1952).

$$\eta_C = [(1 + \frac{d}{D}) - \frac{1}{(1 + \frac{d}{D})}] \, 100 \tag{2}$$

The collection efficiency predicted by equation (2) is plotted in Figure 4.

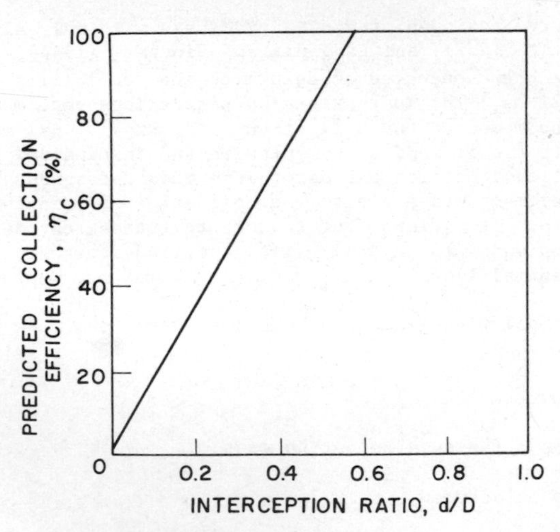

Fig. 4 Theoretical aerosol collection efficiency by cylinders - interception in potential flow

Brownian Diffusion. Very small particles, less than one (1) micron (μm), are rarely captured by inertial impaction or interception because their inertia is insufficient to cause them to diverge from the streamlines and their size limits the number that are intercepted. However, because of their size, they become subject to bombardment by gas molecules and as a result they exhibit Brownian motion. The random movement of Brownian motion allows the aerosols to cross the streamlines and impact on the collector.

The collection of particles on circular cylinders with Brownian diffusion as the controlling mechanism has been studied by a number of workers, using two different approaches. Langmuir (1942) and Bosanquet (1950) both started with well-known models of Brownian diffusion to surfaces in non-flow situations. They then adapted these to the particular case of diffusion to a collector in a flow situation. Ranz (1953), Landt (1956) and Davies (1952) used an emperical approach.

Ranz (1953) modified the turbulent heat transfer correlation for cylinders relating the Nusselt number, Reynolds number and Schmidt number to predict the efficiency due to diffusion for cylinders. Landt (1956) suggested that (π/Pe), while Davies (1952) suggested that (1/Pe) be used as the collection efficiency due to diffusion.

Four of the models predicting the collection efficiency by diffusion on a cylinder are presented below along with the general assumptions upon which they were based. Figure 5 is a plot of the predicted collection efficiency by Brownian diffusion as a function of the Peclet number for the various models.

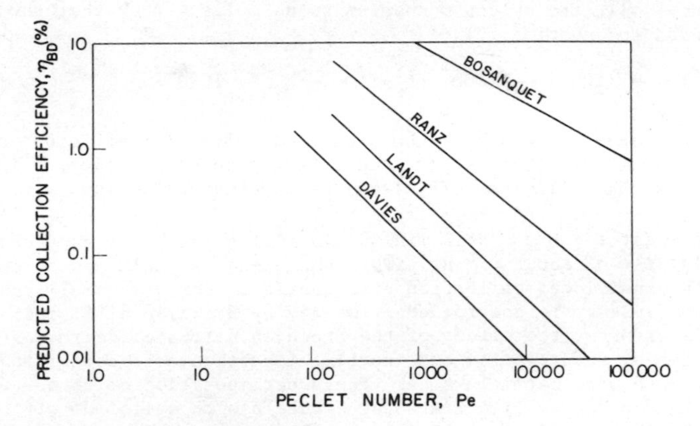

Fig. 5. Theoretical aerosol collection efficiency by a cylinder due to Brownian diffusion

(a) Bosanquet (1950)

$$\eta_{BD} = [\frac{8}{Pe}]^{1/2} 100 \tag{3}$$

(i) non-flow situation

(ii) diffusion to a flat plate

(iii) Brownian motion

(b) Ranz (1953)

$$\eta_{BD} = \frac{\pi}{Pe} \left(\frac{1}{\pi} + 0.55\ Re_c^{1/2} Sc^{1/3}\right) 100 \qquad (4)$$

(i) based on the turbulent mass transfer analogy

(ii) diffusion to a cylinder

(iii) Brownian motion

(c) Landt (1956)

$$\eta_{BD} = \left[\frac{\pi}{Pe}\right] 100 \qquad (5)$$

(d) Davies (1952)

$$\eta_{BD} = \left[\frac{1}{Pe}\right] 100 \qquad (6)$$

Combination of effects. The aerodynamic capture of particles does not proceed by the isolated mechanisms discussed in the previous section but does so by a combination of these mechanisms, with one of the mechanisms often predominating.

A number of equations, most of which are emperical, have been proposed to account for the combination of mechanisms. A common approach has been to allow only the particles not collected by one mechanism to be collected by the others. This approach leads to equation (7):

$$\eta_{ICBD} = \left[(10^6 - (100 - \eta_I)(100 - \eta_C)(100 - \eta_{BD})\right] 10^{-4} \qquad (7)$$

Using this approach, we can plot the predicted collection efficiencies for various particle sizes of unit density particles. Fig. 6 is such a plot, using equation (3) to predict the collection efficiency by Brownian diffusion.

It is of interest to note that a minimum in collection efficiency exists at a particle diameter of about 0.7 μm. This phenomena is explained by examining the effect of the individual collection efficiencies. For very small particles, the collection efficiency is controlled primarily by Brownian diffusion; as the particle size increases, the effect of the Brownian diffusion decreases and, although the mechanisms of interception and inertial impaction are insignificant, their effects increase with particle size. The superimposition of these two effects yields the kind of curve illustrated by Figure 6 with a minimum efficiency occurring where both the diffusive and inertial/interceptive forces are relatively ineffectual.

Streamlined Surfaces

In cases where the dust laden air flows parallel to the collector surface, as in the example of flow over a flat plate, there is no obvious mechanism by which the particles deposit on the surface. In fact, if one assumes streamline flow, as was the case in the development of the inertial impaction and interception models, then one predicts, by definition, no deposition at the surface. If the added complexity of a boundary layer is added, then one actually predicts that the aerosol velocity in the vicinity of the flat plate is directed away from the surface due to the thickening of the boundary layer. Only by inclusion of turbulent velocity fluctuations and eddy diffusion can one account for the observed deposition.

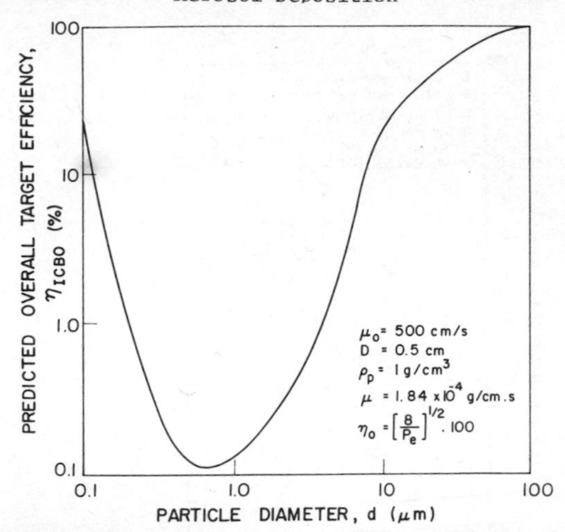

Fig. 6 Predicted overall target efficiency versus particle diameter

Most of the research in this field has been directed towards the problem of aerosol deposition in pipe flow, which is important in determining line losses when sampling aerosols through long tubes. The most important work in this area of aerosol mechanics was conducted by Freidlander and Johnstone (1957); other contributions have been made by Lui and Agarival (1974), Lui and Ilori (1974), Gudmundsson and Bott (1977), Davies (1966) and Beal (1968).

The predominant theory of aerosol deposition on pipe walls in turbulent flow is what has become known as the 'free flight diffusion model' developed by Freidlander and Johnstone (1957). In simplest terms, it is assumed that turbulent velocity fluctuations or eddy diffusion in the radial direction transport the aerosol up to within one stopping distance, x_s, of the wall.

$$\text{where } x_s = \frac{C\rho_p d^2 U_0}{18\mu} \tag{8}$$

With their inertia, the aerosol particles are then able to travel to the wall and deposit on the wall.

Freidlander applied Fick's Law to predict the flux of aerosol particles diffusing from the turbulent core to one stopping distance from the wall. Fick's First Law for eddy diffusion can be written as:

$$N = -\varepsilon(y) \frac{\partial c}{\partial y} \tag{9}$$

Using the hydrodynamic correlations of Laufer (1954), and Lin, Putnam and Moulton (1953) and equation (9), Freidlander predicted the aerosol deposition rate.

Figure (7) is a plot of the deposition rate in terms of a mass transfer coefficient called the deposition velocity, against the duct Reynolds number.

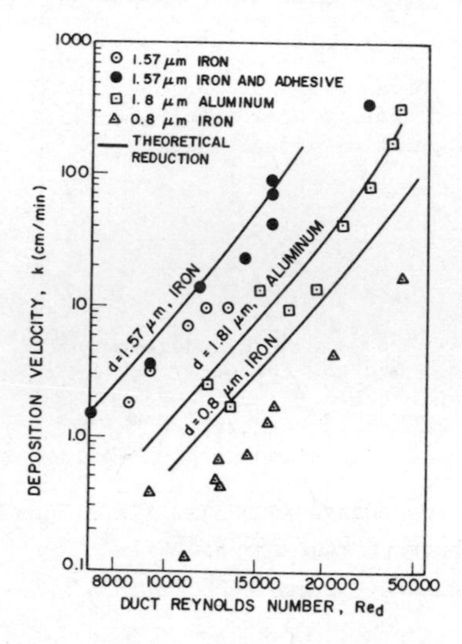

Fig. 7 Deposition velocity versus duct Reynolds number

Freidlander found that the rate of deposition increased with increasing duct Reynolds number; he verified his predictions with experiments, see Fig. 7, and explained the intuitively-acceptable results by showing that the ratio of the stopping distance to the boundary layer thickness, x_s/δ, also increases with the duct Reynolds number. It is then easier to transport an aerosol particle to within one stopping distance of the wall resulting in their deposition, especially when the stopping distance is large.

A brief summary of the literature follows.

The deposition of aerosols on nonstreamlined surfaces, such as a cylinder, in ideal flow (viscous or potential flow) has been studied in some detail. In general, carefully designed experiments to test these models have shown the models to be adequate for the conditions under which they were developed. The deposition of aerosols onto streamlined surfaces in turbulent flow has not been investigated to the same degree, although significant contributions towards understanding the deposition mechanism have been made.

The most significant omission in the literature is the deposition of aerosols onto nonstreamlined surfaces under turbulent flow conditions. As indicated in our first paper, Dullien and Spink (1978), some preliminary experiments have been performed and the deposition results are significantly higher than are predicted by the classical mechanisms of inertial impaction, interception and Brownian diffusion.

The remainder of the paper is devoted to theoretical and experimental investigation of the deposition of an aerosol onto a cylinder in turbulent cross flow.

THEORY

The diffusional deposition approach was used to develop a model to predict the deposition rate onto a cylinder in turbulent flow. The continuity equation for the particles can be written in orthoganal co-ordinates as:

$$\frac{\partial c}{\partial t} + \frac{\partial N_x}{\partial x} + \frac{\partial N_y}{\partial y} + \frac{\partial N_z}{\partial z} = r \tag{10}$$

where y is the radial distance from the surface (cm)
x is the distance measured along the contour from the stagnation point (cm)
z is the logitudinal distance (cm)
N is the total flux in the subscripted direction ($g/cm^2.s$)
r is the rate of production or depletion due to chemical reaction ($g/cm^3.s$)

The total flux is the sum of the convective and diffusive components, and is given by equations (11) and (12) in the x and y directions, respectively.

$$N_x = v_x c - \varepsilon \partial c/\partial x \tag{11}$$

$$N_y = v_y c - \varepsilon \partial c/\partial y \tag{12}$$

The Solution of the Continuity Equation

The solution to the continuity equation given by equation (10) yields a concentration profile c = f(x,y) enabling one to determine the flux at the surface and, hence, the collection efficiency.

When one is solving large complex equations involving many terms, simplifying assumptions are often made, either because the physical realities indicate that certain terms are insignificant, or because the information at hand is insufficient to include others, in which case the experimental programmes to test such models must accommodate the simplifications. Here, the solution of equation (10) is subject to a number of simplifying assumptions and limitations. They are:

(1) $\partial c/\partial t = 0$

(2) $c \neq f(z)$

(3) $\varepsilon \neq f(x)$

(4) $c \neq f(x)$

(5) $r = 0$

Invoking these assumptions and substituting dimensionless variables into equation (10) reduces it to:

$$\frac{\partial^2 \bar{c}}{\partial \bar{y}^2} + \alpha_1 \frac{\partial \bar{c}}{\partial \bar{y}} + \alpha_2 \bar{c} = 0 \tag{13}$$

where

$$\alpha_1 = \frac{1}{\varepsilon}\left(\frac{\partial \bar{c}}{\partial \bar{y}} - Pe_t\ \bar{v}_y\right) \tag{14}$$

$$\alpha_2 = -Pe_t\left(\frac{\partial \bar{v}}{\partial \bar{x}} + \frac{\partial \bar{v}_y}{\partial \bar{y}}\right) \tag{15}$$

$$Pe_t = \frac{U_o \delta}{\varepsilon_o} \tag{16}$$

Equation (13) is a linear, second order, differential equation with variable coefficients. The solution of such an equation requires two (2) boundary conditions. The first is that at the outer edge of the boundary layer, the concentration of the aerosol is equal to the free stream concentration. The second boundary condition is that at a distance d/2 from the surface of the cylinder, the concentration of particles in the gas phase is zero (0). The physical justification for this boundary condition is that when a particle comes in contact with the surface of the cylinder, ideally, the particle is immediately removed from the gas phase. In practice, however, not all particles adhere to the surface of the cylinder once they have contacted the surface.

The boundary conditions, then, for the solution to equation (13) are written as:

$$\text{BC 1:}\quad \bar{y} = 1,\ \bar{c} = 1 \tag{17}$$

$$\text{BC 2:}\quad \bar{y} = \frac{(d/2)}{\delta},\ \bar{c} = 0 \tag{18}$$

The solution of the differential equation was obtained by using the numerical integration technique of Runge-Kutta and a search technique to determine the value of $[\partial \bar{c}/\partial \bar{y}]_{\bar{y}=1}$ such that the second boundary condition was satisfied.

The velocity profile around the cylinder, $(\bar{v}_x, \bar{v}_y) = f(\bar{x}, \bar{y}, Re)$, was determined from the solution to the Navier-Stokes equations, Schlichting (1955). The dimensionless eddy diffusivity profile near the cylinder, $\bar{\varepsilon}$, was assumed to have the same shape as the eddy diffusivity profile near a flat plate: Levich (1962), Schlichting (1955), Laufer (1953), Davies (1966), Alpay (personal communication, 1

The boundary layer thickness was calculated from the solution to the Navier-Stokes equations, Schlichting (1955), and can be estimated at the stagnation point to be:

$$\delta = \frac{1.4\ D}{\sqrt{Re}} \tag{19}$$

Figure (8) is a typical plot of the concentration profile for a Peclet number of 1.12. The concentration profile is not a function of the angle, because $\partial \bar{c}/\partial \bar{x}$ was assumed to be equal to zero.

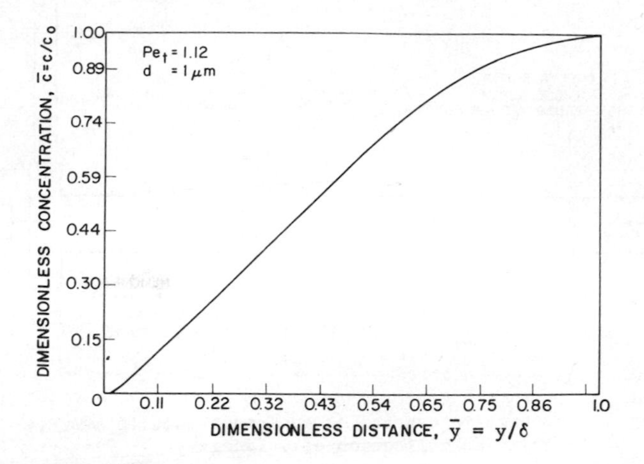

Fig. 8 Dimensionless aerosol concentration versus dimensionless distance within boundary layer

Once the concentration profile is obtained, one can calculate the value of particle deposition rate at the surface to be:

$$N_y' = -\left(\frac{\pi D \ell}{2}\right)\left(\varepsilon\, \partial c/\partial y\big|_{y=d/2}\right) \tag{20}$$

A collection efficiency can now be defined by dividing N'_y by the deposition rate based on the conditions far from the cylinder, N_o', where:

$$N_o' = (\pi D \ell)\,(U_o c_o) \tag{21}$$

The collection efficiency, η_{TD}, due to turbulent diffusion is therefore expressed in terms of percentage by equation (22) as:

$$\eta_{TD} = \frac{N_y'}{N_o'}\,100 = -\left(\frac{\pi}{2}\right)\left(\frac{\varepsilon\, \partial c/\partial y\big|_{y=d/2}}{U_o c_o}\right)100 \tag{22}$$

EXPERIMENTAL

A large number of experiments were performed to measure the deposition rate of aerosols on cylinders in turbulent cross flow.

Figure 9 is a schematic diagram of the equipment used in the experimental portion of this study.

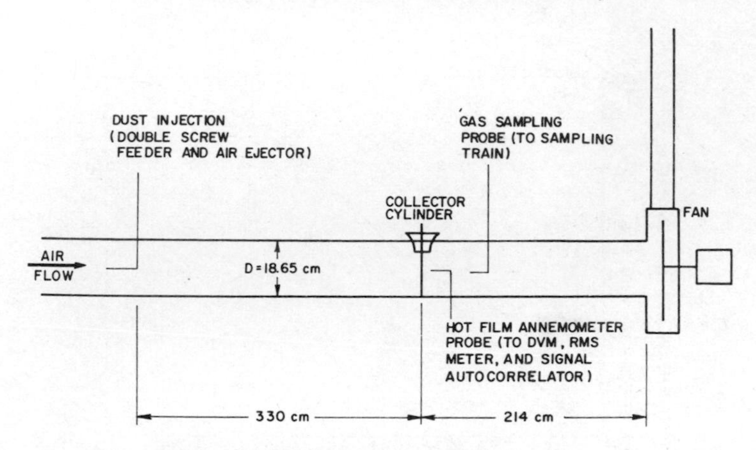

Fig. 9 Schematic diagram of equipment used to measure cylinder collection efficiencies.

The prime mover of air is the fan downstream from the sampling probe and dust injection site. The operating temperature of the equipment was ambiant, 20°C, and the operating pressure was slightly below barometric due to the action of the fan.

The aerosol used in this study was an aluminum silicate pigment. The dust was redispersed and injected into the duct through an air ejector operating at 25 psig. A double screw feeder or vibrating feeder fed the dust to the air ejector. The size distribution of the aerosol was measured using an 8-stage Andersen Cascade Impactor and found to have a log-normal size distribution with a mass median diameter of 1 μm and a geometric standard deviation of 2.2. The dust loading in the duct was measured using a Joy Manufacturing type sampling train.

The collector cylinders were made of mild steel and had diameters ranging from 0.01 to 1.27 cm.

The gas velocity, turbulence intensity and eddy diffusivity were measured using a hot film annemometer.

After a suitable cylinder residence time, "τ", of up to 4 minutes, the cylinder was withdrawn from the duct and the accumulated dust carefully removed from the cylinder using a delicate artist's brush. The dust was collected in small, lightweight aluminum foil envelopes for weighing.

With the dust loading in the air, the residence time in the duct, and the amount of dust on the cylinder the mass percent collection efficiency was calculated.

The duct Reynolds number, over which the experiments were conducted, ranged from 20,000 - 140,000. The turbulence intensity was found to vary from 0.5% to 1.5% in the 18.65 cm diameter duct, and the eddy diffusivity, a function of the duct Reynolds number, was found to range from 6 cm^2/s at Re_d = 20,000 to 25 cm^2/s at

$Re_d = 140,000$.

COMPARISON OF EXPERIMENTAL RESULTS WITH THEORETICAL PREDICTIONS

Figure 10 is a plot of our experimental data versus the inertial impaction parameter.

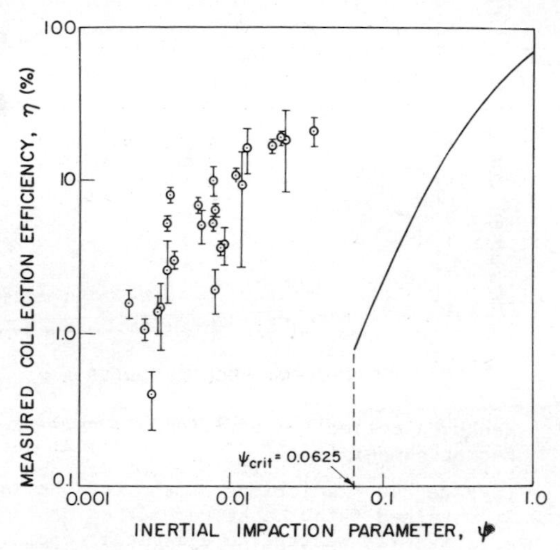

Fig. 10 Measured collection efficiency versus inertial impaction parameter

Although the data indicate a lot of scatter, the results clearly show that the collection efficiency increases with increasing inertial impaction parameter. This trend is substantiated by the theoretically predicted collection efficiency using the mechanism of inertial impaction. The solid line shows the theoretically predicted collection efficiency due to inertial impaction for potential flow.

All of the experiments were performed at values of the inertial impaction parameter less than 0.0625, which has been shown, theoretically, to be a critical value of the inertial impaction parameter below which no deposition occurs, Freidlander (1966). Nevertheless, the observed collection efficiencies are much greater than zero (0) for values of Ψ less than 0.0625.

In the development of the new model, turbulent diffusion and the concentration profile within the boundary layer near the cylinder were taken into account. As a result of this analysis, a turbulent Peclet number, $Pe_t = U_0\delta/\varepsilon_0$, with the boundary layer thickness as a characteristic length, was found to be the most significant parameter in defining the collection efficiency.

Figure 11 is a plot of the measured efficiency versus the turbulent Peclet number ($Pe_t = U_0\delta/\varepsilon_0$). The solid line superimposed on the data is the prediction made for particle diameter of 1 μm, from equation (21). Our aerosol is poly-dispersed and consists of particles which have diameters ranging from 0.1 μm to 3.0 μm and a mass median diameter of about 1 μm. The comparison between predicted and measured results is very good, much better than when the inertial impaction model was used.

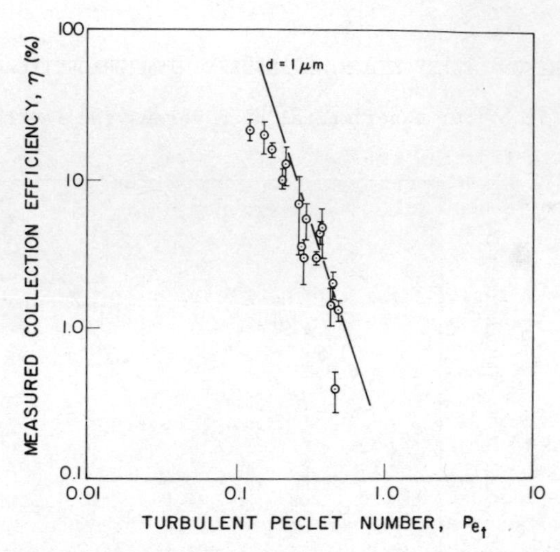

Fig. 11 Measured collection efficiency versus turbulent Peclet number

It is of interest to compare the predictions of the two theoretical models with the experimentally obtained data. A better method of comparison between the predicted collection efficiencies using the classical deposition models and the experimentally obtained collection efficiencies, is to plot the experimental collection efficiency using a combined mechanism approach. Figure 12 is such a plot.

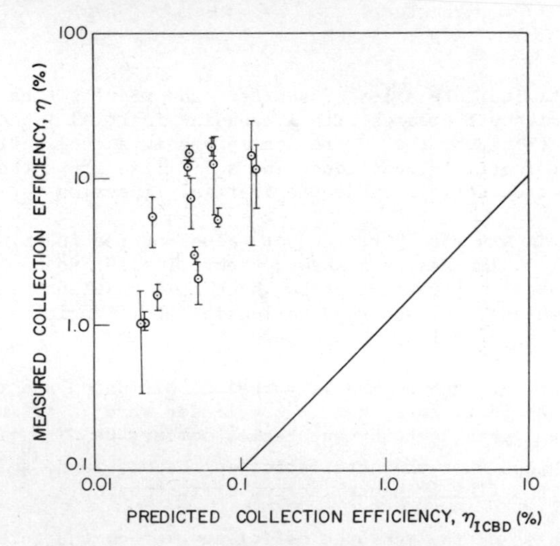

Fig. 12 Measured collection efficiency versus predicted collection efficiency (classical deposition models)

The abcissa contains the combined predicted collection efficiency given by equation (7). The various terms in equation (7) were evaluated as follows:

η_I predicted from the model of inertial impaction (Figure 2)

η_C predicted from the model of interception (equation (2) and Figure 4)

η_{BD} predicted from the Brownian diffusion model. Equation (4), plotted in Figure 5, was chosen to represent this mechanism. The ordinate contains the experimental collection efficiencies.

It is clear from Figure 12 that the predictive model does not adequately represent the experimental results. The most immediate reason for this failure is that the flow regime in the experimental equipment is not that which was assumed in the development of the classical models. The flow in the duct is turbulent, the eddy diffusivity of the gas, which is imparted to the particles, is significant and a boundary layer with a concentration gradiant exists near the cylinder.

Although the classical model appears inadequate, it should be emphasized again that the experiments specifically designed to test the classical model took into consideration the assumptions that were made in their development. As a general rule, those carefully designed experiments confirmed the predictions of the theoretical models.

Figure 13 is a plot of the measured efficiency versus the predicted efficiency from the solution to the continuity equation (13).

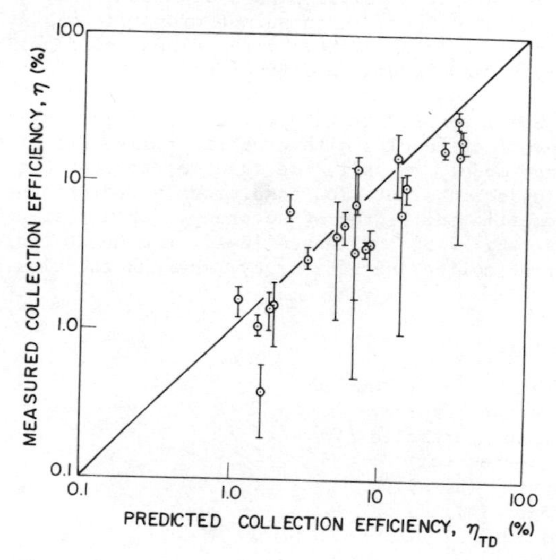

Fig. 13 Measured collection efficiency versus predicted collection efficiency (turbulent diffusion model)

The solution to the continuity equation tends to overestimate the measured collection efficiency. The simplest explanation for this trend is that the theoretical model, unlike the experiments, does not take particle-cylinder adhesion, or particle-particle cohesion into account.

When comparing the classical deposition model and the new model presented here, we should note that both models are theoretical models, unlike empirical models which are fitted to the data using the least squares technique and hence both models have built-in or bias errors which are often avoided when emperical models are used. As a result, the residuals (the differences between the measured and predicted values) are not normally distributed about a mean of zero as would be the case if the model had been fitted using a least squares technique. The errors are composed of errors in the experiment and the errors in the model. The errors in the experiment, which are significant in such work, can be estimated by making replicates at various constant operating conditions. The error in the model is then estimated from the difference in the total error and the experimental error.

From an analysis of variance, it has been shown at the 97.5% confidence level that the new turbulent diffusion model shows no lack of fit in representing the experimental data, while the classical deposition model shows a lack of fit (Douglas 198 Therefore, the new model, although not perfectly representing the data, is a bette predictor of the collection efficiency under the conditions imposed here, than the classical deposition model.

CONCLUSIONS

The collection efficiency of a polydispersed aluminum silicate aerosol, with a mas median diameter of about 1.0 μm, in turbulent cross flow, has been measured and is significantly higher than is predicted by the classical deposition models of inert impaction, interception and Brownian diffusion.

A new theoretical model to predict the collection efficiency has been tested by comparing the theoretical results with experimentally determined collection efficiency. The new model, which is the solution to the continuity equation for aerosol particles under tubulent flow conditions, predicts the observed collectio efficiency with a sufficient degree of accuracy. On a statistical basis, it has been shown that, at the 97.5% confidence level, the new turbulent deposition mode predicts the observed collection efficiency, whereas the classical deposition mod does not.

NOMENCLATURE

c Concentration of aerosol (g/cm^3)
C Cunningham correction factor
d Diameter of aerosol particles (cm)
D Diameter of collector (cm)
k deposition velocity (cm/s)
ℓ Length of cylinder (cm)
N Flux of aerosol to collector surface ($g/cm^2.s$)
Pe Peclet number
r Particle production or depletion rate ($g/cm^3.s$)
Re Reynolds number
Sc Schmidt number
t Time (s)
U Velocity (cm/s)
v Velocity component (cm/s)
x Distance in x direction (cm)
y Distance in y direction (cm)
z Distance in z direction (cm)

Greek letters

$\alpha_{1,2}$ Defined by equations (14) and (15)
δ Boundary layer thickness (cm)
ε Eddy diffusivity (cm^2/s)
η Collection efficiency (%)
μ Viscosity of gas (g/cm.s)
ρ Density (g/cm^3)
τ Residence time of cylinder in duct (s)
Ψ Inertial impaction parameter

SUBSCRIPT

BD Brownian diffusion
c Cylinder
C Interception
d Duct
I Inertial impaction
ICBD Inertial impaction, interception and Brownian diffusion
O Freestream value
p Particle
s Stopping distance
t Turbulent
TD Turbulent diffusion

SUPERSCRIPTS

— Normalized value

ACKNOWLEDGEMENTS

The authors gratefully acknowledge the Natural Sciences and Engineering Research Council for their financial support in the form of an operating grant.

We would also like to thank Mr. I. Buist and Mr. F. Leung for their contribution to the experimental portion of this study.

REFERENCES

Albrecht, F. (1931). Physik, Z., 32, 239.

Beal, S.K. (1968). Transport of particles in turbulent flow to channel or pipe walls. Bettis Atomic Power Laboratory, Westinghouse Corp., Pittsburgh, Pa., Tech. Report No. UC-80.

Bosanquet, C.H. (1950). Trans. Inst. Chem. Engrs. (London), 28, 130.

Davies, C.N. (1952). Proc. Inst. Mech. Engng. 1B, 185.

Davies, C.N. (1966). Deposition of aerosols from the turbulent flow through pipes. Proc. Royal Soc. (London), A 289, 235-246.

Douglas, P.L. (1980). The deposition of aerosols on cylinders in turbulent cross flow. PhD Thesis, Department of Chemical Engineering, University of Waterloo, Waterloo.

Dullien, F.A.L. and D.R. Spink (1978). The Waterloo Scrubber Part I: the aerodynamic capture of particles. In M. Moo-Young and G.J. Farquhar (Eds.), Waste Treatment and Utilization, Pergamon Press, Toronto, 469-486.

Freidlander, S.K. (1966). Smoke, Dust and Haze: Fundamentals of Aerosol Behaviour. John Wiley & Sons, New York.
Freidlander, S.K. and H.F. Johnstone (1957). Deposition of suspended particles from turbulent gas streams. Ind. Eng. Chem., 47 (7), 1151-1165.
Gudmundsson, J.S. and T.R. Bott (1977). Particle diffusivity in turbulent pipe flow. J. Aerosol Science, 8, 317-319.
Landt, E. (1957). Staub Reinaltung der Luft, 48, 9.
Langmuir, I. (1942). Office of Scientific Research and Development. Report No. 86
Langmuir, I. and K.B. Blodgett (1944). Mathematical Model of Water Droplet Trajectories. General Electric Research Laboratory, Schnectaty, New York, Report No. RL. 225.
Laufer, J. (1954). National Advisory Committee Technical Report Number 1174.
Levich, V. (1962). Physico-chemico Hydrodynamics, Prentice-Hall Inc., Englewood Cliffs, U.S.A.
Lin, C.S., R.W. Moulton and G.L. Putnam (1953). Ind. Eng. Chem. Process Design Develop., 45, 636.
Lui, B.Y. and J.K. Agarival (1974). Experimental observation of aerosol deposition in turbulent flow. Aerosol Science, 5, 145-155.
Lui, B.Y.H. and T.A. Ilari (1974). Aerosol deposition in turbulent pipe flow. Environ. Sci. Technol., 8, (4), 351-356.
Ranz, W.E. (1953). Tech. Report No. 8, Univ. Illinois Eng. Exptl. Station.
Ranz, W.E. and J.B. Wong (1952). Impaction of dust and smoke particles. Ind. Eng. Chem., 44, (6), 1371-1381.
Schlichting, H. (1955). Boundary Layer Theory, McGraw Hill, New York.
Sell, W. (1931). Ver. Deut. Ing. Forschungsheft, 347.
Strauss, W. (1975). Industrial Gas Cleaning, Pergamon Press, Oxford.
Wong, J.B. and H.F. Johnstone (1953). Report No. 11, Univ. Illinois Eng. Exptl. Station.

33

A THEORETICAL STUDY OF THE FACTORS AFFECTING THE PERFORMANCE OF A MULTICOMPONENT ADSORBER

R. J. Litchfield*, A. I. Liapis and F. A. Farhadpour***

**Department of Chemical Engineering, Surrey University, Surrey GU2 5XH, UK*
***Department of Chemical Engineering, University of Missouri-Rolla, Rolla, Missouri 65401, USA*

ABSTRACT

Equilibrium theory and constant pattern analysis can be used to predict the performance of adsorbers. Depending upon the validity of the assumptions involved, the answers will be more, or less, accurate. However, only the solution to the full model equations can reliably predict the performance of real plant. Cases are considered where equilibrium conditions and constant patterns are absent.

Temperature effects are also examined; two systems are considered whereby break-through times are substantially increased by, in one case, decreasing and in the other increasing the influent temperature.

KEYWORDS

Adsorption; adiabatic adsorption; gas adsorption; multi-component adsorption; adsorber design; adsorber operation.

INTRODUCTION

Adsorbers are of increasing importance for the removal and separation of chemical mixtures dissolved in fluid streams, not only because of their parsimonious energy requirements, but also because they often provide the only effective way to meet pollution control standards. The use of adsorption equipment is generally diverse and examples range from the removal of trace components from an effluent stream before discharge to the environment to the separation of two or more components from a rich feed stream.

Adsorption systems may be broadly classified into isothermal or adiabatic, both of which may operate with single or multicomponent feeds. The isothermal case is mainly concerned with liquid phase adsorption, where energy release is small compared to the thermal capacity of the liquid, and has been dealt with in considerable depth by Balzli and co-workers (1978). Adiabatic columns mainly operate with gas phase feedstocks, and, since the thermal capacity of the gas is low and the heat liberated (especially from large diameter industrial columns) cannot be removed effectively, the solid and gas phase temperatures may change considerably, affecting the relative and absolute affinities of each adsorbate to the adsorbant.

Reported temperature changes in adsorption columns range from 8 - 10°C (Needham and colleagues, 1966 and Meyer and Weber, 1967) up to about 60°C in the case of CO_2 adsorption on a molecular sieve (Leavitt, 1962). Such temperature changes, as has been shown by Harwell and co-workers (1980), and as will be discussed later, may or may not have a deleterious effect on column performance. It should be noted that these temperature changes often create a distinct thermal wave in the column which may advance ahead of the concentration waves (Pan and Basmadjian, 1967), changing adsorptivity and making the design and the selection of appropriate operating conditions rather difficult.

Several methods are available which can be used to predict the performance of adsorbers under various conditions, and can therefore be used in design and control studies. The earliest, equilibrium theory, is due to Devault (1943), and has been used in the analysis of a multicomponent isothermal system by Shen and Smith (1968) and for a single component adiabatic study by Pan and Basmadjian (1971). A later method, which assumes that constant pattern concentration waves travel relatively slowly along the column, has been used by Pan and Basmadjian (1967, 1970) to analyse their experimental results and those obtained by Leavitt (1962) for a CO_2 - molecular sieve system. However, both equilibrium theory and constant pattern analysis suffer from the fact that the range of validity of the assumptions implicitly made cannot be determined *a priori*. Because of finite resistances to mass transfer, equilibrium theory will always be too optimistic. Constant pattern profiles under isothermal asymptotic conditions have been proved to exist (Cooney and Lightfoot, 1966). However, the length of the bed required to generate these profiles may be much longer than those encountered in practical applications. Consequently, and as has been pointed out by Sweed (1978), real systems can only be reliably treated by the numerical solution of the complete column model equations.

In this work we present a detailed mathematical model of the process, which under appropriate assumptions reduces to either equilibrium or constant pattern conditions. Using the experimental data of Thomas and Lombardi (1971) and Leavitt (1962) and the isotherm data given in Sweed (1978) and Harwell and co-workers (1980), we will use equilibrium theory and, where possible, solutions to the full model to investigate the effects of changing the operating pressure and temperature on the column performance.

THEORETICAL

The mathematical model of the gas adsorber is composed of material and energy balances for the gas and solid phases. The continuity equation for each gas phase solute is

$$\frac{\partial C_i}{\partial t} + V\frac{\partial C_i}{\partial x} + \frac{\rho_s(1-\varepsilon)}{\rho_g\,\varepsilon}\frac{\partial q_i}{\partial t} = 0;\ i = 1,2,\ldots n \tag{1}$$

These equations contain accumulation terms for both the solid and fluid phases as well as a fluid phase convection term. Radial gradients in concentration have been shown by Acrivos (1956) to be negligible. Axial diffusion is also often small, and is neglected, although the term could be incorporated into the model, as indicated later. As temperature changes affect gas densities, the most appropriate units of concentration are mols of solute/mol of gas. Similarly, suitable units of q_i are mols of solute/gram of solid.

The energy balance for the system is

$$\frac{\partial T_g}{\partial t} + V\frac{\partial T_g}{\partial x} + \left[\frac{\rho_s C_{ps}}{\rho_g C_{pg}}\right]\left[\frac{1-\varepsilon}{\varepsilon}\right]\frac{\partial T_s}{\partial t} + \left[\frac{\rho_s}{\rho_g C_{pg}}\right]\left[\frac{1-\varepsilon}{\varepsilon}\right]\sum_{i=1}^{n}\Delta H_i \frac{\partial q_i}{\partial t} = 0 \qquad (2)$$

and the mass and energy balances for the solid phase are respectively:

$$\frac{\partial q_i}{\partial t} = \frac{K_{pi}a}{\rho_s}\left[q_i^* - q_i\right]; \qquad i = 1, 2, \dots n \qquad (3)$$

$$\frac{\partial T_s}{\partial t} = \frac{ha}{\rho_s C_{ps}}\left[T_g - T_s\right] - \frac{1}{C_{ps}}\sum_{i=1}^{n}\Delta H_i \frac{\partial q_i}{\partial t} \qquad (4)$$

In the above equations, mechanisms of heat and mass transfer are expressed as linear functions of overall driving forces. The quantities $K_{pi}a$ and ha are lumped parameters that include both external and internal diffusional resistances. Such lumped expressions are generally accepted as fairly accurate (Balzli and coworkers,1978;Cooney,1974).The relevant resistances can be estimated from empirical correlations (Perry and Chilton, 1974) or by experimentation.

Axial diffusion of mass and energy may be incorporated into the analysis by changing the values of K_{pi} and h in equations (3) and (4) according to the formulae developed by Klinkenberg and Sjenitzer (1956), for example, without changing the hyperbolic form of the equations.

The superficial fluid velocity, V, is assumed constant in the above equations. This assumption is valid for dilute solutions, which are of most interest industrially. However, for non-dilute solutions, the following equation should be included:

$$\rho_g \frac{\partial V}{\partial x} = -\sum_{i=1}^{n} K_{pi}a\,(q_i^* - q_i) \qquad (5)$$

By substituting equations (3) and (4) for corresponding terms in equations (1) and (2), the following set of hyperbolic equations is obtained for the case of dilute solutions

$$\frac{\partial C_i}{\partial t} + V\frac{\partial C_i}{\partial x} = \left[\frac{\rho_s}{\rho_g}\right]\frac{(\varepsilon-1)}{\varepsilon}\frac{K_{pi}a}{\rho_s}(q_i^* - q_i);\ i = 1, 2, \dots n \qquad (6)$$

$$\frac{\partial T_g}{\partial t} + V\frac{T_g}{\partial x} = \frac{ha(1-\varepsilon)}{\partial_g C_{pg}\varepsilon}(T_s - T_g) \qquad (7)$$

$$\frac{\partial T_s}{\partial t} = \frac{ha}{\rho_s C_{ps}}(T_g - T_s) - \frac{1}{C_{ps}}\sum_{i=1}^{n}\Delta H_i \frac{K_{pi}a}{\rho_s}(q_i^* - q_i) \qquad (8)$$

$$\frac{\partial q_i}{\partial t} = \frac{K_{pi}a}{\rho_s}(q_i^* - q_i);\ i = 1, 2, \dots n \qquad (9)$$

The associated initial and boundary conditions are as follows:

for $0 < x < L$, initial conditions at $t = 0$ are,

$$q_i = q_{i,0}(x) \; ; \; T_s = T_{s,0}(x) \; ;$$

$$C_i = C_{i,0}(x) \; ; \; T_g = T_{g,0}(x) \; ;$$

for $t > 0$, boundary conditions at $x = 0$ are,

$$C_i = C_{i,in}(t) \; ; \; T_g = T_{g,in}(t)$$

The quantities $C_{i,in}$ and $T_{g,in}$ may be functions of time when inlet disturbances occur. If equation (5) is included in the model to account for non-dilute solutions, the appropriate boundary condition is

$$V = V_0(t) \text{ at } x = 0, \; t > 0$$

where $V_0(t)$ is a function of time such that $V_0(t) > 0$.

Although the individual equations are not highly complex, they are coupled through the transfer terms and equilibrium expressions. It is through the equilibrium expressions that non-linear terms are introduced into the partial-differential equations.

For this study, solid phase adsorbate concentrations in equilibrium with the fluid phase at the solid phase temperature were represented by a Langmuir type model:

$$q_i^* = \frac{Q_i K_i C_i}{1 + \sum_{i=1}^{n} K_i C_i} \; ; \; i=1, 2, \ldots n \tag{10}$$

where Q_i is the saturation value of q_i^* for component i. The value of Q_i depends on the surface area occupied by one molecular layer of i and, hence, is independent of temperature. The value of K_i is the reciprocal of C_i when half the surface is occupied by component i and the rest of the surface is vacant. The theoretical expression for the quantity K_i is

$$K_i = K_{0i} T^{-\frac{1}{2}} P_T e^{-\Delta H_i / RT_s} \; ; \; i=1, 2, \ldots n \tag{11}$$

in which K_{0i} is a constant, ΔH_i (<0) represents the molar heat of adsorption, P_T is total pressure, and R is the gas law constant. The Langmuir type model was chosen because it has both theoretical and empirical basis in contrast to other adsorption equilibrium models. It has been shown by comparisons with experimental data to represent adsorption equilibria of dilute solutions very well (De Boer, 1968).

Appropriate choices of dimensionless groups permit the model equations to be simplified as follows:

$$\frac{\partial C_i'}{\partial \tau} + \frac{\partial C_i'}{\partial z} = \eta_i(q_i^{*\prime} - q_i') \; ; \quad i=1,\, 2,\, \ldots\, n \tag{12}$$

$$\frac{\partial T_g'}{\partial \tau} + \frac{\partial T_g'}{\partial z} = \theta(T_s' - T_g') \tag{13}$$

$$\frac{\partial q_i'}{\partial \tau} = \xi_i(q_i^{*\prime} - q_i') ; \quad i=1,\, 2,\, \ldots\, n \tag{14}$$

$$\frac{\partial T_s'}{\partial \tau} = \beta(T_g' - T_s') - \lambda \sum_{i=1}^{n} \delta_i(q_i^{*\prime} - q_i') \tag{15}$$

The dimensionless independent and dependent variables are $\tau = tV/L$; $z = x/L$; $q_i^{*\prime} = q_i^* /(q_{i,in}^*(0))$; $C_i' = C_i/C_{i,in}(0)$; $T_s' = (T_s - T_{g,in}(0))/\left[T_{g,in}(0)\right]$; $T_g' = (T_g - T_{g,in}(0))/\left[T_{g,in}(0)\right]$; $q_i' = q_i/(q_{i,in}^*(0))$. The dimensionless parameters necessary to determine solutions of the above equations are $\eta_i = \dfrac{q_{i,in}^* K_{pi} a(\varepsilon-1)L}{\rho_g \varepsilon C_{i,in}(0)V}$;

$\theta = \dfrac{Lha(1-\varepsilon)}{\rho_g C_{pg} \varepsilon V}$; $\xi_i = \dfrac{K_{pi} aL}{\rho_s V}$; $\beta = \dfrac{Lha}{V\rho_s C_{ps}}$; $\lambda = \dfrac{L}{C_{ps}(T_{g,in}(0))V\rho_s}$ and

$\delta_i = q_{i,in}^* \Delta H_i K_{pi} a$.

The partial differential equations (12) - (15) can be reduced to a set of ordinary differential equations by solving along characteristics.

consider $C_i' = C_i'(z,\tau)$

then
$$\frac{dC_i'}{dz} = \frac{\partial C_i'}{\partial z} + \frac{\partial C_i'}{\partial \tau}\frac{d\tau}{dz} \tag{16}$$

Now, provided $d\tau/dz = 1$, the left hand side of equation (12) can be simply written:

$$\left.\frac{dC_i'}{dz}\right|_{\frac{d\tau}{dz}=1} = \eta_i \left[q_i^{*\prime} - q_i' \right] ; \; i=1,\, 2,\, \ldots\, n \tag{17}$$

where $d\tau/dz = 1$ is the characteristic for this equation. Similarly, along this characteristic, equation (13) reduces to

$$\left.\frac{dT_g'}{dz}\right|_{\frac{d\tau}{dz}=1} = \theta\,(T_s' - T_g') \tag{18}$$

In an analogous manner, writing

$$\frac{dq_i'}{d\tau} = \frac{\partial q_i'}{\partial \tau} + \frac{\partial q_i'}{\partial z}\frac{dz}{d\tau}$$

and, choosing $\frac{dz}{d\tau} = 0$ (or z = constant),

$$\left.\frac{dq_i'}{d\tau}\right|_{z\,=\,\text{constant}} = \frac{\partial q_i'}{d\tau}$$

Hence, equations (14) and (15) become

$$\left.\frac{dq_i'}{d\tau}\right|_{z\,=\,\text{constant}} = \xi_i\,(q^{*\prime}_i - q_i') \tag{19}$$

and

$$\left.\frac{dT_s'}{d\tau}\right|_{z\,=\,\text{constant}} = \beta(T_g' - T_s') - \lambda \sum_{i=1}^{n} \delta_i\,(q_i^{*\prime} - q_i') \tag{20}$$

These equations can be solved in the τ - z plane as shown in Fig. 1.

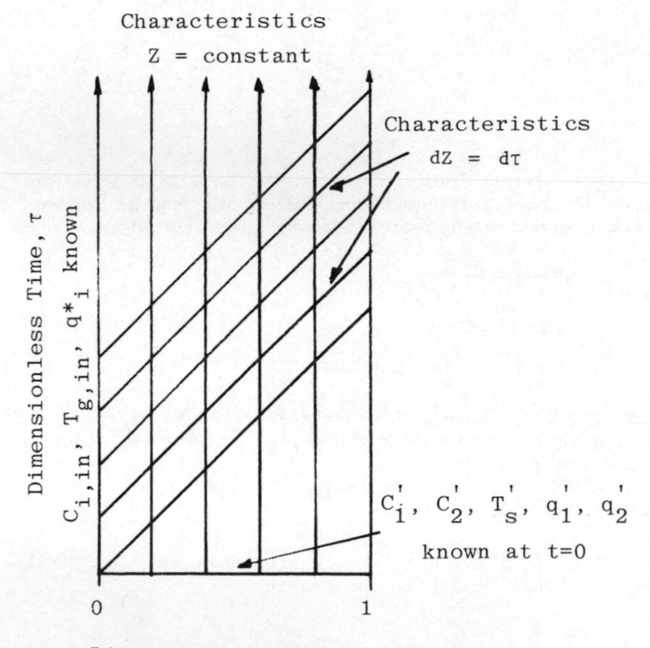

Fig. 1. Integration grid for method of characteristics.

Equations (17) and (18) propagate along the $\frac{d\tau}{dz} = 1$ characteristic, and equations (19) and (20) are evaluated along the z = constant lines. Note that $\Delta z = \Delta\tau$. The above set of equations were solved along the chosen characteristics using the modified Euler predictor-corrector technique.

Reduction of the Model to other forms

Equilibrium model. Under the assumptions of zero resistance to mass transfer, $\xi_i \rightarrow \infty$, and zero resistance to heat transfer, $\theta \rightarrow \infty$, the model reduces to the equilibrium theory form and equations (12) - (15) may be written as:

$$\frac{\partial C_i'}{\partial \tau} = \left[\frac{q^*_{i,in}}{C_{i,in}}\right]\left[\frac{\rho_s}{\rho_g}\right]\left[\frac{\varepsilon-1}{\varepsilon}\right] \frac{\partial q^{*\prime}_i}{\partial \tau} \tag{12a}$$

$$T_g' = T_s' \tag{13a}$$

$$q_i' = q_i^* \tag{14a}$$

$$\frac{\partial T'}{\partial \tau} = - \lambda \sum_{i=1}^{n} \delta_i \frac{\partial q_i^*}{\partial \tau} \tag{15a}$$

Constant Pattern Model. Here it is assumed that time invarient patterns are set up in both the solid and gas phase temperature and concentration. It is further assumed that all the waves travel along the column at a constant velocity γ.

This is equivalent to the statement:

$$\left[\frac{\partial z}{\partial \tau}\right]_C = \left[\frac{\partial z}{\partial \tau}\right]_q = \left[\frac{\partial z}{\partial \tau}\right]_T = \gamma \tag{21}$$

Integrating, it is seen that $f(c) = z - \gamma\tau$, $f(q) = z - \gamma\tau$, and $f(T) = z - \gamma\tau$. Thus, under the constant pattern assumptions, c, q and T are functions of a single variable $x = z - \gamma\tau$.

Therefore, we may write

$$\left[\frac{\partial c}{\partial z}\right]_\tau = \frac{dc}{dx}\left[\frac{\partial x}{\partial z}\right]_\tau = \frac{dc}{dx} \tag{22}$$

$$\left[\frac{\partial c}{\partial \tau}\right]_z = \frac{dc}{dx}\left[\frac{\partial x}{\partial \tau}\right]_z = -\gamma\frac{dc}{dx} \tag{23}$$

$$\left[\frac{\partial q}{\partial z}\right]_\tau = \frac{dq}{dx}, \left[\frac{\partial q}{\partial \tau}\right]_\tau = -\gamma\frac{dq}{dx} \tag{24}$$

$$\left[\frac{\partial T}{\partial z}\right]_{\tau} = \frac{dT}{dx}\ , \quad \left[\frac{\partial T}{\partial \tau}\right]_{z} = -\ \gamma\ \frac{dT}{dx} \tag{25}$$

Hence, for a single component, the original dimensionless equations (12) - (15) reduce to

$$\frac{dC'}{dx} = \frac{\eta}{1-\gamma}\ (q^{*\prime} - q') \tag{26}$$

$$\frac{dT'_g}{dx} = \frac{\theta}{1-\gamma}\ (T'_s - T'_g) \tag{27}$$

$$\frac{dq'}{dx} = -\ \frac{\xi}{\gamma}\ (q^{*\prime} - q') \tag{28}$$

$$\frac{dT}{dx} = \frac{\beta}{\gamma}\ (T'_g - T'_s) - \frac{\lambda}{\gamma}\ \delta\ (q^{*\prime} - q') \tag{29}$$

The solution of the above equations require the determination of the wave velocity γ. To facilitate this, it is assumed that behind each wavefront an equilibrium zone (plateau) exists, as shown in Fig. 2.

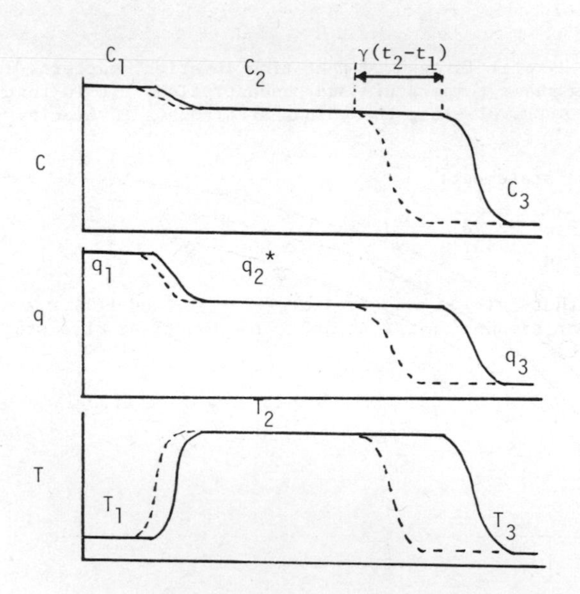

Fig. 2. Asymptotic concentration and temperature profiles.

A material balance calculation over a differential element of the plateau zone would then yield the value of the velocity, γ. Thus, this theory is similar to equilibrium theory, except that leading and lagging concentration and temperature profiles are taken into account. Further details of this method and its application to a CO_2 - molecular adsorption column may be found in Pan and Basmadjian (1967).

BENZENE-TOLUENE ADSORPTION

Isothermal Equilibrium Theory

Thomas and Lombardi (1971) studied the adsorption of benzene and toluene onto activated carbon pellets. They used conditions which were unrealistic for industrial purposes, namely of extremely small inlet concentrations and bed length, but nevertheless their results will be analysed. The Langmuir equilibrium relationship for this binary system was

$$q_B^* = \frac{6.74 \times 10^{-4} \times 0.1246PT^{-\frac{1}{2}}\exp\left[\frac{43542.7}{RT}\right] C_B}{1 + 0.1246PT^{-\frac{1}{2}}\exp\left[\frac{43542.7}{RT}\right] C_B + 163.402PT^{-\frac{1}{2}}\exp\left[\frac{24283}{RT}\right] C_T} \quad (30)$$

and

$$q_T^* = \frac{8.404 \times 10^{-4} \times 163.4025PT^{-\frac{1}{2}}\exp\left[\frac{24283}{RT}\right] C_T}{1 + 0.1246PT^{-\frac{1}{2}}\exp\left[\frac{43542.7}{RT}\right] C_B + 163.4025PT^{-\frac{1}{2}}\exp\left[\frac{24283}{RT}\right] C_T} \quad (31)$$

In these expressions, C_B and C_T are expressed as mole fractions. Note that the isotherm equation for toluene, equation (31) gives values about 20% lower than the isotherm quoted by Thomas and Lombardi (1971), but nevertheless fits their data quite accurately as shown in Figs. 3 and 4.

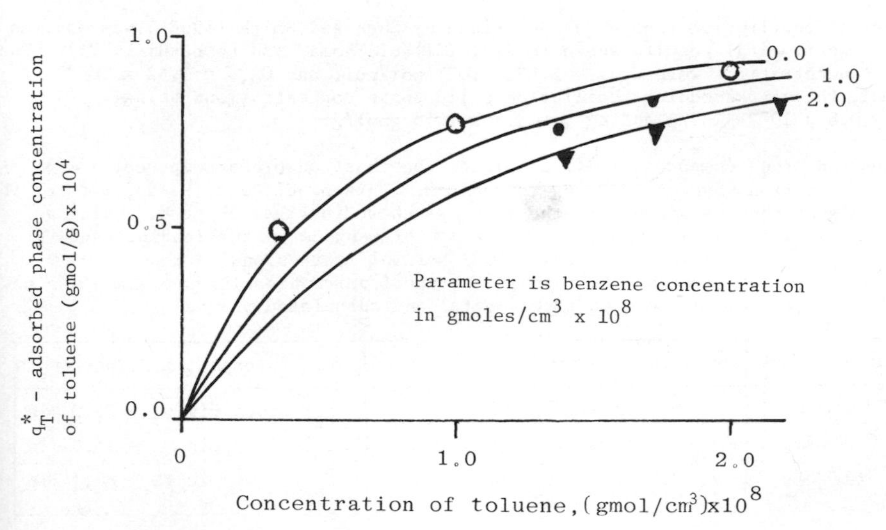

Fig. 3. Langmuir adsorption isotherm at 150°C for benzene - toluene system.

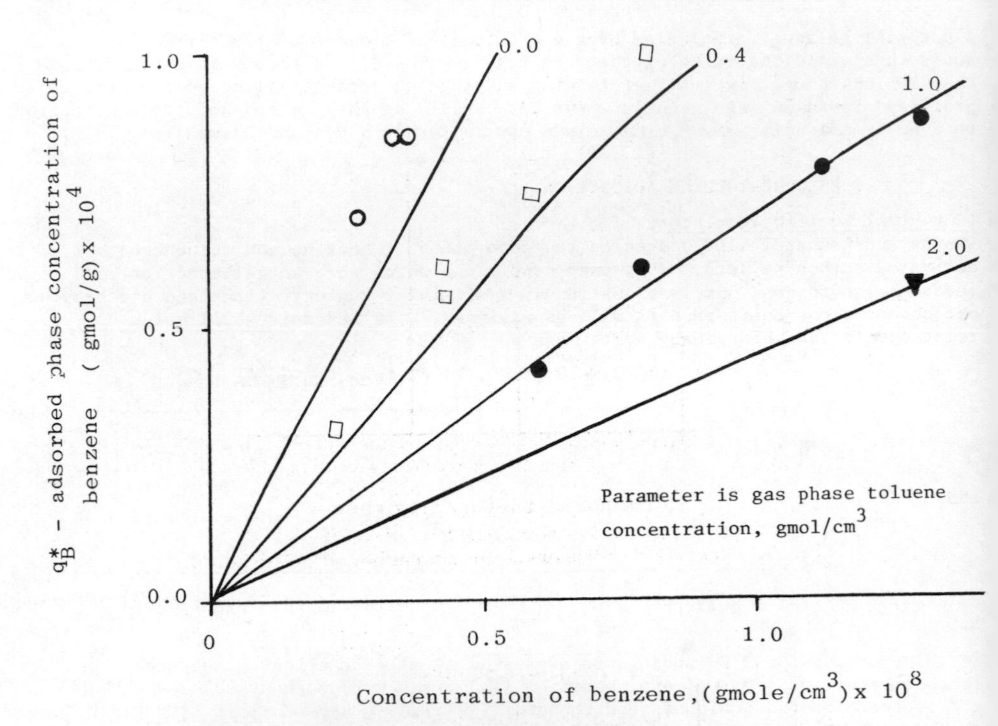

Fig. 4. Langmuir adsorption isotherm at 150°C for benzene - toluene system.

Isothermal equilibrium theory, as described by Shen and Smith (1968), was applied to the experimental results shown in Fig. (21) of Thomas and Lombardi (1971). The inlet concentrations were $C_{o,B} = 2.52 \times 10^{-4}$ mol/mol, and $C_{o,T} = 3.42 \times 10^{-4}$ mol/mol, the corresponding equilibrium solid phase concentrations being $q^*_{o,B} = 0.6 \times 10^{-4}$ gmol/g and $q^*_{o,T} = 5.6 \times 10^{-4}$ gmol/g.

The most adsorbed component (MAC) displaces the least adsorbed component (LAC), as the concentration wave moves along the column. This produces a leading band of the LAC of higher concentration than the feed, as shown in Figs. 5 and 6. Using a material balance, the gas phase concentration of benzene in the leading band is calculated to be approximately 3.54×10^{-4} mol/mol, corresponding to $q^*_{max,B} = 2.3 \times 10^{-4}$ gmol/g. From this, the equilibrium breakthrough times can be calculated. The table compares experimental and calculated values:

Length of bed	Breakthrough time ; calculated	Breakthrough time; experiments
0.012 m	t_B = 32.2 min; t_T = 82.1 min	t_B = 10.5 min; t_T = 25.0 min
0.024 m	t_B = 64.5 min; t_T =164.2 min	t_B = 23 min; t_T = 55.0 min
0.036 m	t_B = 96.8 min; t_T =246.3 min	t_B = 37 min; t_T = 88.0 min

These results indicate that a substantial discrepancy exists between experimental

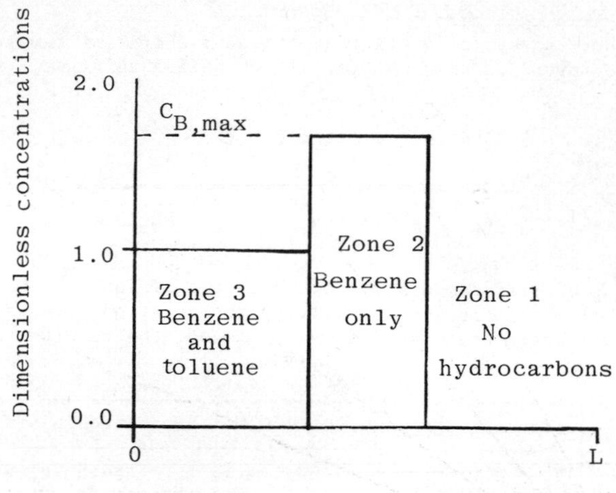

Fig. 5. Equilibrium theory representation of adsorption.

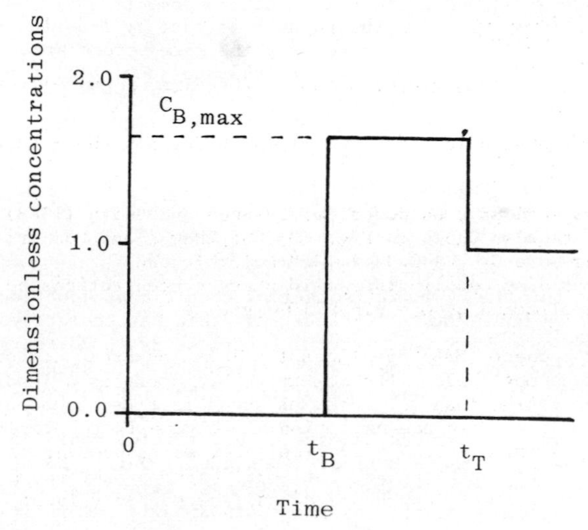

Fig. 6. Outlet concentration profiles according to equilibrium theory.

breakthrough times and those obtained using equilibrium theory.

This is rather surprising in view of the fact that under the experimental conditions reported by Thomas and Lombardi (1971), the time constants associated with gas phase adsorption indicate that equilibrium should be very closely approached. This throws some doubt upon the accuracy of the experimental data, particularly as closer agreement has been obtained by other workers, for instance Shen and Smith (1968).

Isothermal Operation at the Inversion Temperature

Now let us consider isothermal operation, but at a different temperature, and, in particular at the inversion temperature. If we define relative affinity as

$$\psi = \frac{q_B^*/C_B}{q_T^*/C_T} = \frac{6.74 \times 10^{-4} \times 0.1246\exp\left[\frac{43542.7}{RT}\right]}{8.404 \times 10^{-4} \times 163.4025\exp\left[\frac{24283}{RT}\right]} \tag{32}$$

then we can find the solid temperature at which the affinity of both solutes for the carbon is equal; that is, $\psi = 1$. Solving the equation, $T_{INV} = 39.93^oC$. Reworking the equilibrium breakthrough calculations, the results show:

Adsorber length	Breakthrough times at 150°C	Breakthrough times at 39.9°C
0.012 m	t_B = 32.2 min ; t_T = 82.1 min	t_B = 49.5 ; t_T = 62.4
0.024 m	t_B = 64.5 min ; t_T =164.2 min	t_B = 98.9 ; t_T = 124.8
0.036 m	t_B = 96.8 min ; t_T =246.3 min	t_B = 148.4 ; t_T = 187.27

Had the feed concentrations been the same, both components would have broken through at the same time. The conclusion here is that by altering the operating temperature, one can influence advantageously the time before breakthrough. Although these results are only qualitative, they do show that, depending upon the system requirements (e.g. whether for separation or merely removal of all components), quite different operating conditions are required.

Affect of Stiffness Ratio on Computation Time

The stiffness ratio is the ratio of the time constants in the simultaneous ordinar differential equations which comprise the system model, i.e., equations (17) and (Here, the stiffness ratio is equal to η_B/ξ_B.

The solution to the full model under isothermal conditions, equations (12) and (14), has proved to be numerically difficult for this particular system. Here, $\xi_B = 0.6 \times 10^{-3}$, and $\eta_B = 19.5$, so that the stiffness ratio for this system is about 32500, indicating that extremely small space and time steps would be required in the numerical solution. Computation times for this system were estimated to be greater than 100 hours on the powerful Surrey University Multi-Prime multi-access computer network, so that these results could not at present be calculated. Research on the numerical techniques required to overcome this problem is underway, since the problem of stiffness originates in the ratio ρ_s/ρ_g, and will affect calculations on all gas phase atmospheric adsorption columns.

CARBON DIOXIDE-HELIUM ADSORPTION

We will now turn to the experimental results of Pan and Basmadjian (1971) who considered the adsorption of CO_2 - He on a 5A molecular sieve. These authors analysed their results using adiabatic equilibrium theory, and extremely good agreement was found, showing that under appropriate conditions equilibrium theory provides almost exact answers. The same authors (Pan and Basmadjian, 1967) have also analysed Leavitt's (1962) experimental work on CO_2 - N_2 adsorption on a molecular sieve, using constant pattern assumptions. Their predicted breakthroug times agreed closely with the experimental results. Breakthrough times of 0.8 hr were predicted, compared with the experimental value of 0.7 hrs. (Compare this with a simplified equilibrium analysis of this system, which predicts a breakthrough time of 1.02 hrs.) Further work by Pan and Basmadjian (1970) shows, however, that circumstances can exist - particularly with low inlet concentrations - where constant patterns do not exist, and neither equilibrium theory nor constant pattern assumptions could predict accurately the performance of such a system (see Fig. 7).

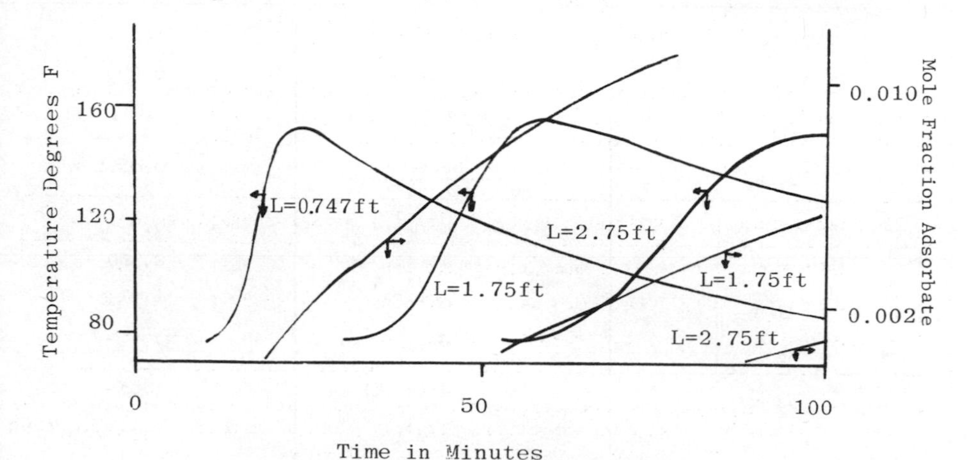

Fig. 7. Experimental results of Pan and Basmadjian (1970) showing absence of constant patterns and plateaus

HIGH PRESSURE ADSORPTION OF BENZENE AND CYCLOHEXANE

We now turn to a binary adsorption column where a mixture of benzene and cyclohexane dissolved in nitrogen is adsorbed onto activated carbon at a pressure of 34 atm. The system equations now can be solved numerically, because the increased gas density reduces the stiffness ratio to a manageable level of about 250. The data pertaining to this system is given in Table 1, the equilibrium relationships having been obtained by James and Phillips (1954). Some typical results, shown in Fig. 8 for different times indicate that both constant profiles and plateaus are absent.

The solid phase concentrations, shown in Fig. 9, also show that equilibrium conditions are not found anywhere along the bed, even, as in this case, at breakthrough. Thus, we conclude that in a typical high pressure system, neither equilibrium theory nor constant pattern analysis could be used with any degree of confidence. Finally, we have calculated the inversion temperature of this system to be 449.9°K, and operating the column with feed at this temperature approximately doubles the time to breakthrough, when compared with feed and initial bed temperatures of 20°C. (See Fig. 10)

TABLE 1. Parameter Values used in the Binary Adsorption of Benzene and Cyclohexane.

Parameter	Units	Numerical Values Used
L	m	1.0
V	m/s	0.0015
$C_{1,in}(t)$	mols of cyclohexane / mol of gas	0.01, t > 0
$C_{2,in}(t)$	mols of benzene / mols of gas	0.05 t > 0
a	m^2/m^3	60.0
ε	-	0.360
$\rho_{g,in}$	$mols/m^3$	1.41×10^3
C_{pg}	$J/(mol\ ^oK)$	37.7
C_{ps}	$J/(g\ ^oK)$	1.05
ρ_s	g/m^3	1.30×10^6
h	$J/(m^2 .s)$	41.9
K_{p1}	$g/(m^2 \cdot s)$	1.00
K_{p2}	$g/m^2 \cdot s)$	0.800
ΔH_1	J/mol	-3.27×10^4
ΔH_2	J/mol	-4.35×10^4
$T_{g,in}(t) = T_{g,0}(x) = T_{s,0}(x)$	^oK	$293.15,\ t > 0,\ 0 \leq x \leq L$
$q_{1,0}(x) = q_{2,0}(x)$	mols of i / g of solid	$1 \times 10^{-15},\ 0 \leq x \leq L$
$C_{1,0}(x) = C_{2,0}(x)$	mols of i / mol of gas	$1 \times 10^{-15},\ 0 \leq x \leq L$
P_T	atm	34.
Q_1	mols of 1/(g of solid)	2.90×10^{-3}
Q_2	mols of 2/(g of solid)	3.67×10^{-3}
K_{01}	$(^oK)^{\frac{1}{2}}/atm$	1.26×10^{-2}
K_{02}	$(^oK)^{\frac{1}{2}}/atm$	4.73×10^{-4}

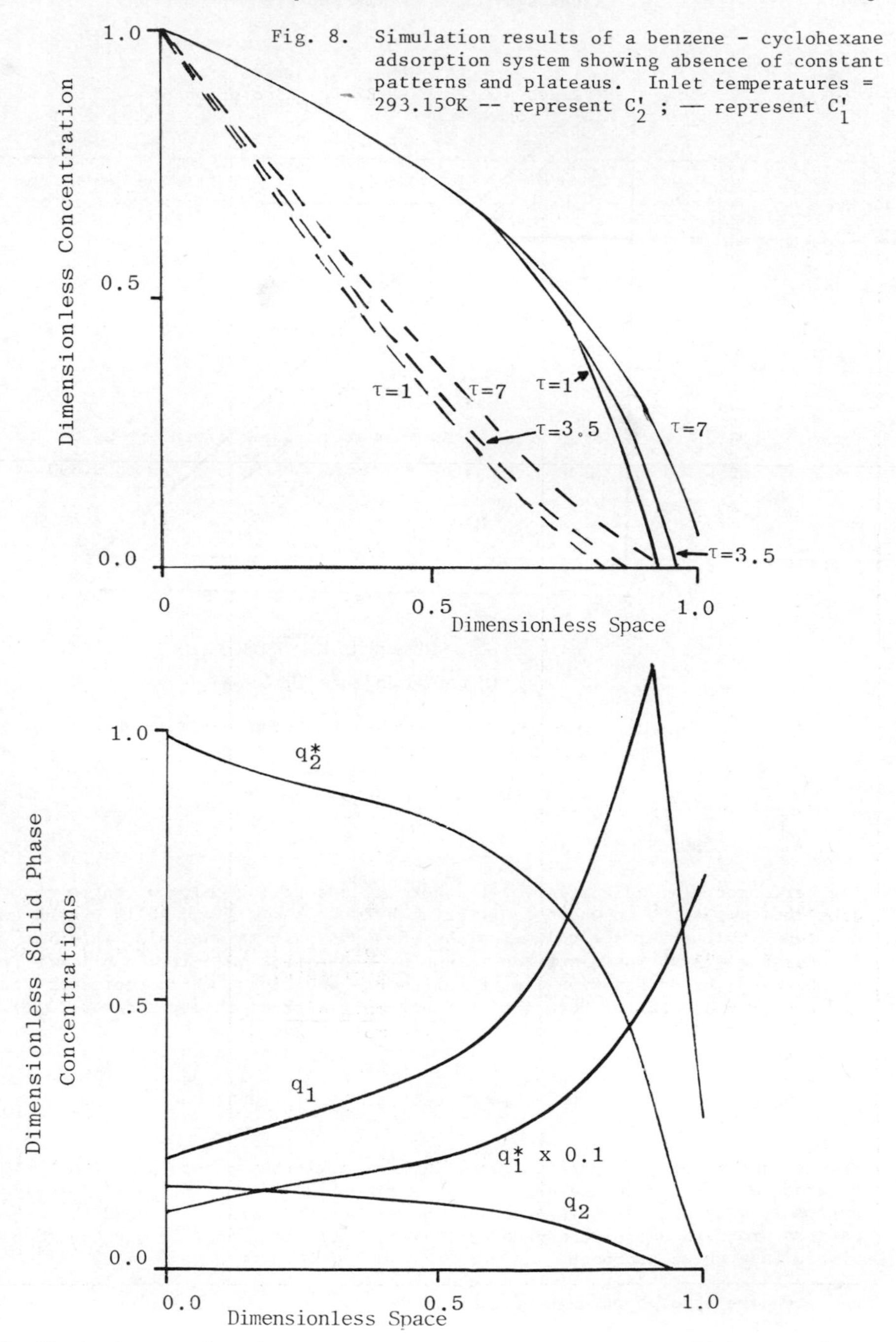

Fig. 8. Simulation results of a benzene - cyclohexane adsorption system showing absence of constant patterns and plateaus. Inlet temperatures = 293.15°K -- represent C'_2 ; — represent C'_1

Fig. 9. Simulation results showing solid phase concentrations; conditions as for Fig. 8, $\tau = 7$

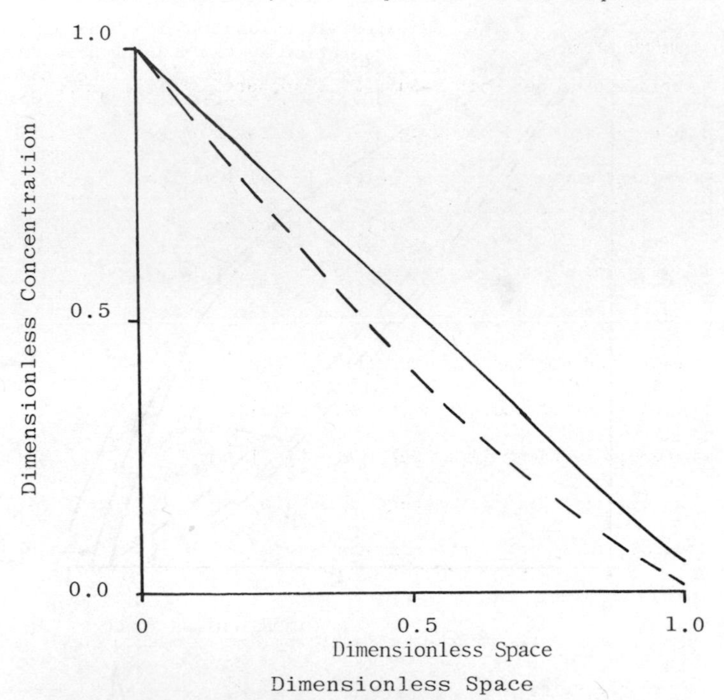

Fig. 10. Simulation results of a benzene - cyclohexane adsorption system showing breakthrough at $\tau = 13$; inlet temperature = $445^{o}K$ -- represent C_2'; — represents C_1'

CONCLUSIONS

The performance of adiabatic (or isothermal) adsorption columns can be analysed by equilibrium theory, in which equilibrium between the gas and solid phases exists a all times throughout the column; or by constant pattern analysis, which assumes ti invarient waves moving along the column at a constant velocity much lower than tha of the feed; or by solution of the full model equations, which represent the mater and energy balances for both phases. Both equilibrium theory and constant pattern analysis can give accurate results provided that the assumptions implicitly made a true in practice. However, there is no way at present in which these assumptions can be guaranteed to be true *a priori*. In general, only the solution to the full model equations can give reliable and accurate solutions, but in some cases comput tion times become excessive.

Results shown here illustrated cases where equilibrium theory produced either accurate or inaccurate answers. In some situations, constant pattern analysis agr extremely well with experimental results; in other cases, we showed that constant pattern profiles and plateaus do not exist, indicating that little confidence can placed in such an approach.

Results were also shown whereby times to breakthrough for different systems may b substantially increased by, in one case decreasing, and in another increasing the influent temperature.

NOMENCLATURE

- a — surface area per unit volume of adsorbate particle, m^2/m^3
- C_i — moles of solute i per mole of gas in the bulk gas
- $C_{i,in}$ — influent concentration of solute i, mol i/mol gas
- $C_{i,O}$ — initial solute concentration, mol i/mol gas
- C_i' — dimensionless solute concentration, $C_i/(C_{i,in}(0))$
- C_{pg} — heat capacity of the gas, J/mol·K
- ΔH_i — heat of adsorption of component i, J/mol
- h — overall heat transfer coefficient, $J/m^2 \cdot s$
- K_i — Langmuir isotherm parameter, $K_i = K_{Oi}\, T^{-\frac{1}{2}}\, p_T e^{-\Delta H/RT}$
- K_{Oi} — Langmuir isotherm parameter, $(^oK)^{\frac{1}{2}}/atm$
- K_{pi} — overall mass transfer coefficient, $g/m^2.s$
- L — column length, m
- n — number of solutes in the influent subject to adsorption
- p_T — total pressure, atm
- Q_i — Langmuir isotherm parameter, mol i/g solid
- q_i — solid-phase concentration of component i, mol i/g solid
- q_i^* — solid-phase concentration of i at equilibrium with fluid phase at the solid phase temperature, mol i/g solid
- $q_i^{*'}$ — dimensionless solid-phase concentration of i at equilibrium with fluid phase at solid phase temperature, $q_i^*/[q_{i,in}^*(0)]$
- $q_{i,o}$ — initial concentration in the bed, mol i/g solid
- q_i' — dimensionless solid phase concentration of component, $q_i/\ [q_{i,in}^*(0)]$
- R — universal gas constant
- T — temperature, K
- T_{INV} — temperature of inversion in relative adsorptivity, K
- T' — dimensionless temperature, $T' = [T - T_{g,in}(0)]/[T_{g,in}(0)]$
- t — time, s
- v — fluid phase velocity, m/s
- x — axial coordinate, m
- z — dimensionless axial corodinate, $z = x/L$

Greek letters

β — $\beta = (haL)/(V \rho_s C_{ps})$

γ — wave velocity

λ — $\lambda = L/[C_{ps}(T_{g,in}(0))V\rho_s]$

ε — void fraction in column

ξ — $\xi_i = (K_{pi}aL)/(\rho_s V)$

η_i — $\eta_i = [q^*_{i,in}K_{pi}a(\varepsilon-1)L]/(\rho_g \varepsilon C_{i,n} V)$

θ — $\theta = [Lha(1-\varepsilon)]/(\rho_g C_{pg} \varepsilon V)$

ρ_s — density of the solid, g/m^3

ρ_g — density of the gas, mol/m^3

τ — dimensionless time, $\tau = tV/L$

Superscripts

' — dimensionless variable

* — at equilibrium

Subscripts

B — benzene

g — gas phase

i — component i

in — evaluated at the inlet conditions

s — solid phase

T — total or toluene

REFERENCES

Acrivos, A. (1956). Method of Characteristics Techniques, Ind. Eng. Chem. Process Des. Develop., 48 (4), 703-710.

Balzli, M.W., A.I. Liapis and D.W.T. Rippin (1978). Applications of Mathematical Modelling to the Simulation of Multi-component Adsorption in Activated Carbon Columns, Trans. I.Chem.E., 56, 145-156.

Cooney, D.O. and E.N. Lightfoot (1966). Multicomponent Fixed-Bed Sorption of Interfering Solutes, Ind. Eng. Chem. Process. Des. Develop., 5 (1), 25-32.

Cooney, D.O. (1974). Numerical Investigation of Adiabatic Fixed-Bed Adsorption, Ind. Eng. Chem. Process. Des. Develop., 13 (4), 368-373.

De Boer, J.H. (1968). The Dynamical Character of Adsorption, 2nd ed., Oxford, Clarendon Press, U.K.

DeVault, D. (1943). The Theory of Chromatography, J. Amer. Chem. Soc., 65, 532-5

Harwell, J.H., A.T. Liapis, R.J. Litchfield and D.T. Hanson (1980). A Non-Equilibrium Model for Fixed-Bed Multi-Component Adiabatic Adsorption, Chem. Eng. Sci., 35, 2287-2296.

James, D.H. and C.S.G. Phillips (1954). The Chromatography of Gases and Vapours, Part III. The Determination of Adsorption Isotherms. J. Chem. Soc., 1066-1070.

Klinkenberg, A. and F. Sjenitzer (1956). Holding-time Distribution of the Gaussian Type, Chem. Eng. Sci., 5, 258-270.

Leavitt, F.W. (1962). Non-Isothermal Adsorption in Large Fixed Beds, Chem. Eng. Progress, 58 (8), 54-59.

Meyer, O.A. and T.W. Weber (1967). Nonisothermal Adsorption in Fixed Beds, AIChE J. 13 (3), 457-465.

Needham, R.B., J.M. Campbell and H.O. McLeod (1966). Critical Evaluation of Mathematical Models used for Dynamic Adsorption of Hydrocarbons, Ind. Eng. Chem. Process Des. Develop., 5 (2), 122-128.

Pan, C.Y. and D. Basmadjian (1967). Constant - Pattern Adiabatic Fixed-Bed Adsorption, Chem. Eng. Sci., 22, 285-297.

Pan, C.Y. and D. Basmadjian (1970). An Analysis of Adiabatic Sorption of Single Solutes in Fixed-Beds: Pure Thermal Wave Formation and its Practical Implications, Chem. Eng. Sci., 25, 1653-1664.

Pan, C.Y. and D. Basmadjian (1971). An Analysis of Adiabatic Sorption of Single Solutes in Fixed-Beds: Equilibrium Theory, Chem. Eng. Sci., 26, 45-47.

Perry, R.H. and C.H. Chilton (Ed.) (1974). Chemical Engineers Handbook, 5th Edition, McGraw-Hill, N.Y. U.S.A.

Shen, J. and J.M. Smith (1968). Rates of Adsorption in the Benzene - Hexane System, Ind. Eng. Chem. Fund., 7 (1), 106-114.

Sweed, N.H. (1978). Nonisothermal and Nonequilibrium Fixed-Bed Sorption, NATO Advanced Study Institute on Percolation Processes, July 17-29, Espinho, Portugal.

Thomas, W.J. and J.L. Lombardi (1971). Binary Adsorption of Benzene - Toluene Mixtures, Trans. I. Chem. E., 49, 240-250.

SECTION VII

WASTE DISPOSAL ON SOIL

34
COPPER-SOIL INTERACTIONS IN WASTEWATER LAND TREATMENT SYSTEMS

M. Sadiq and D. S. Tarazi

Research Institute, University of Petroleum & Minerals, Dhahran, Saudi Arabia

ABSTRACT

'wenty-one mineral soils were used in this study. Some of the selected physico-
:hemical properties of these soils were correlated to Cu sorption. The sorption
of Cu was followed for 50 days at constant temperature and pressure. Cation exchange capacity of test soils was poorly correlated to Cu sorption. No significant
:orrelation was observed between adsorption characteristics; like surface area,
'lay, organic matter, etc; and Cu Sorption. Soil suspension and soil saturated
paste pH were significantly correlated to Cu sorption suggesting that chemisorption
was more important than physical adsorption of Cu in the experimental soils.

'hermodynamic stability diagrams of Cu minerals and solution species were prepared
nd compared with the experimental data of 21 mineral soils. The most probable
nineral which can precipitate in soils was copper-iron compound like cupric ferrite.
he significance of these results in reference to designing a land treatment system
nd their environmental implications were also discussed.

KEYWORDS

u-Fe solid phase formation; Cu-Fe interaction; Cu sorption; Cu precipitation;
hermodynamic solubility diagrams of Cu.

INTRODUCTION

ublic awareness of environmental pollution and socio-legal restrictions on
isposing of pollutants into lakes, rivers and seas have revived interest in land
reatment systems for sewage effluents and other wastes. Recently, because of the
uality and variety of wastes and variations in the methods of disposal, leachate
nd runoff from treatment sites are posing a potential problem to the aquatic
iomass, quality of surface and underground waters, and soil itself. (Apgar and
angmuir, 1971; Bouwer and Chaney, 1974; Cunningham, Ryan, and Keeney, 1975; Weber,
972). In view of this problem, the designers of land treatment systems are often
eeking information for safe disposal of wastes on land.

oxic metals are one of the concerned pollutants present in municipal and industrial
astes. These metals, on reaching the soils, interact with soil constituents in a

way characteristic of habitat in which they enter. The dynamic equilibria of such interactions is presented in Fig.1. Soil solution is the soul of these interactions. Numerous soil characteristics; like texture, structure, cation exchange capacity, lime, organic matter, clay, oxides and hydroxides of alluminum, iron, and manganese, and pH; have been suggested as controlling the solution concentration and mobility of metals in soils (Hildebrand and Blum, 1974; Jones, 1971; Kinniburgh, 1974; U.S. Environmental Protection Agency, 1977). Many times the cation exchange capacity of soils is speculated as the indicator of edaphological behavior of heavy metals (Chaney, 1974; Chumbley, 1973; Garrigan, 1977; Haghiri, 1974).

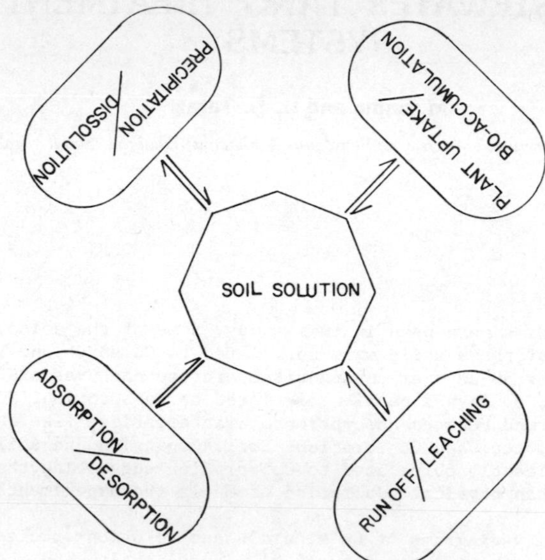

Fig.1. The dynamic equilibria of copper in soils.

Copper(Cu), Phyto- as well as zoo-toxic element, is one of the concerned metals. It ranges from 0.07 to 5.9 ppm in municipal wastewaters (Buzzel, 1972; Driver, 1972; Menzies and Chaney, 1974) and 84 to 10,400 ppm in sewage sludges. (McCalla, Peterson, and Lue-Hing, 1977). As it is obvious in Fig.1, copper on reaching a soil enters into numerous simultaneously occurring interactions.

The concentration of Cu in a leachate or runoff is mainly governed by adsorption, desorption and precipitation/dissolution processes. These processes, in a soil system, are least understood. Therefore, a study was designed to investigate the processes. The objectives of this study were:

1. to study the interactions between some selected physicochemical characteristics of soils and Cu sorption,
2. to develop thermodynamic solubility models to predict probable forms of Cu precipitate and solution complexes in soils, and
3. to examine, experimentally, the precipitation of probable Cu mineral

The information gathered in this study will help in understanding edaphological behavior of Cu in a land treatment system.

EXPERIMENTAL

Twenty-one mineral soils were used in this study. The average and range of some selected physicochemical properties of these soils are listed in Table 1.

TABLE 1. Some Selected Physicochemical Properties of Experimental Soils*

Soil Characteristics	Average	Range
Surface area (m^2/g of soil)	69.95	1.7-203
Clay (percent)	23.33	4.54
Organic matter (percent)	1.26	0.22-4.2
Cation exchange capacity (meq/100g of soil)	24.76	2-71
Total Fe (percent)	1.96	0.4-8.0
Total Al (percent)	4.32	1.0-8.75
Total P (ppm)	337.85	14.7-9.00
Total Ca (percent)	1.30	0.7-8.4
Exchangeable Ca (ppm)	2100.00	184-7020
Total Mg (percent)	0.55	0.02-1.52
Exchangeable Mg (ppm)	298.6	6,960
Saturated paste pH	6.72	4.5-8.5
Saturated paste EC (mmoh/cm)	797	90-6000

* After Enfield and Bledsoe (1975)

Forty grams of each soil, particle size less than 2 mm, was suspended in polyethelene bottles (15cm x 12cm x 7cm) containing 200 ml distilled water. Activated carbon black, 0.5 grams, was added to adsorb organic complexes. To equilibrate the soil-water system, suspensions were shaken for 60 hours prior to copper addition. One hundred parts per million of Cu as $CuCl_2$, on suspension basis, was added to each soil. A blank treatment was similarly prepared. To maintain oxidized conditions in the suspensions, water saturated air was continuously bubbled throughout the experiment. The soil suspensions were shaken for 50 days on an oscillatory shaker at constant speed of 200 rpm and temperature of $25\pm1^{o}C$.

The suspensions were sampled after 1,7,28 and 50 days of shaking. The redox, pH, and electrical conductivity of the suspensions were measured at each sampling. The sampled suspension was centrifuged at 40,000xg for 15 minutes. To remove any undissolved material, the supernatant solution, after centrifuging, was filtered through 0.45μ pore size filter. A portion of the filterate was fixed in 5% nitric acid for subsequent metal analyses. The remaining portion was analyzed for orthophosphate-P (colorimetrically). Chloride, by ion specific electrode, and suffate, by turbidiometrically, were determined after one day of shaking and were assumed constant throughout the experiment.

The ionic strength of the suspension was calculated from the electrical conductivity measurements, using the relationship (Griffin and Jurinak, 1973);

$$I = 0.013 \text{ X } EC \quad (1)$$

where I is the ionic strength and EC is the electrical conductivity in millimohs/cm of soil suspension at $25^{o}C$.

The analytical results of Cu and Fe analyses, and pH, redox, orthophosphate-P, chloride, and sulfate determinations, along with ionic strength calculations, were used in calculating the activity of Cu^{2+} in soil suspensions.

RESULTS AND DISCUSSION

Physicochemical Characteristics of Soils and Cu Sorption

The correlation coefficients between physicochemical properties of soils and Cu sorption were determined by simple linear regression and are given in Table 2.

TABLE 2. Correlation Coefficients between Selected Physicochemical Properties of Soils and Cu Sorption.

Soil Characterists	Correlation Coefficients			
	One day shaking	7 days shaking	28 days shaking	50 days shaking
Surface Area	0.23	0.08	0.21	0.19
Clay Content	0.08	-0.11	0.13	0.07
Organic Matter	0.02	0.14	0.08	0.06
Cation Exchange Capacity	-0.04	-0.11	-0.29	-0.11
Total Fe	0.26	0.17	0.32	0.29
Exchangeable Fe	0.21	0.16	0.14	0.16
Total Al	0.30	0.18	0.38	0.28
Total Ca	0.37	0.27	0.34	0.32
Exchangeable Ca	0.39	0.25	0.39	0.32
Total Mg	0.43*	0.28	0.41	0.36
Exchangeable Mg	0.25	0.08	0.22	0.21
Saturated Paste pH	0.79**	0.66**	0.70**	0.69**
Suspension pH	0.83**	0.77**	0.68**	0.75**

* Significant at 95% level of confidence
**Significant at 99% level of confidence

Cation exchange capacity and other adsorption related soil properties like surface area, clay content, organic matter, total iron and aluminum were not significantly related to Cu sorption. The results of this study disagree with the speculation of Chaney (1974), Chumbley (1973), and Garrigan (1977) who suggested cation exchange capacity as the deciding soil parameter for determining the sludge load. According to their suggestion, the heavy metal loading capacity of soils is a function of cation exchange capacity and can be calculated from a linear mathematical relationship. In authors' opinion, such suggestions are misleading and need to be carefully re-evaluated.

Unlike the soil adsorption capacity parameters, the soil suspension and saturated paste pH were significantly correlated to Cu sorption. The data given in Table 2 suggest that adsorption capacity of soils does not but precipitation probably plays a significant role in Cu sorption in soils. This possibility will be explored in the following pages.

Solid Phase Formation of Cu in Soils

As mentioned earlier, precipitation was probably the main process involved in Cu sorption. Solubility approach was used to speculate what mineral(s) of Cu has been precipitating. In this approach thermodynamic solubility models have been formulated to show the probable minerals and important solution complexes of Cu. The calculated activity of Cu was superimposed on these models to draw inferences.

Thermodynamic Solubility Models of Cu Minerals and Solution Complexes

Most important copper minerals, from an oxidized soil system viewpoint, are CuO(tenorite), $CuFe_2O_4$(cupric ferrite) and $Cu_3(PO_4)_2.2H_2O(c)$(Lindsay, 1979). The equilibrium reactions of these and other minerals, required to develop a representative thermodynamic solubility model for a soil system, are listed in Table 3. The solubility relationships of these minerals are depicted in Fig.2.

TABLE 3. Equilibrium Constant (logK°) for Various Chemical Reactions Used in This Paper

S.NO	Chemical Reactions		LogK^{o}*
1	CuO(tenorite) + $2H^+$	$\rightleftarrows Cu^{2+} + H_2O$	7.66
2	$CuFe_2O_4$(cupric ferrite) + $8H^+$	$\rightleftarrows Cu^{2+} + 2Fe^{3+} + 4H_2O$	10.13
3	Soil-Cu + $2H^+$	$\rightleftarrows Cu^{2+}$	2.80**
4	$Cu_3(PO_4)_2.2H_2O(C) + 4H^+$	$\rightleftarrows 3Cu^{2+} + 2H_2PO_4^- + 2H_2O$	0.34
5	$Cu^{2+} + H_2O$	$\rightleftarrows CuOH^+ + H^+$	-7.70
6	$Cu^{2+} + 2H_2O$	$\rightleftarrows Cu(OH)_2^o + 2H^+$	-13.78
7	$Cu^{2+} + 3H_2O$	$\rightleftarrows Cu(OH)_3^- + 3H^+$	-26.75
8	$Cu^{2+} + 4H_2O$	$\rightleftarrows Cu(OH)_4^{2-} + 4H^+$	-39.59
9	$Cu^{2+} + Cl^-$	$\rightleftarrows CuCl^+$	0.40
10	$Cu^{2+} + 2Cl^-$	$\rightleftarrows CuCl_2^o$	-0.12
11	$Cu^{2+} + CO_2(g) + H_2O$	$\rightleftarrows CuHCO_3^+ + H^+$	-5.73
12	$Cu^{2+} + CO_2(g) + H_2O$	$\rightleftarrows CuCO_3^o + 2H^+$	-11.43
13	$Cu^{2+} + NO_3^-$	$\rightleftarrows CuNO_3^+$	0.5
14	$Cu^{2+} + H_2PO_4^-$	$\rightleftarrows CuH_2PO_4^+$	1.59
15	$Cu^{2+} + H_2PO_4^-$	$\rightleftarrows CuHPO_4^o + H^+$	-4.00
16	$Cu^{2+} + SO_4^{2-}$	$\rightleftarrows CuSO_4^o$	2.36
17	$H_2PO_4^-$	$\rightleftarrows HPO_4^{2-} + H^+$	-7.20

S.NO	Chemical Reactions		LogK°
18	$FePO_4.2H_2O$(strengite) + $2H^+$	⇄ $Fe^{3+} + H_2PO_4^- + 2H_2O$	3.11
19	$CaHPO_4$(monetite) + H^+	⇄ $Ca^{2+} + H_2PO_4^-$	0.30
20	$Fe(OH)_3$(amorp) + $3H^+$	⇄ $Fe^{3+} + 3H_2O$	3.54
21	$Fe(OH)_3$(soil) + $3H^+$	⇄ $Fe^{3+} + 3H_2O$	2.74***
22	α-FeOOH(goethite) + $3H^+$	⇄ $Fe^{3+} + 3H_2O$	-0.02
23	Soil-Ca + $2H^+$	⇄ Ca^{2+}	-.2.50
24	$CaCO_3$(calcite) + $2H^+$	⇄ $Ca^{2+} + CO_2(g) + H_2O$	9.74

* Calculated from ΔG_f^o recommended by Sadiq and Lindsay (1979)

** After Lindsay and Norvell (1969)

*** After Lindsay (1979)

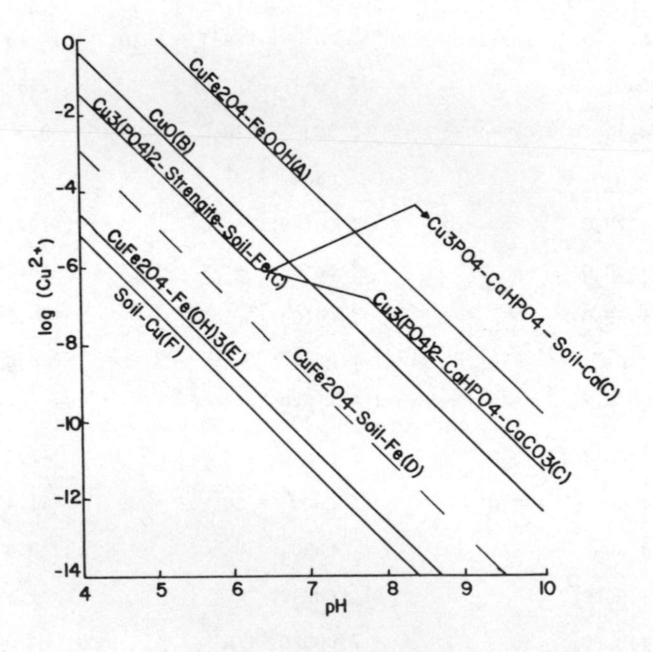

Fig.2. The solubility relationships of various Cu minerals in a soil system.

The procedure for developing this diagram is explained by considering the isotherm of $Cu_3(PO_4)_2.2H_2O(c)$. The assumptions made were:

1. the activity of P, in acidic soils, was controlled by strengite mineral,
2. the activity of Fe^{3+} was in equilibrium with soil-Fe,
3. above pH 6.5, calcium phosphates were stable,
4. since wastewaters contain P, dicalcium phosphate was assumed to control P activity in alkaline soils,
5. between pH 6.5 and 7.3, the activity of Ca^{2+} was assumed $10^{-2.5}\underline{M}$ and
6. above pH 7.3, calcite($CaCO_3$) was controlling Ca^{2+} activity.

The logic of these assumptions in reference to a soil system has been discussed by Lindsay (1979). The solubility isotherm of $Cu_3(PO_4)_2.2H_2O(c)$, in equilibrium with strengite and soil-Fe, was developed by considering the reactions:

Chemical Reactions	LogK^o
$Cu_3(PO_4)_2.2H_2O(c) + 4H^+ \rightleftarrows 3Cu^{2+} + 2H_2PO_4^- + 4H_2O$	0.34
$2Fe^{3+} + 2H_2PO_4^- + 4H_2O \rightleftarrows 2FePO_4.2H_2O(c) + 4H^+$	2(6.85)
$2Fe(OH)_3(soil) + 6H^+ \rightleftarrows 2Fe^{3+} + 6H_2O$	2(2.70)
$Cu_3(PO_4)_2.2H_2O(c)+2Fe(OH)_3(soil)+6H^+ \rightleftarrows 3Cu^{2+} + 2FePO_4.2H_2O(c)+6H_2O$	19.44

The solubility relationship for the above reaction can be modified as:

$$(Cu^{2+})^3 \quad (H^+)^6 = 10^{19.44} \qquad (2)$$

$$\log\ (Cu^{2+}) = 6.48 - 2pH \qquad (3)$$

Equation 3 is plotted in Fig.2. The solubility isotherm of $Cu_3(PO_4)_2.2H_2O(c)$, in equilibrium with $CaHPO_4$ and $CaCO_3$, can similarly be developed. Also plotted in Fig.2 is soil-Cu. Soil-Cu does not refer to any particular Cu mineral but to the activity of Cu^{2+} in soil suspension determined by Lindsay and Norvell (1969).

Although soil-Cu seems most stable mineral but the solubility data on which this isotherm was based is limited Lindsay and Norvell (1969) and Lindsay (1972) reported pK = 3.2 for the reaction:

$$Soil\text{-}Cu + 2H^+ \rightleftarrows Cu^{2+} \qquad (4)$$

whereas Lindsay (1979) adopted pK=2.7 for the above reaction without any explanation. If we assume pK = 3.2 for Reaction(4), the soil-Cu isotherm will superimpose on $CuFe_2O_4$ solubility isotherm in equilibrium with $Fe(OH)_3$(amorphous). It is possible that what Lindsay and Norvell (1969) had misnamed as soil-Cu might be $CuFe_2O_4$. As pointed out that the solubility data of Lindsay and Norvell (1969) is not reliable, therefore, we have not considered this as a solid phase of Cu.

As shown in Fig.2, the stability of curpric ferrite is dependent on Fe^{3+} activity. In a soil system, where the activity of Fe^{3+} is generally governed by Soil-Fe, $CuFe_2O_4$ is the most stable mineral. Since the wastewaters contain Fe which can precipitate as $Fe(OH)_3$, the possibility of $CuFe_2O_4$ formation, in a land treatment system, is further improved. The precipitation of tenorite and $Cu_3(PO_4)_2.2H_2O$ seems feasible too.

The solution complexes of Cu are presented in Fig.3. No matter which mineral controls the activity of Cu^{2+}, the relative significance of Cu complexes will remain unchanged. For simplicity, the equilibrium was assumed with soil-Cu. In acidic soils, $CuSO_4^o$ and Cu^{2+} are important, whereas, in alkaline soils, the complexes $Cu(OH)_2^o$ and $CuCO_3^o$ (depending upon CO_2(g) pressure) are significant.

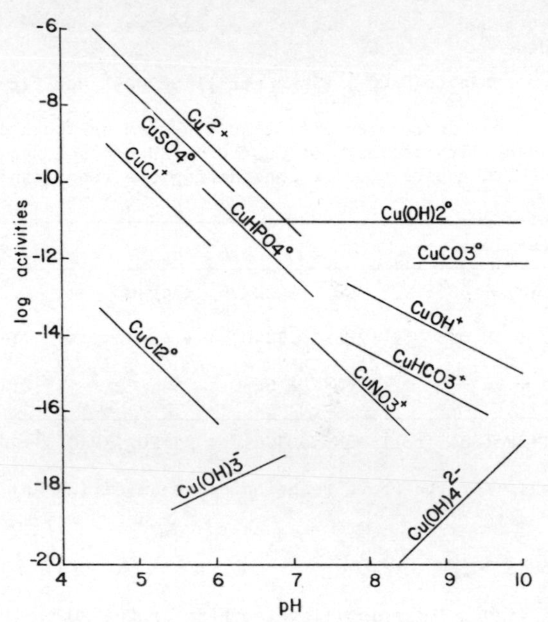

Fig.3. The solution complexes of Cu in equilibrium with Soil-Cu, and 10^{-3}M of Cl^-, NO_3^-, SO_4^{2-}, HPO_4^{2-}, and $10^{-3.52}$ atmosphere partial pressure of CO_2(g).

In a land treatment system, the total soluble Cu can be presented as:

$$\{\text{Total Cu}\} = \{\text{Cu(inorganic)}\} + \{\text{Cu(organic)}\} \quad (5)$$

where brackets indicate concentration. The organic interactions of Cu will be dealt somewhere else (Sadiq, Al-Mashhady, and Idrees, in press). In this paper, only inorganic species of Cu are considered. The total inorganic Cu in solution can be expressed as:

$$\{\text{Total Cu(inorganic)}\} = \{Cu^{2+}\} + \{CuSO_4^o\} + \{Cu(OH)_2^o\} + \{CuCO_3^o\} \quad (6)$$

Substituting activities gives:

$$\{\text{Total Cu(inorganic)}\} = (Cu^{2+}) \;/\; \gamma Cu^{2+} + (CuSO_4^o) + (Cu(OH)_2^o) + (CuCO_3^o) \quad (7)$$

where parentheses represent activities and γ indicates activity coefficient. when each term on the right side is expressed in terms of its equilibrium constant, given in Table 3, Eq.(7) can be rearranged to give:

$$(Cu^{2+}) = \frac{\{\text{Total Cu(inorganic)}\}}{1/\gamma Cu^{2+} + 10^{2.36}(SO_4^{2-}) + 10^{-13.78+2pH} + 10^{-11.43+2pH}(CO_2(g))} \quad (8)$$

Knowing the pH, SO_4^{2-} content, Cu concentration, and CO_2(g) partial pressure, the activity of Cu^{2+} can be calculated by using Eq.(8).

Precipitation of Cu in Soils

Substituting the analytical data of each soil suspension in Eq.8, the activity of Cu^{2+} in soil suspensions was calculated and is plotted in Fig.4 to 6. Superimposed on these solubility plots are the stability isotherms of some important Cu minerals developed in the previous section.

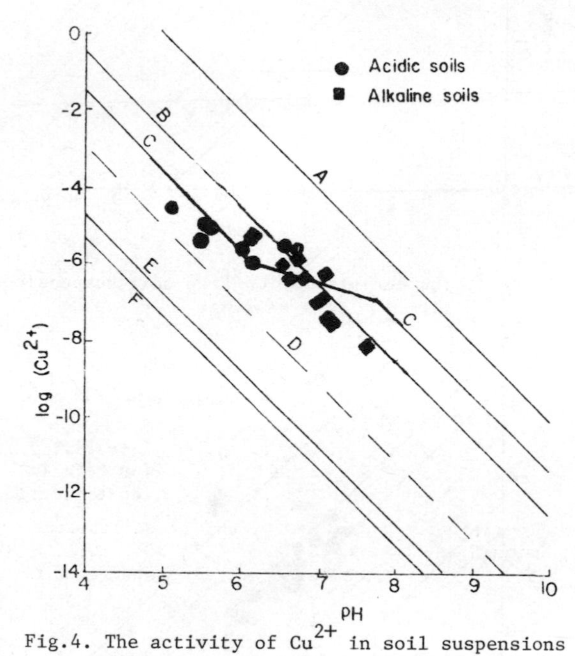

Fig.4. The activity of Cu^{2+} in soil suspensions after one day shaking. Isotherms (A)-(F); see corresponding isotherms on Fig. 2.

The activity of Cu^{2+} in soil suspensions after 1 day shaking is scattered within the stability region of $CuFe_2O_4$, in equilibrium with amorphous $Fe(OH)_3$ and FeOOH (goethite). Some of the solubility data of alkaline soils conform to the solubility isotherm of CuO(c). In acidic soils, the possibility of copper phosphate formation cannot be overlooked either (Fig.4).

The activity of Cu^{2+}, maintained in soil suspensions after 7 days of shaking, followed similar trend as observed in case of suspensions shaken for one day. The analytical data, presented in Fig.5, though fall within the stability region of $CuFe_2O_4$ but also conform to CuO. Observing Figs.4 and 5 closely, one may conclude that CuO, $Cu_3(PO_4)_2$, and $CuFe_2O_4$ were precipitating simultaneously or only $CuFe_2O_4$ was forming in soil suspensions.

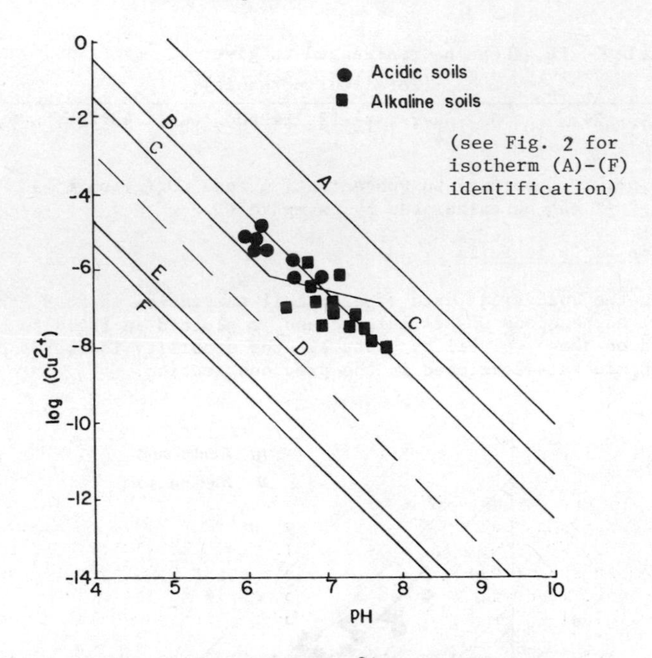

Fig.5. The activity of Cu^{2+} in soil suspensions after 7 days shaking.

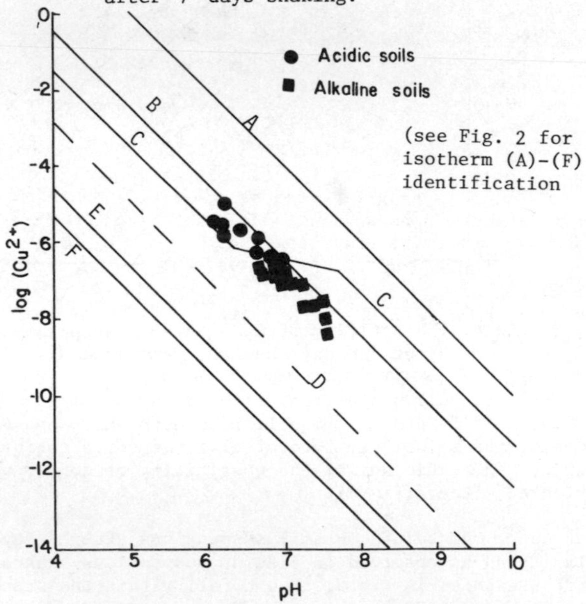

Fig.6. The activity of Cu^{2+} in soil suspensions after 50 days shaking.

The activity of Cu^{2+} observed after 50 days shaking are presented in Fig.6. Almost all the activity data were below the solubility isotherm of CuO(c). Had the Cu been precipitated as CuO, the activity of Cu^{2+} would have been above the solubility isotherm of CuO. Therefore, it is concluded that CuO was not forming in these experimental soils.

In all the soil suspension, the activity of Cu^{2+}, after 1,7,28, and 50 days shaking, was within the levels maintained by $CuFe_2O_4$, in equilibrium with amorphous $Fe(OH)_3$ and goethite. The stability region of $Cu_3(PO_4)_2.2H_2O$ mineral is also overlaping the solubility data of this study. The data presented in Fig.5 and 6 are inconclusive regarding the formation of $CuFe_2O_4$ and $Cu_3(PO_4)_2$. Further treatment of analytical data is necessary to determine which one of these two minerals was precipitating in soils.

First we examine the possibility of phosphate formation. The precipitation of Cu as $Cu_3(PO_4)_2.2H_2O$ in soils can be presented as:

$$3Cu^{2+} + 2HPO_4^{2-} + 2H_2O \rightleftarrows Cu_3(PO_4)_2.2H_2O(c) + 2H^+ \qquad (9)$$

The activity of HPO_4^{2-} ion is controlled by several crystaline and amorphous comounds (Lindsay and Moreno, 1960; Lindsay and Vlek, 1976; Lindsay and Stephenson, 1962; Williams and Patrick 1973). It is very difficult to determine exactly which phosphate was controlling the activity of HPO_4^{2-} in the experimental soil suspensions. To overcome this handicap, the activity of HPO_4^{2-} was calculated from the analytical information of this study. Equation (9) can be modified as:

$$(Cu^{2+})^3(HPO_4^{2-})^2 \;/\; (H^+)^2 = K(\text{equilibrium constant}) \qquad (10)$$

From equation (10), it may be concluded that the quantity $(Cu^{2+})^3(HPO_4^{2-})$ is a linear negative function of pH. No matter what solid phase controls the activity of HPO_4^{2-} in the experimental soil suspensions, the quantity $(Cu^{2+})^3(HPO_4^{2-})^2$ must be linearly related to pH if $Cu_3(PO_4)_2.2H_2O$ or some other phosphates were precipitating.

To test this hypothesis, the quantity $(Cu^{2+})^3(HPO_4^{2-})^{2-}$ was calculated and is plotted in Figs.7 and 8. The observations of these figures indicate that copper phosphate was not precipitating in the experimental soils. So far, we have concluded that neither Cu phosphate nor oxide was forming. This left with the only probability of Cu ferrite precipitation.

As mentioned earlier, the activity of Cu^{2+} was within the levels maintained by $CuFe_2O_4$ in equilibrium with $Fe(OH)_3$ and goethite. The precipitation of $CuFe_2O_4$ can be expressed as:

$$Cu^{2+} + 2Fe^{3+} + 4H_2O \rightleftarrows CuFe_2O_4(c) + 8H^+ \qquad (11)$$

Equation (11) involves Fe^{3+}. The activity of Fe^{3+} in soils is controlled by several simultaneously occurring oxides and hydroxides of iron and is very difficulty to comprehend (Lindsay, 1979; Sadiq, 1977). To overcome this handicap, the activity of Fe^{3+} was calculated from the analytical information of this study and Eq. (11) was rearranged to give:

$$(Cu^{2+})(Fe^{3+})^2 \;/\; (H^+)^8 = K(\text{equilibrium constant}) \qquad (12)$$

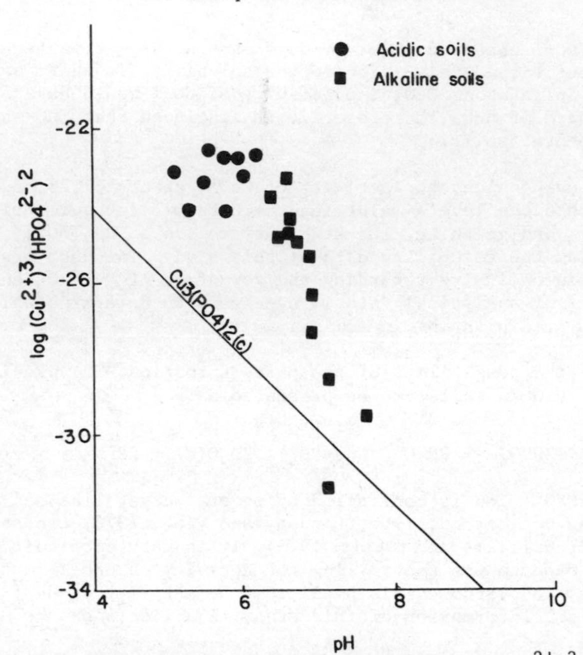

Fig.7. The distribution of the quantity $(Cu^{2+})^3(HPO_4^{2-})^2$ in soil suspensions after 1 day shaking.

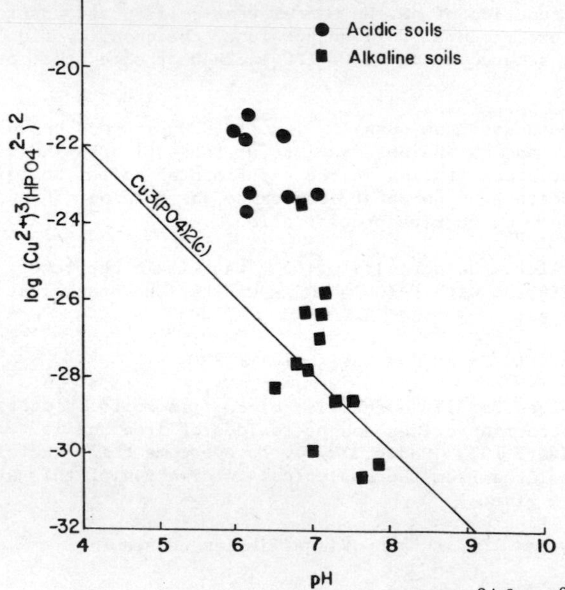

Fig.8. Distribution of the quantity $(Cu^{2+})^3(HPO_4^{2-})^2$ in soil suspensions after 50 days shaking.

The quantity $(Cu^{2+})(Fe^{3+})^2$ was calculated from the activities of Cu^{2+} and Fe^{3+}. This quantity is independent of Fe solid phase and must linearly correlate to pH if $CuFe_2O_4$ or other similar compounds were precipitating in soil suspension. The quantity $(Cu^{2+})(Fe^{3+})^2$, as a function of pH, is plotted in Figs. 9 to 11.

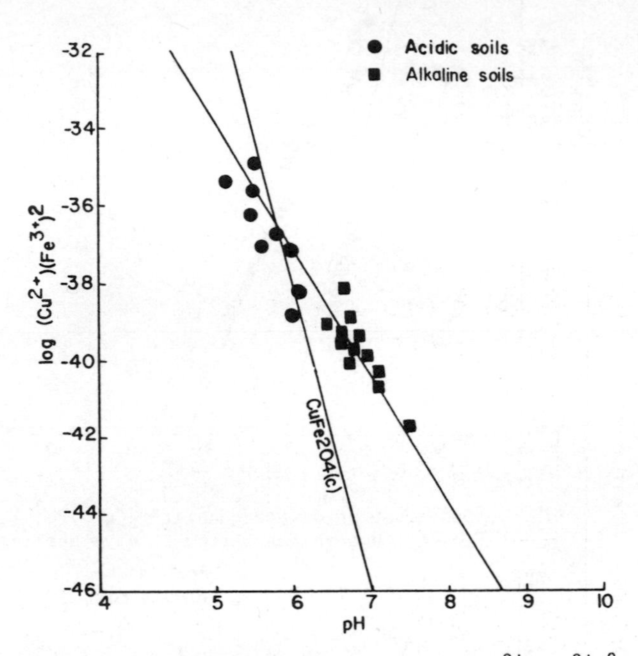

Fig.9. Distribution of the quantity $(Cu^{2+})(Fe^{3+})^2$ in soil suspensions after 1 day shaking.

The distribution of $(Cu^{2+})(Fe^{3+})^2$ was significantly correlated to suspension pH after shaking for 1 day. Although a linear relationship was exhibited between $(Cu^{2+})(Fe^{3+})^2$ and suspension pH, the slope of the best fit line was different from that of $CuFe_2O_4$ mineral. Initially different forms of Cu-Fe mineral was probably precipitating resulting a variable slope as compared to $CuFe_2O_4$. As shaking proceeded, the slope of distribution of $(Cu^{2+})(Fe^{3+})^2$ was becoming more or less comparable to Cu ferrite. This fact is very evident in Fig.11 where the experimental data is almost parallel to the $CuFe_2O_4$ solubility isotherm. The observations of Figs.9 through 11 support the formation of Cu-Fe solid phase formation. The initial precipitate was probably crystalizing to $CuFe_2O_4$. The supersaturation of experimental data, especially in Figs.10 and 11 with respect to $CuFe_2O_4$ isotherm probably reflects the amorphousness of precipitate.

The conclusion of Cu-Fe solid phase formation was also speculated by Jenne (1968) who suggested that Fe oxides and hydroxides surfaces in soils and sediments were more involved than only adsorption in Cu sorption. Lindsay and Norvell (1969) also postulated Cu solid phase controlling the activity of Cu^{2+} which they termed as Soil-Cu. Recently, Lindsay (1979) theorized that $CuFe_2O_4$ can precipitate in soil and is in conformity with the observations of this study.

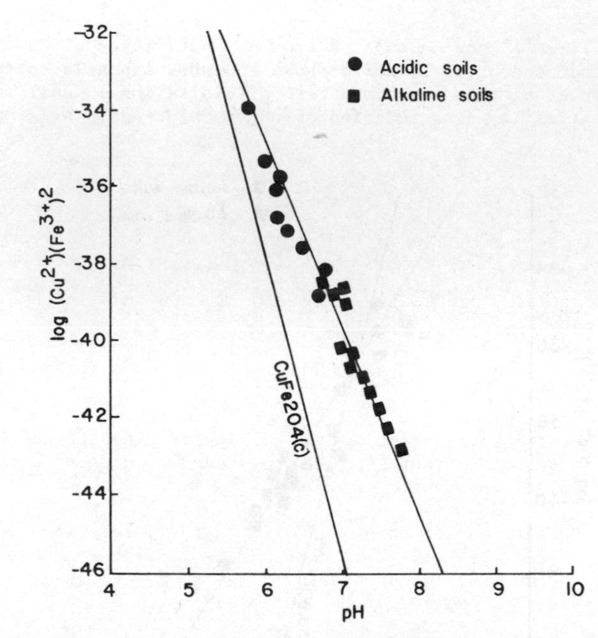

Fig.10. Distribution of the quantity $(Cu^{2+})Fe^{3+})^2$ in soil suspensions after 28 days shaking.

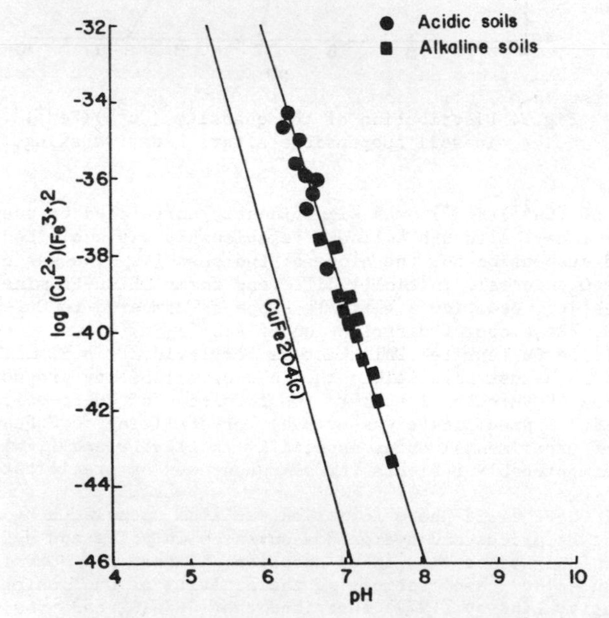

Fig.11. Distribution of the quantity $(Cu^{2+})(Fe^{3+})^2$ in in soil suspensions after 50 days shaking.

CONCLUSIONS

Bases on the observations of this study the following conclusions were drawn:

1. the adsorption capacity parameters of soils were not significantly correlated to Cu sorption,

2. no significant correlation was found between cation exchange capacity of soils and Cu sorption. The guidelines suggested for the land application of sludges using cation exchange capacity needs to be revised and re-evaluated,

3. precipitation process seems more active in Cu sorption than adsorption,

4. phosphate and oxides/hydroxides of Cu were not forming in soil suspensions,

5. copper-iron solid phase was probably precipitating in the experimental soils but further studies are needed to determine the form of such solid phase,

6. the observations of this study proves the interaction between Cu and Fe. In land treatment system of wastewaters or land application of sludges, considerations must be given to the kinetic of Fe supply for effective control of Cu contimination, and

7. further studies are recommended to determine the form Cu-Fe solid phase, the effect of redox, organic complexation and kinetics of soils Fe minerals dissolution.

The scope of this study is very wide. In addition to plant nutrition, the observations can be utilized in designing a land treatment system or determining the load of sludges for land application.

REFERENCES

Apgar, M.A. and D. Langmuir (1971). Ground Water, 9, 76-96.

Bouwer, H. and R.L. Chaney (1974). Adv. Agron.,26, 133-176.

Buzzel, T. (1972). Secondary treatment processes. In Wastewater Management by Disposal on Land. Cold Region Res. Eng. Lab. Corps of Engs., U.S. Army Special Rept. No. 171, 35-47.

Chaney, R.L. (1974). Recommendations for Management of Potentially Toxic Elements in Agriculture and Municipal Wastes. P. 97-120. Factors Involved in Land Application of Agricultural and Municipal Waters. Nat. Program Staff, USDA-ARS. Soil, Water, and Air Services. Beltsville, Maryland, USA.

Chumbley, C.G. (1973). Permissible Levels of Toxic Metals in Sewage Used on Agricultural Land. A.D.A.S. Advisory Paper No. 10.

Cunningham, J.D., A.R. Ryan and D.R. Keeney (1975). J. Environ. Qual. 4, 455-460.

Driver, C.H. (1972). Assessment of the Effectiveness and Effects of Land Disposal Methodologies of Wastewater Management. Corps of Engs. U.S. Army Wastewater Manage. Rept. 72-1, 147.

Enfield, C.G. and B.E. Bledsoe (1975). Kinetic Model for Orthophosphate Reactions in Mineral Soils. U.S. Environmental Protection Agency, Tech. Ser. EPA-660/2-75-022. P.133.

Garrigan, G.A. (1977). J. Water Pollut. Cont. Fed., 2380-2389.

Griffin, R.A. and J.J. Jurinak (1973). Soil Sci., 116, 26-30

Haghiri, F. (1974). J. Environ. Qual., 3, 180-183.
Hildebrand, E.E. and W.F. Blum (1974). Naturwissenschaften, 61, 169-170.
Jenne, E.A. (1968). Adv. Chem., 73, 337-387.
Jones, M.K. (1971). Environ. Letters, 2, 173.
Kinniburgh, D.G. (1974). Cation Absorption by Hydrous Metal Oxides. Ph.D. Dissertation, University of Wisconsin, Madison, USA.
Lindsay, W.L. (1979). Chemical Equilibria in Soils. John Wiley and Sons, New York, p. 222-237.
Lindsay, W.L. (1972). Inorganic phase equilibria of micronutrients in soils. In J.J. Mortvedt, P.M.Giordano and W.L. Lindsay (Eds.), Micronutrients in Agriculture, Soil Science Society of America, Inc., Madison, Wisconsin, U.S.A. pp.41-57.
Lindsay, W.L. and E.C. Moreno (1960). Soil Sci. Soc. Am. Proc., 24, 177-182.
Lindsay, W.L. and W.A. Norvell (1969). Soil Sci. Soc. Am. Proc., 33, 62-68.
Lindsay, W.L. and P.L.G. Vlek (1976). Phosphate minerals. In J.B. Dixon and S.B. Weed (Eds.), Minerals in Soil Environment, Soil Science Society of America, Inc., Madison, Wisconsin, USA, pp. 639-672.
Lindsay, W.L., A.W. Frazier and H.F. Stephenson (1962). Soil Sci. Soc. Am. Proc., 26, 446-452.
McCalla, T.M., J.R. Peterson and C. Lue-Hing (1977). Properties of Agricultural and Wastewaters. Am. Soc. Agron, Inc.
Menzies, J.D. and R.L. Chaney (1974). Waste Characteristics in Factors Involved in Land Application of Agricultural and Municipal Wastes. USDA - Nat. Program Staff Agri. Res. Serv., Soil, Water and Air Sci., Beltsville, Md. p.200.
Sadiq, M. (1977). The Use of Electron Titration to Study Fe and Mn in Soils. Ph.D. Thesis, Dept. of Agronomy, Colorado State University.
Sadiq, M. and W.L. Lindsay (1979). The Selection of Standard Free Energies of Formation for Use in Soil Chemistry. Colorado State Univ. Tech. Bull. 134, p. 1075.
Sadiq, M., Al-Mashhady and M. Idrees (in press). Chelation of copper in contaminated soils. Soil Sci. Soc. Am. J. (1981).
U.S. Environmental Protection Agency (1977). Process Design Manual for Land Treatment of Municipal Wastewaters. EPA-625/1-77-008 p.42.
Weber, J. (1972). J. Water Pollut. Control. Fed., 71, 404-413.
Williams, B.G. and W.H. Patrick (1973). Soil Sci. Soc. Am. Proc., 37, 33-36.

35

EFFECTS OF SEWAGE EFFLUENTS ON CARBON AND NITROGEN MINERALISATION OF A SOIL

T. Sekar and A. K. Bhattacharyya

School of Environmental Sciences, Jawaharlal Nehru University, New Delhi-110 067, India

ABSTRACT

Raw sewage and sewage effluents from a domestic sewage treatment plant serving Delhi were analysed for their physical, chemical characteristics in view of evaluating their suitability for irrigation to agricultural lands. In a laboratory experiment, these effluents were added to a soil and incubated at 30 ± 1°C. Microbial populations, CO_2 evolution and NH_4^+ - N and NO_2^- + NO_3^--N concentrations were determined at 10 day intervals for a period of sixty days and were compared with tap water treatment. Chemical properties of the soil applied with sewage were also determined after 60 days of incubation.

Sewage addition caused significant increase in soil bacterial, fungal and Azotobacter populations. CO_2 production rates of the soil were enhanced by raw sewage and effluents both without and with glucose amendment. Raw sewage and effluent treatment resulted in significant ammonium and increases (nitrite + nitrate) nitrogen concentrations of the soil with respect to the tap water control. Raw sewage treatment increased NO_2^- + NO_3^- -N levels in soil over the effluents also. Soil pH was not found to be altered while sewage application caused substantial changes in electrical conductivity, C.E.C., organic matter and available nitrogen.

KEYWORDS

Raw Sewage; Effluents; Nitrification; CO_2 evolution.

INTRODUCTION

The practice of returning liquid wastes including domestic sewage to land systems is not recent. The multifarious advantages of land treatment of domestic sewage over surface water discharge are water pollution control, water resource reuse and nutrient recycling. The important merits and potential problems of land application of wastewaters have been elaborately discussed in the reviews of Bouwer and Chaney (1974) and Lee (1976). The success and life span of a disposal site and the environmental consequences from land treatment are to a large extent dependent upon the wastewater characteristics (Ellis, 1978), site selection (Sheaffer, 1979) etc. Serious environmental problems which include contamination of ground water with nitrate, buildup of biotoxic elements in soils and plants and health hazards from pathogenic microorganisms are likely to occur depending upon the sewage characteristics and management practices (Dunlop, 1968; Chaney, 1973; Johnson, Jr., 1979).

As domestic sewage carries chemical constituents of known and unknown character and concentrations, its application to soil over a period is expected to cause changes in soil microbial communities and their activities. The contribution of NO_3-N in sewage treatment soil to ground water nitrate nitrogen levels will definitely be influenced by nitrogen transformation processes in soil with subsequent leaching of nitrate (Lance, 1975). As such ammonification, nitrification and denitrification rates in soil are likely to be affected by the nature and properties of the wastewater applied. In the present study, raw sewage and effluents from a treatment plant are characterised. Effects of these effluents on microbial population, CO_2 evolution, nitrogen mineralisation rate and selected chemical properties of a soil are also reported.

MATERIALS AND METHODS

Grab samples of raw sewage and effluents leaving various treatment units at Okhla sewage treatment plant, New Delhi were collected at two periods, viz., 4.9.1979 and 25.11.1979. The samples collected were (i) Raw sewage; (ii) Primary settling tank effluent; (iii) Clarifier effluent (both (ii) and (iii) represent primary treatment); (iv) Activated sludge plant effluent; (v) Final clarifier effluent (both of which represent secondary treatment) and (vi) Irrigation channel effluent. The influent to this plant is mainly from domestic sewage of New Delhi Municipal Corporation area.

The soil used was a sandy loam and was collected from 9 inches top layer of a field receiving sewage effluents as a source of irrigation since 1940. The properties of the soil are furnished in Table 1.

TABLE 1 Properties of Experimental Soil

Property		
1. Particle size analysis	% sand	71.9
	% silt	21.3
	% clay	6.8
2. Water Holding capacity	%	37.0
3. pH		8.5
4. Electrical conductivity (mmho/cm)		0.5
5. C.E.C. (meq/100 g)		8.0
6. Organic carbon (%)		0.6
7. Total nitrogen (%)		0.1
8. Total phosphorus (ppm)		48.0
9. Available nitrogen (ppm)		99.1
10. Available phosphorus (ppm)		28.5
11. Available potassium (ppm)		276.7

About 500 g of air-dried soil sieved to 2 mm was placed in polyethylene beakers and was added with required amounts of raw sewage and sewage effluents separately to maintain the soil moisture at the level of 50% of maximum water holding capacit Three replications were kept for each treatment and the treatment which received tap water served as control. Samples were drawn at the interval of 10 days for a period of 60 days. Bacteria, fungi, actinomycetes and *Azotobacter* populations, CO_2 evolution (both without and with glucose), NH_4^+ - N and NO_2^- + NO_3^- -N concentrations were determined. Samples drawn at the end of incubation period were used for other chemical determinations. During the experimental period, the beakers were covered with polyethylene sheets and incubated at 30 $\pm$ 1°C and loss in moisture was adjusted with fresh additions of the respective effluents.

Methods outlined in Standard methods for examination of Water and Wastewater (Anon., 1975) were employed to determine the properties of sewage vis., total solids, pH, electrical conductivity (EC), D.O., BOD_5, COD, chloride, total N, NH_4^+ - N, NO_2^- - N, NO_3^- - N, phosphate, potassium, sodium, hardness and boron. Organic carbon of the sewage and soil was analysed using Walkley and Black method as detailed by Allison (1965). Populations of bacteria, fungi, actinomycetes and Azotobacter were enumerated using serial dilution plate technique (Pramer and Schmidt, 1966). Soils drawn from different treatments were amended with and without 2% glucose and CO_2 evolution was measured as described by Terry and co-workers (1979). 10 g of soild samples drawn at different intervals were extracted with 50 mL of 2 M KCl and the steam distillation method (Bremner, 1965a) employing MgO with and without Devarda's alloy was used to determine NO_2^- + NO_3^- -N and NH_4^+ -N, respectively,in the extracts. Particle size analysis and cation exchange capacity of the soil were determined using methods outlined by Dewis and Frietas (1970). Total nitrogen in the initial soil was determined employing the method as given by Bremner (1965b). Total phosphorus in the soil was determined by the method detailed by Murphey and Riley (1962) after the perchloric acid digestion. pH, electrical conductivity (EC) of the initial and incubated soils were determined in 2:5 (w/v) soil water suspension. Available nitrogen, phosphorus and potassium of the soils were determined as per methods of Subbiah and Ajija (1956), Olsen and coworkers (1954) and Toth and Prince (1949), respectively. The significance of The results was analysed statistically employing the factorial Randomized Block design as given by Panse and Sukhatme (1978).

RESULTS AND DISCUSSION

Results on the composition of raw sewage and effluents are shown in Table 2. The data show that there is considerable variation in the characteristics of the individual plant effluent at the two sampling periods. The observations also tend to explain that main characteristics like BOD, COD, organic carbon and other constituents including total nitrogen, ammonium nitrogen, phosphate, and boron were higher in raw sewage while parameters like dissolved oxygen, nitrite-nitrogen and nitrate-nitrogen were maximum either in activated sludge plant effluent or final clarifier effluent. It is, therefore, evident that physical, chemical quality of raw sewage entering the plant was modified to a significant extent when it was subjected to primary and secondary treatments. However, parameters like pH, EC, Mg hardness failed to reveal significant changes among raw sewage and effluents. These observations are in agreement with the report of Tebbutt (1965) who demonstrated that conventional biological treatment processes were able to deal fairly effectively with BOD, COD, NH_4^+ - N, NO_2^- - N, NO_3^- - N, $PO_4^{3-}$and SO_4^{2-}. The author also reported that chlorides and pH were little changed by conventional treatment process.

The chemical properties of raw sewage and effluents are to be assessed not only to evaluate their suitability for irrigational purposes but also for managing soil productivity factors. Hence values obtained for various parameters for raw sewage and effluents in the present study were compared with the stipulated standards for those parameters (Table 3). It has been noticed that the range of total solids, pH, EC, BOD, chloride, magnesium and boron in the sewage were well below the tolerance limits as prescribed by Scofield and Wilcox (1931), Scofield (1935), Okun and Ponghis (1975) and Indian Standards Institution (1977). The present observation, therefore, suggest that raw sewage and effluents from Okhla Treatment Plant

TABLE 2 Physical, Chemical Characteristics of Raw Sewage and Effluents

Characteristics	Raw Sewage		Primary Settling Tank Effluent		Clarifier Effluent		Activated Sludge Plant Effluent		Final Clarifier Effluent		Irrigation Channel Effluent	
	4.9. 1979	26.11. 1979	4.9 1979	26.11. 1979	4.9. 1979	26.11. 1979	4.9. 1979	26.11. 1979	4.9. 1979	26.11. 1979	4.9. 1979	26.11. 1979
Total solids (mg/L)	980.0	880.0	760.0	680.0	720.0	600.0	760.0	620.0	680.0	580.0	840.0	820.0
pH	7.5	7.7	7.3	7.7	7.2	7.7	7.4	7.9	7.4	7.9	7.4	7.8
EC (mmhos/cm)	1.2	1.1	1.3	1.1	1.3	1.1	1.3	1.1	1.3	1.1	1.4	1.2
D.O. (mg/L)	0	0	0	0	0	0	0	1.1	0	1.1	0	1.0
B.O.D. (mg/L)	143.4	110.0	37.8	48.0	59.5	29.4	18.2	6.3	27.2	10.9	39.3	24.0
C.O.D. (mg/L)	240.0	231.0	92.5	102.6	82.7	65.0	23.1	17.6	47.7	27.0	72.9	51.0
Organic carbon (%)	0.1	0.1	0.1	0.1	0.1	0.1	0.1	0.1	0.1	0.1	0.1	0.1
Chloride (mg/L)	130.2	101.6	133.0	105.5	139.3	105.5	138.2	113.3	137.5	113.3	141.6	105.5
Total N (mg/L)	33.3	29.4	26.3	16.8	24.9	16.1	13.9	11.2	23.5	15.4	15.9	18.9
NH_4^+ N (mg/L)	3.8	3.6	3.2	3.2	3.4	3.0	2.3	2.7	2.5	3.0	3.2	3.2
NO_2^--N (ug/L)	40.0	18.0	64.0	24.0	52.0	18.0	54.0	26.0	68.0	26.0	48.0	18.0
NO_3^--N (mg/L)	0.6	0.5	0.9	0.5	0.8	0.6	1.1	1.0	0.9	0.6	0.8	0.6
Phosphate (mg/L)	16.0	14.8	16.1	13.4	13.3	11.0	7.3	9.4	10.7	12.2	10.2	14.8
Potassium (mg/L)	27.5	31.0	28.0	29.0	30.0	28.0	29.0	28.5	28.0	28.0	25.0	27.0
Sodium (mg/L)	140.0	160.0	148.0	154.0	136.0	146.0	152.0	148.0	146.0	152.0	126.0	146.0
Total Hardness (mg/L)	164.8	148.0	154.8	148.0	156.8	136.0	156.8	140.0	156.8	136.0	160.2	140.0
Ca Hardness (mg/L)	132.7	124.0	124.6	116.0	128.5	120.0	120.6	116.0	116.6	112.0	124.6	120.0
Mg Hardness (mg/L)	32.1	24.0	30.3	32.0	28.3	16.0	36.2	24.0	40.2	24.0	35.6	20.0
Boron (mg/L)	0.4	0.4	0.2	0.2	0.2	0.2	0.2	0.2	0.2	0.2	0.2	0.2

TABLE 3. Comparison of Raw Sewage and Effluent Characteristics with Standard Permissible Limits for Wastewater Effluents and Water to be Irrigated on Land

S. No.	Parameter	Sample Range	Scofield (1935)	Scofield and Wilcox (1931)	Okun and Ponghis (1975)	I S I (1977)	Remarks
1.	Total solids (mg/L)	580-980	-	-	800-1200	-	-
2.	pH	7.2 - 7.9	-	-	6.8 - 7.2	5.5-9.0	-
3.	Electrical Conductivity (mmhos/cm)	1.05 - 1.35	0.75-2.00	-	-	-	EC values less than 0.75 down to 0.25 mmho/cm are excellent for irrigation (Scofield, 1935).
4.	B.O.D. (mg/L)	6.3 - 143.0	-	-	-	500	-
5.	Chloride (mg/L)	101.6-141.6	-	-	70 - 200	600	-
6.	Magnesium (mg/L)	16.0 - 40.2	-	-	40.9-65.4	-	Standard magnesium converted to mg hardness.
7.	Boron (mg/L)	0.16-0.40	-	0.67-3.00	-	2.00	The range quoted by Scofield and Wilcox (1931) for Boron covers ranges for sensitive, semi-tolerant and tolerant crops

can be utilized for irrigation depending upon their physical, chemical and biological characteristics.

Results on the influence of sewage on microbial load are presented in Tables 4-7. Addition of different effluents and raw sewage resulted in significant increases in bacterial numbers at different intervals of enumeration (Table 4).

TABLE 4 Effect of Sewage Effluents on Bacterial Population in the Soil (Population expressed as 10^6/g oven-dry soil)

Treatment	Initial	10	20	30	40	50	60
				Interval in Days			
Tap water control	35.0	73.3	281.1	38.2	34.1	34.1	33.4
Raw sewage		92.1	325.0	44.0	45.0	43.4	42.6
Primary settling tank effluent		90.8	554.0	45.8	47.8	46.6	41.5
Clarifier effluent		60.9	365.0	48.2	38.2	36.9	36.2
Activated sludge plant effluent		79.1	426.0	48.4	38.2	38.2	41.9
Final clarifier effluent		79.5	345.0	44.8	37.3	37.3	35.4
Irrigation channel effluent		100.0	393.0	45.3	43.4	41.8	40.0

	S_{ED}	CD (P=0.05)
Treatment	4.14	8.23
Interval	3.83	7.62
Treatment vs. Interval	10.14	20.16

The results also indicate that the introduction of raw sewage to soil caused a significant increase in the fungal population at various periods of incubation (Table 5). However, when the individual effluent treatments were compared, they failed to show significant changes in bacterial as well as fungal populations.

TABLE 5 Effect of Sewage Effluents on Fungal Population in the Soil (Population expressed as 10^3/g oven dry soil)

Treatment	Initial	10	20	30	40	50	60
				Interval in Days			
Tap water control	24.0	39.0	36.1	39.8	33.0	42.6	51.0
Raw sewage		42.0	50.5	52.7	44.0	42.2	49.5
Primary settling tank effluent		40.6	46.7	43.7	42.3	43.8	46.0
Clarifier effluent		49.0	40.9	49.4	38.6	36.5	48.0
Activated sludge plant effluent		40.6	41.4	48.0	41.0	42.6	51.0
Final clarifier effluent		44.0	39.4	46.7	42.0	41.3	54.9
Irrigation channel effluent		48.3	42.6	50.4	41.6	37.8	43.0

	S_{ED}	CD (P=0.05)
Treatment	1.45	2.87
Interval	1.34	2.67
Treatment vs. Interval	3.54	7.04

With regard to changes in bacterial numbers with period, maximum number of bacteria were recorded at 20th day followed by 10th day, which gradually declined toward the end of 60 days (Fig. 1a). Statistical analysis of the actinomycetes population did not show any significant effect of sewage on this group of microbes (Table 6), although in all the treatments the population declined from the 10th day towards the end of incubation period. The effluent treatments resulted in an increase in Azotobacter counts over the control (Table 7). The population obtained for soil which received secondary sewage effluents were maximum over the other treatments. The population pattern did not reveal any definite trend of enhancement or reduction at different periods.

TABLE 6 Effect of Sewage Effluents on Actinomycetes Population in the Soil (Population expressed as 10^4/g oven-dry soil)

Treatment	Initial	Interval in Days					
		10	20	30	40	50	60
Tap water control	44.0	61.3	24.8	20.7	17.7	18.0	17.1
Raw sewage		65.7	30.3	32.1	19.7	20.7	17.9
Primary settling tank effluent		48.0	29.4	28.0	20.7	20.3	16.7
Clarifier effluent		58.1	26.0	24.7	18.0	18.0	16.8
Activiated sludge plant effluent		64.7	27.6	28.0	17.0	16.0	16.0
Final clarifier effluent		54.7	27.0	27.6	22.7	20.3	16.8
Irrigation channel effluent		54.0	33.0	29.2	20.0	19.0	17.9

	S_{ED}	CD (P=0.05)
Treatment	-Non significant-	
Interval	1.77	3.51
Treatment vs. Interval	-Non significant-	

TABLE 7 Effect of Sewage Effluents on Azotobacter Population in the Soil (Population expressed as 10^3/g oven-dry soil)

Treatment	Initial	Interval in Days					
		10	20	30	40	50	60
Tap water control	6.0	8.8	10.8	11.8	13.2	14.1	12.4
Raw sewage		12.1	9.2	14.6	14.9	15.3	16.4
Primary settling tank effluent		12.7	13.6	12.6	14.9	16.5	10.3
Clarifier effluent		13.4	18.0	17.4	14.9	14.1	12.5
Activated sludge plant effluent		12.5	12.4	19.4	21.7	19.3	17.1
Final clarifier effluent		12.7	13.2	21.4	15.6	18.5	15.1
Irrigation channel effluent		10.4	11.2	22.7	18.0	15.7	13.7

	S_{ED}	CD (P=0.05)
Treatment	0.75	1.49
Interval	0.70	1.38
Treatment vs. Interval	1.85	3.69

The increases in microbial numbers can be explained due to the presence of a wide range of heterotrophic microbial population in raw sewage and effluents, which might have also offered a considerable amount of nutrient elements for their growth and multiplication in soil environment. Observation by Pillai and co-workers (1955) revealed higher bacterial counts on nutrient agar from the soil under sewage irrigation for a number of years in comparison with a virgin soil. Cairns and co-workers (1978) maintained that 15 months sewage irrigation failed to induce any significant variation in the microbial population of ten pasture soils. It seems probable that the enhancement of inhibition of soil microbial communities are largely influenced by the quality and quantity of wastewater and additional source of organic matter through plant residues.

Carbon-dioxide evolution rate was found to be stimulated in soil (both without and with glucose amendment) due to effluent treatments (Fig. 1b). Cumulative values of CO_2 calculated from evolution rate also indicated increases in total CO_2 production due to effluent addition (Tables 8 and 9).

TABLE 8 Effect of Sewage Effluents on the Cumulative CO_2 Evolution from the Soil

Treatment	Initial	CO_2 produced during 10 days in mg/100 g						Total
		d a y s						
		10	20	30	40	50	60	
Tap water control	59	40	28	33	40	52	41	234
Raw sewage		56	60	40	66	78	56	356
Primary settling tank effluent		56	58	40	51	55	44	304
Clarifier effluent		69	53	40	53	67	50	332
Activated sludge plant effluent		58	47	33	60	64	41	303
Final clarifier effluent		58	40	44	51	61	47	301
Irrigation channel effluent		68	57	41	51	58	41	316

The enhanced CO_2 evolution rates in effluent treated soil may be due to the constant addition of organic matter through wastewater application and the subsequent degradation by native microbial populations. Moreover, effect of the sewage was more pronounced when the treated soil was amended with 2% glucose. This might be due to the decomposition of added organic carbon by more number of microbial cells as well as the amplified activity of the population by way of utilization of glucose (easily available and additional carbon source).

A critical observation of the figures for bacterial population and rate of CO_2 evolution from soil amended with 2% glucose at 20th day of incubation indicated a direct correlation between them which was not observed in case of soil without glucose amendment (Fig. 1a and 1b). The activity of the fraction of bacterial population contributing to the decomposition of organic matter (Vela and Eubanks, 1973) might have been amplified by the addition of ready source of organic carbon.

TABLE 9 Effect of Sewage Effluents on the Cumulative CO_2 Evolution from Glucose Amended Soil

Treatment	Initial	Cumulative CO_2 produced during 10 days in mg/100 g						Total
		d a y s						
		10	20	30	40	50	60	
Tap water control	772	716	1223	965	904	870	892	5570
Raw sewage		712	1258	1038	959	1003	939	5909
Primary settling tank effluent		740	1249	1012	928	919	936	5784
Clarifier effluent		790	1241	1009	1003	954	926	5923
Activated sludge plant effluent		785	1267	1010	966	873	854	5755
Final clarifier effluent		804	1267	1038	986	948	950	5993
Irrigation channel effluent		604	1258	1031	994	942	929	5758

Fig. 2a gives NH_4^+ - N values from the soil receiving different treatments. Addition of raw sewage and effluents resulted in significant variations in the ammonium concentration of soil in comparison with tap water control. Raw sewage recorded an overall enhancement in the levels of NH_4^+ - N than other treatments.

Accumulations of NO_2^- + NO_3^- - N during the incubation period are presented in Fig. 2b. A striking increase in the levels of NO_2^- + NO_3^- - N at different intervals was observed in the soil which received raw sewage. When primary and secondary treated effluents were added to the soil, it also recorded varying degrees of enhancement in NO_2^- + NO_3^- - N over tap water control. In general while raw sewage application caused significant increases in these fractions of nitrogen over tap water and other effluents, statistical scrutiny of the data indicated that the variation in nitrogen levels in soil added with different effluents were not significant among themselves.

It is clear from the results that raw sewage and effluents addition caused significant increase in nitrogen mineralisation rate. Increases in ammonium and nitrate levels in effluent treated soil might be due to (i) direct contribution of the above fractions from effluents as they were found to be potential sources of NH_4^+ - N and NO_3^- - N; and (ii) ammonification and nitrification processes. Such enhancement of ammonium nitrogen might be more significant in case of raw sewage treatment. It is likely that fresh additions of raw sewage and effluents at various period during the incubation introduced additional supply of ammonium nitrogen directly as well as through ammonification. Significantly higher levels of NO_2^- + NO_3^- - N at different intervals in all the treatments including tap water control are probably because of the continuous conversion of NH_4^+ - N into NO_2^- + NO_3^- - N. Such higher levels of nitrate+ nitrite nitrogen were earlier demonstrated in soil that received different levels of domestic and industrial sewage sludges by Wilson (1977).

While several reports appeared to illustrate the influence of sewage sludge incorporation on the nitrogen mineralisation processes in soil, observations on similar effects of short term application of effluents are scanty. Nevertheless, critics of land treatment systems stress the need for understanding the interaction of several forms of nitrogen in do stic sewage and plant effluents when they are applied to soil (Bower and Chaney, 1974; Lance, 1975; Broadbent and co-workers, 1977). Thus, results from the study of present nature serve to ascertain the effect of sewage effluents applied to soil on important nitrogen transformation processes including ammonification and nitrification.

Changes in certain chemical properties of soil due to effluents addition are shown in Table 10. Addition of effluents did not cause any perceptible change in soil pH after 60 days which could be explained in terms of buffering capacity of soil. The increase in soil EC values can be well correlated due to the presence of more soluble salts in raw sewage and effluents than the initial soil solution (Tables 1 and 2). Bole (1979) reported that electrical conductance of surface 90 cm soil that received wastewater for 6 years raised from a low initial value of 0.4-0.8 to 0.9-1.6 mmhos/cm. Although the cation exchange capacity of the soil was found to increase in all of the treatments over the control at the end of the experiment, no firm conclusion should be drawn from this observation. Hence, there is a need for further investigation in this aspect on long term basis.

The results indicate (Table 10) that organic carbon of the soil maintained almost its initial value after the incubation period in all the treatments except when tap water was added. Organic carbon contribution by the effluents might have been balanced by the quantity which had been decomposed during 60 days period. In case of tap water control, the decrease in organic carbon of the soil at the end of experi-

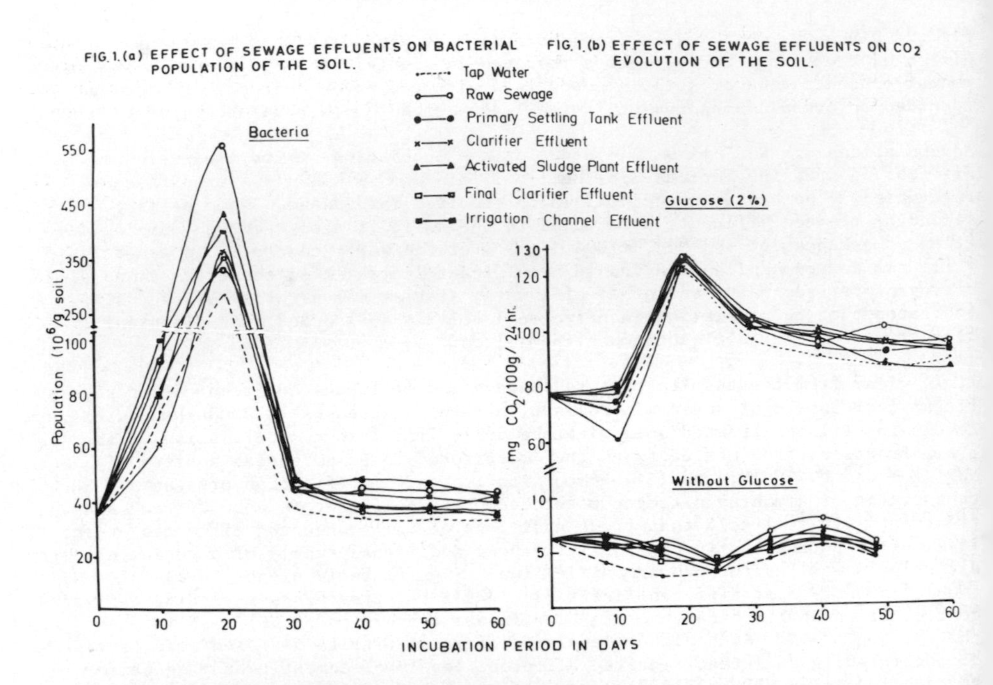
FIG. 1. (a) EFFECT OF SEWAGE EFFLUENTS ON BACTERIAL POPULATION OF THE SOIL.
FIG. 1. (b) EFFECT OF SEWAGE EFFLUENTS ON CO2 EVOLUTION OF THE SOIL.
Tap Water
Raw Sewage
Primary Settling Tank Effluent
Clarifier Effluent
Activated Sludge Plant Effluent
Final Clarifier Effluent
Irrigation Channel Effluent
Bacteria
Glucose (2%)
Without Glucose
Population (10^6/g soil)
mg CO2/100g/24 hr.
INCUBATION PERIOD IN DAYS

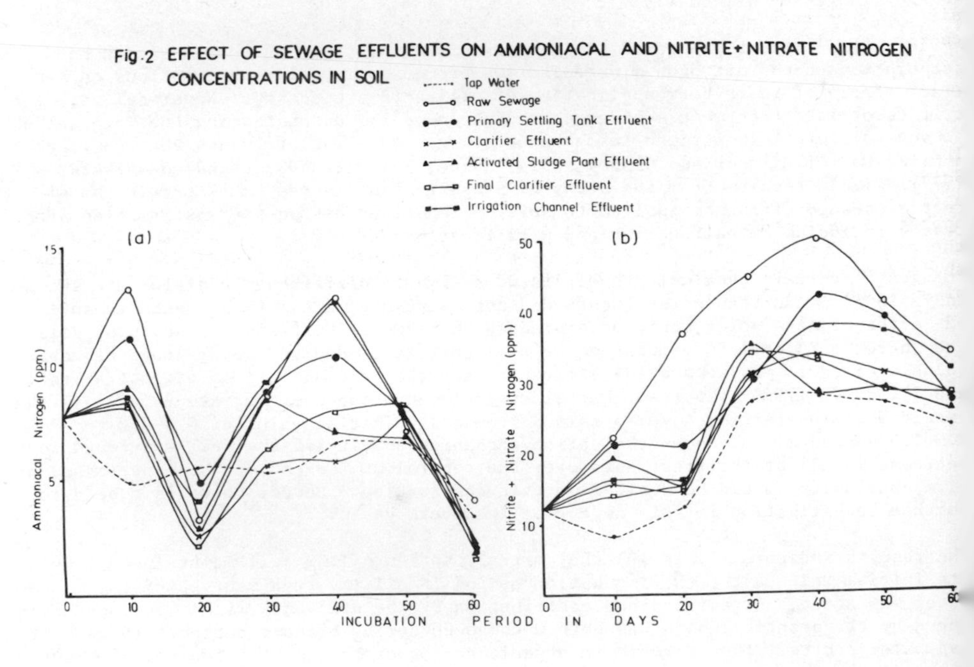
Fig. 2 EFFECT OF SEWAGE EFFLUENTS ON AMMONIACAL AND NITRITE+NITRATE NITROGEN CONCENTRATIONS IN SOIL
Tap Water
Raw Sewage
Primary Settling Tank Effluent
Clarifier Effluent
Activated Sludge Plant Effluent
Final Clarifier Effluent
Irrigation Channel Effluent
(a)
(b)
Ammoniacal Nitrogen (ppm)
Nitrite + Nitrate Nitrogen (ppm)
INCUBATION PERIOD IN DAYS

ment can be assessed by the fact that it did not receive any organic effluent during the period but loss of organic matter might have occurred through normal microbial decomposition processes. A little increase in organic matter build up in case of raw sewage treatment might be because of its relatively higher organic carbon content (Table 2).

Enhancements in soil available nitrogen, phosphorus and potassium levels, after the incubation period of 60 days, can be related to the effluents application which were rich in nitrogenous compounds, phosphate and potassium (Table 2). Quin and Woods (1978) reported substantial increases in total N. total and available phosphorus levels at different layers of soil irrigated with municipal sewage effluent for a longer period over tap water irrigated soils. Soil potassium was reported to be reduced in field in case of application with domestic sewage for 4 years when crop uptake exceeded the amount supplied by the wastewater (Palazzo and Jenkins, 1979). However, in the present study, in the absence of any crop removal reasonable increases in soil K level due to effluents addition has been noticed as per expectation.

CONCLUSION

The results of the present investigation serve to conclude that raw sewage and sewage effluents from Okhla Plant can be used to irrigate agricultural lands subject to further characterization of sewage for heavy metals, pathogenic bacteria and viruses apart from other parameters. Effluents treatments showed pronounced enhancement in total microbial population, CO_2 evolution and $NO_2^- + NO_3^-$ - N levels in soil, thus hinting that carbon and nitrogen mineralisation processes were significantly influenced by the nature of effluents. However, further work with crop systems under sewage irrigation is needed before the environmental significance of such enhance C- and N-mineralisation rates in sewage applied soils can be specified.

ACKNOWLEDGEMENTS

The authors express their gratitude to Prof. J.M. Dave, Dean, School of Environmental Sciences, Jawaharlal Nehru University, New Delhi for his keen interest and encouragement in this work. The authors also wish to acknowledge the kind help and cooperation extended by the authorities of Okhla Sewage Treatment Plant, Delhi in the collection of effluent samples. One of the authors (TS) gratefully acknowledges the financial assistance by Jawaharlal Nehry University, New Dalhi, in the form of a research fellowship.

TABLE 10 Effect of Sewage Effluents on Chemical Properties of the Soil during 60 days of Incubation

Treatment	pH	EC (mmho/cm)	CEC (meg/100g)	Organic carbon (%)	Available nitrogen (ppm)	Available phosphorus (ppm)	Available potassium (ppm)
Initial	8.5	0.5	8.0	0.57	99.1	28.5	276.7
Tap water control	8.6	0.5	8.2	0.53	100.7	33.8	276.7
Raw sewage	8.7	0.6	9.7	0.59	149.5	35.0	301.7
Primary settling tank effluent	8.6	0.6	8.9	0.57	137.8	39.5	288.3
Clarifier effluent	8.7	0.6	9.0	0.57	123.8	35.0	301.7
Activated sludge plant effluent	8.6	0.6	8.8	0.55	145.0	40.3	291.7
Final clarifier effluent	8.6	0.6	8.8	0.55	135.0	37.5	308.3
Irrigation channel effluent	8.7	0.6	9.2	0.57	143.1	35.7	301.7

REFERENCES

Allison, L.E. (1965). Organic Carbon. In Methods of Soil Analysis, Part 2, C.A. Black et al. (Eds.), American Society of Agronomy, Madison, Wisconsin p. 1367-78.

Anonymous (1975). Standard Methods for the Examination of Water and Wastewater, 14th edition. American Public Health Association, American Water works Association and Water Pollution Control Federation, New York. p. 1193.

Bole, J.B. (1979). Land application of municipal sewage wastewater: Changes in soil chemistry. Agron. Abst. (Environ. Qual.), p. 41. 71st Annual Meeting of American Society of Agronomy held at Fort Collins, Colorado, on August 5-10, 1979.

Bouwer, H. and R.L. Chaney (1974). Land treatment of wastewater. Adv. Agron., 26, 133-176.

Bremner, J.M. (1965a). Inorganic forms of nitrogen. C.A. Black and others (Eds.), Methods of Soil Analysis, Part 2, American Society of Agronomy, Madison, Wisconsin Wisconsin. pp. 1179-1237.

Bremner, J.M. (1965b). Total nitrogen. C.A. Black and others (Eds.), Methods of Soil Analysis, Part 2, American Society of Agronomy, Madison, Wisconsin, pp.1149-1176.

Broadbent, F.E., D. Pal and K. Aref (1977). Nitrification and denitrification in soils receiving municipal wastewater. F.M. D'Itri (Ed.), Wastewater Renovation and Reuse. Marcel Dekker, Inc., New York. pp. 321-348.

Cairns, A., M.E. Dutch, E.M. Guy and J.D. Stout (1978). Effect of irrigation with municipal water of sewage effluent on the biology of soil cores - Introduction. Total microbial populations and respiratory activity. N.Z. J. Agric. Res., 21, 1-10.

Chaney, R.L. (1973). Recycling Municipal Sludges and Effluents on Land, National Association State University and Land Grant Colleges, Washington, D.C. pp.129-141.

Dewis, J. and F. Freitas (1970). Physical and Chemical Methods of Soil and Water Analysis. Soils Bulletin No. 10, FAO, Rome. p. 275.

Dunlop, S.G. (1968). Survival of pathogens and related disease hazards. In C.W. Wilson and F.E. Beckett (Eds.)Louisiana Tech. Alumini Foundation Tech. Station, Ruston, Louisiana. pp. 107-122.

Ellis, B.G. (1978). Analysis and their interpretation for wastewater application on agricultural land. B.D. Knezek and R.H. Miller (Eds.). Applications of Sludges and Wastewaters on Agricultural Land - A Planning and Educational Guide, USEPA, Office of the Water Program Operations, Municipal Corporation Division, Washington, D.C. pp. 6.1-6.6.

Indian Standards Institution (1977). Indian Standard - Tolerance Limits for Industrial Effluents Discharged on Land for Irrigation Purposes (First Revision), ISI, New Delhi. p. 8.

Johnson, Jr. C.C. (1979). Land application of waste - An accident waiting to happen. Ground Water, special issue, Proceedings of the fourth NWWA-EPA National Ground Water Quality Symposium. Vol. I, 69-72.

Lance, J.C. (1975). Fate of nitrogen in the sewage effluent applied to soil. J. of the Irrigation and Drainage Div., 101 (IR3), 131-144.

Lee, G.F. (1976) Potential problems of land application of domestic wastewaters. In R.L. Sankds and T. Asano (Eds.), Land Treatment and Disposal of Municipal and Industrial Wastewater, Ann Arbor Sci. Publ. Inc., Ann Arbor, Mich. pp. 179-1921

Murphey, J. and J.P. Riley (1962). A modified single solution method for the determination of phosphate in natural waters. Anal. Chem. Acta, 27, 31-36

Okun, D.A. and G. Ponghis (1975). Community Wastewater Collection and Disposal. World Health Organisation, Geneva. pp. 211-220.

Olsen, S.R., C.L. Cole, F.S. Watonable and D.A. Dean (1954). Estimation of available phosphorus in soils by extraction with sodium bicarbonate. U.S.D.A. Circ., 939.

Palazzo, A.J. and T.F. Jenkins (1979). Land application of wastewater: Effect on soil and plant potassium. J. Environ. Qual., 8, 304-309.

Panse, V.G. and P.V. Sukhatme (1978). Statistical Methods for Agricultural Workers. Third Revised Edition. Indian Council of Agricultural Research, New Delhi. p. 347.

Pillai, S.C., R. Rajagopalan and V. Subrahmanyam (1955). Sewage farming and sewage-grown crops in relation to health. Bull. of the Natl. Inst. of Sciences of India, No. 10, 160-179.

Pramer, D. and E.L. Schmidt (1966). Experimental Soil Microbiology. Burgess Publ. Co., Minneapolis, Minnesota. p. 107.

Quinn, B.F. and P.H. Woods (1978). Surface irrigation of pasture with treated sewage effluent I. Nutrient status of soil and pasture. N. Z. J. Agri. Res., 21, 419-426.

Scofield, C.S. (1935), as quoted by R.P. Mishra (1965). Factors affecting the reuse of effluents in agriculture. Proc. Symp. Water Pollut. Control., 3, 78-87.

Scofield, C.S. and L.V. Wilcos (1931),as quoted by R.P. Mishra (1965). Factors affecting the reuse of effluents in agriculture. Proc. Symp. Water Pollut. Control, 3, 78-87

Sheaffer, J.R. (1979). Land application of waste - important alternative. Ground Water, special Issue, Proceedings of the Fourth NWWA-EPA National Ground Water Quality Symposium. vol. I, 69-72.

Subbiah, B.V. and G.L. Ajija (1956). A rapid procedure for estimation of available nitrogen in soils. Curr. Sci., 25, 259-260.

Tebbutt, T.H. (1965). Reuse of sewage effluent. In Proc. Symp. Water Pollution Control, 3, 88-96.

Toth, S.J. and A.L. Prince (1949). Estimation of cation exchange capacity and exchangeable calcium, potassium and sodium contents of soils by flame photometer techniques. Soil Sci., 67, 439-445

Vela, G.R. and E.R. Eubanks (1973). Soil microorganism metabolism in spray irrigation. J. Water Pollut. Contr. Fed., 45, 1789-1794.

Wilson, D.O. (1977). Nitrification in soil treated with domestic and industrial sewage sludge. Environ. Pollut., 12, 73-82.

36

POLISHING INDUSTRIAL WASTE STREAM EFFLUENTS USING FLY ASH NATURAL CLAY SORBENT COMBINATION

P. C. Chan, J. W. Liskowitz, A. Perna, Mung-Shium Sheih, R. Trattner and M. K. Stinson*

New Jersey Institute of Technology, Newark, New Jersey, USA
**U.S. Environmental Protection Agency, N.J., USA*

ABSTRACT

The combined use of acidic and basic fly ashes, bentonite, bauxite, illite, zeolite, and vermiculite was evaluated in the laboratory for polishing a 1 mgd industrial waste stream which contains significant concentrations of fluoride, iron, lead, chromium, and cadmium. Activated alumina was included in this study for comparison purposes.

A combination consisting of illite-basic fly ash using lime to achieve the optimum pH (6.3) of the combination was found to be the most effective for reducing the fluoride and iron in the waste water from concentrations 17.5 mg/L and 4.5 mg/L to 1 mg/L, respectively. In addition, the lead, chromium, and cadmium concentrations were reduced from 0.12 mg/L, 0.05 mg/L, and 0.15 mg/L to 0.013 mg/L, 0.015 mg/L and 0.010 mg/L, respectively. The lysimeters were operated in the up-flow expanded bed mode to overcome the low permeability exhibited by clay and fly ash. The reduction of the fluoride concentrations in these industrial waste streams controlled the treatment process.

The regeneration of the spent sorbents was successfully accomplished by passing through the lysimeter column H_2SO_4 on the basis of 5 ml of 1% H_2SO_4 per gram of sorbent. A 20% loss was observed in sorbent capacity after the first regeneration. Subsequent regenerations resulted in no further loss of sorbent capacity. Estimated treatment with acid regeneration of the sorbent case for reducing the fluoride ion to 1.0 mg/L was 13 cents per 1000 gallons.

KEYWORDS

Industrial waste water treatment; pH control; flow-rate control.

INTRODUCTION

The discharge of industrial waste streams that contain heavy metals, toxic anions, and organics into receiving waters can severely limit its reuse. The availability of an inexpensive waste treatment and effluent polishing process could be an economic incentitive for industry.

For the past two and one-half years, we have been examining the use of waste products such as acidic and basic fly ashes in combination with clays such as illite, kaolinite, vermiculite, and natural zeolite, commonly found in soils to inexpensively treat leachate that is generated from industrial sludges deposited in landfills. The effectiveness of the above sorbents for removing the hazardous and toxic constituents in these leachates was established and the results have been reported (4).

It is the intent of this investigation to apply the fly ash-clay combination technology for the inexpensive treatment of industrial waste streams.

A number of investigators have shown that organics (6-9) (e.g. phenols, surfactants, pyridine, other organics characterized by COD) pesticides (10-12) (e.g. parathion, DDT, dieldria and heptachlor, herbicidesley, paraguet), heavy metals (13-16) (e.g. lead, cadmium, mercury, zinc, mangenese, copper), and toxic anions (17) (e.g. chromium VI, arsenic and selenium) in leachates and waste streams are attenuated by clay minerals, soils and waste products. However, virtually none have explored the use of fly ash-clay sorbent combinations for polishing industrial waste streams for the removal of heavy metals, toxic anions, and organics, and identified the parameters which control the removals of heavy metals, toxic anions and organics present in these waste streams.

Thus, the objective of this investigation is to develop an inexpensive treatment technology to reduce the concentrations of heavy metals, toxic anions, and organics to acceptable levels for discharge into receiving waters and to identify the parameters that governs this sorbent treatment approach.

Materials and Methods

Sorbents Description

The sorbent materials that were selected for this investigation are fly ash, zeolite, vermiculite, illite, kaolinite, bauxite, bentonite, and activated alumina. The selection of these materials was based on the consideration of economics, availability, and removal potential. Activated alumina was included for comparison purposes. Details of these materials have been published previously (4).

All sorbent materials were used as received. Sorbent materials which were not obtained as a powder (illite and vermiculite) were ground in a laboratory hammer mill (Weber Bros., and White Metal Works, Inc. Type 22) and passed through an eighty mesh A.S.T.M. standard sieve. All sorbents were dried to constant weight at 103°C in accordance with "Standard Method" procedures and stored in a desiccator until used.

Methods

The investigation was undertaken in three different phases of experimentation, viz., static studies, dynamic studies, and sorbent regeneration.

1. Static studies: The primary objective of the static studies were to evaluate those parameters affecting the waste water-sorbent materials interactions under batch equilibrium conditions. These parameters include the effect of pH on the sorbent capacity , the relationship between sorbent capacity and the desired effluent concentration, and contact time.

A measured amount of dried sorbent material was placed into a tared one liter screw capped polypropolene Erlenmeyer flask. The flask was sealed and agitated for 24 hours, with the exception of the contact time study, at ambient temperature. The mixture was then filtered through a glass fiber filter and filtrate was analyzed.

2. Dynamic studies: Lysimeter studies were conducted to evaluate hydraulic conditions and removal capacity of the sorbents for the specific contaminants present

in significant concentration in the waste water under dynamic conditions. In order to evaluate feasibility of the sorbent approach to handle large volume of discharge, two hydraulic systems, viz., gravitational flow and upflow expanded bed, were investigated.

Gravitational flow - Lysimeters were constructed of plexiglass tubing, supported in a vertical position. The laboratory arrangement of the lysimeters are shown in Figure 1. Waste water was fed to the top of the column through a valved manifold that distributed the waste water to the lysimters simultaneously from a central reservoir. A constant hydraulic head was maintained in the lysimters at all time, and the volume of waste water through the packed sorbent material was continuously monitored. Samples of waste water effluent were collected at intervals and analyzed to determine the concentration of all measurable constituents remaining in the effluent after a known volume of waste water had passed through the lysimeter. This was continued until breakthrough for all measurable contaminants had occurred or excessively low permeabilities were encountered. Breakthrough is defined as that condition when the concentration of the species of concern in the collected effluent sample approached or exceeded that in the influent.

Upflow Expanded Bed Operation

An upflow expanded bed approach was investigated to overcome low permeability exhibited by the sorbents combination. Using the same lysimeter and sorbent bed layout as in the gravitational flow study, the waste water was fed from the reservoir through the bottom of the bed at a velocity sufficient to expand the bed without loss of the sorbent in the overflow.

A flow of 6×10^{-3} gal/min was achieved in lysimeter containing 750 grams of the illite-basic fly ash mixture and 10 grams of lime without loss of sorbent. The effluent was collected from the overflow outlet and volume monitored continually. The collected effluent samples were analyzed for all measurable pollutants until breakthrough had occurred.

3. Regeneration. Regeneration of the sorbents was investigated to reduce the costs associated with the use of the clay-fly ash sorbent combinations still further, as well as eliminate the need for stock-piling of the sorbents. Since sulfuric acid is considered fairly inexpensive, our earlier studies reveal that leaching of heavy metals and anions are enhanced under strongly acidic conditions, regeneration using sulfuric acid was examined (4,18).

The regeneration studies were carried out under batch conditions on spend sorbent combinations. Different volumes of a one percent solution of sulfuric acid was added to 10 grams of spent sorbents and shaken for six hours. A series of tests showed that six hours is required for maximum regeneration of the spent sorbent for each specific volume of sulfuric acid solution used. After each regeneration step, the 10 grams of sorbent combinations was filtered, washed, and 0.2 g of lime added. The sorbent combination was then mixed with the 0.5 liters of feldspar waste water according to the static studies described earlier. This was repeated for different volumes of one percent sulfuric acid solution until maximum recovery of the spent sorbent as indicated by its sorbent capacity was achieved.

Analytical procedures for the experimentation were conducted in accordance with the Standard Methods (19). The hydraulic conditions of waste water through sorbent bed were based on the Darcy's law of hydraulic conductivity. Details of experimentation have been described previously (4).

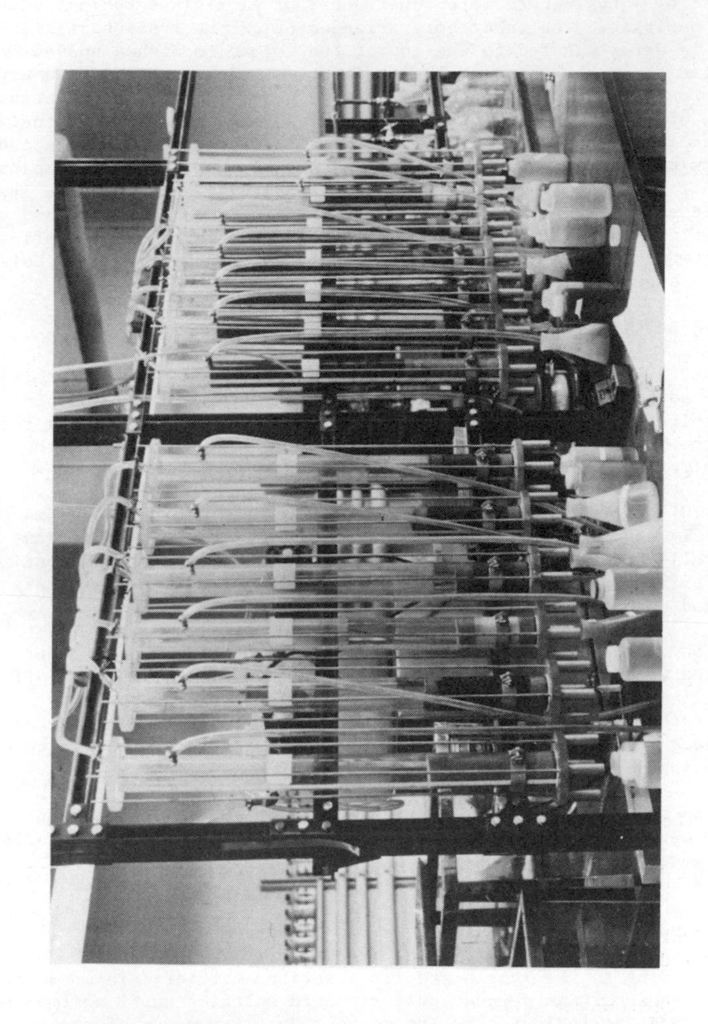

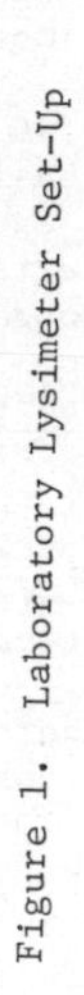

Figure 1. Laboratory Lysimeter Set-Up

RESULTS AND DISCUSSION

The results of the analysis of industrial waste water samples show significant concentrations of heavy metals and toxic anions such as iron and fluoride, respectively (see Table 1).

TABLE 1. Analysis of Industrial Waste Water Samples

Contaminants	Concentrations (mg/L)
F	4.0 - 17.5
Cl	27 - 39
CN	0.0 - 0.009
SO_4	57 - 82
Ca	5.2 - 11.3
Cd	0.007 - 0.015
Cr	0.007 - 0.05
Cu	0.035 - 0.02
Fe	0.3 - 4.5
Mg	1.7 - 3.7
Ni	0.044 - 0.15
Pb	0.013 - 0.12
Zn	0.054 - 0.18
COD	14 - 20
pH	4.9 - 6.3

Evaluation of Sorbents

The means to reduce the most concentrated contaminants (F, Fe) encountered in waste water samples by the selected sorbent materials was examined under batch equilibrium condition.

Among the sorbent materials examined, both illite and kaolinite showed comparable removals in reducing the fluoride concentrations to below 1.5 mg/L. Bentonite, bauxite, acidic fly ash, zeolite, and vermiculite were unable to reduce fluoride concentration to below 1.5 mg/L even when sorbent concentrations of 100 g/L were used (Figure 2).

Pertaining to the removal of iron, basic fly ash was found to be the most effective sorbent, followed by kaolinite, illite, and acidic fly ash. A sorbent concentration of 1 g/L was required to reduce the iron concentration from 1.8 mg/L to 0.02 mg/L (Figure 3).

These findings are in agreement with earlier reported results for waste samples obtained from industrial sludge leachates (4).

Effect of pH on Sorbent Capacity

The effect of pH upon the sorbent capacity of the fly ashes and clay for the removal of fluoride ion was investigated. Here, the sorbent capacity represents the amount of contaminant that will be removed in a waste stream by one gram of sorbent. The results indicate for an increase of pH of the waste water from 3 to 7, the sorbent capacity dependence upon pH for illite and kaolinite differed markedly. The sorbent capacity for the kaolinite exhibited a fairly braod maximum in the pH range from 3.0 to 4.5, but decreased with an increasing pH above 4.5 (see Figure 5). The sorbent capacity for the illite, however, exhibited a convex-shaped curve which

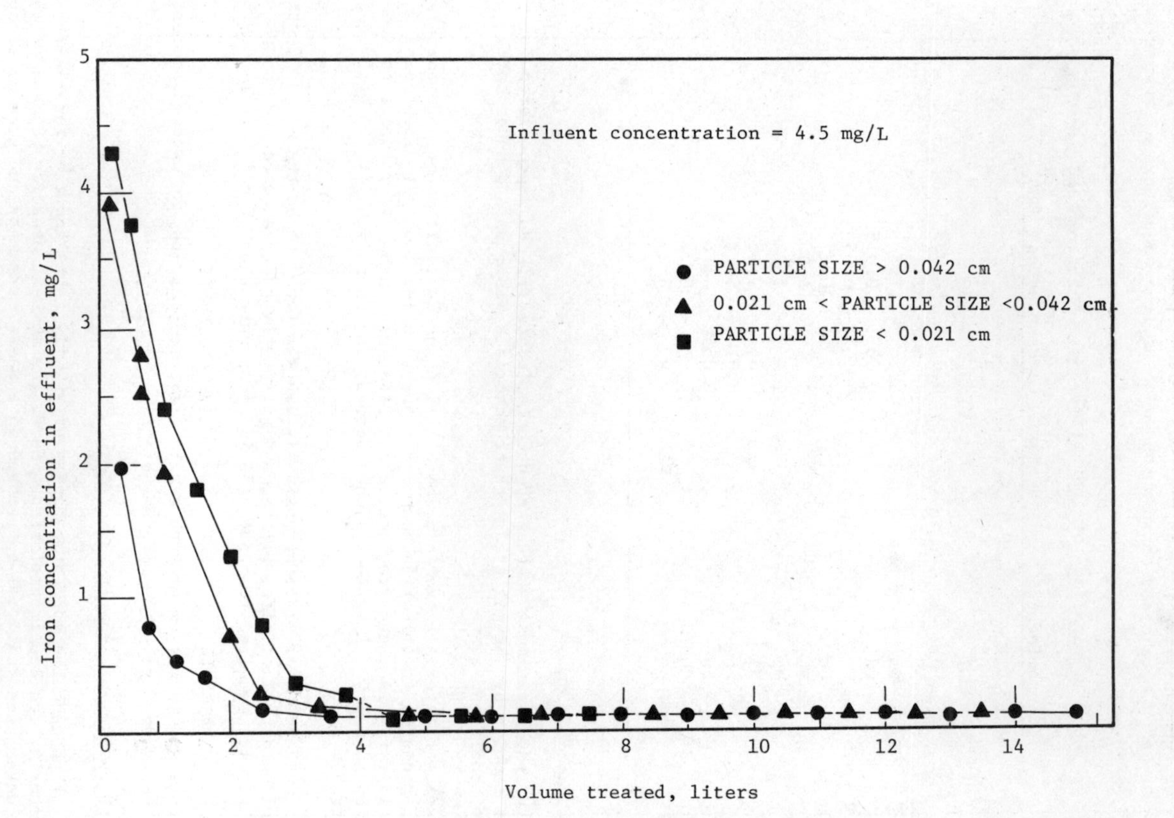

Figure 2. Iron removal dependence on sorbent particle size (gravitational flow).

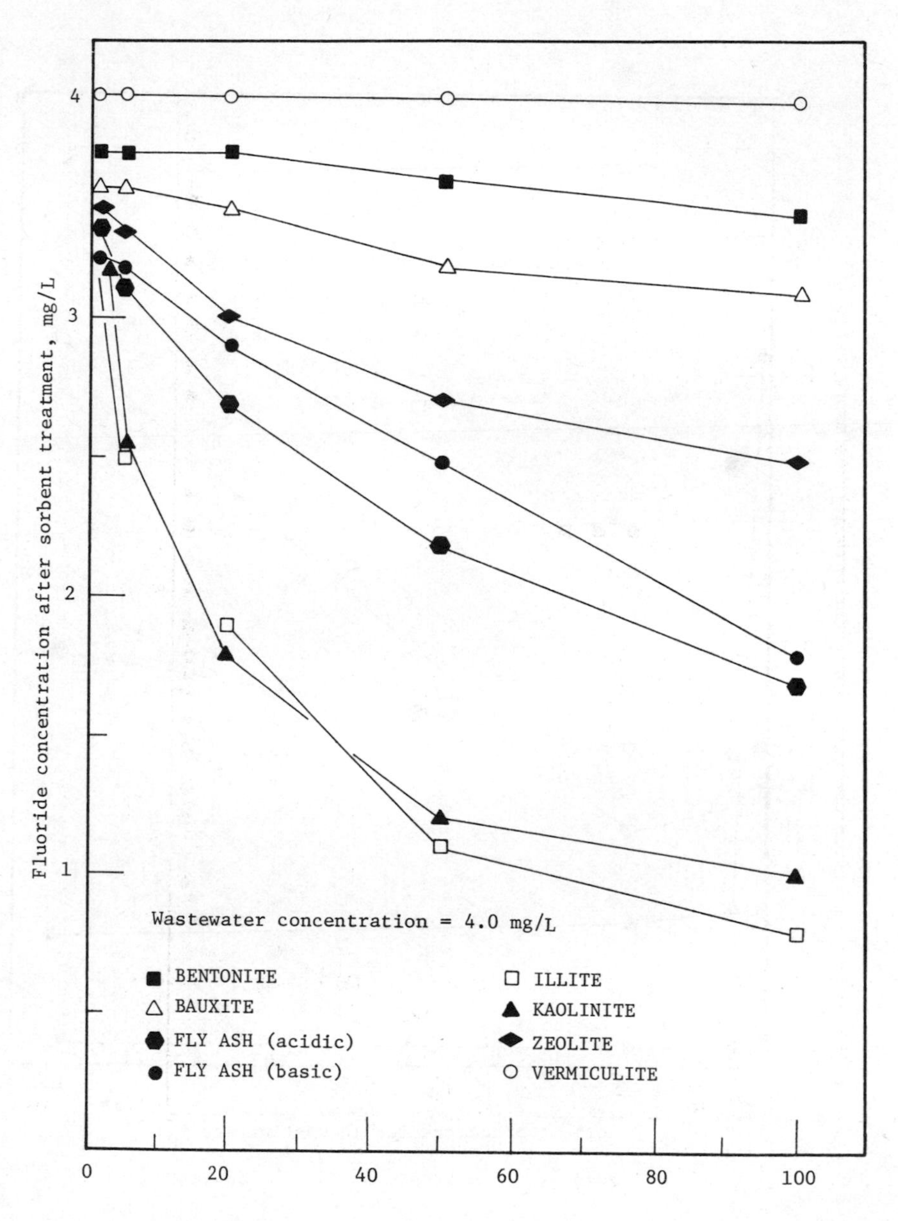

Figure 3. Fluoride treatment (batch conditions)

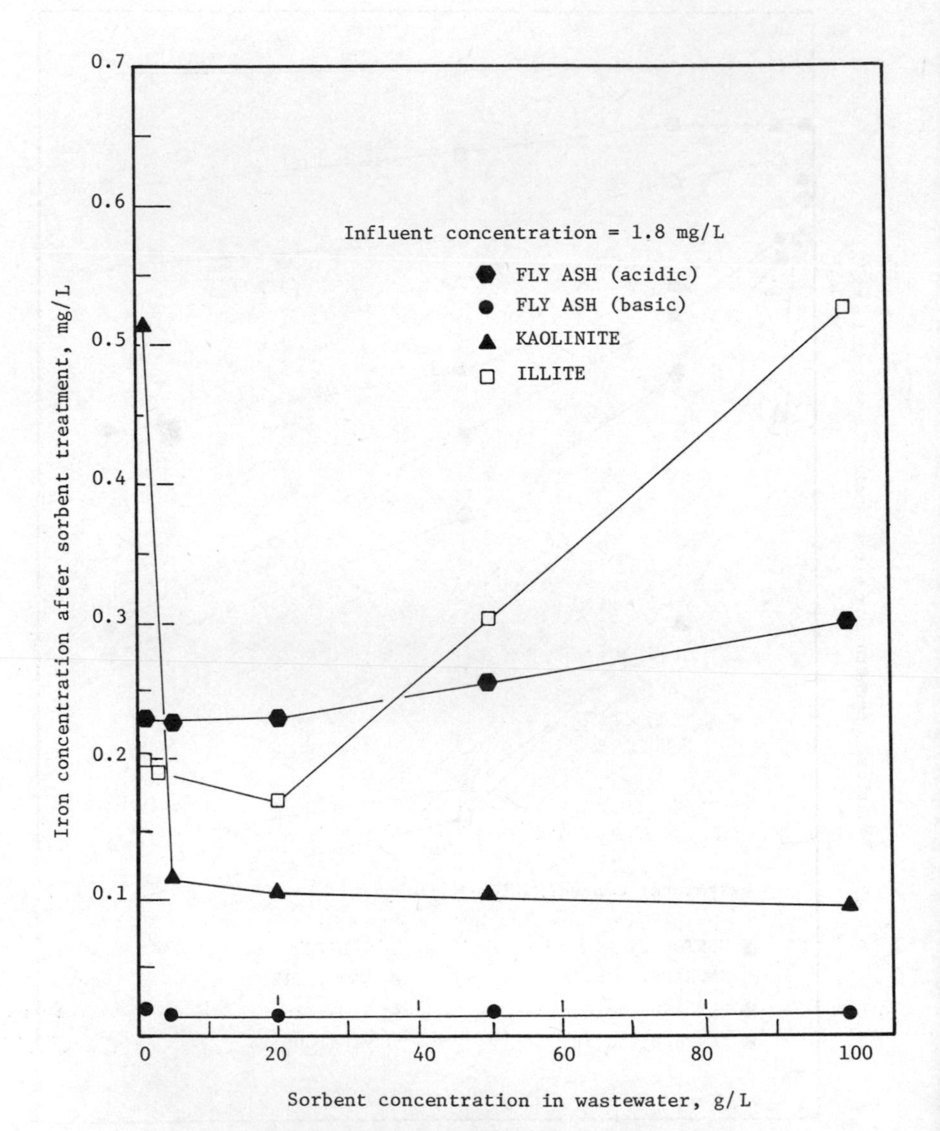

Figure 4. Iron treatment (batch conditions)

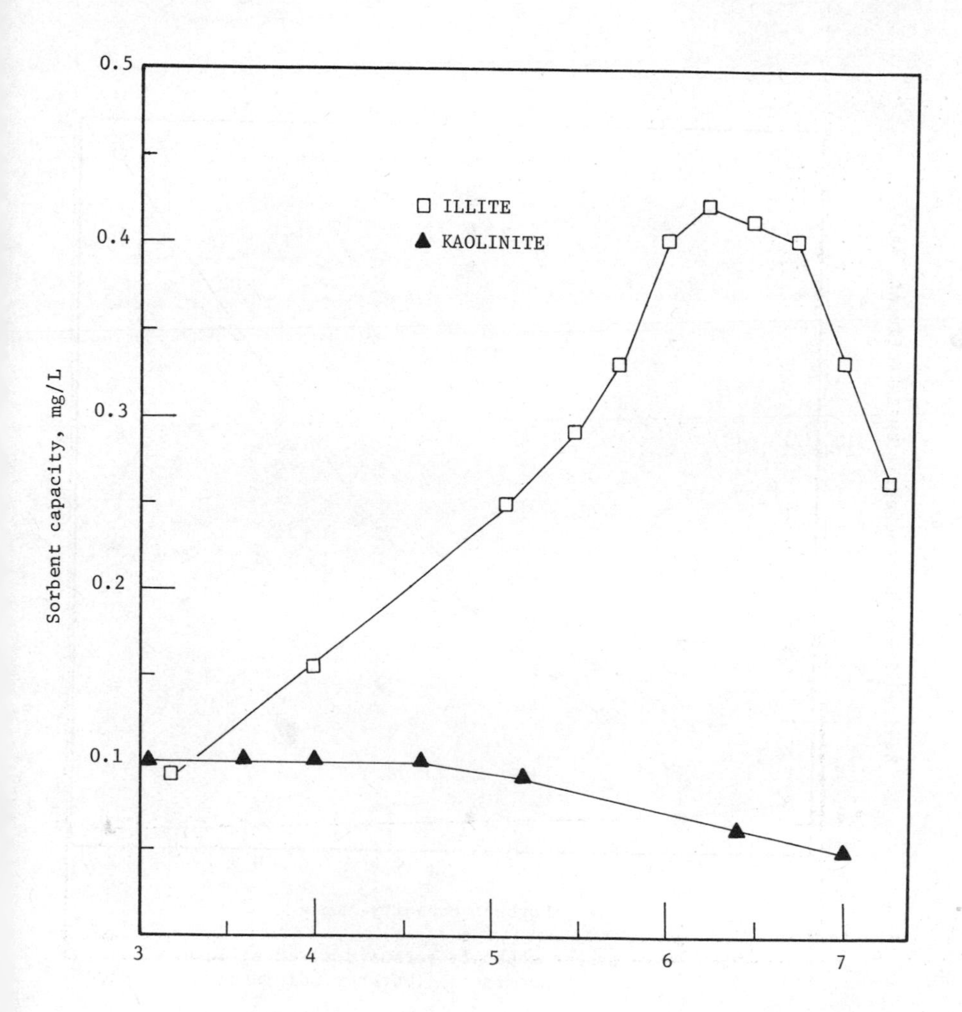

Figure 5. pH effect on sorbent capacity (batch conditions)

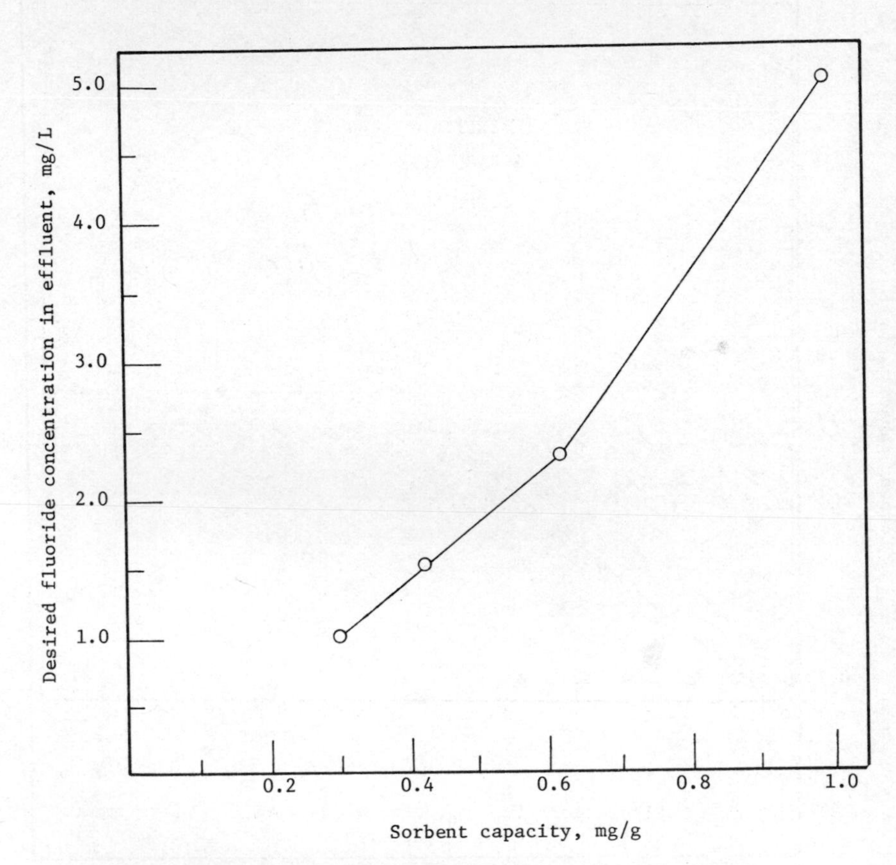

Figure 6. Sorbent capacity versus desired effluent concentration (batch conditions)

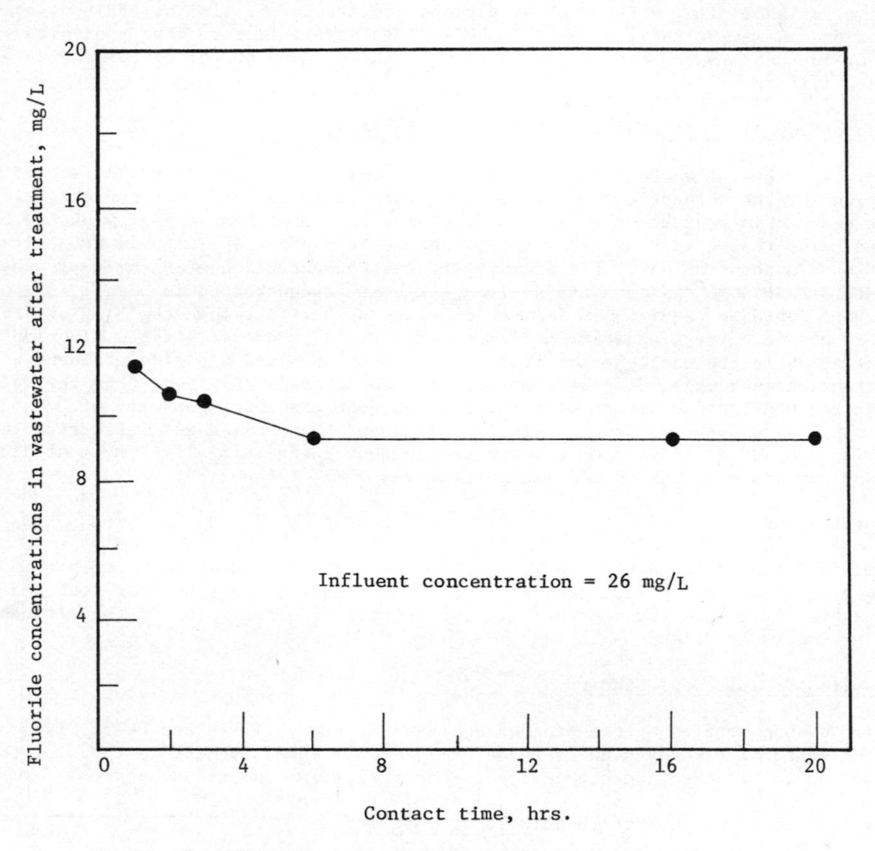

Figure 7. Fluoride influent concentration versus contact time (batch conditions)

increased to an optimal pH of 6.3 followed by a decrease at higher pH. It should be emphasized that the sorbent capacity at optimal pH exhibited by the illite is about 4 times than that of kaolinite.

Thus, it is of paramount importance to control the pH of the waste water to achieve the optimal removals of a contaminant. For example, given the same volume of waste water sample to be treated, it requires approximately the same amount of illite or kaolinite, without pH adjustment, to attain the similar effluent concentration. However, the illite would be 4 times more effective than kaolinite if the pH of the waste water were adjusted to 6.3, the optimum pH for fluoride removal using illite.

Sorbent Capacity and Desired Effluent Concentration

Given a volume of waste water, the amount of pollutant required to be removed is defined by the sorbent capacity. However, the lower the effluent concentrations are desired in a given waste stream, the smaller the sorbent capacity should be used. The reason is that the sorbents become less effective with lower desirable effluent concentrations. For example, if the concentration of fluoride in the waste stream is 8 mg/L and the desired effluent concentration is 5 mg/L, 3 g of sorbent would be required in 1 liter of waste stream to remove the 3 mg of fluorid based upon a sorbent capacity of 1 mg/g (Figure 5). However, if the fluoride concentration in the waste stream is 4 mg/L, and the desired fluoride effluent concentration is 1 mg/L, 3 mg of fluoride will be removed. However, this removal wil require 10 g instead of 3 g of sorbent based upon a sorbent capacity of 0.3 mg/g. In other words, the sorbent combination is about three times more effective in removing 3 mg of fluoride from a waste stream containing initially 8 mg/L of fluorid than from waste stream containing initially 4 mg/L of fluoride.

Contact Time

A minimum contact time of 6 hours between the sorbents and waste water sample is required to ensure optimal removal of the fluoride. Contact periods less than 6 hours result in greater concentration of the fluoride remaining in the effluent after treatment (Figure 7).

Results of Lysimeters Studies

Based on the results of the preceeding investigation, illite and basic fly ash we selected for lysimeter studies to determine their effectiveness in reducing the fluoride and iron to acceptable levels under flowing conditions.

Particle Size and pH Effects on Sorbent Capacity

Different sorbent particle sizes were examined in the lysimeters to determine whether increasing the particle size would be capable of handling flows of the order of 3.8×10^6 liters per day (1 mgd) by sorbent beds under gravitational fee

Three different particle sizes of fly ash - illite mixture - 1) greater than 0.42 mm; 2) between 0.21 and 0.42 mm; and 3) less than 0.21 mm - were selected to eval ate their effect upon sorbent capacity. The results indicate that the particle size range appear to influence the reduction of fluoride, but not the iron. For the treatment of the fluoride, the two smaller particle size ranges reduce the fluoride to below 1 mg/L in the treatment of 8 liters of waste stream; the partic greater than 0.42 mm only reduce the fluoride to these levels in the treatment of 5 liters of waste stream (Figure 8). However, the iron concentration was reduced from 4.5 mg/L to 0.12 mg/L with identical behavior after the initial treatment of 4 liters amongst the three different size ranges (Figure 9). Pertaining to the

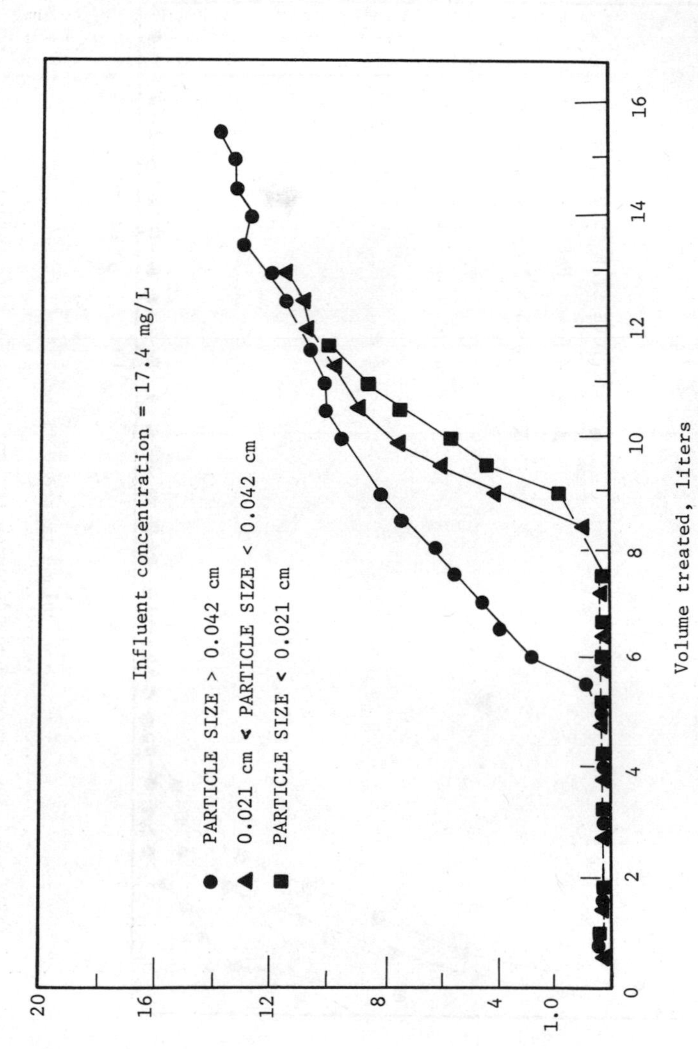

Figure 8. Fluoride removal dependence on sorbent particle size (gravitational flow)

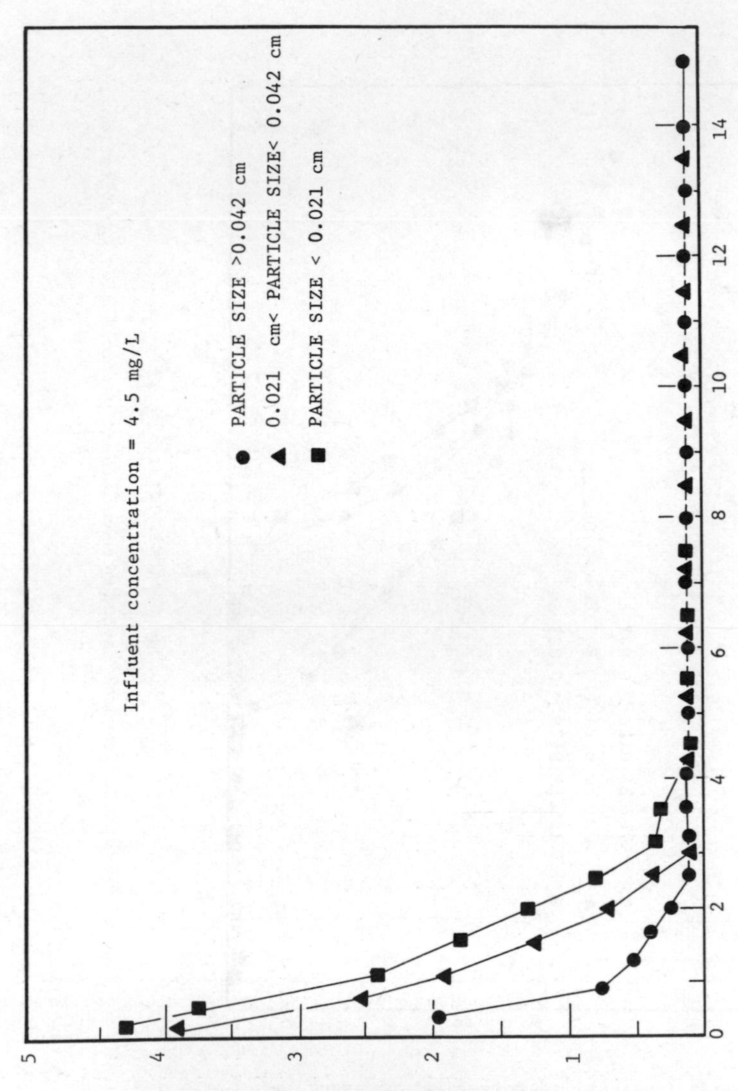

Figure 9. Iron removal dependence on sorbent particle size (gravitational flow)

initial treatment, the sorbent whose particle sizes are less than 0.21 mm exhibited the poorest removal followed by the particles in the 0.21 mm to 0.41 mm range, and those greater than 0.41 mm in increasing order, respectively. The initial acidic pH of 3.6 encountered from the lysimeter containing the smallest particles is probably responsible for the poorest removal of iron in the initial stage of treatment (Figure 9). From the treatment point of view, this situation can be corrected rather easily. Adjustment of the pH of the influent with lime, for example, to provide a pH of 4 in the initial effluent from the lysimeter containing the smallest particle results not only in iron removals that are comparable with the larger particles, but also results in improved removals of fluoride. Since the initial volume of untreated effluent is rather small, an alternative solution is to recycle this volume to the system for subsequent treatment.

Breakthrough of fluoride in the lysimeters where the pH of the influent was adjusted to 6.3 occurs after 7 liters of waste water was passed through these lysimeters. In comparison, breakthrough for the influent where the pH was not adjusted occurs after only 5.5 liters of waste water was passed through this lysimeter (Figure 10). These results are as expected since it was shown earlier that the maximum removal of the fluoride occurs at a pH of 6.3.

Particle Size Effect on Flow

The permeability exhibited by the sorbents increases with particle size. The sorbents in size range less than 0.21 mm, between 0.21 mm and 0.42 mm, and greater than 0.42 mm increased from 1.5×10^{-4} cm/sec, to 3×10^{-4} cm/sec and to 6.5×10^{-4} cm/sec, respectively (Figure 11). These permeabilities, however, limit the flows through the lysimeters to only 11.4×10^{-5} L/min, 15.2×10^{-5} L/min, and 30.4×10^{-5} L/min, respectively. These flows provide a contact time longer than the 6 hours which was shown earlier to be required for maximum removal of the fluoride ion under batch conditions.

The above results imply that an actual sorbent bed of six feet deep would only be able to handle gravitational flows that range from 132 L/m^2/day to 575 L/m^2/day to avoid ponding. On the other hand, a sorbent bed six feet deep with a contact time of 6 hours should be able to handle flows of 7420 L/m^2/day. Two alternatives may be used: (1) by increasing the size of the particles in the bed; and (2) by upflow expanded bed treatment. However, an increase in particle size results in a significant decrease in the effectiveness of the sorbent bed for removal of fluoride in specific volumes of waste water (Figure 8). Discussion of the latter approach follows.

Upflow Expanded Bed Treatment

The operation of the upflow expanded bed treatment is based on the concept of fluidized bed. The use of this mode of treatment is to take advantage of the hydraulic situation: (1) A significantly large volume of waste water can be passed through due to the increase in pore volume of the expanded bed; and (2) the fluidized bed can provide an adequate contact time for the sorbent to interact with the waste water even at a higher effluent flow rate than that of gravitational feed.

The maximum possible upflow that could be achieved in the lysimeters which contained a mixture of 750 grams of illite and basic fly ash, and 10 grams of lime without the loss of sorbent was 22.8×10^{-3} L/min. This is at least two orders of magnitude greater than the above gravitational flows obtained with the same sorbent combination. In an actual six feet deep sorbent bed, a flow of 22.8×10^{-3} L/min in the lysimeter would be equivalent to a treatment rate of about 2.8×10^4 L/min^2/day.

In these experiments, 13 mg of lime was mixed with one g of 50 percent illite-basic

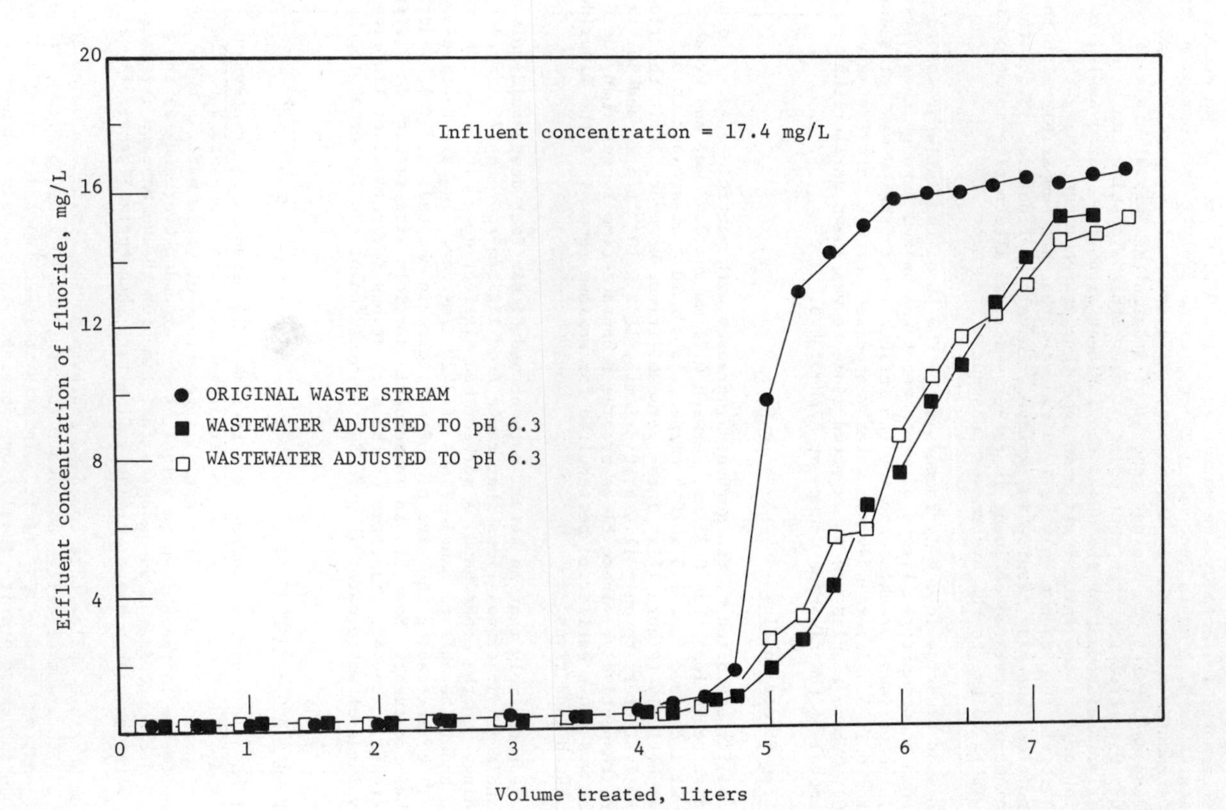

Figure 10. Fluoride treatment with pH adjustment (gravitational flow)

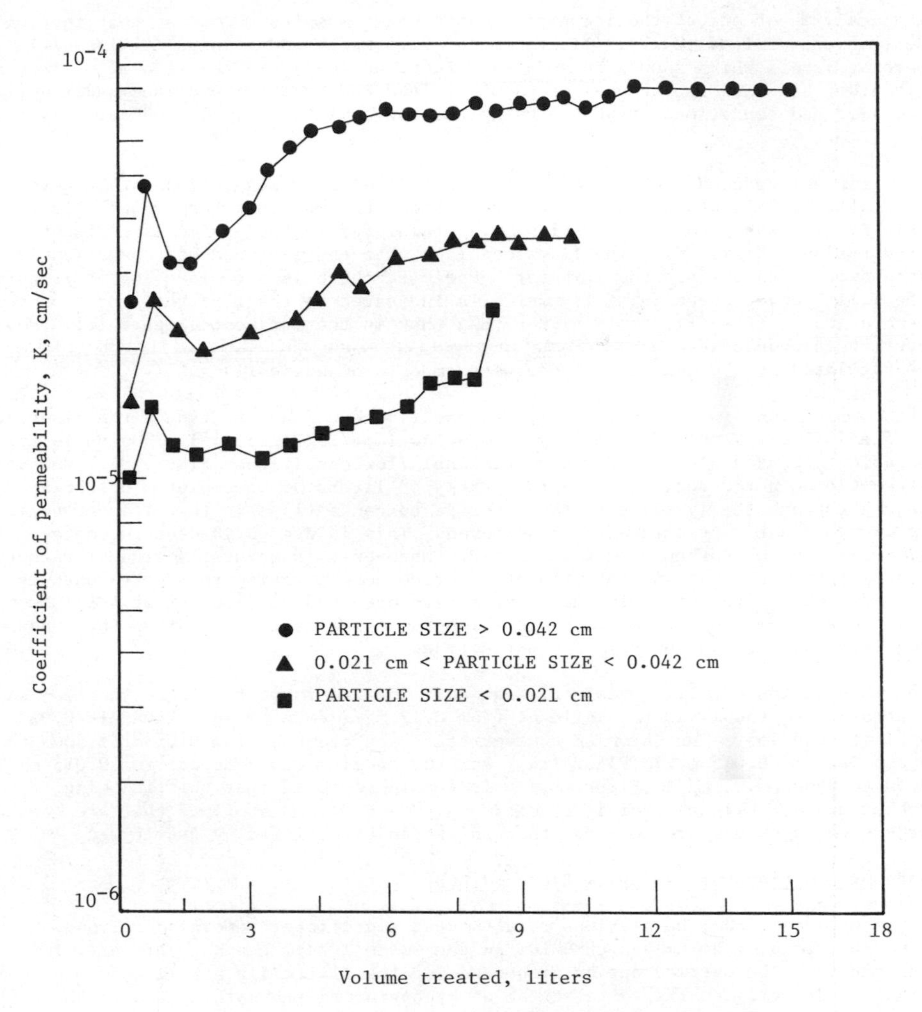

Figure 11. Sorbent particle size effect on permeability

fly ash to neutralize the acidic properties of the clay. It was also added to the lysimeter influent to provide a pH of around 6.3 for maximum removals of the iron and fluoride (Figure 12).

Removal of Iron, Fluoride, Lead, Chromium, and Cadmium

The analysis of one of the industrial waste water samples revealed that the concentrations of lead (0.12 mg/L), chromium (0.05 mg/L) and cadmium (0.015 mg/L) were at levels which should be reduced. Thus, not only was the removal of iron and fluoride, but also the removal of cadmium, lead and chromium examined under upflow expanded bed conditions at pH 6.3 using the above mixture of illite-basic fly ash and lime.

The iron was reduced from 0.3 mg/L to 0.12 mg/L with no suggestions of breakthrough even after 140 liters of waste water were added to the lysimeters (Figure 13). This is comparable to concentration levels achieved in the lysimeter effluent under gravitational flow. Yet, the flows obtained with the expanded bed operation is some two orders of magnitude greater. However, there is some leaching of iron from the illite encountered initially which is indicated by the fact that the iron concentration in the effluent is higher than that in the influent (Figure 13). However, this should pose no problems in practice since this initial leaching can be regulated by the addition of adequate amounts of basic fly ash (4).

The combination of illite-basic fly ash and lime was found to reduce the fluoride in the influent sample from 5.8 mg/L to below 1 mg/L (Figure 14) of which is comparable to that achieved under gravitational flow conditions (Figure 8). However, it should be noted that after approximately 40 liters of waste water has been passed through the lysimeters, the fluoride concentrations in the effluent begin to increase until breakthrough is achieved. This is due to inadequate contact time between the sorbent and waste water. Whenver this occurs, a gradual rather than a rapid increase in the influent fluoride concentration to breakthrough is observed. The contact time in these lysimeters operated at flows of 22.4×10^{-3} L/min was determined to be 1.6 hours which is significantly less than the 6 hours required for maximum removal of the fluoride.

The combination of illite-basic fly ash and lime was found to reduce the lead concentration in the lysimeter influent from 0.12 mg/L down to approximately 0.013 mg/L (Figure 15). The chromium concentration was reduced from 0.05 mg/L down to approximately 0.015 mg/L (Figure 16), and the cadmium was reduced from 0.015 mg/L down to about 0.01 mg/L (Figure 17). It should be noted that this is being achieved under slightly acidic conditions (pH = 6.3) and at flows that are some 2 orders of magnitude greater than that which can be achieved by gravity.

Sorbent Addition Versus Upflow Expanded Bed

It is apparent from the previous results that significant removals of iron, fluoride, lead, cadmium and chromium in the waste stream can be achieved. This treatment can be carried out by either adding the illite/fly ash/lime sorbent combination directly to the waste stream or by operating the sorbent bed in the upflow expanded bed mode. The addition of sorbent directly to the waste stream leads to better control of the concentration of pollutants in the treated effluent. The amounts of the sorbent combination that is added directly to the waste stream defines the pollutant concentration levels after treatment. However, maximum removals of the pollutants such as iron and fluoride occur with the use of sorbent beds. This results in a specific weight of sorbent treating less volume of waste water in a bed than that which can be treated by adding the same amount of sorbent mixture to the waste stream. Thus, the use of a sorbent bed can lead to greater removals than which may be desired. For example, the addition of 10 grams of

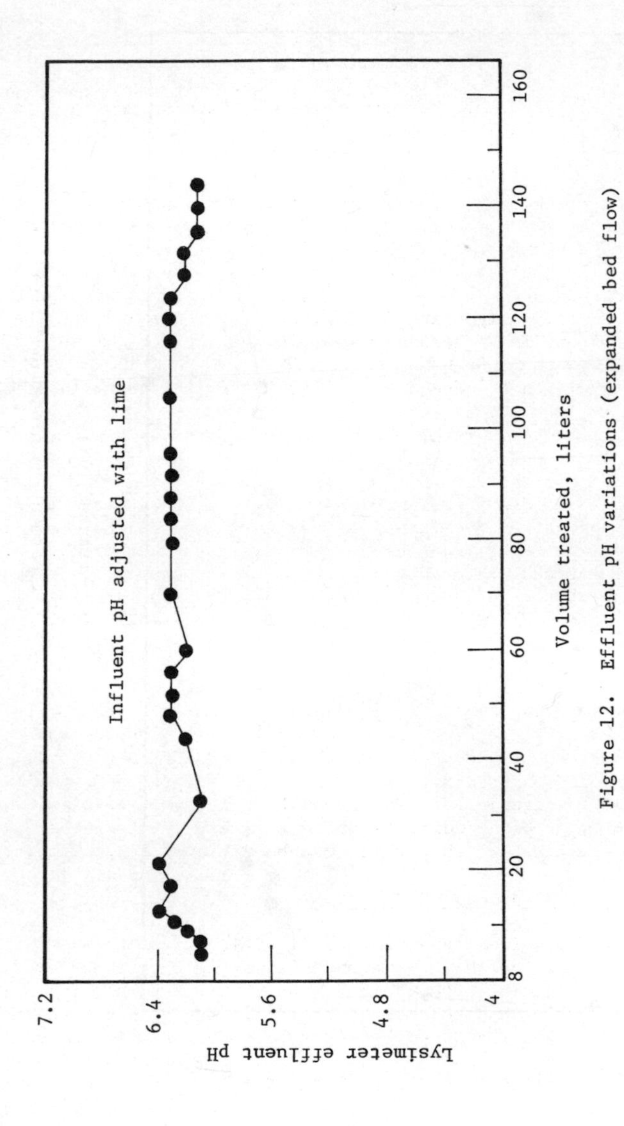

Figure 12. Effluent pH variations (expanded bed flow)

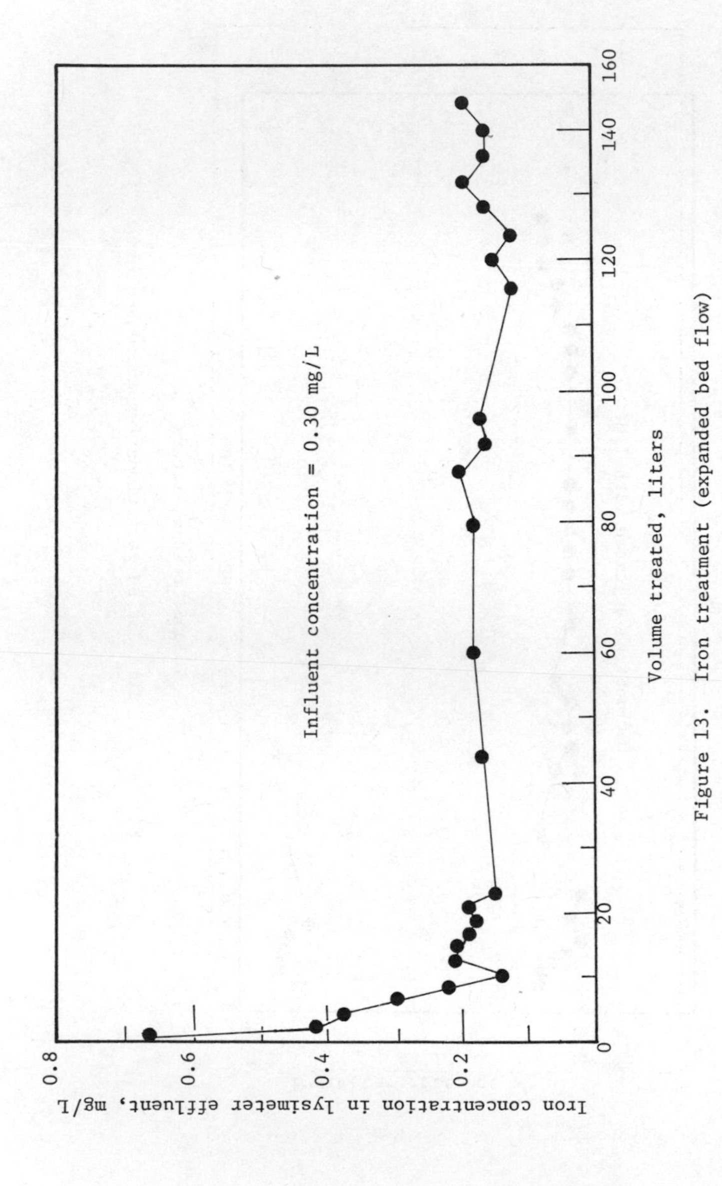

Figure 13. Iron treatment (expanded bed flow)

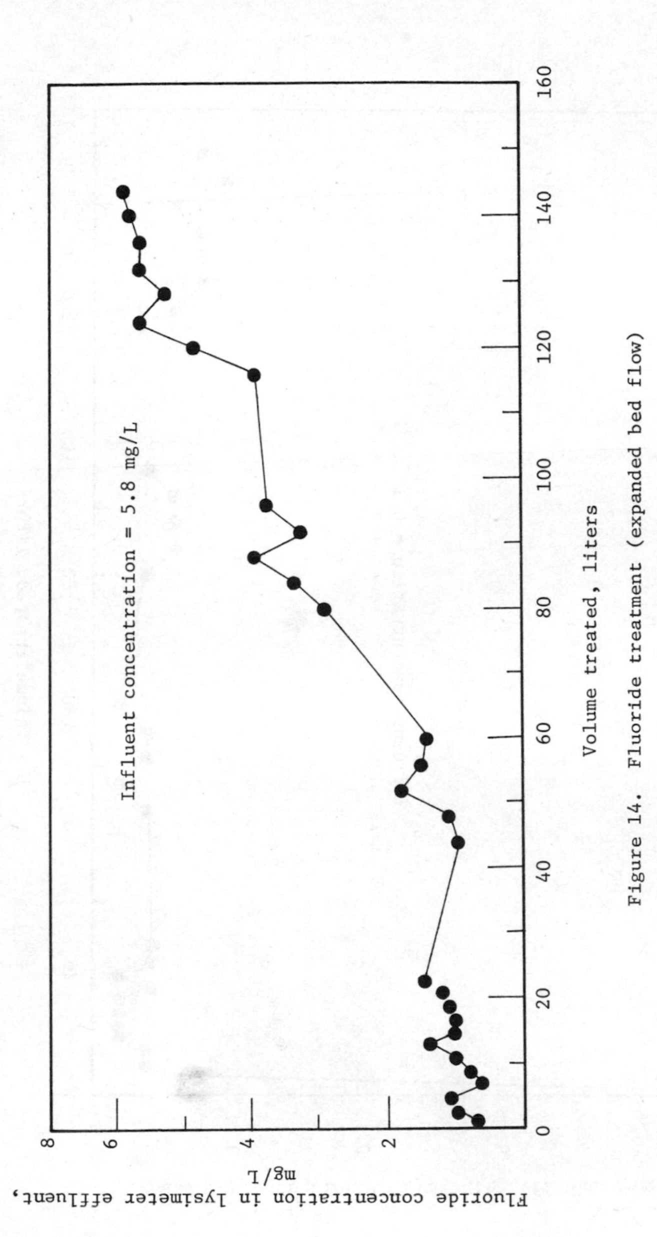

Figure 14. Fluoride treatment (expanded bed flow)

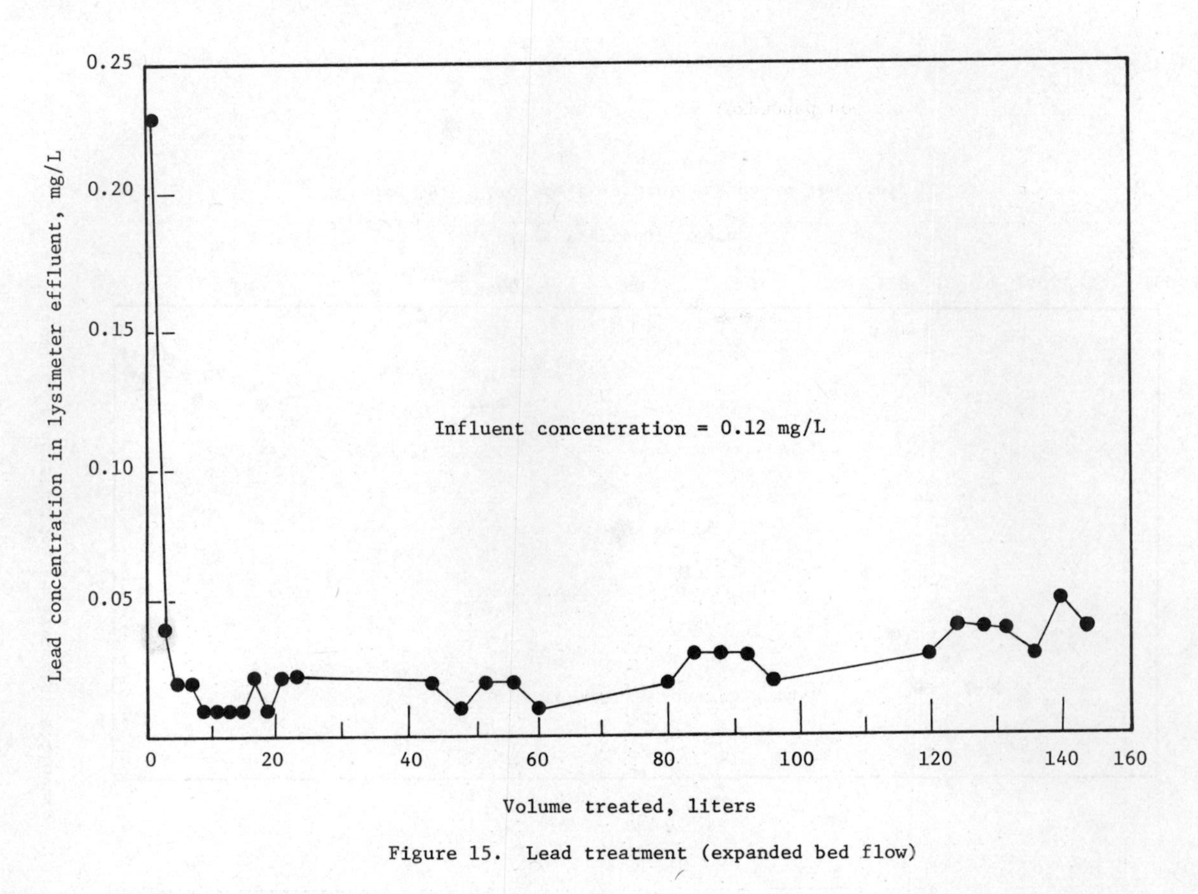

Figure 15. Lead treatment (expanded bed flow)

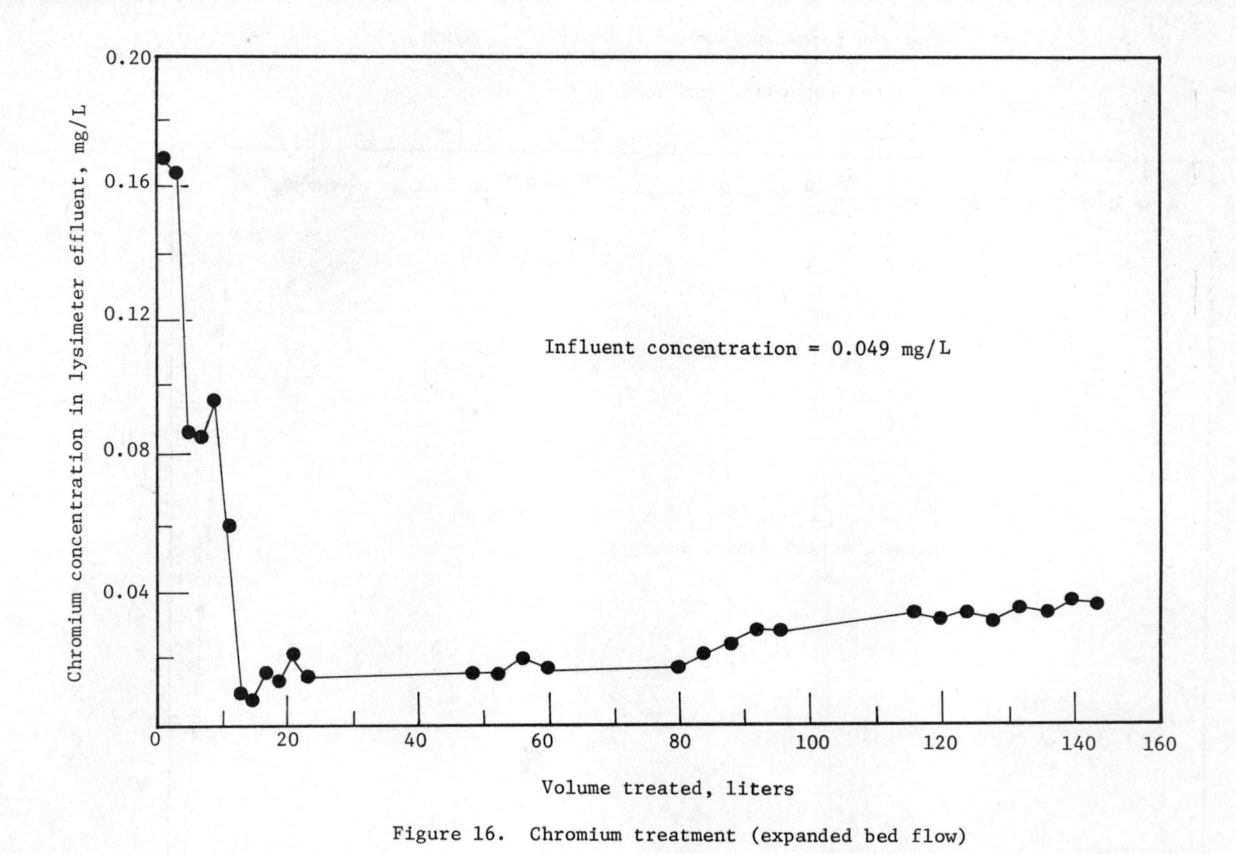

Figure 16. Chromium treatment (expanded bed flow)

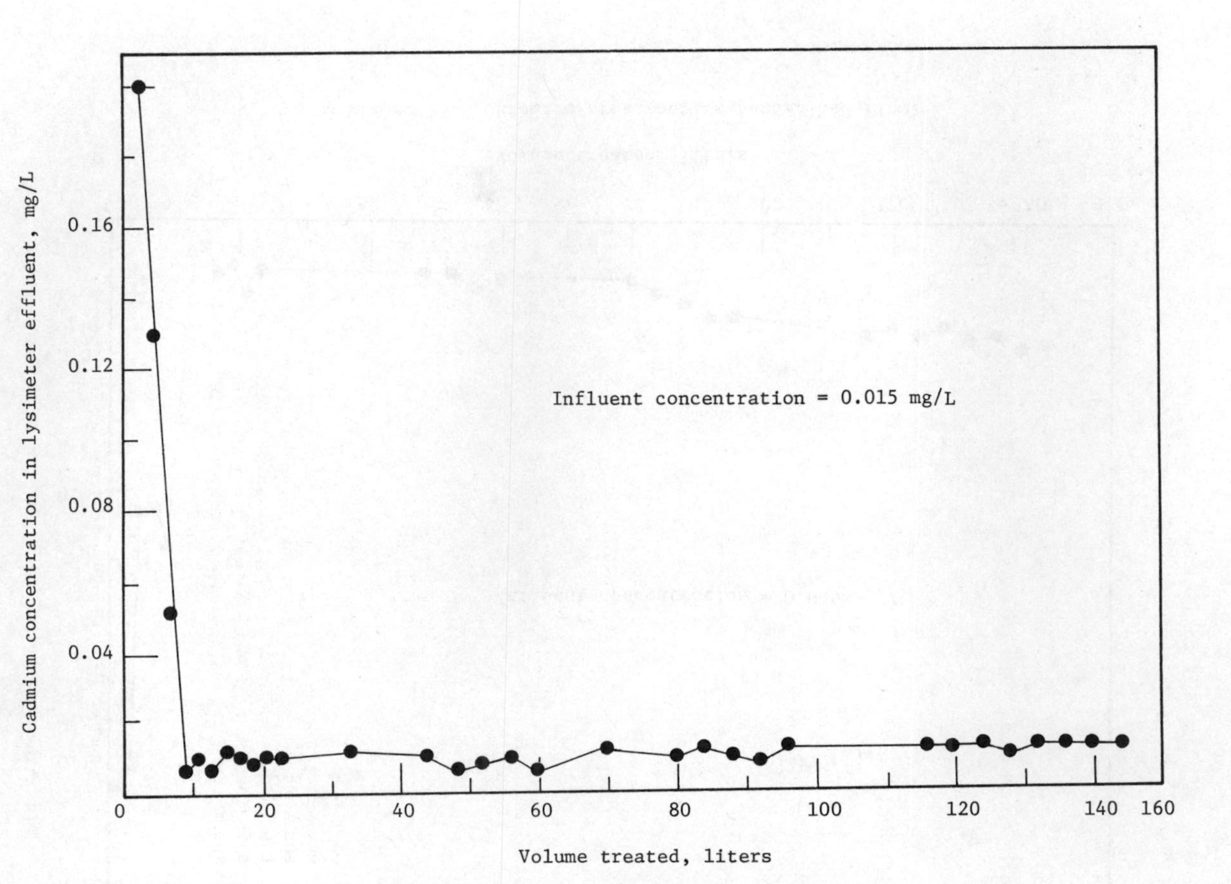

Figure 17. Cadmium treatment (expanded bed flow)

sorbent directly to 1 liter of waste water will reduce the fluoride from 5.8 mg/L to a desired fluoride effluent concentration of 1.5 mg/L, whereas the same amount of sorbent in a bed will reduce the effluent combination to below 1.0 mg/L but treat only 0.9 liter of waste water.

Settling Characteristics of Sorbent

The sorbents added to the waste stream can be removed in sedimentation basins provided that adequate settling rates are encountered. Settling tests carried out on the illite/basic fly ash/lime sorbent combination by measuring the interface between the suspension and clear liquid versus time results in a straight line (Figure 18). These results translate into a loading ratio of 1.78×10^4 $L/m^2/day$ which is comparable to those generally encountered in the settling basins used in conventional municipal waste treatment systems.

Cost Comparison Between Illite and Activated Alumina

In reducing the fluoride concentration from 25 mg/L to 5 mg/L, both activated alumina and illite exhibit optimal removal capacity at pH = 6.3 approximately. However, the removal capacity at this pH is 5 mg/g for activated alumina as compared to 1 mg/g for illite.

In terms of material cost for treatment of fluoride in the case of without regeneration, the use of illite is far less expensive than that of activated alumina. The cost of illite is about 5 cents per pound, whereas the cost for activated alumina is about 30 cents per pound. Thus, the material costs associated with the use of illite to obtain a fluoride effluent concentration of 5 mg/L would be some 12 times less than that associated with the use of activated alumina, even though the activated alumina is five times more effective for removing the fluoride.

The use of activated alumina for the treatment of fluoride generally requires regeneration because of the high cost of alumina. The regeneration involves the treating of the spent activated alumina with sodium hydroxide followed by sulfuric acid to reduce the pH to 6.4 for maximum removal of fluoride. A calcium fluoride sludge is generated as a result of the treating of sodium hydroxide with calcium ion to pricipitate the fluoride as CaF_2. Material cost associated with the addition of the illite/basic fly ash/lime combination to the waste stream without regeneration to reduce the fluoride concentrations from an average of 6.4 to 1.5 mg/L is 45 cents per 3.8×10^3 liters (1000 gallons). The spent sorbents can be discarded because the sorbed pollutants in the spent sorbents are held and released to the environment when in contact with water at levels that pose no threat to ground and surface waters. For example, repeated washing of a 10 gram and 20 gram of spent sorbent sample with 250 ml of water results in soluble fluoride concentrations in the initial washing that does not exceed 1.1 mg/L. Continued washing of the spent sorbent results in a decrease in the fluoride concentration in the rinse to below 1.0 mg/L which is generally considered acceptable for potable water supplies.

Regeneration of the spent sorbents using sulfuric acid results in a reduction in material cost to 13.3 cents per 3.8×10^3 liters. However, the regeneration process results in the generation of a metal hydroxide, CaF_2, $CaSO_4$ sludge that must be disposed of when the spent sulfuric acid containing the fluoride and heavy metals is neutralized.

In comparison, the material cost associated with the use of activated alumina with regeneration to reduce the fluoride concentration in potable water supplies is twice as much. In 1953, the chemical cost associated with the removal of fluoride in the concentration range encountered in the feldspar waste stream using activated

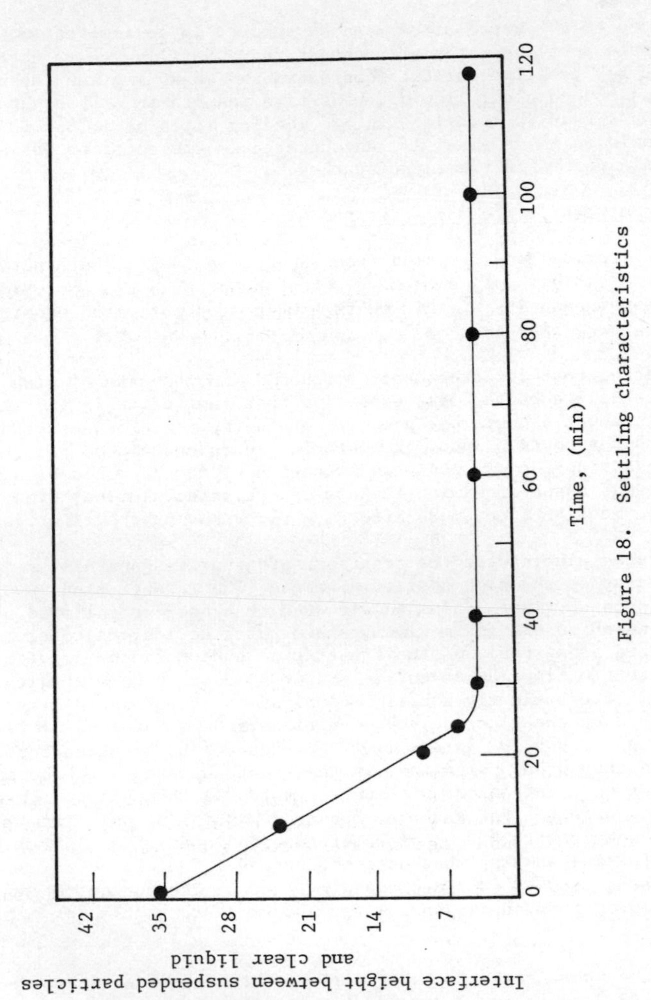

Figure 18. Settling characteristics

alumina with regeneration was reported to be 5.2 cents per 3.8×10^3 liters. If it is assumed that the average inflation rate for 26 years is 6 percent, the present chemical cost for this treatment is estimated to be 25 cents per 3.8×10^3 liters. A sludge also results that must be disposed of when the spent activated alumina is regenerated. The use of activated alumina without regeneration to avoid the sludge disposal problems would cost $4.95 per 3.8×10^3 liters.

CONCLUSIONS

Based on findings summarized in this investigation, the following conclusions are made:

1. The sorbent approach has been shown to be a simple industrial waste water treatment alternative capable of providing a high quality effluent. The illite-basic fly ash-lime sorbent combination is effective for industrial waste stream that contain heavy metals, toxic anions, and organics with a discharge rate in the order of 1 million gallons per day. The sorbent combination can be added directly to the waste stream and the spent sorbent removed by sedimentation at a loading rate of 17,700 L/m^2/day (432 gal/ft^2/day). Alternatively, the waste stream can be treated in a sorbent bed operated in an upflow expanded bed mode. A six-foot deep sorbent bed can treat the above flows at a loading rate of 7400 L/m^2(180 gal/ft^2). The sorbent combination reduced the iron, lead, chromium, cadmium, and fluoride concentration levels that are generally acceptable for potable water.

2. Reliance upon gravitational flow through the sorbent bed is impractical for treatment of a 3.8×10^6 L/day waste stream. The permeability of the sorbents limit the treatment to such low loadings as 575 L/m^2/day (14 gal/ft^2/day). This loading would require a bed surface area of 6.7×10^3 m^2 (7.2×10^4 ft^2) to avoid ponding.

3. Increasing the gravitational flow through a sorbent bed by increasing the particle size in the bed is not practical for treating large volumes of waste water. An increase in the particle size results in a decrease in the volume of waste water that can be treated with a given weight of sorbents. Apparently, the advantage of the increasing in bed pore volume by increasing the particle size is offset by a reduction in the sorbent capacity due to the decrease in particle surface area encountered with the larger particles.

4. The conditions required for maximum removal of the fluoride also provide effective treatment of iron, lead, chromium, and cadmium present in the industrial waste stream. The maximum sorbent capacity for the removal of fluoride occurs at a pH of 6.3 with a contact time of 6 hours between the sorbents and waste water.

5. The material costs associated with the use of the illite/basic fly ash/lime sorbent for treating 3.8×10^3 liters (1000 gallons) of industrial waste stream with regeneration is three times less than that associated with the use of activated alumina. These costs amount to 13 cents per 3.8×10^3 L and 45 cents per 3.8×10^3 L for the sorbent combination and activated alumina, respectively. The difference is primarily associated with the use of sodium hydroxide in the regeneration of activated alumina.

6. The sorbent combination may be disposed of when spent to eliminate the problem associated with the disposal of a metal hydroxide CaF_2 sludge from the regeneration process. The spent sorbent contaminants do not pose any threat to ground or surface waters. Repeated washing of different amounts of sorbents did not indicate any fluoride concentration above 1.1 mg/L in the rinse water. However, replenishing the spent sorbents with unused sorbents raises the treatment costs to 45 cents per

3.8×10^3 liters. Disposal of spent activated alumina without regeneration would amount to $4.95 per 3.8×10^3 liters.

REFERENCES

American Public Health Association, American Water Works Association and Water Pollution Control Federation (1976). Standard Methods for the Examination of Water and Wastewater, 14th Edtion. American Public Health Association, Washington, D.C.

Babich, H. and G. Stotzky (1977). Reduction in toxicity of cadmium to microorganisms by clay minerals. Appl. Environ. Microbiol., 3, 696-705.

Baker, R. and M. Hah (1971). Pyridine sorption from aqueous solutions by montmorillonite and daolinite. Water Research, 5, 839-848.

Bhargava, R. and M.P. Khanna (1974). Removal of detergents from wastewater by adsorption on fly ash. Indian J. Environ. Health, 16, 109-120.

Bittell, J.G. and R.J. Miller. Lead, cadmium and calcium selectivity coefficients on a montmorillonite, illite and kaolinit. J. Environ. Qual., 3, 250-253.

Chan, P.C., R. Dresnack, J.W. Liskowitz, A. Perna and R. Trattner (March 1978). Sorbents for Fluoride Metal Finishing and Petroleum Sludge Leachate Contaminant Control. EPA - 600/2-78-024, U.S. Environmental Protection Agency, Washington, D.C.

Culp, R.L. and H.A. Staltengerg (1958). Fluoride reduction of LaCrosse, Kan. A.W.W.A., 50, 423-431.

Damanakia, M., D.S.H. Drennan, J.D. Fryer and K. Holly (1970). The adsorption and mobility of paraquet on different soils. Weed Research, 10, 264-277.

Emig, D.D. (1973). Removal of heavy metals from acid batch plating wastes by soils Diss. Abst., B, 2661.

Griffin, R.A., K. Cartwright, N.F. Shrimp, J.D. Steele, R.R. Ruch, W.A.White, G.M. Hughes and R.H. Gilkeson (1976). Aternation of pollutants in municipal landfill leachate by clay minerals: column leaching and field verification. Environ. Geol. Notes, 78, 1-34.

Griffin, K.A., R.R. Frost, A.K. Au, G.D. Robbinson and N.F. Shrimp (1977). Attenuation of pollutants in municipal landfill leachate by clay minerals: heavy-metal adsorption. Environ. Geol. Notes, 79, 1-47.

Kliger, L. (1975). Parathion recovery from soils after a short contact period. Bull. Environ. Contamin. Toxicol., 13, 714-719.

Liao, C.S. (1970). Adsorption of pesticides by clay minerals. A.S.C.E. Sanitary Eng. Div., 96, 1057-1078.

Lub, M. and R. Baker (1971). Sorption and desorption of pyridine-clay in aqueous solution. Water Research, 5, 849-858.

Meier, F.S. (1953). Defluoridation of municipal water supplies. A.W.W.A., 45, 79-83.

Nelson, M.D. and C.F. Quarino (1970). The use of fly ash in wastewater treatment and sludge conditioning. J.W.P.C.F., 42, R-125-135.

Rios, C.B. Removing phenolic compounds from aqueous solutions with absorbents. U.S. Patent #2, 937, 142.

Savinelli, E.A. and A.P. Black (1968). Defluoridation of water with activated alumina. A.W.W.A., 50, 33-44.

37

A SIMULATION OF THE EFFECTS OF BACTERIA, ORGANICS AND SALT SOLUTIONS ON URANIUM MINE MILL TAILINGS FROM ELLIOT LAKE, ONTARIO

M. Silver and G. M. Ritcey

Process Metallurgy Section, Extractive Metallurgy Laboratory, Mineral Sciences Laboratories, Canada Centre for Mineral and Energy Technology, Energy, Mines and Resources, 555 Booth Street, Ottawa, Ontario K1A OG1, Canada

ABSTRACT

The effect of iron-oxidizing bacteria and organic solvent extraction reagents on uranium mine mill tailings from Elliot Lake, Ontario were examined to establish a model to predict acid formation and radium-226 leaching. In a series of lysimeter tests simulating abandoned tailings, accelerated by a nine-fold exposure to sunlight and darkness, wind and the amount of precipitation normally associated with the area at 18±4°C, the elution of radium-226, thorium, uranium, sulfate, total iron and the pH were monitored. The tests were conducted in the presence of bacteria either in the absence or presence of a mixture of kerosene, a tertiary amine and isodecanol applied to about one tonne of fresh tailings; the bacteria in a control lysimeter box were inhibited with hypochlorite bleach. The patterns of the elution of radium-226, total iron, sulfate and uranium and the pH changes with time were similar in both lysimeter boxes containing bacteria but were different in the lysimeter box to which the bleach was applied. The proportion of radium-226 eluted in about 750 days (or 19 simulated years) was 2.3-2.4% from the lysimeter boxes in which the bacteria were active and 3.4% in which the bacteria were inhibited with bleach. The tertiary amine was completely eliminated within 91 days (or 2.3 simulated years) and the kerosene within 1½ years (or 10.8 simulated years) from the lysimeter box to which these were added.

In a separate experiment using tailings from the same source, about half the radium-226 was removed by treatment with salt solutions and water. Radium-226 in the effluents of columns packed with these treated tailings was decreased by about 90% compared with columns packed with untreated tailings.

KEYWORDS

Iron-oxidizing bacteria; uranium tailings; radium-226; lysimeter; solvent extraction reagents.

INTRODUCTION

Sulfuric acid leaching is the principal process that is used in Canada for the extraction of uranium from ores. Although satisfactory for the recovery of uranium this process results in the discarding of a lime neutralized (pH 9-10) residue containing thorium, radium and often pyrite into surface disposal areas. Subsequent oxidation and long term weathering results in the neutralization of the lime and, in the presence of pyrite, the eventual formation of acid conditions. This acid generation and subsequent seepage from the tailings is a source of contamination by sulfate and environmentally objectionable metals such as iron, thorium and uranium. Radium, solubilized and reprecipitated during the leaching process in the uranium extraction circuit, is deposited with the solid residues in the tailing dumps from which it is slowly leached releasing unacceptable levels of radium-226 (Itzkovitch and Ritcey, 1979).

In the presence of dissolved oxygen and iron-oxidizing bacteria, sulfide minerals such as pyrite are oxidized according to the following series of reactions:

$$48\ FeS_2 + 180\ O_2 + 24\ H_2O \xrightarrow{\text{bacteria}} 24\ Fe_2(SO_4)_3 + 24\ H_2SO_4 \qquad (1)$$

$$12\ FeS_2 + 12\ Fe_2(SO_4)_3 \longrightarrow 36\ FeSO_4 + 24\ S^\circ \qquad (2)$$

$$24\ S^\circ + 36\ O_2 + 24\ H_2O \xrightarrow{\text{bacteria}} 24\ H_2SO_4 \qquad (3)$$

$$36\ FeSO_4 + 9\ O_2 + 18\ H_2SO_4 \xrightarrow{\text{bacteria}} 18\ Fe_2(SO_4)_3 + 18\ H_2O \qquad (4)$$

$$30\ Fe_2(SO_4)_3 + 120\ H_2O \longrightarrow 20\ H\ Fe(SO_4)_2.2Fe(OH)_3 + 50\ H_2SO_4 \qquad (5)$$

$$60\ FeS_2 + 225\ O_2 + 150\ H_2O \longrightarrow 20\ H\ Fe(SO_4)_2.2Fe(OH)_3 + 80\ H_2SO_4 \qquad (6)$$

with the formation of sulfuric acid and basic iron sulfate precipitates, such as jarosite (Lundgren and Silver, 1980). Ferric sulfate, formed as an intermediate in this series of reactions can oxidize residual uranium thus releasing it into solution.

When solvent extraction is used for the selective recovery of a high grade uranium product, some of the reagents are lost from the process and are discarded with the tailings (Ritcey, Lucas and Ashbrook, 1974). These reagents, consisting of tertiary amine, kerosene and a long-chain alcohol modifier, can be present due to various types of solvent losses (Ritcey, Slater and Lucas, 1973). Each has a separate solubility level in the aqueous effluent and thus their proportions in the effluent differ from the original solvent composition. Such organic component can constitute an environmental problem to aquatic life, and bioassay results have indicated a wide variation of toxicity depending upon the reagent, its concentration, and the type of fish tested (Ritcey, Lucas and Ashbrook, 1974). Previous research (Gow and colleagues, 1970; Parsons, 1974; Torma and Itzkovitch, 1976) have indicated that these and similar compounds altered the metabolism of the iron-oxidizing bacteria. To study further the influence of these reagents on the interactions of iron-oxidizing bacteria with freshly produced tailings, an experiment was devised in which this could be monitored in an accelerated test simulating natural conditions of abandoned tailings, the results of which would aid in future experimental work on the impoundment and treatment of tailings. Thus it would be possible to predict the production of sulfate and the solubilization of metallic elements by the action of iron-oxidizing bacteria, the effects of organic uranium milling reagents on these bacteria, and the leaching of radium-226 and other chemical constituents caused by water percolation through a tailings area.

To substantially decrease the problem with radium-226 pollution in seepage from uranium mine mill tailings, a number of add-on processes have been suggested. Amongst these is the removal of radium-226 by sequential washings with water and salt solutions. Silver and Andersen (1980) demonstrated that 50-75% of this radionuclide could be removed from Elliot Lake uranium tailings by variations of this procedure. In this communication, it is shown that the radium-226 concentrations in the effluents of the tailings treated in this manner is decreased by 90% in comparison to untreated tailings.

LYSIMETER STUDIES

Experimental

Three reinforced plexiglass lysimeter boxes (Fig. 1) with approximate dimensions

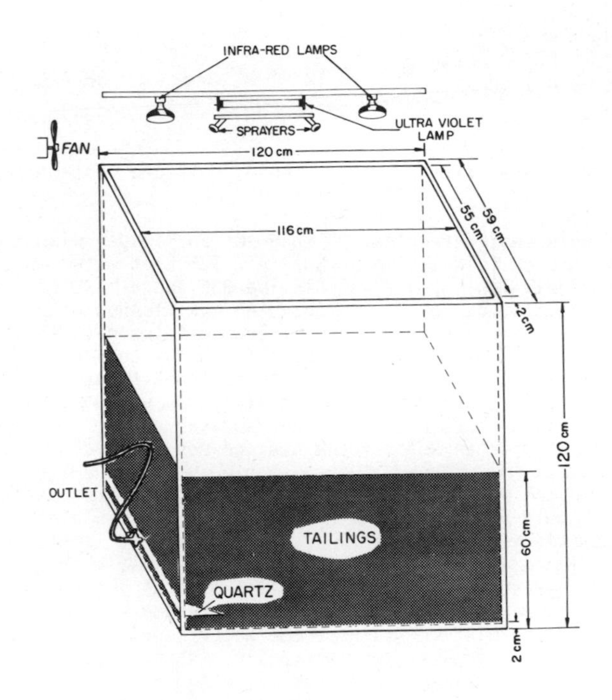

Fig. 1. Lysimeter Box Details.

of 120 cm by 55 cm by 120 cm (length, width and height) were constructed and loaded to a depth of approximately 60 cm with fresh solid uranium mine mill waste. The tailings were obtained by decantation from the tailing neutralization discharge of the uranium milling circuit of the Rio Algom Quirke Lake Mine at Elliot Lake, Ontario, immediately after addition of lime. A portion of the finer tailings was lost during this procedure and is the reason for the relatively low radium-226 content. This material contains, on a dry weight basis, 2.5-3.5% pyrite, 0.72% sulfate (as sulfur), 2.82% total iron, 0.007% uranium, 0.025% thorium and 170 pCi/g radium-226. The contents of each lysimeter box are described in Table 1.

TABLE 1 Lysimeter Box Contents

	Lysimeter Box 'A' Bacteria and Organics	Lysimeter Box 'B' Bacteria Only	Lysimeter Box 'C' Bleach
Dimensions (cm)	124.5 x 59.5 x 58.0	124.5 x 59.5 x 58.0	116.5 x 55.5 x 59.5
Volume (cm^3)	429 650	429 650	384 712
Weight (kg)*	1 047	1 047	938
Radium-226 (μCi)	178	178	159
Sulfate (g as S)	7 539	7 539	6 75
Uranium (g)	73	73	66
Total iron (g)	29 528	29 52	26 440
Thorium (g)	262	262	234

*Calculated dry weight. Specific gravity - 2.6548, Moisture content 8.8%

The lysimeter boxes were designated 'A', 'B' and 'C' and loaded with approximately 1 tonne of tailings in each. To lysimeter box 'A', 175 L of water containing 90 mL/L of kerosene (Shell 140), 5 gm/L of Alamine 336 and 5 mL/L of isodecanol was added. After equilibration for 2 days, box 'A' was flushed with 175 L of water and inoculated with a washed 72-hour culture of iron-oxidizing thiobacilli isolated at Elliot Lake. Boxes 'B' and 'C' were flushed with 175 L of water, and box 'B' was inoculated similarly. Modified simulated conditions were automatically controlled with exposure to infra-red and ultra-violet lamps and a fan which were activated every 1½ hours, and 5.2 L of water was sprayed on each box every 8 hours, resulting in an application rate 9 times the average annual precipitation of the Elliot Lake region (966.1 mm). Thus, if the leaching of the chemical constituents were reliant only upon the water applied to the tailings, a chronological augmentation of nine fold should be achieved and 40 days of experimental time would be equivalent to 1 simulated year. The water supplied to box 'C' contained 1-3 mL/L of hypochlorite solution (Javex bleach) containing 5% available chlorine to inhibit bacterial action and thus sulfate generation whilst maintaining oxidizing conditions. The experiment was conducted at ambient temperature (18±4°C) with extraneous light being excluded by black curtains. The water level in the lysimeter boxes was maintained within 10 cm (or less) of the tailings surface by a syphon tube from the effluent port. Liquid samples, taken from the effluent ports at intervals of approximately 2 weeks, were analyzed for total iron, pH, sulfate, uranium, thorium and radium-226. The effluent and core samples taken from the centre and each corner of lysimeter box 'A' were periodically analyzed by gas liquid chromatography for kerosene. Due to interference of the kerosene on the chromatographic spectrum of isodecanol, this latter compound could only be detected but not quantitated. Alamine 336 was analyzed by a colorimetric procedure (Barkley, 1970).

Note that all results are reported on the simulated time basis of 1 year = 9 years.

Results

Figure 2 shows the pH and concentrations of radium-226, sulfate and total iron in the effluents of the three lysimeter boxes over a period of about 750 days, simulating about 19 years in which time 11.5 tonnes of water were applied. Should the assumption that a chronological acceleration of nine-fold be valid, then each 40-day period of the experiment would correspond to one year under natural conditions. No correction to these results is applied to compensate for evaporation of 9.3±2.3%. The total amount and proportion of each constituent initially present in each of these three lysimeter boxes removed during the course of this experiment is summarized in Table 2. Radium-226 is constantly being eluted, the concentrations in the effluents from lysimeter boxes 'A' and 'B' being between 100 and 300 pCi/L for about the first 7.5 simulated years, increasing thereafter and remaining around 500 pCi/L. The radium-226 content of the effluent of lysimeter box 'C' is between 85 and 240 pCi/L. The amounts and proportions of the radium-226 eluted from lysimeter boxes 'A', 'B' and 'C' are 4.0 μCi and 2.3%, 4.1 μCi and 2.4%, and 5.4 μCi and 3.4% respectively.

After an initial release of sulfate, the elution patterns of this anion and total iron of the effluents from each of the three lysimeters are similar, with the pH decrease correlating to the cumulative sulfate elution. These elution patterns are similar in boxes 'A' and 'B', whereas those of box 'C' are distinctly different. The amounts and proportions of sulfate eluted from lysimeters 'A', 'B' and 'C' are 1473 g (45.9 moles) and 19.8%, 1346 g (42.0 moles) and 17.8%, and 908 g (28.3 moles) and 13.4% respectively. The amounts and proportions of total iron eluted from these three boxes are 923 g (16.5 moles) and 3.1%, 1002 g (18.0 moles) and 3.4%, and 249 g (4.5 moles) and 0.9% respectively. The molar ratios of total iron to sulfate in the effluents of box 'A' is 1:2.8, of box 'B' is 1:2.3 and of box 'C' is 1:6.4.

Thorium was not extracted from the tailings contained in any of the lysimeters, the concentrations in the effluents being consistantly less than 0.1 mg/L. After an initial release, uranium was not found in the effluents of lysimeters 'A' and 'B' until after 7.5 simulated years and from lysimeter 'C' not until after 17.5 simulated years. When uranium is detected in the effluents, it is rarely found in concentrations greater than 1 mg/L. The total amounts and proportions of the uranium of the tailings eluted from lysimeter boxes 'A', 'B' and 'C' are 3.9 g and 5.3%, 2.5 g and 3.5%, and 1.1 g and 1.5% respectively.

The mixture of organic compounds originally introduced into lysimeter box 'A' was almost completely eliminated, either by voidage through the effluent or by evaporation within the first 11 simulated years. Kerosene, the largest component in the mixture of organic compounds, was eliminated from the tailings, its concentration in the tailings decreasing almost by 80% in the first 4 simulated years and by more than 95% in the first 10.8 simulated years (Table 3). Examination of the core samples from the centre and at the corners of lysimeter box 'A' indicated migration toward the effluent outlet. Isodecanol, which could be detected but not quantified due to the interference of the chromatographic spectrum of kerosene, was always found with the kerosene. The elution of Alamine 336 was observed to be much more rapid (Table 4); of the original 875 mg of this reagent applied to the tailings, 30.3% was voided during the initial 0.25 simulated years, 36.1% during the next 0.7 simulated year, 31.2% during the next 0.9 simulated year and 5.1% during the next 0.5 simulated year. No amine could be detected in the effluent after 2.3 simulated years.

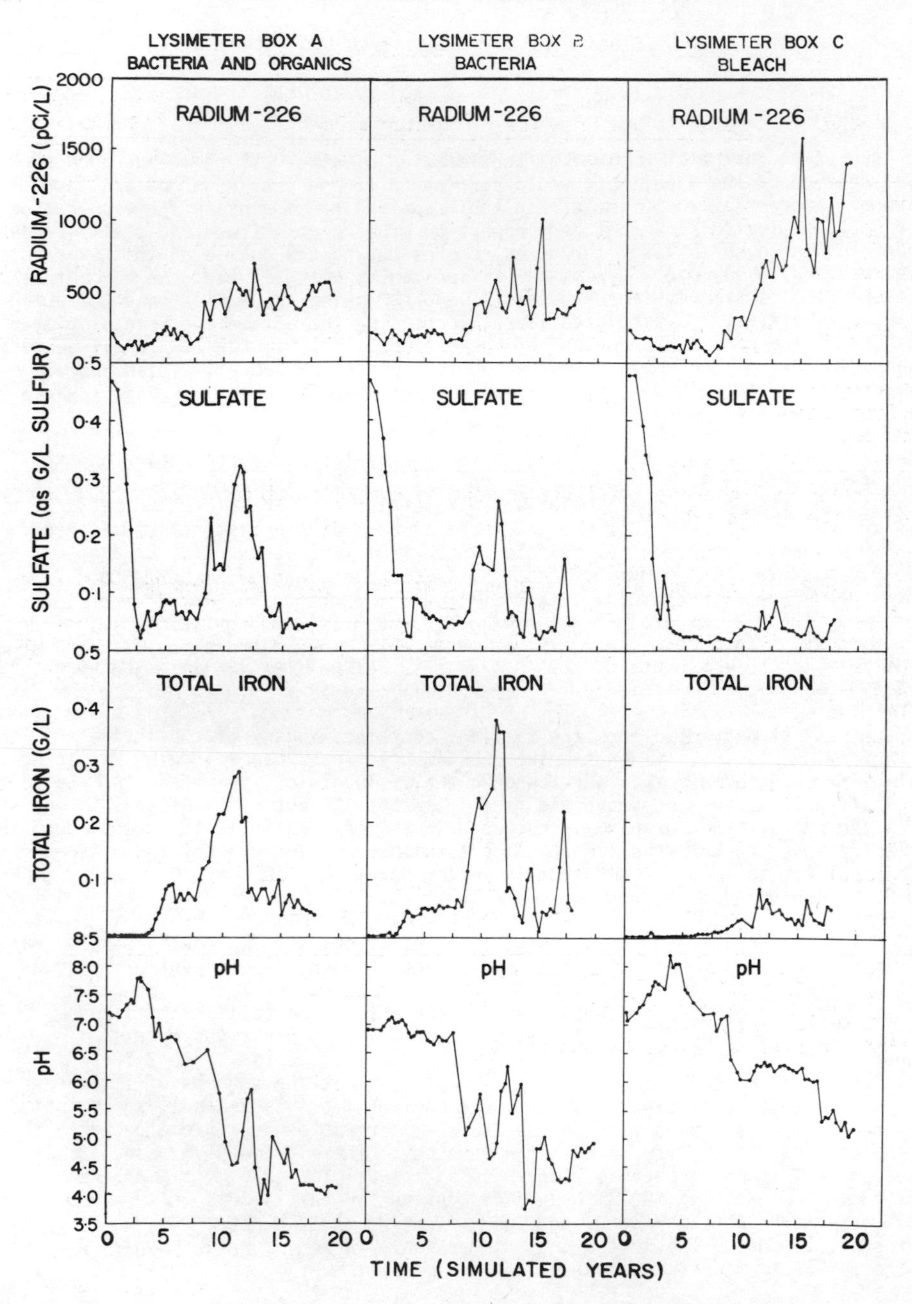

Fig. 2. Elution of radium-226, sulfate and total iron and change in pH in the lysimeter box effluents with respect to simulated time.

TABLE 2 Elution of Components from Lysimeter Boxes

Component	Lysimeter Box 'A' Bacteria and Organics		Lysimeter Box 'B' Bacteria Only		Lysimeter Box 'C' Bleach	
	Amount	Proportion*	Amount	Proportion*	Amount	Proportion*
Uranium	4.3 g	5.8%	3.1 g	4.3%	1.1 g	1.5%
Radium-226	4.0 μCi	2.3%	4.1 μCi	2.4%	5.4 μCi	3.4%
Sulfate (as S)	1.5 kg 45.9 moles	19.8%	1.4 kg 42.0 moles	17.8%	0.9 kg 28.3 moles	13.4%
Total iron	923 g 16.5 moles	3.1%	1002 g 18.0 moles	3.4%	249 g 4.5 moles	0.9%
Total iron: sulfate ratio molar	---	1:2.8	---	1:2.3	---	1:6.4

*Proportion of component initially present in the original tailings leached during the course of the experiment.

TABLE 3 Elution of Kerosene and Isodecanol from Lysimeter Box 'A'

Test Period (days)	Simulated Time (years)	Concentration of Kerosene	Presence of Isodecanol*
14**	0.35	1.1 mg/L	+
23**	0.58	0.5 mg/L	+
35**	0.88	0.2 mg/L	+
105**	2.63	0.025 mg/L	+
163**	4.08	0.025 mg/L	+
163***	4.08	2.81 ng/g (20.2%)****	+
431***	10.78	0.81 ng/g (3.5%)****	-

* + = detectable; - = undetectable
** liquid effluent samples
*** solid core samples
**** % remaining in the tailings

TABLE 4 Elution of Alamine 336* from Lysimeter Box 'A'

Test Period (days)	Simulated Time (years)	Volume of Water (L)	Average Concentration (mg/L)	Total (mg)	% Eluted
10	0.25	331**	0.80	264.8	30.0
37	0.93	421	0.75	315.9	36.1
72	1.80	546.0	0.50	273.0	31.2
91	2.28	296.4	0.15	44.5	5.1
				898.2	102.7

*875 mg Alamine 336 initially applied to the tailings.
**Includes 175 L applied to equilibrate the lysimeter box.

REMOVAL OF RADIUM-226 FROM TAILINGS BY SALT WASHING

Experimental

For the removal of radium-226 by salt washing, a quantity of fresh Rio Algom Quirke Lake tailings was treated with water and a KCl solution as described previously (Silver and Andersen, 1980); 30 kg of tailings containing 200 pCi/g radium-226, 0.23% sulfate sulfur and 0.0065% uranium were exposed to 150 L of tap water with stirring for 30 min at ambient temperature (20±2°C) and filtered under 1 atmosphere of pressure through three thicknesses of Whatman #3 filter paper. The residue was retreated in the same manner with 150 L of 2 M KCl, filtered and washed with 150 L of tap water. After recovery by filtration, the residue contained 100 pCi/g radium-226, less than 0.02% sulfate sulfur and 0.0058% uranium. Two polycarbonate columns (10 cm by 135 cm) were each loaded to a height of 1 m with 12.5 kg (dry weight) of untreated or treated tailings and filled with water. Weekly samples of 1.75 L were withdrawn from the bottom of these columns while replenishing the water, and the pH concentrations of radium-226, total iron, sulfate and uranium in the effluents were determined.

Results

Table 5 shows that treating the tailings with 2 M KCl, preceeded and followed by washing with water, removed virtually all of the sulfate, very little of the uranium and iron, 50% of the radium-226 and 87.5% of the calcium. Of the 200 pCi/g of the radium-226 initially present, 0.6% was removed by the first water wash, 47.6% by the salt leach, and 1.6% by the second water wash.

Figure 3 shows the mobility of the remaining radium-226 in the column with 12.5 kg of unleached tailings containing 2.5 mCi of radium-226 and in the column with 12.5 kg of leached tailings containing 1.25 mCi of radium-226. The column containing the untreated tailings is, in effect, a small scale version of lysimeter box 'B' and is similar except for the size and the amount of water applied, which is 30 fold the annual precipitation at Elliot Lake, Ontario. About 2500 pCi of radium-226 was easily removed during the first two weeks with 3.5 L of water from each of these two columns. Although radium-226 continued to be eluted from the treated tailings, this was at the rate of approximately one-tenth the rate of removal from

TABLE 5 Leaching of Tailings with 2M KCl

Fraction	Component				
	Radium-226	Sulfate (as S)	Total iron	Calcium	Uranium
Fresh tailings	200 pCi/g	0.23%	2.75%	0.45%	0.0065%
1st Water Wash	205 pCi/L (0.6%)*	0.46 g/L	0.0002 g/L	0.552 g/L	0.0001 g/L
Residue	180 pCi/g (10.0%)	<0.02%	---	0.10%	0.0053%
KCl Leach	18800 pCi/L (47.6%)	0.036 g/L	0.04 g/L	0.071 g/L	<0.0001 g/L
Residue	105 pCi/g (47.5%)	<0.02%	---	0.08%	0.0060%
2nd Water Wash	640 pCi/L (49.2%)	0.009 g/L	0.0003 g/L	0.027 g/L	<0.0001 g/L
Residue	100 pCi/g (50.0%)	<0.02%	2.65%	0.06%	0.0059%

*Cumulative Extraction of Radium-226.

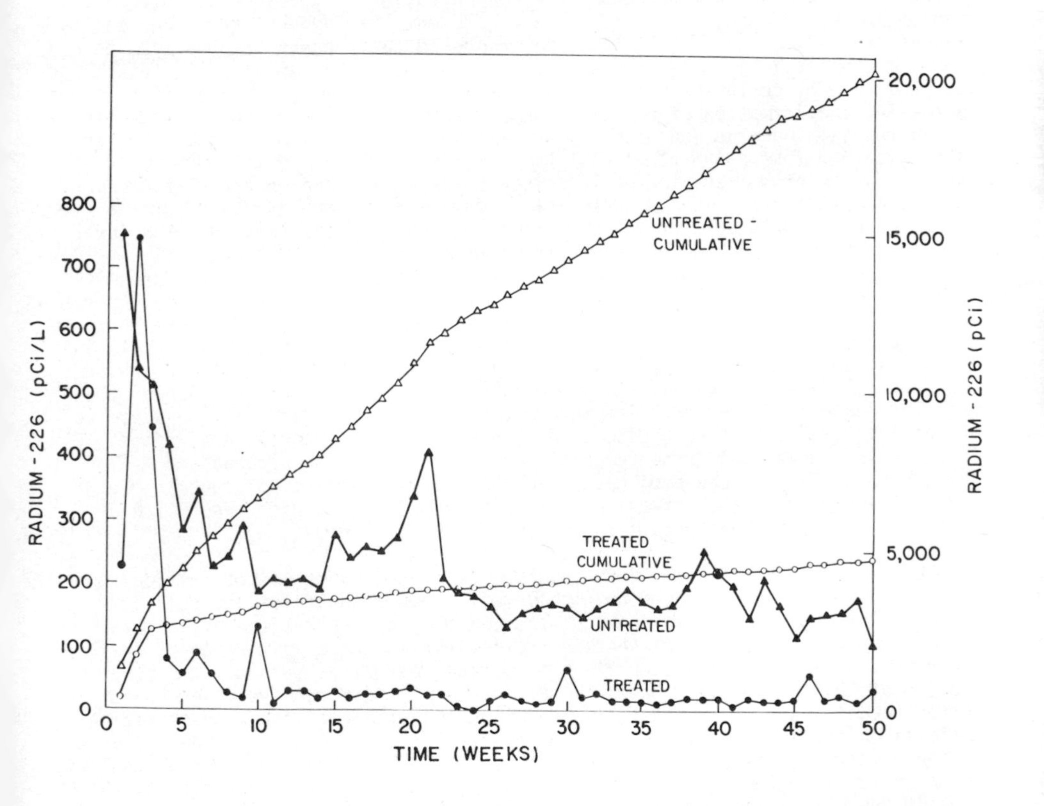

Fig. 3. Mobility (closed symbols) and cumulative removal (open symbols) of radium-226 from untreated (triangles) and treated (circles) uranium mine mill tailings.

the untreated tailings. After 50 weeks elution with a total of 87.5 L of water, 20 100 pCi of radium-226 was removed, which was 0.8% of that originally present in the untreated tailings; 4800 pCi of radium-226 was eluted in the same period from the treated tailings. The sulfate sulfur concentrations in the effluents from the fresh tailings remained around 0.5 g/L for the first 30 weeks and then decreased to around 0.1 g/L by the 45th week; in contrast, the sulfate sulfur from the treated tailings decreased from around 0.2 g/L to 0.01 g/L within the first ten weeks and remained at that concentration. The concentrations of total iron and uranium in the efflunets of both columns remained below 1 mg/L each.

DISCUSSION

The presence of the organic solvent extraction reagents kerosene, isodecanol and Alamine 336 that are used in the recovery of uranium have little effect on the solubilization of iron, sulfate, uranium, thorium or radium-226 from fresh Elliot Lake tailings in the presence of iron-oxidizing bacteria. Conversely, the presence of bleach used as a bacteriostatic agent significantly alters the pattern of radium-226 elution and decreases the concentration of total iron, sulfate and uranium in the effluents. We are aware at present that hypochlorite, due to its oxidant properties, may not be the ideal bacteriostatic agent for these tests and that a metabolic inhibitor, such as sodium azide might be superior for this purpose. Nevertheless, hypochlorite not only inhibits the iron-oxidizing bacteria thus decreasing the formation of soluble iron and sulfate from the iron sulfide minerals in the tailings and consequently preventing uranium solubilization, but this anion also may form soluble salts of the alkaline earth elements, such as radium. With the diminution of sulfate in the lysimeter boxes during the course of the experiment, the competition between these anions may be decreased in favour of the hypochlorite, the radium salt of which is more soluble than the sulfate. The inhibition of bacterial oxidation of iron sulfide minerals is reflected not only in lower concentrations of total iron and sulfate in the effluents in the presence of bleach but also in the slower decline in the pH of these effluents demonstrating less acid production. The solvent extraction reagents added to lysimeter box 'A' are eliminated almost entirely through voidage through the effluent, the Alamine 336 being depleted during the first 2.3 simulated years and the kerosene during the first 11 simulated years. Thus the movement of these reagents through the tailings could be discerned. The interaction of these reagents in the presence of uranium tailings has been discussed previously (Ritcey, Lucas and Slater, 1973), where it was shown that these reagents can be lost from the process and discharged to tailings because of upsets in the solvent extraction process. Such losses are caused by the formation of stable emulsions and high adsorption losses on the solid gangue particles.

As the tailings contain about 8.8% moisture by weight, some of the chemical constituents present in dissolved or highly mobile adsorbed states are readily eliminated from the lysimeter boxes. Thus, higher concentrations of sulfate, radium-226 and uranium are present in the effluents at the initiation of the experiment. With the exception of this initial removal of sulfate, the concentration of this anion and iron in the effluents of the three lysimeter boxes correlate with each other; this may be due to the action of the iron-oxidizing bacteria on the iron sulfide minerals present in the tailings. In the presence of the bacteriostatic reagent in lysimeter box 'C', not only are iron and sulfate found in lower concentrations but also the ratio of the soluble iron to sulfate is greatly decreased. Bacterial action also appears to be responsible for the solubilization of uranium, which after 18.8 simulated years is greater in the absence of bleach than in its presence, but is similar in all three lysimeter boxes during the initial stage of the experiment. During the first two simulated years, the initial removal of uranium is found to be 0.28 g or 0.4% from lysimeter box 'A', 0.27 g or 0.4% from lysimeter box 'B' and 0.49 g or 0.7% from lysimeter box 'C'. Thorium is never found in the lysimeter box effluents in concentrations greater than 0.1 mg/L.

The water applied to the lysimeter boxes on the simulated time cycle of light, dark, rain, is equivalent to about 8100 mm of precipitation per year, or nine-fold that of the Elliot Lake region. In experiments using similar tailing materials, Bryant, Cohen and Durham (1979) applied water equivalent to 492, 984 and 5620 mm of precipitation per year in a number of lysimeter column tests. Despite the differences in the volumes of water passed through these lysimeter columns, which were twice the depth of our lysimeter boxes, the concentrations of radium-226 in the effluents were not dissimilar to those reported in this present communication during the initial 7.5 simulated years. Thus it appears that the radium-226 concentration in these tailing effluents, which is limited by the dissolution properties of the salts of this radionuclide, is independent of the volume of water to which the tailings are exposed initially. Upon continued exposure to water, sulfate, the principal anion of radium salts in the tailings, is removed and the radium is thus rendered more mobile. Activity of the iron-oxidizing thiobacilli, however, results in the replenishment of the sulfate and thus tends to decrease this mobility. Calcium is also removed from the tailings, and the presence of this cation in the effluent might affect the solubility characteristics of radium salts.

Assuming that the elution of radium-226 from the tailings in the lysimeter boxes 'A' and 'B' has reached a steady state of 500 pCi/L of water applied and that all the radium-226 in the tailings are leachable (mobile), then all the radium-226 will be removed from the lysimeter boxes in about 560 simulated years. This does not take into consideration the formation of radium-226 by the nuclear disintegration of thorium-230.

In a study of an actual tailing dump at the abandoned Rio Algom Nordic mine in which the movement of radium-226 was monitored by a network of multilevel piezometers, Blair and colleagues (1980) showed that this radionuclide migrated downwards through the tailings with heavy metal ions, but much more slowly — about 0.05% the rate — than the ground water itself and such mobile ions as sulfate. Soluble radium-226 concentrations found in the tailings below the water table by these investigators ranged from 30 to 230 pCi/L. This concentration is similar to that found (unpublished observations) in the effluents of the two lysimeter boxes without hypochlorite after about 25 simulated years.

In an experiment such as the one presented here with an accelerated water flow, a number of deviations from natural conditions would be expected. An increase in the removal of sulfate and soluble iron, whether initially present in the tailings or formed by bacterial action, would result in a decrease in their concentrations in the tailings waters due to dilution. This would tend to increase the mobility of radium-226 and the concentration of this radionuclide would be elevated in comparison with natural conditions. Because of the lack of vertical distance in the lysimeter boxes, a well developed hydrogeological profile would be prevented from being formed which would also contribute to deviations from actual conditions. Comparison of the results of the present lysimeter study to those of outdoor tailing pit experiments (Murray and Okuhara, personal communication) illustrate some of these deviations. In the outdoor tailing pit experiments, both iron and sulfate were detected in the effluents in gram per litre quantities during the fourth year of these experiments, whereas in the lysimeter box effluents, these were detected in quantities at least an order of magnitude lower after 2.5-3 simulated years correlating well with the results of the lysimeter experiment. In both experiments, sulfate and iron are found in the effluents in concert with each other. As the Nordic tailing dump is in its 23rd year of existance, and as the simulated time of the lysimeter experiment is approximately 19 years, observations over the next year will reveal whether or not this type of test can be used for the prediction of the movement of the tailing constituents. Despite possible shortcomings of this experiment, we believe that it will yield results that will aid in the explanation of this phenomenon.

Various treatments of uranium mine mill tailings with salt solutions and water have been shown to remove up to 75% of the radium-226 (Silver and Andersen, 1980). Radium-226 removal is dependent upon the salt concentration, the solid-liquid ratio, and the number of times the material is treated sequentially. In the experiment illustrated in this communication, 50% of this radionuclide is removed. This treatment also results in the removal of almost 90% of the calcium and virtually all of the sulfate. The KCl solutions could be regenerated by precipitating about 99% of both the radium-226 and the sulfate with $BaCl_2$ and $Ca(OH)_2$. The radium-226 residual in the treated tailings is much less mobile in comparison with that in the untreated tailings; the average concentration of this readionuclide in the effluents from the treated tailings is 21±12 pCi/L compared with 193±58 pCi/L from those of the untreated tailings after the initial purge of the soluble radium-226 from the columns. As the salt treatment removes most of the more mobile radium from the tailings, further dissolution is greatly reduced, the mobility of the residual radium being governed by the slower solid diffusion process (Nathwani and Phillips, 1979). Thus, lower concentrations of radium-226 in the effluents from uranium tailings treated with salt solutions is observed.

CONCLUSIONS

From the results of this study, it is shown that there is no difference in the rates that radium-226, sulfate, iron and uranium are leached from Elliot Lake uranium tailings in the absence or presence of organic solvent extraction reagents. The inclusion of a bacteriostatic agent, hypochlorite, in the irrigation water causes relatively low production of acid and results in lower concentrations of iron, sulfate and uranium and higher concentrations of radium-226 in the lysimeter box effluents. In addition to inhibiting bacteria, hypochlorite might also oxidize the sulfide minerals present in the tailings. The data of this report indicates, however, that the oxidation caused by hypochlorite is considerably less than that caused by bacteria. Thus we can conclude that inhibition of iron sulfide mineral oxidation decreases sulfate generation and results in a condition in which radium solubility in water is increased. Conversel the oxidation of iron sulfide minerals by the bacteria results not only in increa iron and sulfate in the effluents but also in decreased concentrations of radium- in the lysimeter box effluents. At the concentrations used in this experiment, the organic solvent extraction reagents kerosene, Alamine 336 and isodecanol appeared to have no effect on the pH and concentrations of radium-226, sulfate, iron, uranium and thorium of the lysimeter box effluents.

Treatment of uranium mine mill tailings with a 2M solution of KCl preceeded and followed by washing with water is shown to effectively remove much of the more mobile radium-226, resulting in lower concentrations of this radionuclide in the tailing effluents.

Data have been recently generated in laboratory (Bryant, Cohen and Durham, 1979; Silver and Andersen, 1990) and field (Blair and colleagues, 1980; Murray and Okuhara, (personal communication)) experiments on the movement of chemical constituents of abandoned uranium mine tailings, and work in this area continues. Although field experiments are absolutely necessary in accurately describing the present situation with regard to the concentrations of these constituents in the tailings, their utility to predict future changes is limited. The usefulness of accelerated lysimeter experiments in this regard must await further information from projects such as this one. The advantages of these experiments are the controllability and the chronological acceleration that they afford. Their relevance to the actual situations with regard to abandoned and current tailing disposal dumps will be shown in future comparisons with field experiments.

REFERENCES

Barkley, D. (1970). The determination of amine solvent extraction reagents in ore solutions and in aqueous and solid phases of pulps or slurries. Divisional Report EMI 70-5, Mines Branch, Energy, Mines and Resources Canada. 4 pp.

Blair, R.D., J.A. Cherry, T.P.Lim and A.J. Vivyurka (1980). Groundwater monitoring and contaminant occurrence at an abandoned tailings area, Elliot Lake, Ontario. Paper presented at the First International Conference on Uranium Mining Waste Disposal, Vancouver, B.C., 19-21 May, 1980.

Bryant, D.N., D.B. Cohen and R.W. Durham (1979). Leachability of radioactive constituents from uranium mine tailings. Status report (June, 1974 to January, 1977). Technology Development Report EPS 4-WP-79-4, Water Pollution Directorate, Environment Canada. April, 1979. 32 pp.

Gow, W.A., H.H. McCreedy, G.M. Ritcey, V.N. McNamara, V.F. Harrison and B.H. Lucas (1970). Bacteria-based processes for the treatment of low-grade uranium ores. In The Recovery of Uranium, Proceedings of a Symposium on the Recovery of Uranium from its Ores and Other Sources Organized by the International Atomic Agency Held in Sao Paulo, 17-21 August 1970. International Atomic Energy Agency, Vienna. pp. 195-211.

Itzkovitch, I.J. and G.M. Ritcey (1979). Removal of Radionuclides from Process Streams - A Review. CANMET Report 79-21, Energy, Mines and Resources Canada, Ottawa, 172 pp.

Lundgren, D.G. and M. Silver (1980). Ore leaching by bacteria. Ann. Rev. Microbiol., 34, 263-283.

Moffett, D.E. (1976). The disposal of solid wastes and liquid effluents from the milling of uranium ores. CANMET, Energy, Mines and Resources Canada. Scientific Bulletin 76-19. 76 pp.

Nathwani, J.S. and C.R. Phillips (1979). Rate controlling processes in the release of radium-226 from uranium mill tailings. II. Kinetic studies. Water, Air, and Soil Pollution, 11, 309-317.

Parsons, H.W. (1974). Effect of organic solvents on bacterial activity in the bacterial leaching of uranium ores. CANMET, Energy, Mines and Resources Canada. Divisional Report EMA 74-20. 11 pp.

Ritcey, G.M., B.H. Lucas and A.W. Ashbrook (1974). Some comments on the loss, and environmental effects of solvent extraction reagents used in metallurgical process. Intern. Solvent Extn. Conf., Lyon, France, September, 1974. pp. 2873-2884

Ritcey, G.M., M.J. Slater and B.H. Lucas (1972). Chapter 17. A comparison of the processing and economics of uranium recovery from leach slurries by continuous ion exchange or solvent extractions. In D.J.I. Evans and R.S. Shoemaker (Eds.), International Symposium on Hydrometallurgy, Chicago, Illinois, February 24-March 1, 1973. The American Institute of Mining, Metallurgical, & Petroleum Engineers, Inc., New York, pp. 419-474.

Skeaff, J.M. (1977). Distribution of radium-226 in uranium tailings. CANMET, Energy, Mines and Resources Canada. Report MRP/MSL 77-349(J). 20 pp.

Silver, M. and J.E. Andersen (1980). Removal of radium from Elliot Lake uranium tailings by salt washing. Water Pollution Research in Canada. Accepted for publication.

Torma, A.E. and I.J. Itzkovitch (1976). Influence of organic solvents on chalcopyrite oxidation ability of *Thiobacillus ferrooxidans*. Appl. Environ. Microbiol., 32, 102-107.

38

ASSESSING THE IMPACT OF WASTE DISPOSAL ON GROUNDWATER QUALITY

G. F. Farquhar and J. F. Sykes

Dept. of Civil Engineering, University of Waterloo, Waterloo, Ontario, Canada

ABSTRACT

This paper discusses the impact assessment of waste disposal on groundwater quality. Of concern is the extent of contaminant transport in time and space as compared to environmental standards. The factors which affect contaminant transport are examined. They include leachate contaminant loading, patterns of groundwater movement and processes which attenuate contaminant concentrations during transport. Emphasis is placed on the attenuating mechanisms of dispersion, adsorption, precipitation and biological activity. Modelling contaminant migration in groundwater systems is also discussed. Examples of modelling Cl, K and COD are presented together with comparisons to field data.

INTRODUCTION

In the large majority of cases, landfills both produce leachate and discharge it to areas beyond their boundaries. The extent to which this occurs can be reduced through the use of containment and flow control systems. However, to date, these systems are implemented only occasionally and tend to be not totally effective. Consequently, the discharge of landfill leachate to groundwater systems is a frequent occurrence and will probably continue as such for many years.

The problem with leachate discharge is therefore not if but rather to what extent contamination will occur. Cole (1972), Miller and others (1974), Clark (1979) and others have all reported cases in which leachate contamination has had harmful effects upon groundwater aquifers. However, there are also many landfills at which problems have not occurred (Hughes and others, 1971; Farquhar and co-workers, 1972; Rovers and colleagues, 1974).

Whether or not a problem of groundwater pollution exists depends on:

1. the extent of contaminant transport with respect to time and space within the groundwater system,
2. the contaminant concentration at a point of concern within the system and
3. the quality standards imposed upon the system at that point.

Problems tend to occur when the extent of contaminant transport is high, the point is close to the source and/or the quality standards imposed on the system point are stringent. The factors which affect the severity of groundwater contamination problems include:

1. the leachate volume and strength and their variations with respect to time,
2. the patterns of groundwater movement in areas adjacent to the landfill and
3. the processes which modify contaminant concentrations during transport through the groundwater system.

The dominant metallic ions reported in leachates were the alkaline earth metals with calcium (Ca) and magnesium (Mg) exhibiting concentrations often exceeding 1000 mg/L. Potassium (K) and sodium (Na) concentrations were somewhat lower with concentrations occasionally exceeding 1000 mg/L. Of the heavy metals, iron (Fe) exhibited the highest concentrations. It was rarely in excess of 1000 mg/L and more frequently in the range of 100 - 300 mg/L. Zinc (Zn) concentrations were generally less than 100 mg/L. Other heavy metals were mostly less than 1 mg/L in concentration.

Ammonia (NH_3) in solution yielded concentrations up to 1000 mg/L.

The dominant anion reported was the bicarbonate (HCO_3) ion. Its concentrations were found to lie mostly between 10,000 and 20,000 mg/L. Chloride (Cl) and sulphate (SO_4) ions followed,occurring generally between 500 and 1000 mg/L. The remaining species included silicates (H_3SiO_4) and phosphates (PO_4) in concentrations generally less than 10 mg/L.

As with the organic contaminants, the peak concentrations of the inorganic substances occurred 2 to 3 years after placement. Concentrations subsequently reduced with time but did so irregularly. Three landfills, approximately 15 years old, exhibited leachate parameter levels well in excess of domestic sewage levels (SHWRD,1976).

In the assessment of existing landfills and the design of new ones, consideration of leachate strength must centre on contaminant type and concentration and its variation with respect to time. With regard to contaminant strength, there appears to exist sufficient information to make reasonable estimates of peak contaminant strengths. However, with regard to time trends, the situation is not as clear. Sufficient long term data do not appear to exist. Determining the time dependent variation of leachate strength is essential to proper assessment of the landfill's impact on the environment. Hence, further research on this subject is needed.

Groundwater Movement Patterns

Leachate is carried downward from the landfill as part of the groundwater system. As a result, groundwater movement represents the principal mode of sub-surface leachate transport. It determines both the direction and, to some extent, the distance of contaminant migration away from the landfill.

Figure 2 represents the leachate contaminant plume measured at the CFB Borden Landfill in Ontario, Canada. (Waterloo Research Institute, 1978). The landfill was 36 years old at the time of study and was in operation from 1942 to 1970. The soils at the site as shown in the Figure consist of uniform silty sand underlain by poorly permeable clay.

The extent of the plume was traced with a system of wells and piezometers distributed about the site. Measurements of chloride ion (Cl) and conductivity made on samples collected from the piezometers were used to detect the presence of landfill leachate.

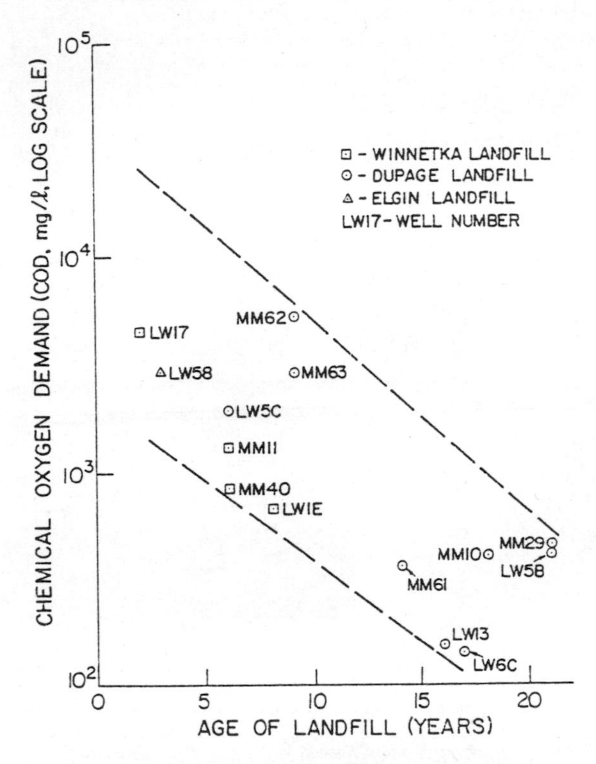

Fig. 1. Variation in leachate COD with time at three Illinois landfill sites (After Rovers and others, 1974)

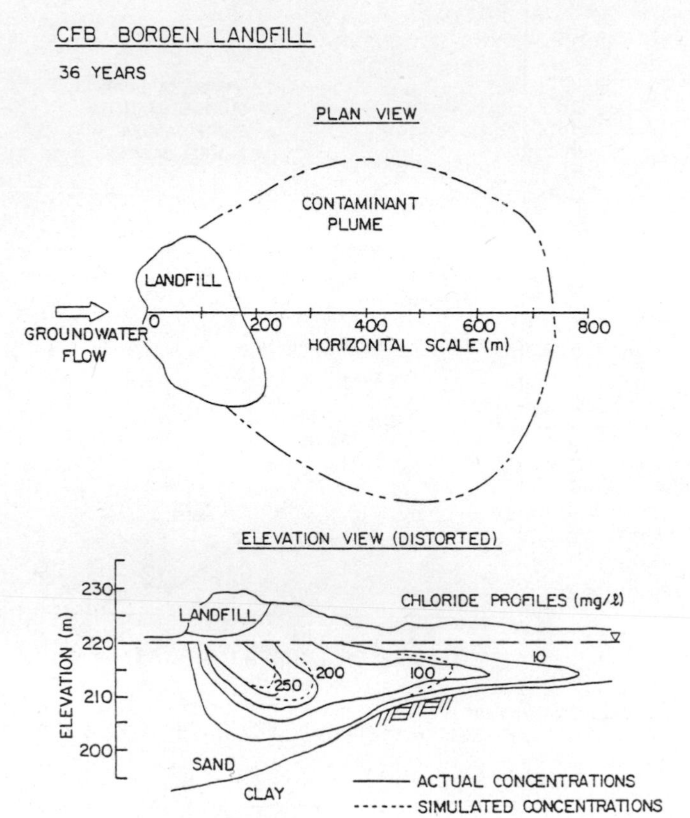

Fig. 2. Contaminant plume at the CFB borden landfill (After Waterloo Research Institute, 1978)

As can be seen from Figure 2, after a period of 36 years, the plume had moved downflow in the groundwater system a distance of 550 m. Some lateral extension of the plume beyond the width of the landfill had also occurred.

Miller and co-workers (1974) described similar plume migration patterns for several of the landfill sites studied in the northeastern United States. Such migration patterns are felt to be typical of finely divided soils such as glacial tills, lacustrine sediments and wind deposits. Although spatial variations must be expected, the groundwater flow can be treated as intergranular.

Groundwater movement direction and velocity are determined by the hydrologic moisture input and the soil's geophysical properties. Of major importance are the properties of permeability and porosity and their variability throughout the soil space. Movement patterns become more irregular as this variability increases. The greatest difficulty occurs when channelling of the groundwater flow occurs through highly permeable soil sections or along cracks and fissures in both rock and unconsolidated soils.

In a recent study by Sudicky and Cherry (1979), point dilution tests were conducted in a sandy soil below a landfill site. A pulse of chloride ion (Cl) was added to the groundwater. Measurements of Cl were made subsequently using a three dimensional piezometer grid placed at 1 m intervals in the groundwater system. The results exhibited substantial variations in Cl transport over distances less than 5 m. Such variations were not anticipated from the soil composition information obtained during instrumentation. A uniform sand had been identified throughout. The authors speculated that such variability may be wide-spread and may serve as an important factor in contaminant dispersion.

Young and others (1978) studied the movement of nitrate (NO_3) ion through unsaturated chalk at several locations in southern England. They observed permeabilities on the order of 10^{-5} to 10^{-6} cm/s, much greater than would be expected for solid chalk. They concluded that groundwater movement was taking place along joints and fissures and not through the body of rock as a whole. This resulted in rates of groundwater movement more rapid than had been expected. Maximum velocities on the order of 0.8 m/d were observed.

In cases where the majority of the groundwater moves through joints, fissures or lenses of sand and gravel, the leachate migration problem is intensified. Not only is the rate of movement rapid but also, the time for reaction and the interfacial contact between contaminants and the soil is reduced. This then reduces the opportunity for attenuating mechanisms to take affect.

Contaminant Attenuation

There exist within soil/groundwater systems, processes which work to lessen or attenuate leachate contaminant concentrations. In some cases, they are effective enough to prevent problems of groundwater pollution. Most of the processes have limited and/or time dependent capacities. These may exist as a minimum contaminant concentration attainable or a minimum reaction time required to achieve a specific contamination concentration reduction. Attention must be paid to these limitations if the processes are to be properly understood.

The following sections examine these processes with respect to their mechanics and capacities. The processes include:

1. dispersion
2. adsorption
3. precipitation and
4. biological activity

While other processes such as filtration and volatilization are known to exist, the four listed above are felt to be the most actively involved and are the focus of consideration here.

1. Dilution Through Dispersion

For certain conservative and/or poorly reactive contaminants such as Cl, Na and certain refractory organic species, dilution through dispersion represents the only means of concentration reduction.

The mechanisms by which dispersion appears to occur are represented schematically in Figure 3.

Groundwater flow generally occurs in the laminar range. Thus the spread of contaminants must involve molecular diffusion. Figure 3a shows a velocity gradient across a soil pore channel. At the migrating contaminant front, the more rapidly moving fluid contains higher contaminant concentrations than the slower fluid nearer to the soil surface. Thus, a concentration gradient along which contaminants will diffuse is set up.

Figure 3b represents the so-called chromatographic effect caused by the presence of dead end pores and other poorly mobile fluid areas within the groundwater system. Contaminants diffuse along concentration gradients into these more slowly moving or stagnant areas. Consequently a slowing and a lessening of the contaminant concentrations results. (Young and colleagues, 1978).

The third and perhaps most effective mechanism of contaminant dispersion is shown in Figure 3c. This involves pore channel mixing where the inter-connected pore channels allow fluid masses to separate and join with new masses thereby setting up concentration gradients for diffusion.

The impact of dispersion as an attenuating process is shown in Figure 2 presented previously (Waterloo Research Institute, 1978). The contaminant plume is marked by the conservative Cl ion. The maximum Cl concentration in the landfill at the time of the study, 36 years after the site was begun, was 830 mg/L. However, based on the previous discussion regarding leachate strength, it is likely that the peak Cl concentration was higher than this value. The reduced Cl concentrations in the groundwater are evidence of the attenuating influence of dispersion. It has also caused the lateral spread of the plume beyond the width of the landfill.

Sykes (Waterloo Research Institute, 1978) used a computer simulation model to predict the 10 mg/L Cl concentration contour as it progressed downflow. His analysis was based on the concept that contaminants, in particular Cl, leached from the landfill were dispersed by means of vertical mixing with infiltrating water. He then compared Cl leaving the landfill with that leaving the 10 mg/L Cl contour of the plume and found them to be similar. A simplified representation of his analysis is presented in Figure 4. The assumptions used in preparing Figure 4 are listed below:

1. The landfill and the plume were represented by flat, cylindrical discs with radii equal to r_i and r_o respectively. The corresponding surficial areas were A_i and A_o.

2. Fluid exited the landfill through the bottom and the plume through the vertical sides. The corresponding exit Cl concentrations were C_i and C_o and were assumed to be uniform across the exit surfaces.

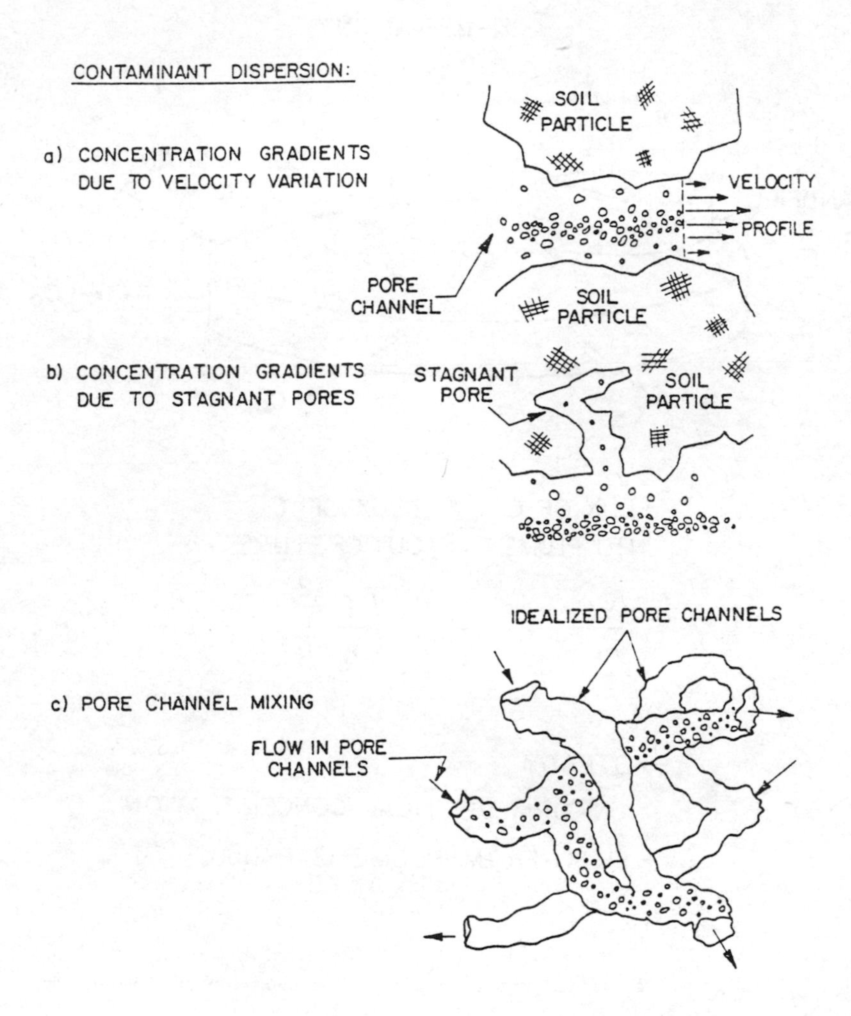

Fig. 3. Mechanisms of dispersion in groundwater

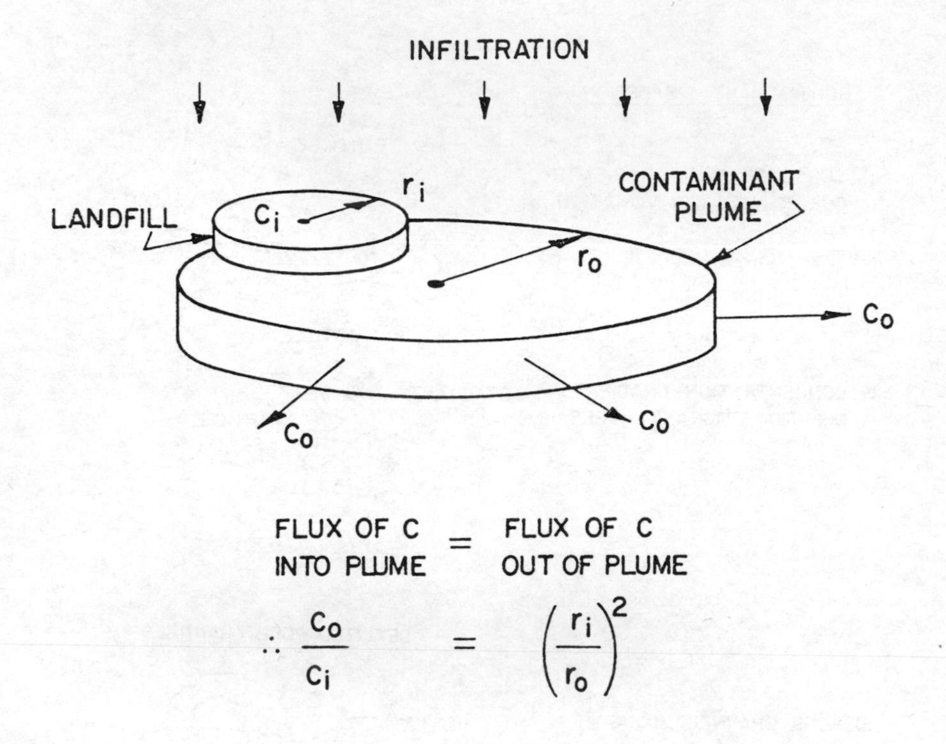

FLUX OF C INTO PLUME = FLUX OF C OUT OF PLUME

$$\therefore \frac{C_o}{C_i} = \left(\frac{r_i}{r_o}\right)^2$$

- STEADY
- UNIFORM VERTICAL CONCENTRATION
- FLUX FROM PLUME IS RADIAL

Fig. 4. Idealized representation of dispersion at CFB borden

3. Steady conditions were used.

Under these conditions, it can be shown that the concentration ratio C_o/C_i is given by:

$$\frac{C_o}{C_i} = \left(\frac{r_i}{r_o}\right)^2$$

If r_i and r_o for CFB Borden are taken as 100 m and 400 m respectively, C_o/C_i at the 10 mg/L contour (see Figure 2) equals 0.06. Given C_i = 830 mg/L, C_o = 50 mg/L as compared to the actual value of 10 mg/L observed in the field. The comparison is reasonable in view of the assumptions made in preparing the figure.

Consequently, dilution of contaminants by means of dispersion into uncontaminated water is a powerful attenuating process.

2. Adsorption

Adsorption represents a collection of processes which remove contaminants into or onto the solid soil surface. Ion exchange, surficial adhesion and internal diffusion are three of the processes involved (Weber, 1972). Some of the processes are highly reversible and therefore result in the return of the contaminant to the fluid phase as conditions change. Ion exchange is of this sort whereas certain adhesive processes are poorly reversible.

Most of the processes experience both capacity and reaction rate limitations and either may control adsorption in the field depending on conditions.

The adsorption of leachate contaminants has been studied in detail by Fuller and Korte (1976), Griffin (1977, 1978), Davidson (1978) and Farquhar (1977, 1978). Collectively, their work covers a wide range of experimental conditions. The result has been a series of equilibrium removal isotherms for many contaminants under many soil conditions. While some contradictions do exist, certain results appear to be common to most of the work:

1. Adsorption capacity appears to increase with increased soil surface area, free lime content, oxide coatings and pH. In some cases increased cation exchange capacity (CEC) and organic content also increase adsorption capacity.

2. Contaminants most readily adsorbed include:
   a) cations; heavy metals plus ammonia
   b) phosphate
   c) numerous pesticides (except dicamba)
   d) polychlorinated biphenyls (PCB's)
   e) most leachate organic matter and monovalent ions tend to adsorb poorly.

3. Heavy metals, several pesticides and PCB's desorb only slightly subsequent to the adsorption phase.

Figures 5 and 6 exemplify the type of equilibrium isotherm data which is available from these studies. Figure 5 exhibits Griffin's (1978) isotherms for the adsorption of PCB's on various materials ranging from sand to coal char. Figure 6 shows data presented by Farquhar (1977) about ammonia (NH_3) adsorption onto seven different soil types. The isotherms are Freundich, reversible and exhibit increased adsorption with increased soil CEC.

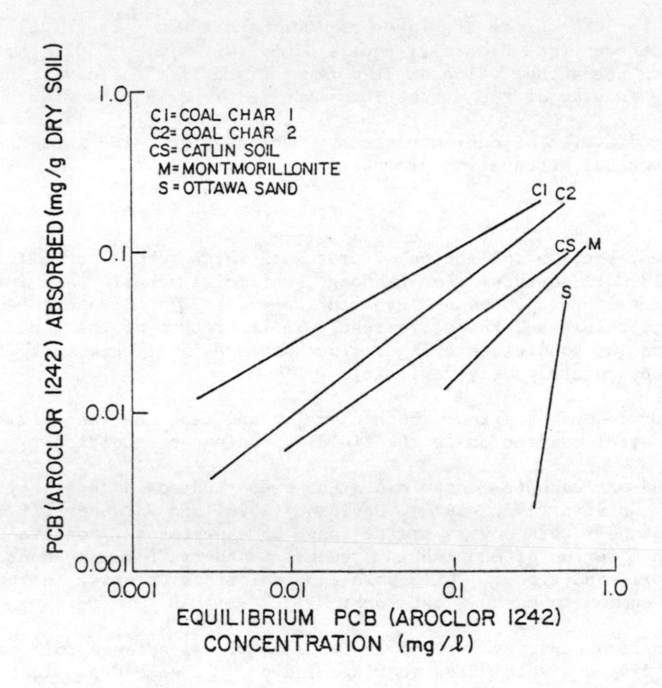

Fig. 5. PCB adsorption isotherms for two solids and two coal chars (After Griffin, 1978).

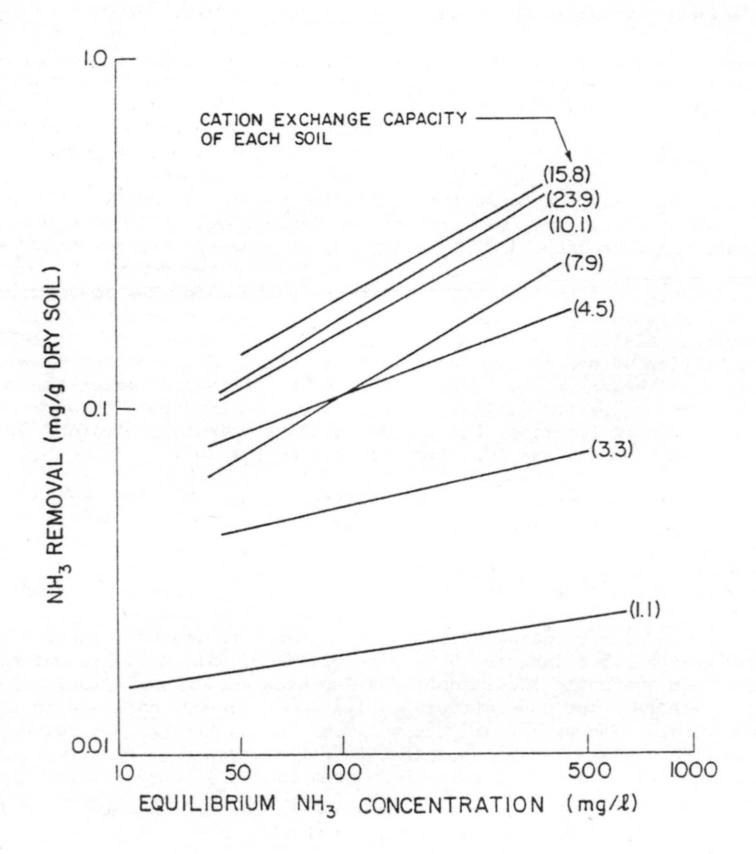

Fig. 6. Ammonia adsorption isotherms for seven soils (After Farquhar, 1977, 1978)

The isotherm format is useful for numerically simulating contaminant transport in groundwater systems. The details of simulation are discussed in a subsequent section. However, the results of such activities are exemplified in Figure 7 which shows both measured and simulated potassium (K) profiles at the CFB Borden Landfill (Waterloo Research Institute, 1978). While the simulations are only approximations, they are fairly close to the measured values particularly as compared to the total distance of plume excursion. The isotherm data used for the simulations were of the same form as those presented for NH_3 in Figure 6 and come from the work of Thompson (1978) and Farquhar (1977, 1978).

As a result, in addition to there being information available on the adsorption of many contaminants onto a variety of soils, the information is in such a form as to be useful in field simulation work.

3. Precipitation

Precipitation involves the formation of insoluble salts of multivalent metallic ions. The major anions involved include carbonate (CO_3), hydroxide (OH), silicate (H_3SiO_4) and phosphate (PO_4). The advantages of precipitation as a mechanism for contaminant removal are its high capacity and its low reversibility. As well, there is no equivalent contaminant release as is the case with ion exchange.

Griffin and others (1977), Fuller and Korte (1976) and Farquhar (1977, 1978) studied the precipitation of metals during contact between landfill leachate and soils of various types. Their results indicate that most multivalent metallic ions except calcium (Ca) and magnesium (Mg) are kept in low concentrations through precipitate formation. The most important factors affecting precipitation appear to be pH, leachate composition and the free lime content of the soil.

The basic chemistry of metallic ion solubility has been well described by Stumm and Morgan (1970). While activities and complex formation are difficult to estimate in liquids as complicated as landfill leachate, reasonable estimates of precipitate formation can be made for specific leachate compositions. Such estimates can be refined through reference to the leachate specific work described above.

Figure 8 shows effluent characteristics in response to leachate passage through a silty sand column 0.5 m long and 0.07 m in diameter. The soil was maintained in a saturated and anaerobic condition. The leachate used was a strong laboratory lysimeter leachate. The concentrations plotted on the ordinate are in the normalized C/C_o format. The number of pore volumes passing through the column are plotted on the abscissa. The chloride ion (Cl) represents a conservative tracer following the fluid flow. The normalized ammonia (NH_3) concentrations lie below the Cl concentrations in the initial part of the curve. This denotes NH_3 removal probably as the exchanged ion.

However, the data of concern are those presented for iron (Fe) and zinc (Zn). Removal of these two ions is essentially complete throughout the experiment. As well, no release of these metals is evident during the subsequent passage of water through the soil. Such a pattern could be expected for precipitation processes.

Figure 9 summarizes the results of a study on soil and groundwater contamination in the area of a zinc smelter in Illinois (Gibb, 1978). The smelter has been in operation in one form or another since 1890. Zinc rich cinders, deposited on the property, had reached depths of 3 m at the time of the study.

Figure 9a is a plan view of the smelter showing Zn concentrations in the groundwater. The concentrations encircle the plant because of a groundwater mound centred near the building. They range from 1 to 10^4 mg/L. The higher concentra-

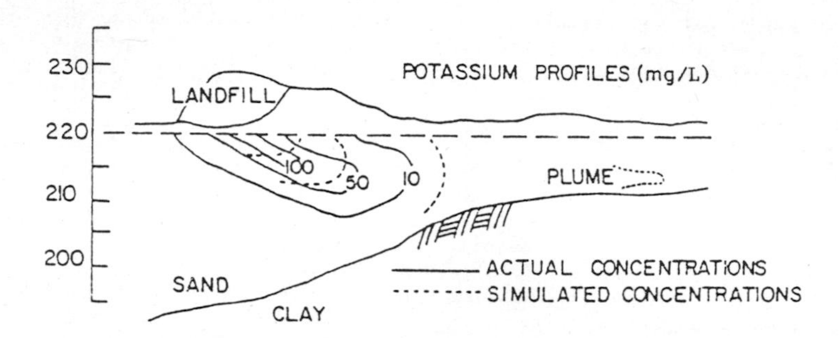

Fig. 7. Concentration contours for potassium migration at CFB borden (After Waterloo Research Institute, 1978)

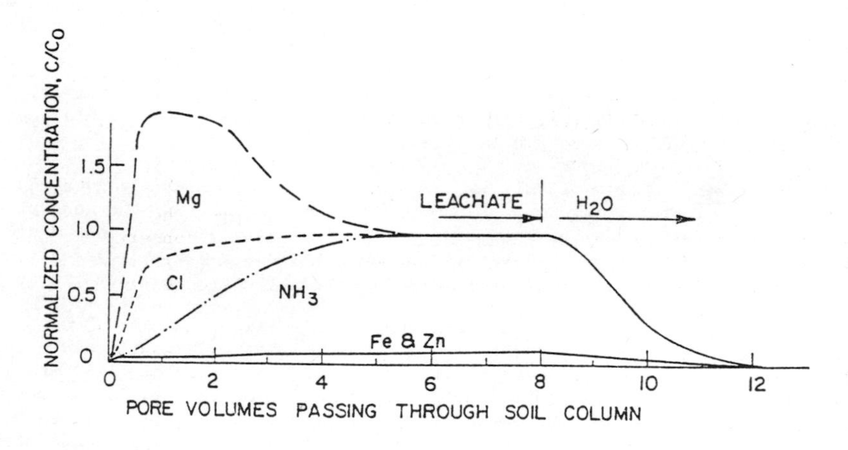

Fig. 8. Effluent characteristics in responses to leachate passage through a silty soil column (After Farquhar, 1977, 1978)

tions exceed the expected solubility limit of Zn and are assumed to represent the presence of Zn complexes and/or colloidal Zn containing precipitates in suspension in the water samples. Similar patterns were also observed for copper (Cu), cadmium (Cd) and lead (Pb), however Zn concentrations were well in excess of the others.

The data show that the groundwater has become contaminated in the area of the smelter. However, it is most important to note that, after approximately 85 years of heavy metal addition to the soil, the zone of pollution is limited to a radius of less than 300 m.

Figure 9b is a cross section taken at A-A through the site. The data presented shown Zn concentrations on the soil up to 10^4 mg Zn/kg soil adjacent to the plant. They result from the transport of Zn from the cinders down into the soil with subsequent deposition. This disposition is likely to have resulted from Zn precipitation with the retention of the precipitate within the soil.

The data demonstrate the effectiveness of precipitation as a contaminant attenuation process in soil. Unfortunately, it is functional for only a limited number of the contaminants found in leachate.

4. Biological Activity

Of all of the contaminants in landfill leachates, particularly young leachates, the highest concentrations are exhibited by organic matter expressed as biochemical oxygen demand (BOD), chemical oxygen demand (COD) and/or total organic carbon (TOC).

These may exist in the range of 10,000 mg/L to 50,000 mg/L and thus represent a significant potential source of groundwater contamination. As noted previously, these organic components tend to consist mainly of volatile fatty acids giving way to humic and fulvic acids as the landfill ages (Chian and DeWalle, 1977).

However, in spite of the high concentrations of organic matter in leachate, incidences of groundwater contamination by these materials occur only rarely. As demonstrated by Soyupak (1979), this results mainly from active microbial decomposition in soils adjacent to the landfill.

The adsorption of leachate organic matter onto soil surfaces was found by Soyupak (1979) to be negligible. Griffin and others (1977) also reported poor adsorption. Farquhar (1977; 1978) observed some adsorption of organics during contact with soil but noted that subsequent desorption into water was rapid and extensive. Consequently, biological degradation represents the primary attenuating process.

Results from Soyupak's (1979) work are summarized in Figure 10. It shows the concentration of TOC in the effluent from a 0.4 m saturated sandy soil column operate anaerobically. The leachate added to the column was a mixture of field and lysimeter leachates with a combined TOC of approximately 1800 mg/L (a COD of approximately 6000 mg/L). The low TOC was chosen to represent an old landfill such as at CFB Borden (Waterloo Research Institute, 1978). The Figure also presents methane (CH_4) production data as a percentage of the TOC added to the column. Fluid velocities were typical of field conditions.

The results show that, for the first few days, the TOC in the effluent increased in a typical breakthrough fashion and remained near the influent concentration for about 10 days. Little CH_4 production occurred indicating a low level of biologica culture had been well established and TOC reduction in the column had reached approximately 90%. Of the TOC removed, approximately 60% was converted to CH_4.

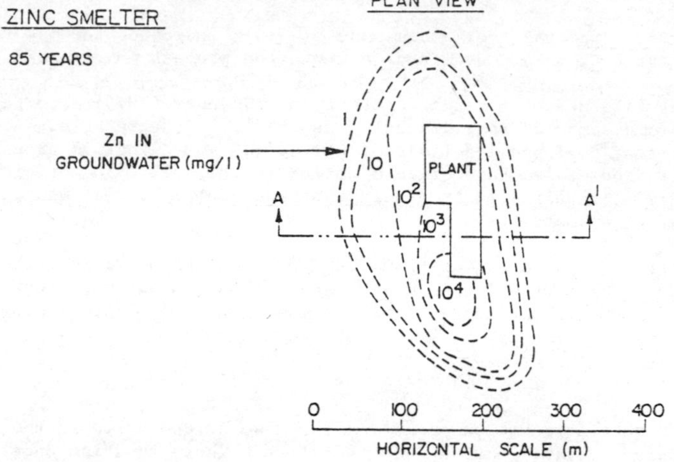

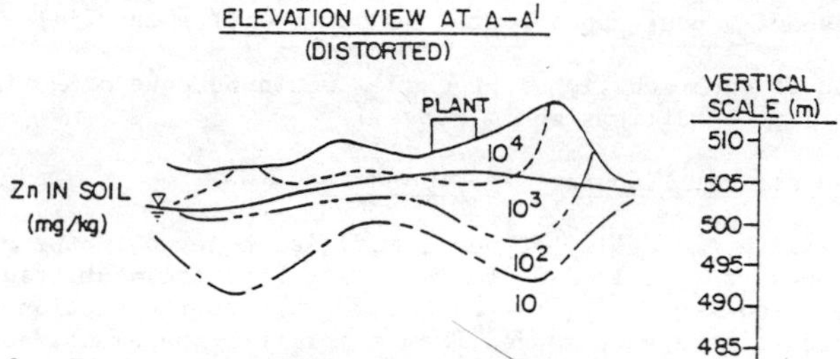

Fig. 9. Zinc conentrations in soil and groundwater near a smelter (After Gibb, 1978)

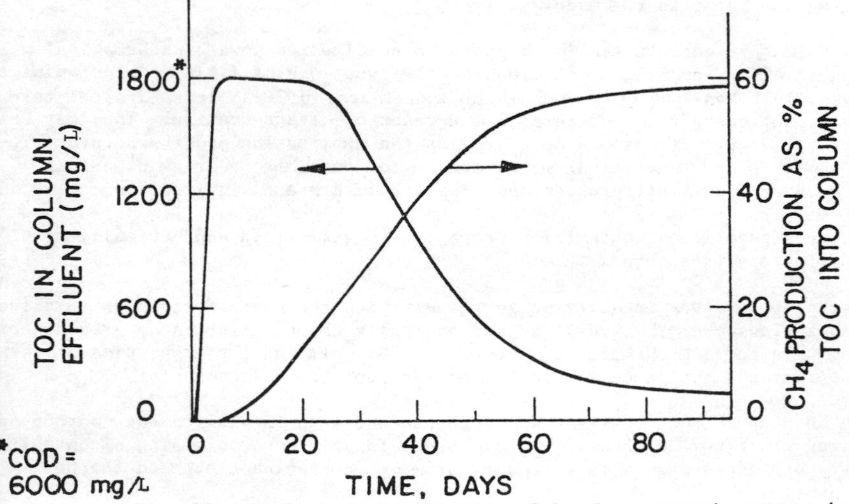

Fig. 10. Biological degradation of leachate organic matter in soil (After Soyupak, 1979)

Such significant removal over a distance of 0.4 m indicates the importance of biological activity as a contaminant attenuation process in the soil environment.

In reviewing literature on leachate treatment, Soyupak (1979) reported that most investigators described their leachates as readily biodegradable. As a result, it is clear that biological activity functions at many landfill sites to reduce the concentration of leachate organic matter in the soil. In addition, based on Soyupak's (1979) results, it is likely that substantial reductions occur over distances of a few meters.

The transport of organic matter from the CFB Borden Landfill into the groundwater below is shown in Figure 11. As can be seen, the COD concentrations reduce from in excess of 3000 mg/L in the landfill to 100 mg/L in a vertical distance of less than 5 m!

5. Other Processes

Other contaminant attenuation processes may also be operative in the soil. Examples include volatilization and filtration. These function under special circumstances. In cases where volatile materials are buried in the landfill, volatilization may serve as an important means of contaminant release. Farmer (1978) observed that hexochlorobenzene (HCB) deposited at a landfill site was dissipated by the release of vapours up through the cover of the landfill.

Consideration of such additional processes would be done on a site by site basis where appropriate conditions would prevail.

Modelling Contaminant Transport in Groundwater

It is essential in the design of new landfills and even in the proper assessment of existing ones that some means for simulating contaminant transport in groundwater be used. Consequently, there has been a recent evolution of models for use in this regard. These have ranged from simplistic judgements based on experience and intuition to sophisticated mathematical expressions. The accuracy of these models tends to increase with increased complexity as well as with the quality of the information input to the model.

Presently, most research attention appears to be directed toward mathematical modelling involving numerical solutions to the equations of fluid and contaminant flux (Sykes, 1975; Van Genuchten, 1978; Fuller and others, 1979). These include terms for contaminant convection, dispersion and reaction or attenuation. They may be written in 1, 2 or 3 dimensions depending on the application and the accuracy required. Solutions by analytical means are seldom possible and thus numerical methods such as finite differences and finite elements are required.

The process of modelling contaminant transport is summarized schematically as a series of 5 steps in Figure 12.

The first step involves identifying and quantifying the basic factors which affect contaminant transport; the hydrology and hydraulic characteristics of groundwater flow, the physical and chemical properties of the soil and the properties of the waste material and the leachate, in particular its chemical composition.

Based on the input from the first step, the second step identifies the contaminant attenuating processes which will be involved. This is done on a single contaminant type basis since the processes will vary from one contaminant type to the next.

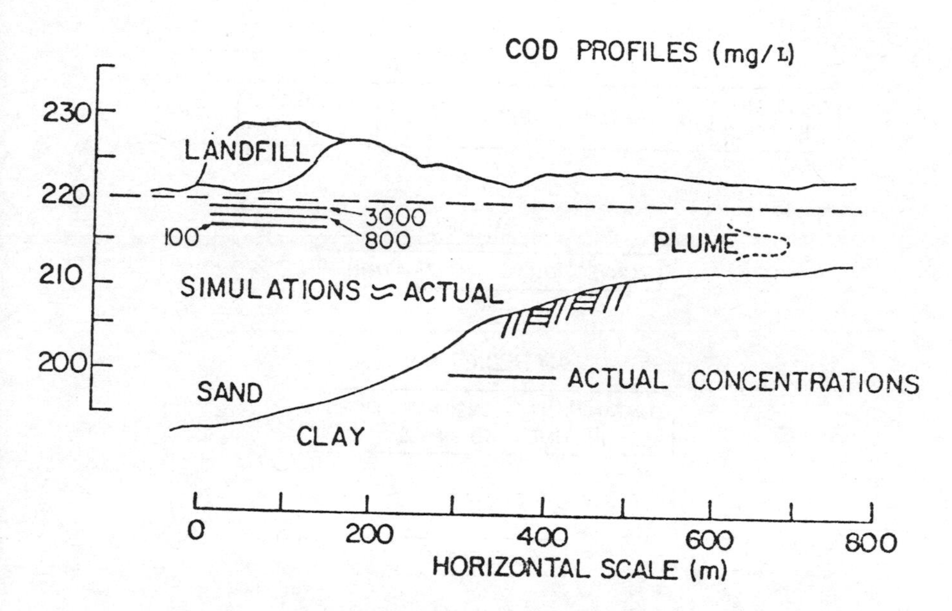

Fig. 11. Concentration contours for organic matter (COD) migration at CFB borden (After Waterloo Research Institute, 1978)

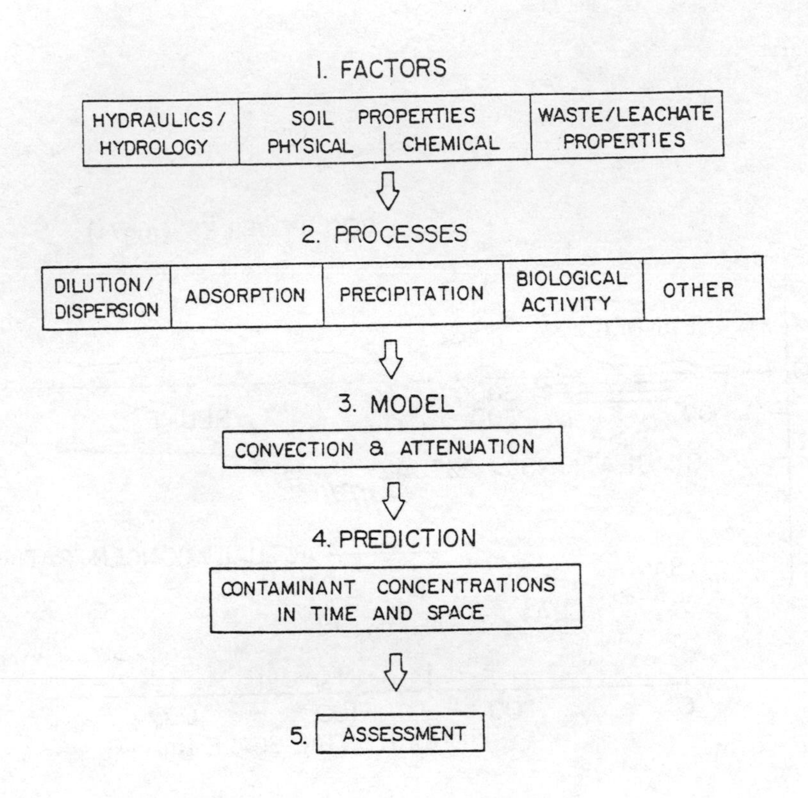

Fig. 12. Steps in modelling landfill contaminant transport in soil

The third step consists of setting up the contaminant transport model including terms to account for the attenuating processes. This includes establishing a method for solving the equations involved.

The fourth step is that of prediction in which contaminant concentrations at various points in time and space are stimulated.

Finally, the fifth step involves an assessment of the modelling process itself (with subsequent revisions if necessary) and of the environmental impact of the predictions.

These 5 steps are applicable to most model types whether simplistic or complex but are clearly more suitable in the case of the latter. While all steps are difficult to accomplish, the first is probably the most difficult since the factors vary from site to site and must be quantified at each. In contrast, the second and third steps can be described in formats which are applicable in general to most cases.

This approach to modelling was undertaken at the CFB Borden Landfill Site (Waterloo Research Institute, 1978). The site was instrumented extensively to identify and qualify the factors described in step 1 or Figure 12. This was accomplished by collecting climatological records, installing wells and piezometers in and around the landfill, collecting and analyzing samples of soil, waste, leachate and groundwater and monitoring groundwater movement over time.

The example to be presented here focuses on three contaminants, chloride ion (Cl), potassium (K) and organic matter expressed as COD. From step 2 in Figure 12 the following attenuating processes were identified:

1. Cl: This is a conservative contaminant with dispersion during transport being the only attenuating mechanism. Dispersion coefficients typical of the CFB Borden soils were used.
2. K: The only significant attenuating processes reported for K are adsorption and, as for all contaminants, dispersion. The work by Thompson (1978) and Farquhar and Constable (1979) was used to quantify a Freundlich adsorption isotherm for the silty sand soil beneath the landfill.
3. COD: Soyupak (1979) developed parameters to quantify a kinetic biological model of the Monod - type for COD reductions at the CFB Borden Landfill. The leachate used by Soyupak was not from the CFB Borden Landfill, although attempts were made to simulate the field liquid as closely as possible. The soil used was taken directly from the site.

In steps 3 and 4 of Figure 12 the model development was based on the work of Sykes (1975). A complete description of the model is beyond the scope of this report. However, its features can be summarized below:

1. The equations accounted for convection, dispersion plus other attenuating processes and continuity, and were expressed in differential form.
2. The principal directions of the 2 - dimensional cartesian co-ordinates were vertical and horizontal in the direction of flow.
3. Appropriate boundary conditions were applied.
4. Solutions were obtained by the Galerkin finite element method.
5. Solutions were used to simulate contaminant transport in space and time.

An assessment of the simulations as set out in step 5 was performed by comparing them with actual measurements made in the field. The comparisons are shown in part in Figures 2, 7 and 11 for Cl, K and COD respectively. While not exact, the comparisons between actual and simulated values are good.

Notwithstanding the fact that modelling of this sort is a complicated process requiring substantial amounts of data and computation, the potential it offers to the waste management engineer is substantial. It provides an opportunity to estimate contaminant excursion as part of the design process to locate and manage landfill sites. In the context of today's need for environmental protection, such capabilities are essential.

REFERENCES

Chian, E.S.K. and F.B. DeWalle (1977). Evaluation of Leachate Treatment. Volume I, Characterization of Leachate. Report No. EPA - 600/2-77-186a, U.S. EPA. Washington, D.C.

Clark, D.C. (1979). Remedial activities for Army Creek Landfill. In Proceedings of the Fifth Annual EPA Research Symposium, EPA Report, U.S. EPA, Washington,D.C.

Cole, J.A. (1972). Groundwater pollution in Europe. Proceedings of a Water Research Association Conference, Reading, England, Water Information Centre, Inc. New York.

Davidson, J.M. (1978). Adsorption, movement and biological degradation of high concentrations of selected pesticides in soils. In Proceedings of the Fourth Annual EPA Research Symposium. EPA Report No. EPA - 600/9-78-016, U.S. EPA, Washington, D.C.

Eagleson, P.S. (1970). Dynamic Hydrology, McGraw-Hill Book Company, New York.

Farmer, W.J. (1978). Land disposal of hexachlorobenzene wastes: controlling vapour movement in soil. In Fourth Annual EPA Research Symposium. Report No. EPA-600/9-78-016, US EPA, Washington, D.C.

Farquhar, G.J. (1977). Leachate treatment by soil methods. In Proceedings of the Third Annual EPA Research Symposium, Report No. EPA - 600/9-77-026, U.S.EPA, Washington, D.C.

Farquhar, G.J., R.N. Farvolden, H.M. Hill and F.A. Rovers (1972). Sanitary Landfill Study Final Report Volume I: Field Studies on Groundwater Contamination from Landfills. Waterloo Research Institute Report, University of Waterloo, Waterloo, Ontario.

Farquhar, G.J. and T.W. Constable (1978). Leachate Contaminant Attenuation in Soil. Report No. ISBN #0-920772-02-1. Waterloo Research Institute, Waterloo, Ontario.

Fuller, W.H. and N. Korte (1976). Attenuation mechanisms of pollutants through soils. In Proceedings of the First Annual EPA Research Symposium, EPA Report No. EPA - 600/9-76-004, U.S.EPA, Washington, D.C.

Fuller, W.H., A. Amoozegar-Fard and G.E. Carter (1979). Predicting movement of selected metals in soils: application to disposal problems. In Proceedings of the Fifth Annual EPA Research Symposium, EPA Report, U.S.EPA, Washington, D.C.

Gibb, J.P. (1978). Field verification of hazardous waste migration from land disposal sites. In Fourth Annual EPA Research Symposium, Report No. EPA - 600/9-78-016, U.S.EPA, Washington, D.C.

Griffin, R.A., R.R. Frost, A.K. Au, G.D. Robinson and N.F. Shimp (1977). Attenuatio of pollutants in municipal landfill leachate by clay minerals: Part 2 - heavy metal adsorption. Environmental Geology Notes, No. 79, Illinois State Geological Survey, Urbana, Ill.

Griffin, R.A. (1978). Disposal and removal of polychlorinated biphenyls in soil. In Proceedings of the Fourth Annual EPA Research Symposium, Report No. EPA - 600/p-78-016, U.S.EPA, Washington, D.C.

Hillel, D. (1971). Soil and Water, Physical Principles and Processes. Academic Pres New York.

Hughes, G.M., R.A. Landon and R.M. Farvolden (1974). Hydrogeology of Solid Waste Disposal Sites in Northeastern Illinois. Report No. SW-12d, U.S.EPA, Washington, D.C.

Miller, D.W., F.A. DeLuca and T.L. Tessier (1974). Groundwater Contamination in the Northeast States. Report No. EPA - 660/2-74-056, U.S.EPA, Washington, D.C.

Rovers, F.A., G.J. Farquhar and J.P. Nunan (1974). Landfill contaminant flux - surface and subsurface behaviour. In Proceedings of 21st Ontario Industrial Waste Conference.

Rovers, F.A. and G.J. Farquhar (1972). Sanitary Landfill Study, Volume II Effect of Season on Landfill Leachate Production. Waterloo Research Institute Report, University of Waterloo, Waterloo, Ontario, Canada.

Rovers, F.A. and G.J. Farquhar (1973). Infiltration and landfill behaviour, ASCE Journal of the Environmental Engineering Division, 99, EE5.

Solid and Hazardous Waste Research Division (SHWRD) (1976). Summary Report: Municipal Solid Waste Generated Gas and Leachate. Draft Report, SHWRD, Municipal Environmental Research Laboratory, Cincinnati.

Soyupak, S. (1979). Modification to Sanitary Landfill Leachate Organic Matter Migrating Through Soil. Ph.D. Thesis, University of Waterloo, Waterloo, Ontario.

Stumm, W. and J.J. Morgan (1970). Aquatic Chemistry, Wiley-Interscience, New York.

Sudicky, E.A. and J.A. Cherry (1979). Field determination of contaminant dispersion in the sandy aquifer below the C.F.B. Borden landfill. Presented at the 14th Canadian Symposium on Water Pollution Research, University of Toronto, Toronto, Ontario.

Sykes, J.F. (1975). Transport Phenomena In Variably Saturated Porous Media. Ph.D. Thesis, University of Waterloo, Waterloo, Ontario.

Thompson, M.K. (1978). Potassium Chloride Transport In Unsaturated Soil Columns. M.A.Sc., Thesis, University of Waterloo, Waterloo, Ontario.

Van Genuchten (1978). Simulation models and their application to landfill disposal siting; a review of current technology. In Proceedings of the Fourth Annual EPA Research Symposium, EPA Report No. EPA - 600/9-78-016. U.S.EPA, Washington, D.C.

Waterloo Research Institute (1978). CFB Borden Landfill Study. Interim First Year Report to Canadian Dept. of Fisheries and Environment, Waterloo, Ontario, Canada.

Weber, W.W. Jr. (1972). Physicochemical Processes for Water Quality Control. Wiley - Interscience, New York.

Young, C.P., E.S. Hall and D.B. Oakes (1978). Nitrate in Groundwater - Studies on the Chalk Near Winchester, Hampshire. Technical Report TR 31, Water Research Centre, Medmentram Laboratory.

39

LYSIMETER STUDIES OF HIGH RATE LAND DISPOSAL OF SLUDGE

J. A. Mueller

Environmental Engineering and Science Program, Manhattan College, Bronx, New York 10471, USA

ABSTRACT

Lysimeter experiments were conducted to determine maximum allowable sludge loading rates which produce a significant nitrification in a northeast U.S.A. climate. Incorporation into total waste treatment system by nitrate recycle is discussed.

KEYWORDS

Sludge Disposal, Land Disposal, Soil, Nitrification, Metals Leaching, Nutrient Leaching.

INTRODUCTION

For wastewater treatment plants with suitable sites, land spreading of sludge is the most attractive ultimate disposal technique presently available. Environmental constraints on the allowable rates of land spreading exist due to the possibility of groundwater contamination and crop uptake of toxicants. Disposal of sludges is also normally considered a liability requiring a significant percentage of treatment plant operating and capital cost. Incorporation of a land spreading operation at a designated site into the total treatment system by leachate recovery and recycle of nitrate back to the plant should minimize the aforementioned environmental constraints limiting application rates and provide some operational benefits at the plant. This paper presents the results of lysimeter studies conducted to evaluate the feasibility of the above concepts and to determine allowable loading rates for a climate similar to that of the relatively wet eastern United States.

CONCEPTUAL DESCRIPTION

A flow diagram of the concept is shown in Figure 1. Anaerobic digestion is utilized since it is not only a low energy consuming system compared to aerobic digestion, but effectively hydrolyzes a substantial portion of the organic nitrogen to ammonia in the digester. Both forms of nitrogen will be held by the soil; the amonia nitrogen will be relatively rapidly nitrified compared to the relatively slow organic nitrogen hydrolysis. Concentrations of nitrate nitrogen in anaerobically digested sludge leachate as high as 400 mg/L have been measured by Hinesly and co-workers (1974) at a solids loading of approximately 34 tons/acre/year, while Andrew and

Troemper (1975) measured concentrations of about 40 mg/L for aerobically digested sludge leachate at a loading of 40 tons/acre/year after two years and about 200 mg/L after 4 years application (Alvey, 1979).

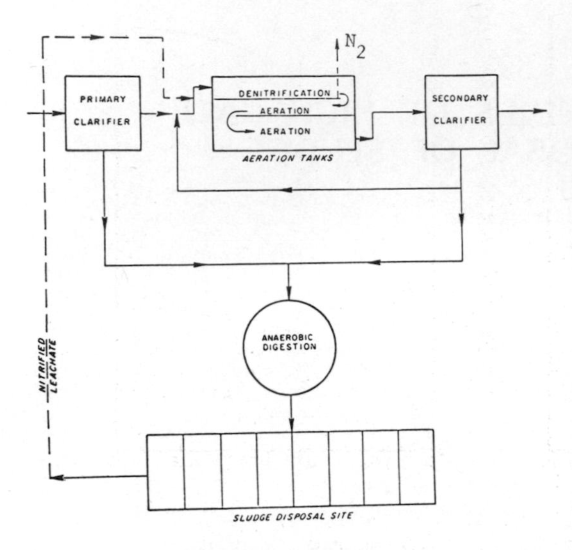

Fig. 1. Flow diagram of total treatment system

An estimate of the effect of the nitrate recycle on treatment plant oxygen requirements is shown in Fig. 2 as a function of the F/M ratio (lb BOD_5 applied/day/per lb MLVSS) using typical characteristics and stoichiometric coefficients for domestic sewage (Eckenfelder,1966). At increasing F/M ratios, less oxygen is required and more secondary sludge is produced. Assuming that the nitrogen in the sludge is ultimately converted to nitrate at a steady state rate equal to the rate of addition, the treatment plant oxygen requirements could be reduced by about 20% to 40% at practical F/M ratios of 0.2 to 0.5. For these estimates, the anaerobic digester supernatant oxygen demand was conservatively taken as 7% of the primary effluent load from data by Malina and DiFilippo,(1971). The benefits to be accrued at a given waste treatment plant will be a function of the specific waste characteristics and existing operation, as well as the degree of nitrification occurring in the soil. The values shown in Fig. 2 are relatively conservative estimates for a typical domestic plant.

The maximum loading rate that can be applied at the sludge site will be a function of soil permeability both initially and after sludge application, soil cation exchange capacity, as well as nitrification rate in the soil. Assuming that the former rate controls, an upper limit can be estimated from sand drying bed data (Eckenfelder, 1966) as 15 lb/ft^2/year or 327 tons/acre/year. If a practical limit is one-half to one-quarter this value, it will be about an order to magnitude greater than the generally accepted values of 10-20 tons/acre/year. Figure 3 indicates that this will reduce land requirements from greater than 7 acres/MGD to 1 to 2 acres/MGD of wastewater flow.

EXPERIMENTAL DESIGN

To evaluate maximum sludge loading rates which would allow high degrees of nitrification without clogging the soil in a northeast climatic condition, in August 1976, fiv lysimeters were set up at the Ridgewood, New Jersey treatment plant outside the control building. As shown in Figure 4, each lysimeter consisted of a 30 gallon heavy duty plastic drum. Added to the bottom were 2" of gravel followed by 17" sand, 4" topsoil and 1 1/2" sod leaving about 6.5" freeboard. A hole was drilled in the side of each drum at the base and a drain installed. Each drum was placed on cinder bloc to allow gravity draining of leachate to a 5 gallon plastic bucket adjacent to the drum. Chemical analyses were conducted on each of the layers prior to startup to establish the initial composition of the lysimeters for mass balance purposes. The study was conducted in two phases. The first phase was conducted for 4 months at 3 diffe ent loading rates using liquid digested sludge and one loading rate using vacuum fil cake (Table 1). The liquid sludge was a well digested combined sludge from the prim and secondary (contact stabilization) clarifiers. The sludge was obtained from the

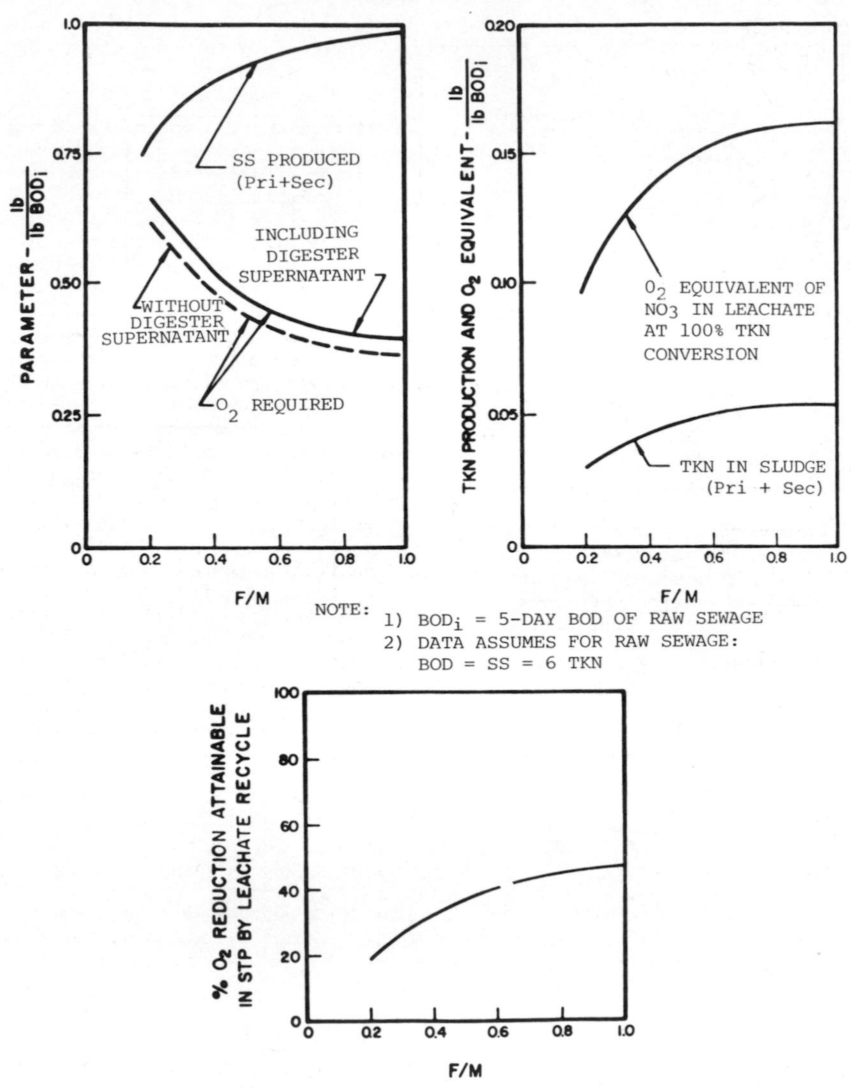

Fig. 2. O_2 reduction

primary digester prior to supernatant separation and thus had a solids content of less than 2%. The vacuum filter cake was pretreated with 6% $AlCl_3$ and 32% lime. Primary digested sludge (PDS) was added once per week to the lysimeter surface and vacuum filter cake (VFC) initially and after 3 months. The above schedule was disrupted from day 49 to 73 when only 1 addition was made instead of 3. Also no sludge was added to the highest loaded unit on days 81 and 87 due to surface ponding.

The total volume of leachate from each lysimeter was measured weekly prior to and normally a few hours following sludge addition. Fresh samples were periodically collected overnight for chemical analysis; emphasis was placed on nitrate and COD analysis.

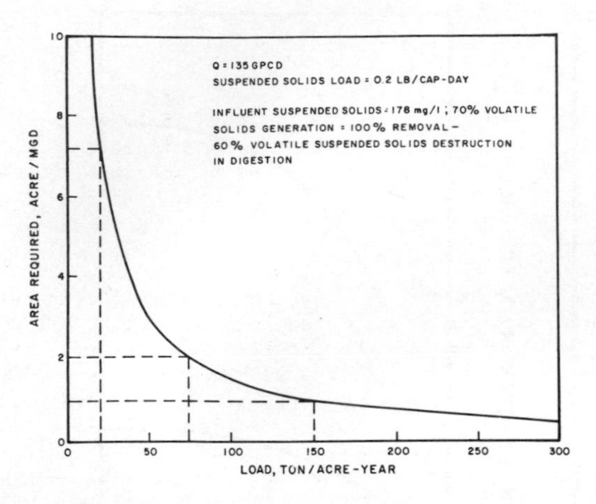

Fig. 3. Area requirements

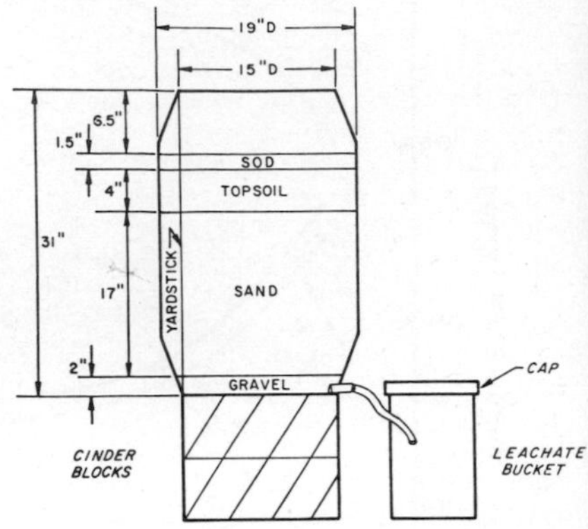

Fig. 4. Lysimeter schematic

Daily rainfall (7 AM - 7 AM) and 8 AM air temperature were monitored by the Ridgewoo Water Company at a site approximately 2 miles from the plant. The chief operator of the Ridgewood STP operated the units and conducted total and volatile solids analyse on the sludge added each week as well as some *E. coli* analyses on leachate. After a schedule was firmed up, operation and monitoring required about two hours per week.

TABLE 1. Experimental Design for Lysimeter Studies

Unit	Sludge % Solids Avg. Conc.	Std. Dev.	Sludge Loading Design Freq.	Vol. 1	Average Loading Rate* Tons/Acre-Year	in./yr.	m-tons h-year	Alkalinity Addition
			EXPERIMENTAL DESIGN PHASE I					
1	0	-	-	0	0	0	0	0
2	1.67	± 0.13	Weekly	1.2	23	12	52	0
3	1.67	± 0.13	Weekly	3.5	68	36	152	0
4	1.67	± 0.13	Weekly	7.0	119	63	267	0
5	17.4(VFC)	-	3 months	1.5	25**	0.9	56	0
			EXPERIMENTAL DESIGN PHASE II					
1	0	-	-	0	0	0	0	0
2	1.83	± 0.31	Weekly	3.5	70	37	157	0
3	1.83	± 0.31	Weekly	3.5	70	37	157	0
4	1.83	± 0.31	Weekly	7.0	125	67	280	10% CaO
5	1.83	± 0.31	Weekly	3.5	70	37	157	10% CaO 25-50% $NaHCO_3$

* To day 160 for Phase II; excludes last addition on day 196
** Based on 6 months interval

In the first phase, sludge was added initially on August 6, 1976 during hot summer conditions with the last addition made on November 22 when the lysimeters were froz The last leachate sample was obtained on December 14, 1976. The second phase of th study, begun the following Spring after thawing of the soil, was conducted to determine the extent of nitrification to be obtained and the ability of the soil to receive sludge at high rates over a longer priod with and without alkalinity addition. Unit #1 was maintained as the control while Units #2 and #3 were increased from 23 to 68 to 70 tons/acre-year with no chemical addition. Unit #5 which had previously received vacuum filter cake at 25 ton/acre-year, was also increased to 70 but with chemical addition, initially utilizing lime followed by sodium bicarbonate. Unit #4 was maintained at a high rate of 125 tons/acre-year but with lime addition. Weekly sludge additions were again utilized unless ponding on the soil surface due to heavy rain prohibited it.

Leachate volume was measured weekly prior to sludge addition and samples refrigerated after collection for the following analyses: ALK, pH, total P, TDS, TKN, NO_3-N, COD and conductivity. Periodic metal analyses were also conducted and samples of grass and plant growth were analyzed. The study duration for the second phase was 7 months beginning April 12, 1977 with the last sludge addition on October 25, 1977 and leachate sample analyzed December 7, 1977. After the winter of 1977-1978, during which the lysimeters were frozen, core samples of each lysimeter were taken on April 13-14, 1978 for chemical analyses. Some difficulties were experienced with leachate collection in Units #1 and #3 due to blockage or clogging in the second phase. Also, after three additions of sludge to Unit #4, the first two with 1% lime and the last with 10% lime, the Unit clogged as evidenced by water ponding on the surface. It remained in this condition for three weeks with no sludge additions made. To prevent this clogging, the sludge mat was removed from Unit #4 on May 17 (Day 35) and the upper six inches of soil turned under. This initial clogging may have been due in part to the previous year's heavy load and to the high application rates with low-lime addition during the cool spring weather. No further clogging problems resulted during the warmer weather with 10% lime addition.

Climatic Conditions

The temperature data over the course of the study are shown in Table 3. Air temperature during the first 64 days of the study (August and September, 1976) were warm to hot, the minimum 8 AM temperatures varying from 45° to 70°F. The remaining months were relatively cold, with freezing temperatures occurring periodically through November and consistently in December. Rainfall was high during the first 3 months of the first phase followed by a month of no rainfall in November. The 19.6" of rainfall that fell during the first 4 month period is equivalent to 58.8"/year, a fairly wet period compared to the mean rainfall for this area of 42"/year. In the second phase, the 8 AM air temperature increased from around 40°F in the spring to 76°F in the summer decreasing to about 36°F in the late fall. Table 3 shows the average temperature to be 10 to 12°F. The 8 AM air temperature in the spring and summer was 10 to 12°F higher than the average but only 1 to 5°F higher in the fall. The November temperature in Phase II was significantly higher than in Phase I (the 1976 Phase I was during an abnormally cold winter). Utilizing the rainfall records from the weather station in New York City at Central Park, the rainfall over the major portion of the Phase II study period from April 12 to September 30, 1977 was near normal. July and May were drier than normal, a departure of -3.84 inches, while the other months were wetter (+2.91 inches). In October and November, 1977 the rainfall was significantly greater than normal, a departure of +10.84 inches, due mainly to a record amount of rainfall of 8.09 inches on November 7 and 8.

RESULTS

Figure 5 shows the volume balance over the Phase II study. Data obtained from April through October 1977 show about 50% evaporation from the units in which leakage or effluent blockage did not occur. A range of 60 to 70% evaporation occurred during the summer months with generally lower values in the wetter spring and fall months.

TABLE2. Digested Sludge Composition at Ridgewood, N.J.

	Phase I Sludge Concentration August 6, 1976 mg/g dry		Phase II PDS Sludge Concentration August 30, 1977	
Parameter	PDS	VFC	mg/L	mg/g dry
Total solids	1.84%	17.4%	13,000	1.3%
VSS	709	–	8,700	670
COD	1,040	330	–	–
Ca	–	–	400	30.8
Mg	–	–	98	7.54
K	–	–	34	2.62
Na	–	–	120	9.23
Cd	0.015	0.013	0.28	0.022
Cr	0.48	0.43	3.2	0.25
Cu	5.4	4.4	50	3.9
Pb	0.38	0.38	4.1	0.32
Zn	2.2	1.75	23.8	1.8
NH_4^+-N	40	1.4	490	38
Org N	79	34	749	58
Total P	23	15.8	–	–
ALK	(182)	–	2,920(2,710)	225(209)
T. Coli, MPN/g dry	270,000	<260		
F. Coli, MPN/g dry	92,000	<260		
				Supernatant
NO_3-N				2
Cl				394
SO_4				220
ALK				2,380
Conductivity, μmho/cm (at 27°C)				4,210
Volatile Acids			(500)	(361)
pH			(7.24)	–

(Average value over study)

In the Phase I study, less (31-33%) evaporation occurred for the units receiving the higher sludge loads due to the colder weather and higher rainfall while only 24% of the total liquid added evaporated from the control.

Leachate Analysis

The leachate nitrate nitrogen concentrations are shown in Figures 6 and 7. In Phase I, the NO_3-N concentrations (Figure 6) increased to 71 mg/L in the control due to the relatively high initial TKN concentrations in the soil. The lightly loaded unit receiving the liquid digested sludge peaked after 60 days to 450 mg/L then dropped to about 150-200 mg/L over the last two months. Temperature also decreased sharply during this period. The 68 PDS unit shows a similar trend but more quickly and at higher concentrations, reaching a peak of 670 mg/L after 50 days. The 119 PDS Unit attained high nitrate concentrations most rapidly and peaked after 40 days at 590

TABLE 3. Phase I and Phase II
8 a.m. and Average Daily Temperatures (°F)

Month	Average Monthly 8 A.M. Air Temperature From Ridgewood Water Company	Average Monthly Daily Temperature From Central Park (NOAA)
1976 PHASE I		
August	65.6	74.3
September	61.8	66.6
October	48.3	52.9
November	40.4	41.7
1977 PHASE II		
April	49.2	59.2
May	54.4	65.0
June	61.7	70.2
July	67.3	79.0
August	66.5	75.7
September	63.4	68.2
October	50.0	54.9
November	46.9	47.3

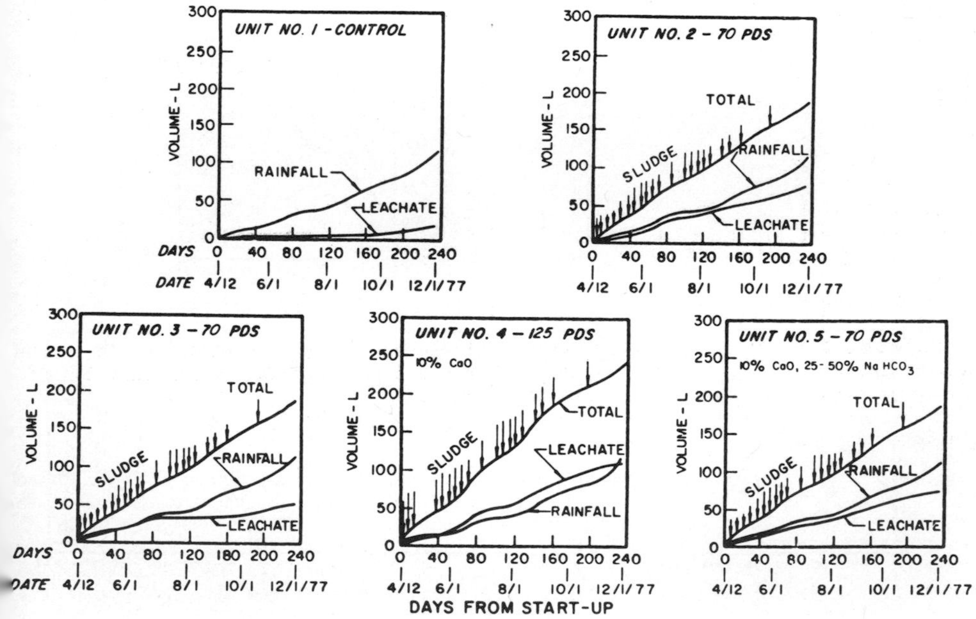

Fig. 5. Volume balance - Phase II

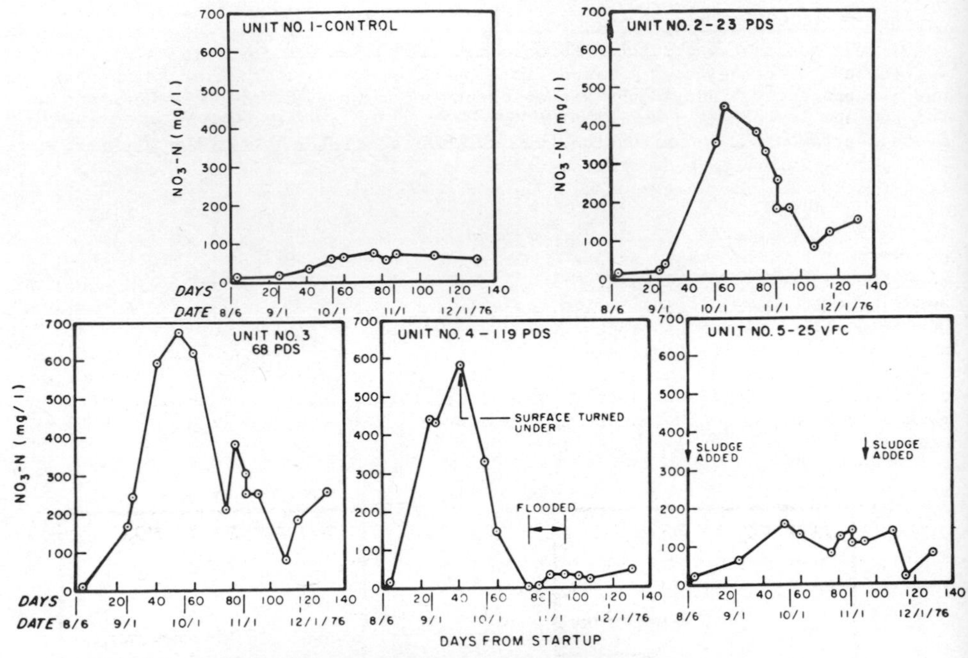

Fig. 6. Leachate NO_3-N - Phase I

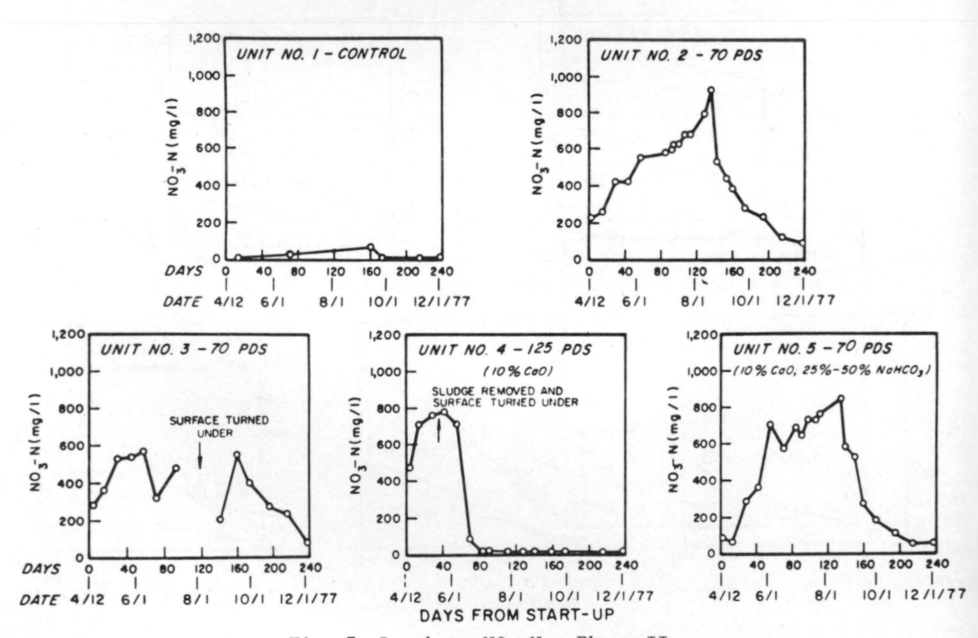

Fig. 7. Leachate NO_3-N - Phase II

mg/L. At this time the sludge was turned under due to lack of surface drying with nitrate concentrations plummetting to zero and surface flooding occurring. Anaerobic conditions apparently persisted throughout this period due to clogging of the surface pores. Turning under was done to attempt to alleviate lack of surface drying but it apparently hindered the situation by causing consolidation and compaction of the surface layer. If a normal plow depth of 6" had been used instead of the 1-2" of the surface sod, the situation may have been reversed. The 25 VFC lysimeter showed lower nitrate concentrations than the 23 PDS since only 4% of the TKN in the filter cake was ammonia compared to 34% in the liquid sludge.

In Phase II, both units #2 and #5 obtained high NO_3-N concentrations during the summer months with a rapid drop off during September and October. The highest loaded unit rapidly obtained high NO_3-N concentrations which dropped off close to zero after 60 days similar to the Phase I study, presumably due to gross overload condition. The high NO_3-N concentrations occurring initially in Unit #4 were apparently the result of nitrate buildup in the lysimeter over the winter months. This is also reflected in the NO_3-N lysimeter profile shown later. After surface clogging, little additional nitrification occurred in Unit #4. The high nitrate values from day 30 to 50 result from washout of the nitrate produced prior to clogging.

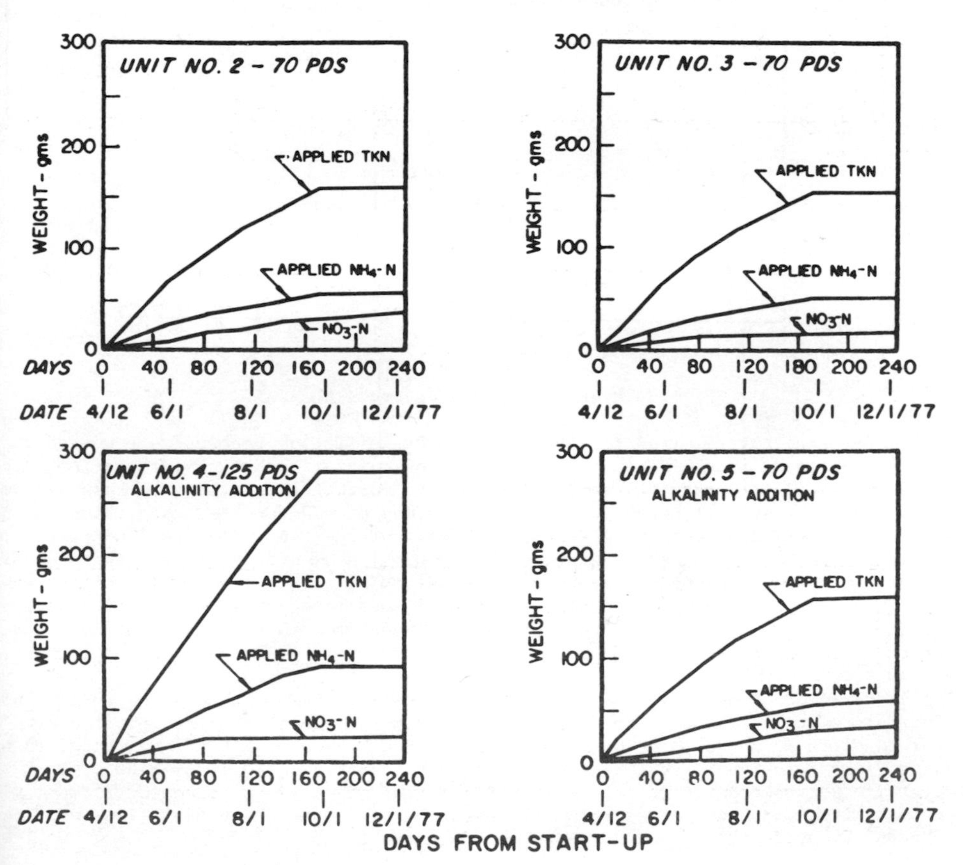

Fig. 8. Cumulative nitrogen balance - Phase II

Figure 8 indicates that the quantity of nitrogen recovered in the leachate in Phase II was less than the ammonia nitrogen applied. Similar results were obtained for Phase I, except for the lowest loaded Unit #2 where the NO_3-N recovery was about equal to the NH_4-N added. Monthly nitrate recoveries (leachate nitrate/total N added in sludge) vary significantly as shown in Figures 9 and 10. For Phase I, the 23 PDS Unit showed 43 to 97% recovery of nitrate after the initial lag period and prior to onset of winter temperature while the 68 PDS Unit yielded 33 to 43% recover

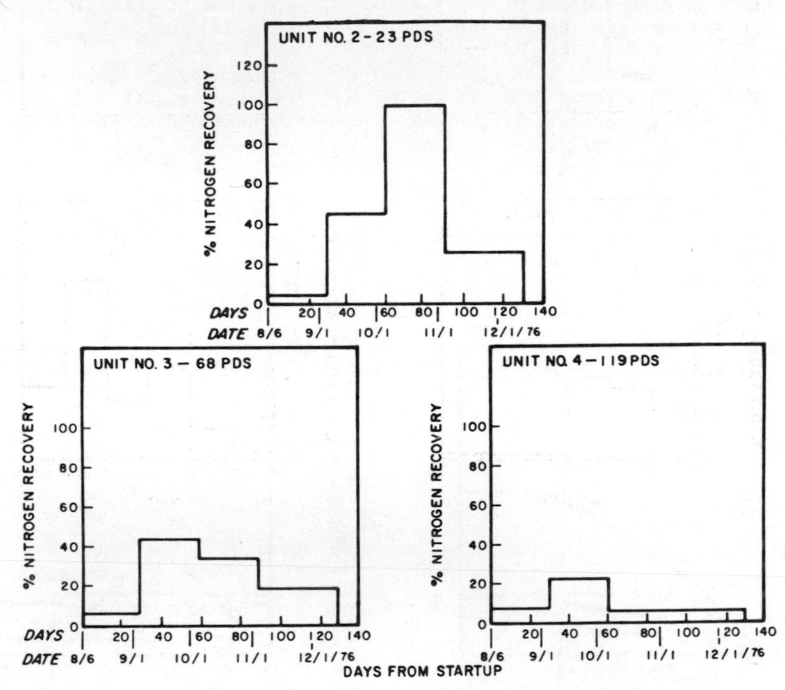

Fig. 9. Monthly NO_3-N recovery - Phase I

Much lower recovery resulted for Unit #4. In Phase II, low recoveries occurred in July for Units #2 and 5. This is due to the low rainfall and high evaporation rate providing little leachate generation during this month. An effluent recycle system during the low rainfall-hot weather conditions may provide greater recoveries in the dry weather similar to the June and August data. The effect on overall nitrate recovery, however, is not known since the accumulation of nitrate in the soil in July may have provided the high recoveries in August. The low recoveries in Unit #3 may have been due to operational problems previously mentioned as well as soil clogging and low pH. The maximum nitrification rates varied from 40 to 50 g/mo-m^2 for Units #2 and #5 in Phase II and from 50 to 60 g/mo-m^2 for Unit #3 in Phase I.

The leachate TDS concentrations in Figure 11 closely resemble the NO_3-N profiles of Figure 7. The process of nitrification mobilizes the previously precipitated or stored solids in the soil. The slope of a TDS-NO_3-N plot of 5.8 for Units #2, #3 an #5 is similar to the Phase I results representing either $Ca(NO_3)_2$ or $NaNO_3$. To determine the suitability of tracking TDS and thus NO_3-N leachate concentrations wi conductivity, measurements at 25°C were begun on Day 86 of the study. The correlat between conductivity and NO_3-N is shown in Figure 12.

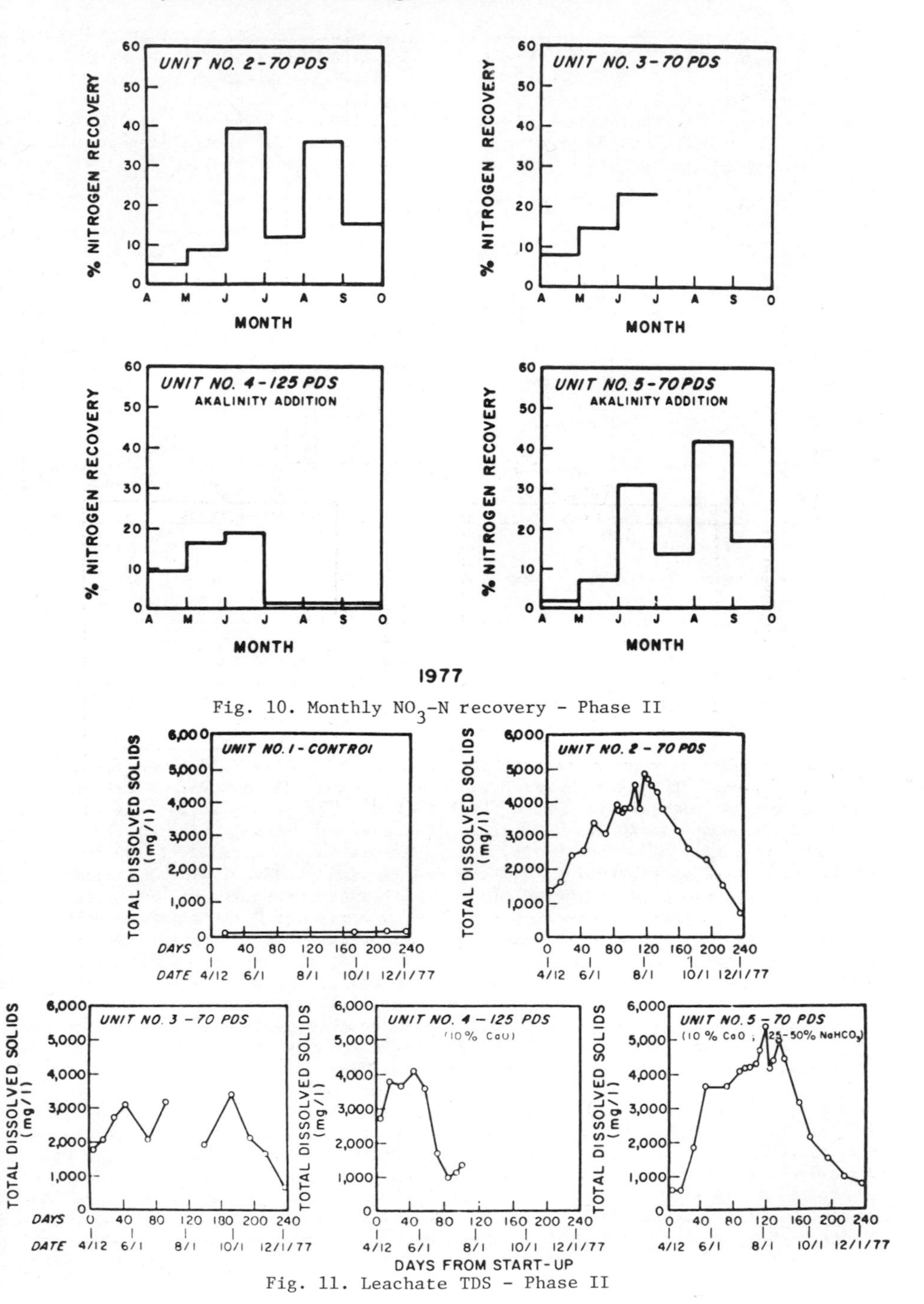

Fig. 10. Monthly NO_3-N recovery - Phase II

Fig. 11. Leachate TDS - Phase II

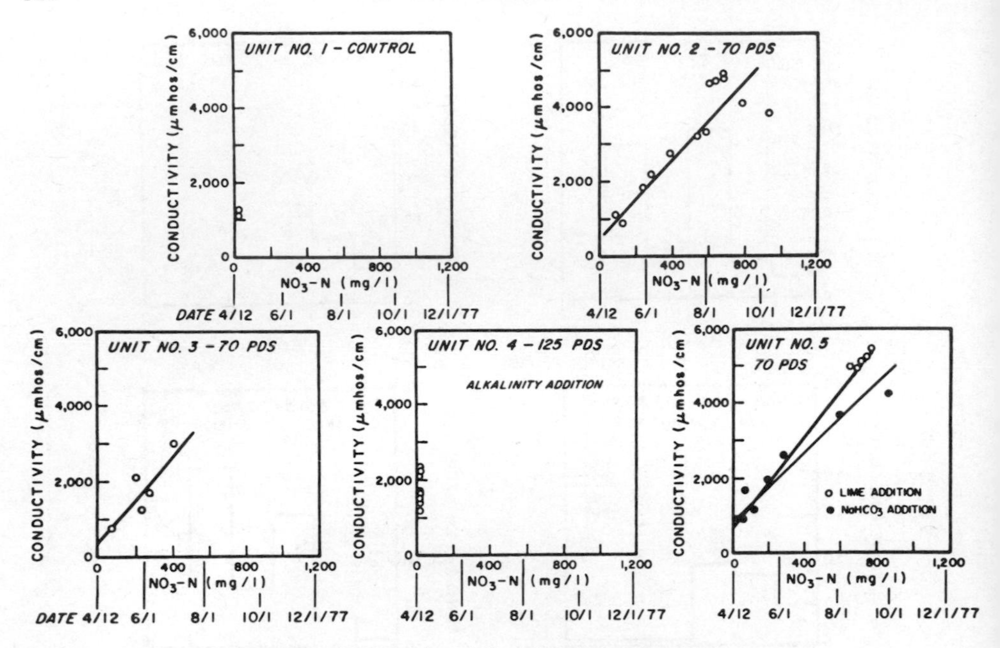

Fig. 12. Conductivity NO_3-N - Phase II

Higher conductivity is obtained for $Ca(NO_3)_2$ in the leachate over $NaNO_3$ as seen from the Unit #5 results. A minimum conductivity of about 1000 exists in the leachate at low NO_3-N concentrations, increasing above 5000 at high leachate concentrations.

Alkalinity analyses near the end of the Phase I study (Day 87) showed very low leachate values (16 - 25 mg/L as $CaCO_3$) for Units #2, #3, and #5. These Units had significant nitrification while Unit #4, which had essentially ceased nitrifying, had a concentration of 220 mg/L. In the second phase, Units #1, #2 and #3 without alkalinity addition had typical leachate values normally below 100 mg/L as $CaCO_3$, as seen in Figure 13. Unit #3 had the lowest alkalinities, at times equal to zero with a resulting pH of 4.5. After the cessation of nitrification in Unit #4, alkalinity of the leachate rapidly increased. The results on Unit #5 were surprising in that the leachate alkalinities were not significantly higher than those in Unit #2 although 10% lime and later 25 and 50% $NaHCO_3$ were added as alkalinity sources. To evaluate the alkalinity removal on the soil surface, 300 mL samples of the digested sludge with various chemical treatments were filtered through 2 inches of soil. With no chemical addition, only 630 mg/L out of the initial 3090 mg/L alkalinity leached through the soil. With 10% lime addition, the major portion (76%) of the added alkalinity was also removed by the soil due to typical softening reactions. Use of soda ash (Na_2CO_3) allowed between 40 and 50% of the alkalinity to leach, while $NaHCO_3$ provided 60 to 80% alkalinity leaching. The last compound raises the sludge pH the least, thus causing the lowest amount of chemical precipitation. Thus, use of lime in Unit #5 did not provide a large increase in leachate alkalinity since most should be stored in the soil. During addition of 25% $NaHCO_3$, a slight increase in leachate alkalinity concentration resulted. This was a period of high nitrate concentration where most of the added alkalinity may have been destroyed. After 50% $NaCHO_3$ addition, although nitrification was significantly reduced, high alkalinity was not attained in the leachate. The reason for this is not known. Also,

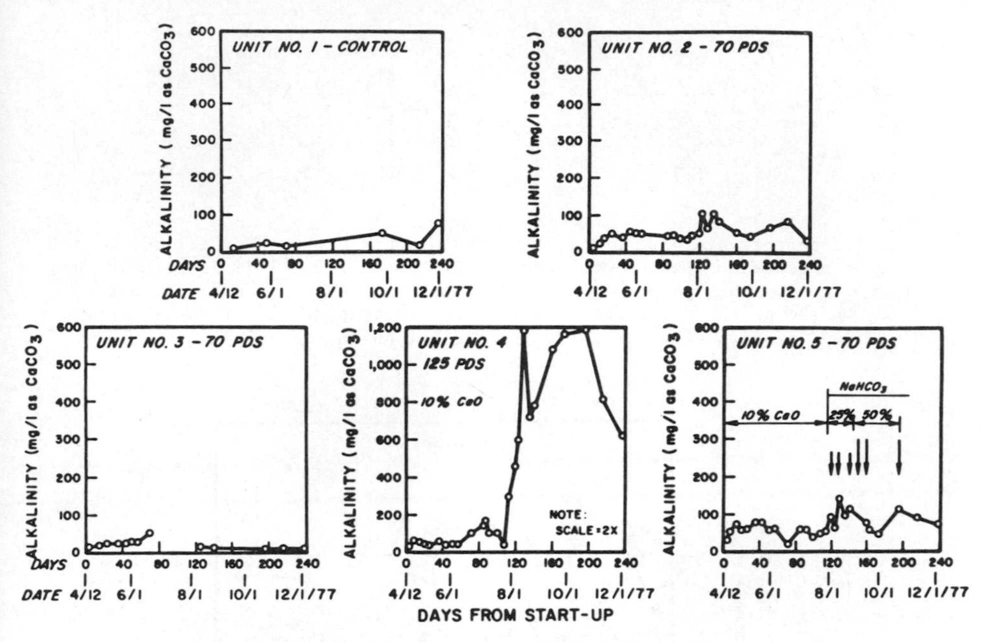

Fig. 13. Leachate alkalinity - Phase II

referring to Figure 7, nitrification in Unit #5 was significantly reduced by Day 160, the last day of weekly sludge addition, due to possible toxic effects of the high $NaCHO_3$ addition.

Figure 14 presents the leachate pH values over the Phase II study. Units #2 and #5, both loaded at 70 PDS, had pH values between 6.5 and 7.5 during the first 160 days of the study and around 6.0 thereafter. Although alkalinity was added to Unit #5, only a slight increase in leachate pH was realized with significant variation occurring from sample to sample. Towards the end of the study during the lower temperature period, the pH was maintained around 6.0 for Unit #2. For Unit #3, also loaded at 70 ton dry solids per acre per year, the pH is significantly lower than the above units: initially between 5 and 6 and at the end of the study, between 4 and 5. This unit was previously (Phase I) loaded at 68 PDS while the other two were loaded at 23 PDS and 25 VFC. The higher loaded unit seemed to have lost its buffering capacity to a greater extent than the previous lower loaded units. Substantially all its alkalinity was exhausted by the end of the study. There was some clogging in this unit, possibly in the effluent line, towards the middle of the study, thus anaerobic conditions may have been present during a portion of this run. The highest loaded Unit, #4, also had pH values of 5 and 6 during the first 60 days of the run even with the 10% lime addition. During this period, the nitrate concentrations were high, 800 mg/L. As nitrification ceased, an immediate increase in pH resulted and remained between 6 and 8 for the remainder of the run while the alkalinity was between 600 and 1200 mg/L. During this period, the pH in Unit #1 varied between 6 and 7.5 after the initial start-up value of 5.2.

Figures 15 and 16 show the leachate phosphorous and TKN concentrations. High TKN (mainly NH_4-N) and P values existed in Unit #4, especially after nitrification ceased.

Based on the CEC values for the various soil layers in the lysimeter, each lysimeter has the capacity to exchange 7750 meq of cations as shown below:

Layer	CEC (meq/100 g)	Dry Weight (g)	Total Cations Exchangeable (meq)
Sod	23.8	2,900	690
Topsoil	17.7	15,600	2,760
Sand	4.7	91,500	4,300
TOTAL			7,750

In Unit #4, approximately 27.5% of the added ammonia has been nitrified, thus leaving an ammonia content of 8000 meq not accounted for. This is greater than the CEC value for the lysimeter, indicating that Unit #4 was probably saturated with ammonia. Although the above comparison is not exact since the amount of ammonia stripping and organic nitrogen hydrolysis is not known, it illustrates the

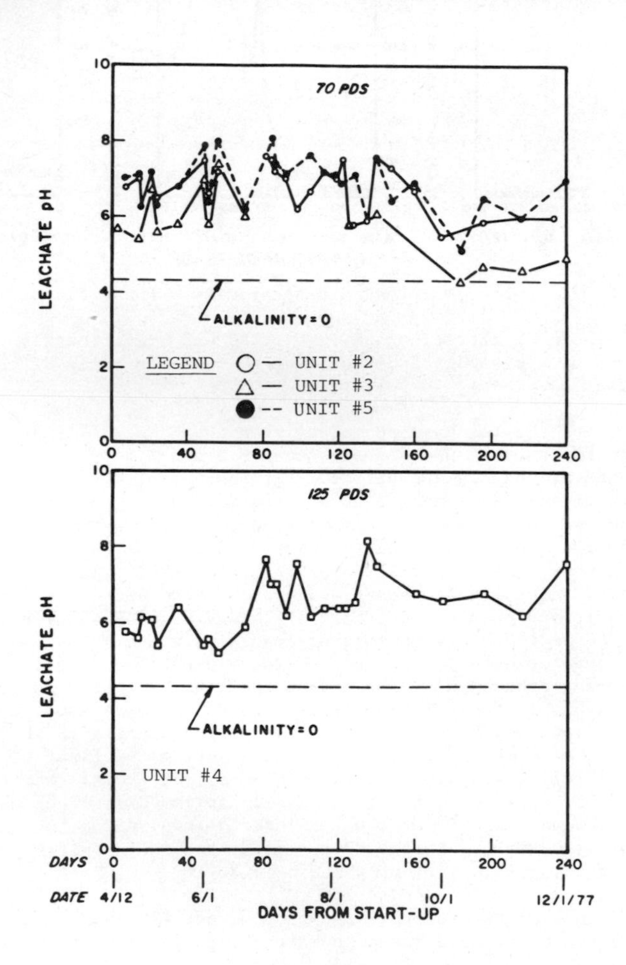

Fig. 14. Leachate pH - Phase II

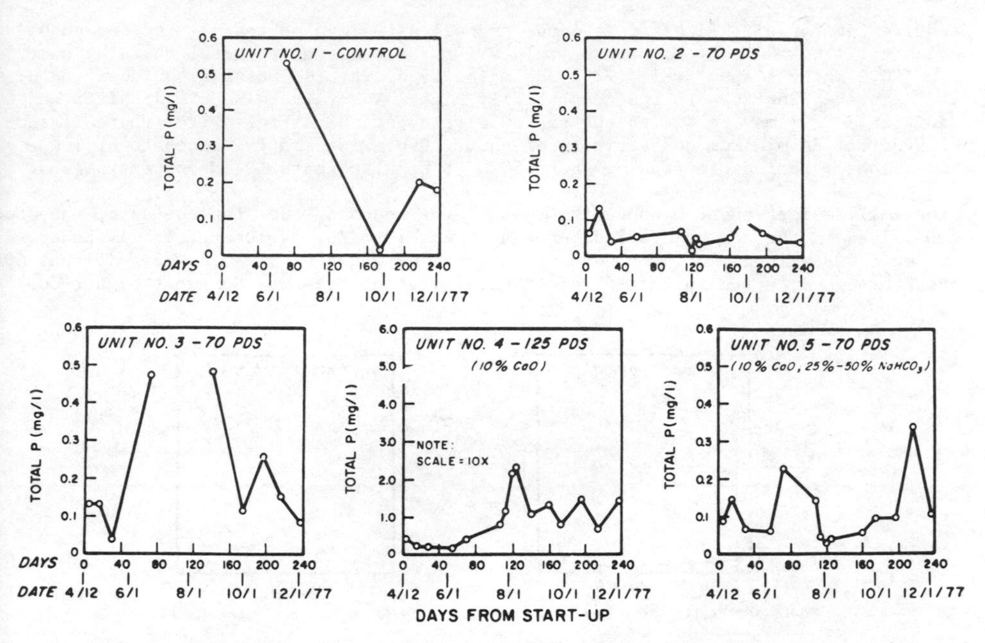

Fig. 15. Leachate total P - Phase II

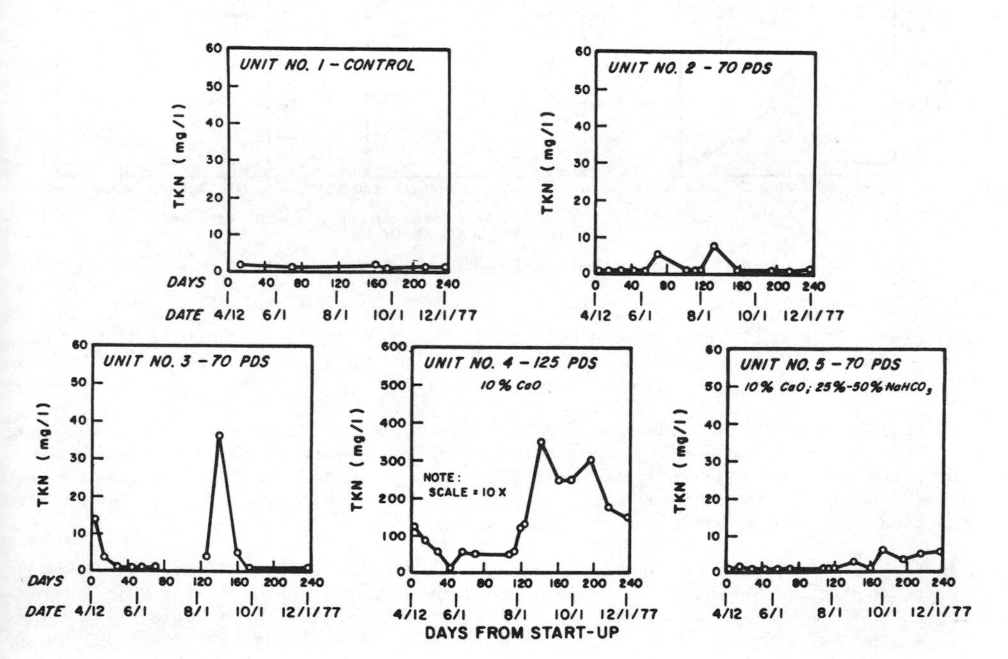

Fig. 16. Leachate TKN - Phase II

requirement for nitrification to occur so soil sites can be regenerated for ammonia exchange. The TKN and phosphorous values for the other units were generally under 5 and 0.2 mg/L, respectively, with the majority of the TKN being organic nitrogen. It is interesting to note the consistently higher TKN values in Unit #5 after $NaHCO_3$ addition. This may be due to the sodium displacing the nitrogen from the soil surface sites. A minimum conductivity of about 1000 exists in the leachate at low NO_3-N concentrations, increasing above 5000 at high leachate NO_3-N concentrations.

Figure 17 depicts the COD concentrations in the leachate over the course of the stud COD values in the control varied between 20 and 40 mg/L. Values generally under 100 mg/L COD existed in the sludge units which were not overloaded. Significant COD breakthrough occurred in Unit #4 after nitrification ceased. Somewhat higher COD

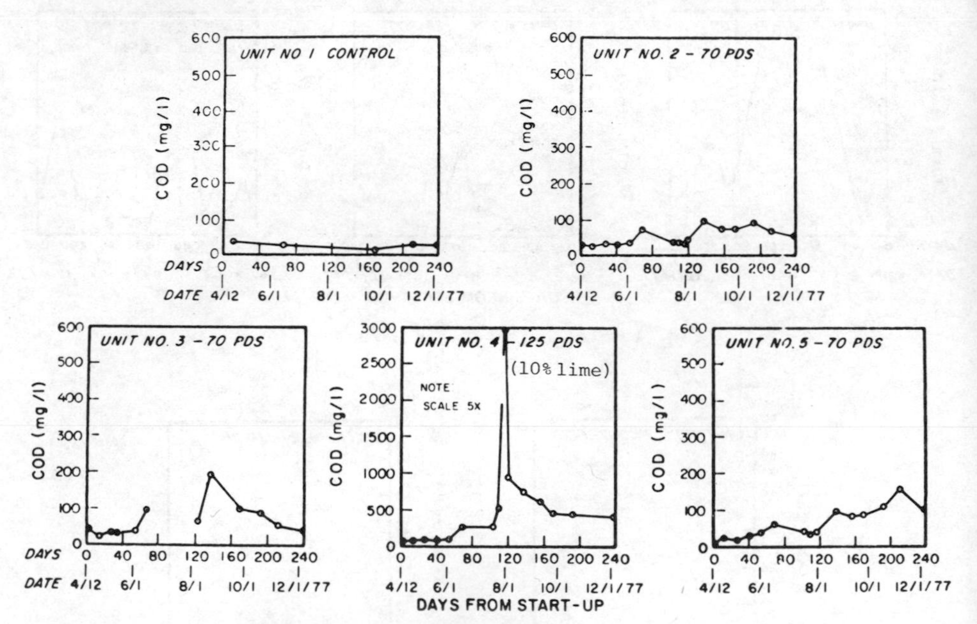

Fig. 17. Leachate COD - Phase II

values occurred in Unit #5 than #2 at the end of the study, possibly due to the addition of the $NaHCO_3$. Some COD breakthrough also occurred in Unit #3 after the soil surface had been turned under and leachate once again collected. At this time, high TKN and phosphorous concentrations also existed.

The phosphorous, TKN and COD leachate concentrations for Phase I were also low excep for Units #2 and #3 toward the end of the first Phase (Figure 18). In the highly loaded unit, higher leachate concentrations became evident after nitrification ceased when flooding of the surface occurred. In the 68 PDS unit, high leachate values resulted when sludge was added to frozen soil.

Figures 19 and 20 present the cation-anion analysis for sludge and leachate samples collected during the Phase II study in the form of bar diagrams of the major chemica constitutents. The major cations of the sludge are calcium and ammonia while alkalinity is the major anion. The leachate samples for Units #1 and #3 (Figure 19) which had no chemical added to them, show a marked contrast between the control and the sludge amended units. For the control, leaching of constitutents through the soil

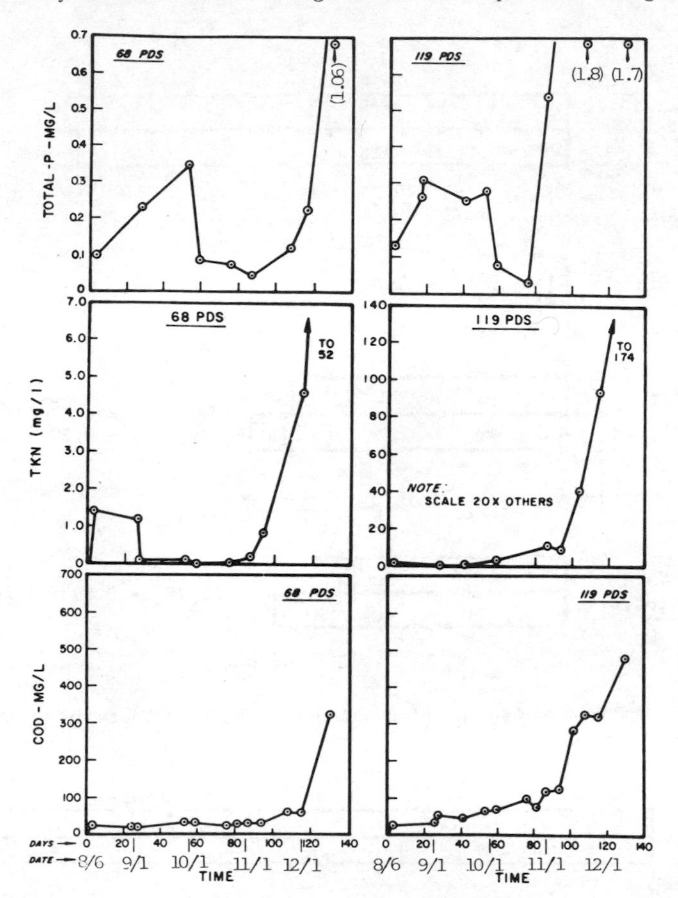

Fig. 18. Leachate P, TKN & COD - Phase I

column is relatively slight compared to the sludge amended soils. The lack of an exact equivalent balance is due to a large extent to analyses on leachate samples collected on different dates. The major cations leaching through Unit #1 are potassium and sodium with much lower concentrations of calcium. The opposite is true for Units #2 and #3, where the cation constitutents in the leachate more closely follow the concentrations in the sludge, except for the heavy metals which were held in the soil column and the ammonia which is oxidized to nitrate. Only a slight amount of alkalinity is present in the leachate due to the portion captured in the soil and that destroyed by nitrification. In Unit #3, no alkalinity existed with the pH at 4.3 in the leachate. The higher Ca^{++} and Mg^{++} concentrations in Unit #2 may be due to the high degree of evaporation (low V_l/V_s) occurring during this period.

With 10% lime addition to Unit #4, the pH is raised and the softening reactions take place to precipitate the Ca^{++} and Mg^{++} hardness as well as the alkalinity. The quantity of lime added to the Units was 1710 mg/L as CaO or 61 meq/L, or 78% of that required for theoretically complete softening. The actual hardness removal was 71%, in close agreement to that anticipated. Due to these softening reactions, the total dissolved solids in the leachate is significantly lower than the other sludge amended units. Although most of the initial alkalinity is precipitated out as $CaCO_3$, a much

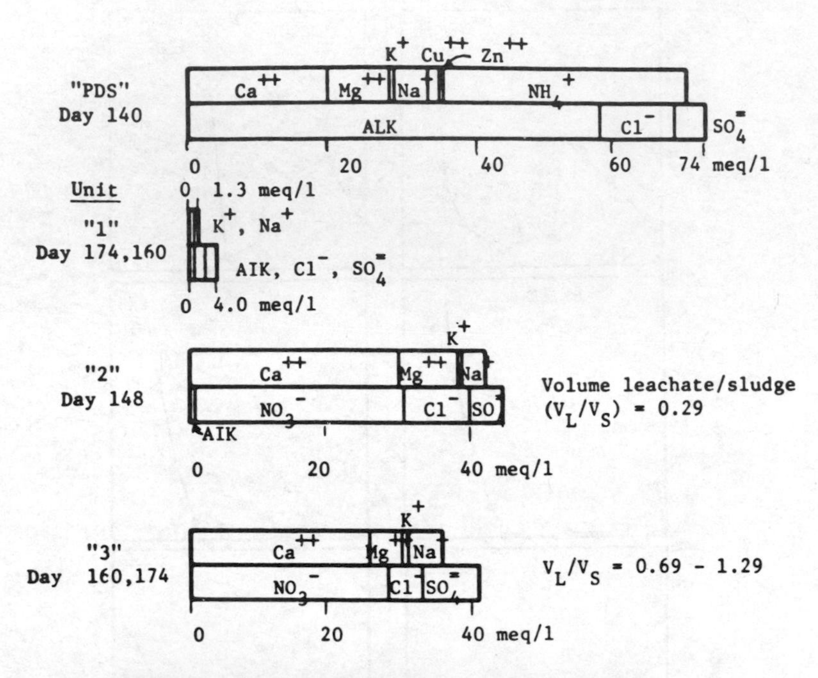

Fig. 19. Cation-anion balance - Phase II Units 1-3

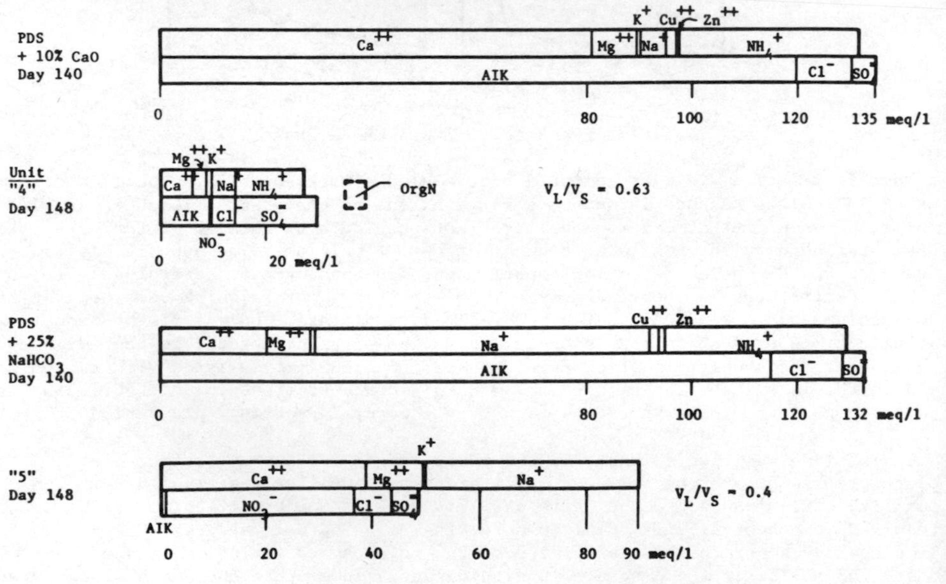

Fig. 20. Cation-anion balance - Phase II - Units 4 and 5

higher alkalinity concentration exists in the leachate of this unit compared to the others due to the lack of nitrification. A significant portion of organic nitrogen also appears in the leachate, the net charge not known. The higher measured sulfate concentrations in the leachate may be due to the solubilization of reduced aulfur in the upper lysimeter layers followed by chemical oxidation to sulfate when exposed to oxygen in the lower layers or in the collection bucket. In Unit #5, sodium bicarbonate addition was used instead of lime in the latter portion of the study. No significant pH increase occurred and no softening reactions as with the previous lime addition. Thus, the leachate reflects the increased sodium concentrations and higher calcium and magnesium concentrations. These latter are actually higher than in the PDS, indicating some displacement of previously exchanged Ca^{++} and Mg^{++} from the soil along with the evaporation effect. The $NaHCO_3$ addition was initiated on Day 119 and should not have been completely washed through the column by Day 148, although it is obvious that its effect is being felt in the above leachate sample. The reason for the lack of an anion-cation balance and low leachate alkalinity is not know, one possibility being NO_2-N which was not measured in the study.

Recovery of Chemicals in Leachate

Table 4 presents the percentage of added nutrients, sludge plus chemical, that leached through the soil during the Phase I and Phase II studies. The total solids recovery is due to the leachate TDS, the greater recoveries in the units obtaining greater degrees of nitrification. Less than 1% COD, alkalinity and phosphorous

TABLE 4. Nutrient Recovergy in Leachate

Parameter	Percent Recovery in Leachate for Unit				
PHASE I	(1)[a] CONTROL	(2) 23 PDS	(3) 68 PDS	(4) 119 PDS	(5) 25 VFC
Total Solids	-	29 (21)	20 (18)	11 (9)	11 (4)
COD	0.04	0.5 (0.2)[b]	0.5 (0.4)	0.8 (0.8)	0.01 (0)
Total N as NO_3-N in Leachate	4.2	35 (28)	23 (21)	11 (10)	34 (18)
Total P	0.06	0.1 (0.01)	0.05 (0.01	0.1 (0.08)	0.07 (0)
PHASE II		(2) 70 PDS	(3) 70 PDS	(4) 125 PDS	(5) 70 PDS
Total Solids		16	9	7[c]	13[c]
COD		0.3	0.2	2	0.4
Total N as NO_3-N in Leachate		20	12	8	18
Total P		0.01	0.02	0.10	0.01
ALK		1	0.3	5[c]	1[c]

a - Based on initial soil quantity
b - () Based on leachate quantity of unit minus control
c - Includes chemical addition

leached through the soil except for Unit #4, the highest loaded unit. The nitrogen recovered as nitrate was generally between 18 and 21% for the 68 and 70 PDS units, while somewhat higher values occurred for the lower loaded units in Phase I. The Phase II Unit #3 at 70 PDS had lower recoveries due to clogging and pH problems. The metals recovery in the leachate of all units over the two phases of the study is given in Table 5. In the Phase I study, many of the leachate concentrations were below the detection limits. With the exception of Pb, the quantity of metals added was greater than initially present in the lysimeters by a factor of 1.4 to 7.8. Lead additions varied from 0.5 to 1.1 times the initial quantity present. For zinc, significantly greater concentrations and loads leached during the second phase than during the first, except for the control. The metals recovery varied from a maximum of 4 to 6% for the cadmium and zinc in Unit #3, which had the lowest pH, to a minimum of 0.4%, excluding copper. Copper recovery was less than 0.1%; however, a copper analysis was not conducted near the end of the study where higher concentrations may be anticipated. Only 0.2% or less of the metals leached out of the controls.

TABLE 5. Percentage Metals Recovery in Leachate over Total Study

	Leachate Concentration Range (mg/L)	Percent Recovery for Unit				
		#1	#2	#3	#4	#5
Cd	0.002 - 0.0222	0.2	0.6	5.8	2.4	1.6
Cr	<0.025 - 0.05	ND	0.7	0.5	0.4	0.6
Cu*	<0.025 - 0.033	ND	0.03	0.07	0.06	0.03
Pb	<0.05 - 0.50	ND	1.1	1.5	0.7	2.9
Zn	0.03 - 3.0	0.2	2.5	4.2	0.5	0.4

* Data only for Phase I and initial portion of Phase II

Grass Analyses

Unit #2 loaded at 23 ton/acre-year (23 PDS) showed about 100% increase in grass growth compared to the control, while Unit #3 (68PDS) showed a 40 to 60% decrease but consistently maintained some grass growth on the surface. The heavily loaded unit (119 PDS) completely eliminated grass growth; the total surface was covered by a layer of black sludge. Application of the dried filter cake initially inhibited grass growth, but it subsequently returned to normal after 2 months. This was apparently due to leaching from the sludge of the lime and $AlCl_3$ conditioning chemicals as well as a reduction in the surface coverage of the cake upon drying. The VFC attained 90% solids after 25 days drying on the lysimeter. Grass and plant samples were obtained from the units in May and June 1977 and analyzed for metals as shown in Table 6.

TABLE 6. Land Disposal Grass Samples - Phase II

Date 1977	Unit	Cd (mg/kg)	Cr (mg/kg)	Cu (mg/kg)	Pb (mg/kg)	Zn (mg/kg)	
5/17	1	0.202	0.84	4.05	6.41	19.9	
5/17	2	0.173	3.67	48	4.97	37.6	
6/27	1	0.40	0.80	3.75	3.20	34.0	) grass
6/27	2	0.60	22.0	250.0	28.0	131.0	) grass
6/27	3	0.30	3.50	38.1	6.75	40.5	) "plants"
6/27	5	0.15	1.60	19.5	5.50	28.5	) "plants"

For Units #1 and #2, metals concentrations in the grass for the sludge amended sample was generally significantly greater than for the control. Large weeds growing in Units #3 and #5 also had significant metals concentrations,but no comparison was available from the control. A significant increase in Unit #2 metals occurred from the grass samples on May 17, 1977 (Day 35) to June 27, 1977 (Day 6), probably due to the significant additional metals load added during this period.

Soil Analyses

Lysimeter profiles were obtained from soil samples taken on April 13 and 14, 1978. The units were frozen most of the winter so the profiles obtained should be representative of the lysimeters at study completion the prior December. Figure 21 shows the volatile solids and, thus, organic content decrease from the surface through the depth of the lysimeters. The control, Unit #1, had about 30% of the dry solids on the surface ("1/4") as volatile due to the initial sod used in the units. The units with sludge addition contained significantly greater organic concentrations on the surface. Unit #3 was an exception,probably due to the surface turned under on Day 119 of Phase II - thus only the pH profile is plotted for this unit. In both Units #2 and #4, organic matter was transported to a depth of about 6 inches in the units. Unit #4 was turned under in both Phase I and II. Unit #2 was not turned under, but still had significantly higher organic concentrations than the control at 2 and 6" contrasted to Unit #5 which only had higher surface concentrations. This may be due to the chemical addition precipitating out the organics on the surface in Unit #5 while the organics are more mobile in Unit #2 which was not chemically treated.

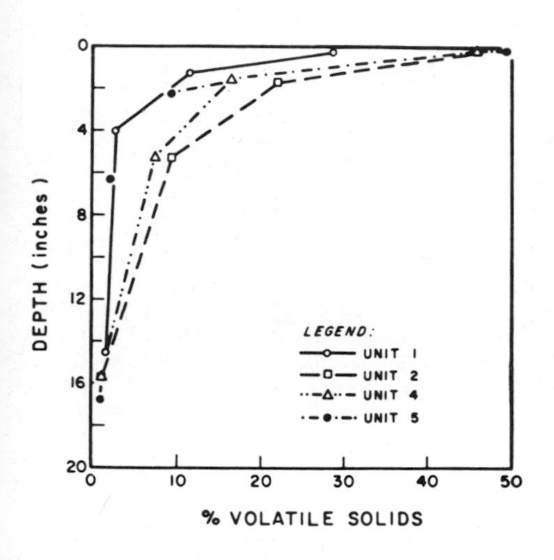

Fig. 21.
Volatile solids profile

Chemical addition to Units #4 and #5 is seen in Figure 22 to significantly affect both the pH and alkalinity distributions in the lysimeters, both having significantly greater values than the other units. The lower pH value of Units #2 and #3 are consistent with the values obtained in the leachate at the end of the study (Figure 14). Inhibition of nitrification would be expected at these lower pH values. The higher alkalinity and pH at the surface of Unit #2 is due to precipitaiton of the alkalinity in the digested sludge. The pH is seen to be higher near the bottom of the lysimeters, thus the pH existing in portions of the soil column may be significantly lower than the leachate value. The alkalinity, pH and nitrate values for these samples were obtained by diluting 10 grams of the soil sample with 100mg/L distilled water. The TKN distribution in Figure 23 is similar to the volatile solids with most accumulated near the surface. Ammonia is similar, except for the highly loaded Unit #4, which had ammonium ion distributed throughout the soil layer. This may be due to overloading of the ammonia adsorption capacity of the soil and lack of nitrification due to anaerobic conditions. The major portion, typically 99%, of the nitrogen in the soil column is organic nitrogen. Unit #2 is seen to have greater nitrogen penetration than Unit #5, again probably due to no chemical addition allowing greater mobility. The nitrate values in Figure 24 show the highest concentrations on the surface for Units #2 and #5

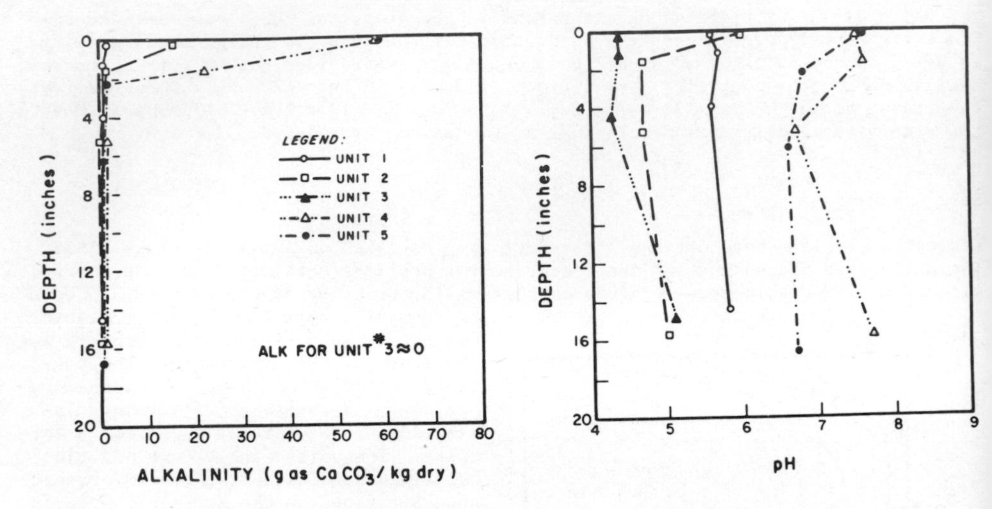

Fig. 22. Alkalinity and pH profiles

decreasing with depth. This is indicative of high nitrate production at the surfac where high ammonia concentrations exist with subsequent washout of the nitrate through the units. For Unit #4, a significant amount of nitrate production occurs within the units due to the higher ammonia concentrations existing in the upper six inches. After freezing over the winter, the surface dried to allow sufficient oxygen penetration for nitrification to occur. The phosphorous values in Figure 25 show the expected trend for Unit #4, significant accumulation at the surface. However, for Units #2 and #5, lower surface values result; the reason for this is not clear.

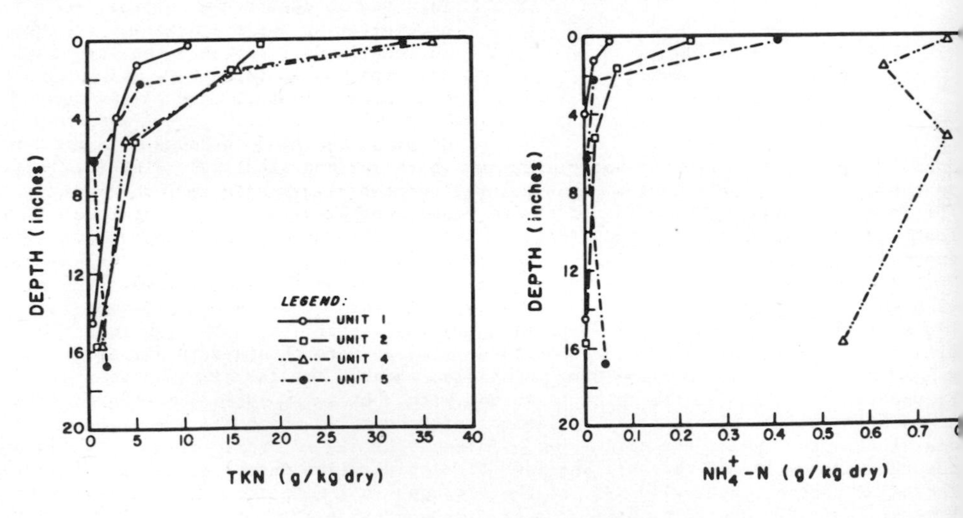

Fig. 23. TKN and NH_4-N Profile

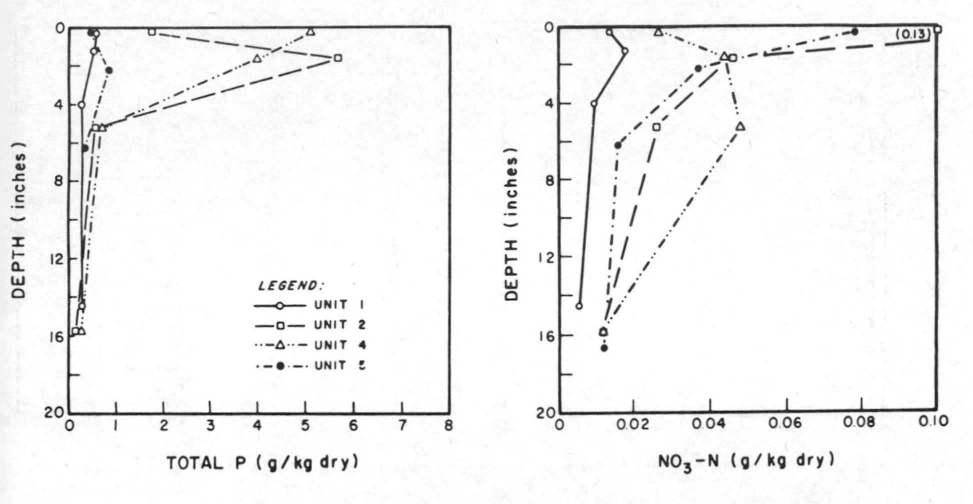

Fig. 24. Total P and NO_3-N profiles

'he metals data also typically show high accumulations at the surface decreasing 'ith depth. For cadmium in Figure 25, however, Unit #5 again shows lower surface 'alues but higher values at the 6" depth. Whether the Na^{++} added to this sample .owards the end of the study caused cadmium mobility is not known. Figures 26 and 27 show the remaining metals profiles. Both copper and zinc have accumulated to the highest surface concentrations of these compounds in the digested sludge. Based on 10% of the soil cation exchange capacity, the sand layer appears to still have some adsorption capacity for metals. The much greater metals concentrations on the sludge amended soils compared to the control provides the possibility of nitrification inhibition due to metals toxicity. Relating the copper content of the soil to the volatile solids shows that the soil in the lower layers has a lower copper concentration per unit volatile solids than the upper layer. The concentration on the surface is greater than the sludge concentration, indicating either volatile solids destruction or transport into the lower layers to a greater extent than the metals. The depths below 6" are substantially equal to the control concentrations.

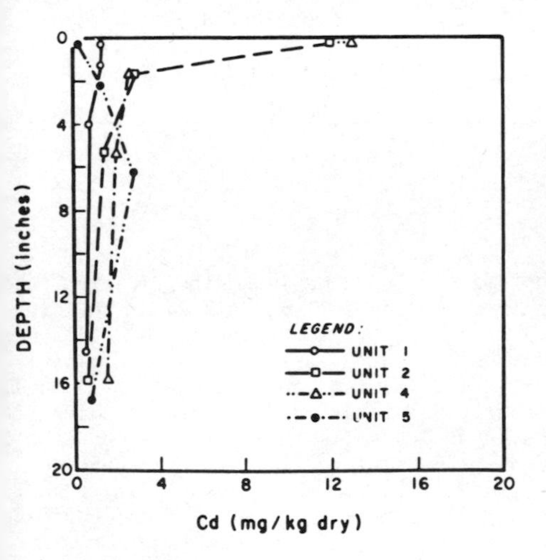

Fig. 25. Cd profile

$_2$ Utilization

n estimate of the O_2 requirements for nitrification can be made as 4.57 times the itrification rate. For maximum nitrification rates of 46 to 60 g/mo-m^2, required $_2$ transfer rates vary from 210 to 270 g/month-m^2. The total O_2 transfer rate from

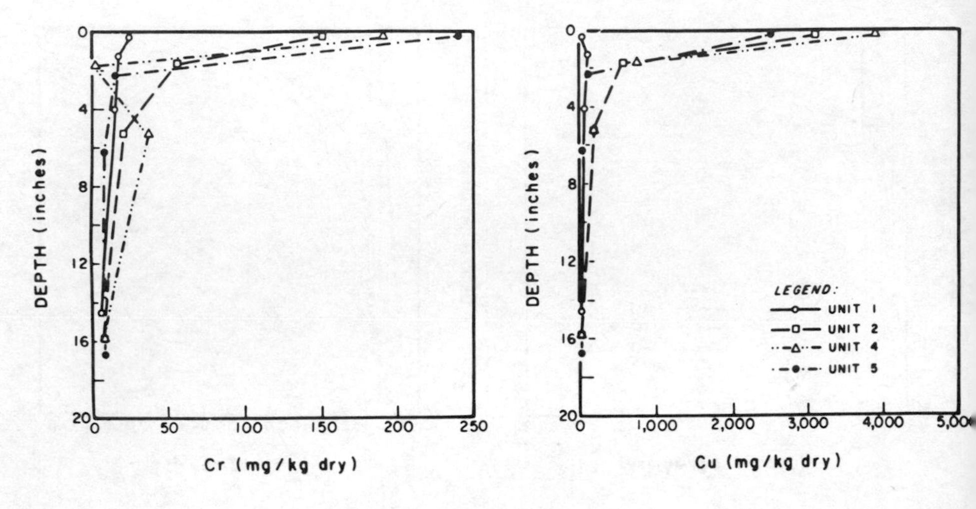

Fig. 26. Cr and Cu profiles

the atmosphere must be greater than this due to organic carbon oxidation. Based o total volatile acids and 20% of the solids COD oxidized along with 40% of the TKN during the maximum month, the organic carbon oxidation rate would be about equal t the nitrification rate:

$$\text{Volatile Acids} = 0.20\ \frac{gVA}{gdry} \times 1.07\ \frac{gO_2}{gVA} \qquad = 0.02\ gO_2/gdry$$

$$\text{COD} = 1.0\ \frac{gCOD}{gdry} \times 0.20\ \text{(assumed)} \qquad = 0.20\ gO_2/gdry$$

$$\text{Total Organic Carbon Oxidation} \qquad = 0.22\ gO_2/gdry$$

$$\text{Maximum TKN Oxidation} = 0.11\ \frac{gTKN}{gdry} \times 0.40 \times 4.57\ \frac{gO_2}{gTKN} \qquad = 0.20\ gO_2/gdry$$

The rate of O_2 transfer into the soil, therefore, during maximum nitrification rat was approximately 500 g/month·m^2. It appears that the soil in Unit #4 at the high loading rate was unable to maintain oxygen transfer at the required rate. This wa probably due to periodic flooded conditions, since the O_2 transfer rate in a wet s is low. The amount of O_2 input from the rainfall is also low, 1-2 g/mo.-m^2. Thus the requirement for nitrification must properly be evaluated in setting loading ra as a function of climatological conditions. For dry soils (0.1 atm water tension) O_2 transfer rates between 2000 to 3000 g/mo.-m^2 have been calculated (Eckenfelder and O'Connor, 1961), based on a nonsteady state diffusion through an infinite soil with no uptake rates, an assumption which appears unrealistic. For trickling filt slimes, maximum rates in the order of 300 g/mo.-m^2 have been measured, while Loehr others (1979) have shown similar rates for typical soils at a 6 inch depth.

Metals Limitation Analysis

The high loading rate utilized in this study causes high metals per unit area of land site to result. If the site were to be utilized for agricultural purposes, allowable metals loadings for a soil with a 10 meq/100 g CEC have been given by

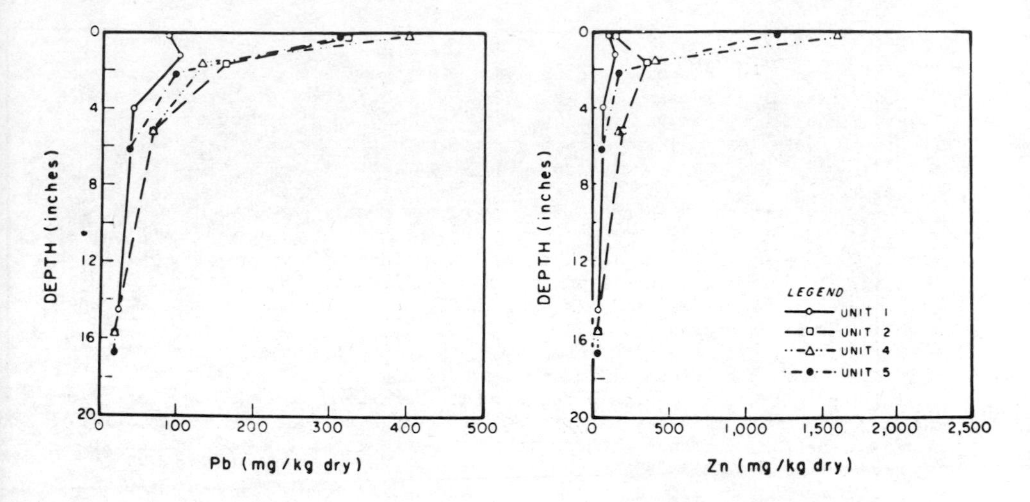

Fig. 27. Pb and Zn profiles

Overcash and others (1977). Based on these data, copper would limit usage of a site to less than one-half year as shown in Table 7, if agricultural constraints controlled site usage. From this analysis, it is obvious that the site would have to be classified as one designated for sludge disposal only and viewed as part of the waste treatment plant. If the site were to be abandoned at a future date, the surface layer may have to be removed and the metals either recovered or disposed of as a hazardous material to prevent ground water contamination and make the site available for agricultural usage.

TABLE 7. Metals Limitation Analysis for Agricultural Usage

Metal	Ridgewood Sludge Concentration mg/g dry solids	Metals Loading Rate lb/acre-year at dry solids loading rate of 70 ton/acre-year	Allowable Metal Load for Agricultural Usage* lb/acre	Time for Allowable Sludge Discharge Based on Agricultural Constraint year
Cd	0.022	3	9	3
Cr	0.25	35	500	14
Cu	3.9	550	225	0.4
Pb	0.32	45	900	20
Zn	1.8	250	450	2

*Based on 10 meq/100 g CEC soil [Overcash and others (1977)]

CONCLUSIONS

1. A high rate land disposal system using a sludge loading rate of 70 ton/acre-year (excluding freezing conditions) would provide 20% yearly nitrogen recovery as nitrate in the leachate with no soil clogging problems.
2. At the above loading rate, low pH values may result unless alkalinity (lime) addition is practiced to insure adequate pH control in the nitrifying soil layer.
3. The nitrification rate in the soil is significantly affected by weather. Lower temperatures and high rainfall not only affect organism reaction rates but the drying of the soil and thus the oxygen transfer capability into the soil. The maximum oxygen transfer rate attained over the course of the study was estimated at 500 g/mo-m^2, easily attainable in dry soil but unattainable in saturated soil.
4. Highest soil nitrification rates (50-60 g/mo-m^2) occurred during summer weather when significant rainfall provided high leachate quantities. To increase nitrate recovery in the leachate, effluent irrigation of the sludge disposal site during low rainfall - hot weather periods may be desirable.
5. Measurement of leachate conductivity is an excellent, simple, and inexpensive test for monitoring leachate nitrate concentrations.
6. A loading rate of 125 ton/acre-year caused soil clogging and resulted in little nitrification due to suspected anaerobic conditions in the soil. Under these conditions, soil cation exchange capacity is quickly (2-3 months) saturated with NH_4 with significant breakthrough occurring. Thus, unless nitrification occurs in the soil, significant ammonia breakthrough to lower layers will result at high loading rates.
7. High concentrations of most sludge constituents (TKN, P, volatile solids, alkalinity and metals) occur in the surface (2 inch) soil layer. This is due to a combination of the following factors: (1) the major constituents are associated with the solids fraction of the sludge which is filtered out on the surface; (2) the chemical precipitation (softening) reactions occurring during lime addition; and (3) the higher cation exchange capacity of the soil and topsoil layers compared to the sand layer.
8. Grass and plants grown on the sludge amended soils at 70 ton/acre-year contain significantly greater metals concentrations than the controls. Based on an estimate of allowable metals loading for agricultural usage, copper would limit the application time to 1/2 to 1 year at a rate of 50-70 ton/acre-year for the municipal sludge used in the analysis. The sludge site may have to be designated for sludge disposal only for applications more than one year. Periodic removal of the upper 2 inches of soil, every two to five years, and disposal in a small controlled landfill site adjacent to the sludge area should prevent soil clogging and allow the larger sludge site to be utilized for other purposes in the future.
9. Maximum metals leaching occurred for the lowest pH unit (#3), resulting in 5.8% of the cadmium and 4.2% of the zinc recovered in the leachate. When lime addition was utilized, less than 1% of the added metals appeared in the leachate, except for cadmium where a maximum of 2.4% appeared.
10. A soil system model should be constructed to evaluate optimum loading frequency as a function of localized weather conditions. Field studies are desired to verify allowable application rates, chemical requirements, and economic viability.

ACKNOWLEDGEMENTS

The analytical and financial assistance of Hydroscience, Inc. is acknowledged along with the cooperation and assistance of the Ridgewood Treatment Plant personnel, especially John La Grosa and Lou Gallo. The work of Douglas Starr and Ajit Singh Pannu in sampling and data analysis is also acknowledged.

REFERENCES

Alvey, R.A. (1979). Effects of aerobic sludge disposal on land. Presented at the 52nd WPCF Convention, Houston, Texas.

Andrew, R.C. and A.P. Troemper (1975). Characteristics of underflow resulting from cropland irrigated with sewage sludge . Presented at the 48th WPCF Conference, Miami Beach.

Eckenfelder, W.W. and D.J. O'Connor (1961). Biological Waste Treatment, Pergamon Press, Oxford.

Eckenfelder, W.W. (1966). Industrial Water Pollution Control, McGraw-Hill, New York.

Hinesly, T.D., O.C. Braids, R.I. Dick, R.L. Jones and J.E. Molina (1974). Agricultural Benefits and Environmental Changes Resulting from the Use of Digested Sludge on Field Crops. Report for Metropolitan Sanitary District of Greater Chicago.

Loehr, and others (1979). Land Application of Wastes, Vol. II, Van Nostrand Reinhold Co., N.Y., pg. 50.

Malina, J.F. and J. DiFilippo (1971). Treatment of supernatants and liquids associated with sludge treatment , Water and Sewage Works, R-30-R-38.

Overcash, M.R. and others (1977). Pretreatment-land application of textile plant wastes. Presented at the 70th Annual AICMG Meeting, New York.

LIST OF SYMPOSIUM PARTICIPANTS

Abou-El-Hassan, M.E., Cairo Univ., Dept. of Chem. Eng., Giza, Egypt
Abraham, G., Univ. of Detroit, Dept. of Chem. Eng., Detroit, Michigan, USA 48221
Allison, D.P., Sutherland-Schultz Ltd., Eng. Mgr., 859 Courtland Ave.E., Kitchener, Ontario, Canada, N2C 1K4
André, G., Univ. of Waterloo, Dept. of Chem. Eng., Waterloo, Ontario, Canada, N2L 3G1

Baldwin, A., Ministry of the Environment, 637 Niagara St.N., Welland, Ontario, Canada, L3C 1L9
Bello, A., Univ. of Waterloo, Dept. of Chem. Eng., Waterloo, Ontario, Canada, N2L 3G1
Belore, R., Univ. of Waterloo, Dept. of Civil Eng., Waterloo, Ontario, Canada, N2L 3G1
Bhattacharyya, A.K., Jawaharlal Nehru Univ., School of Environmental Sciences, P.O. New Delhi - 110067, India
Brown, W.F.M., Walter Brown Associates, 533 Arbor Road, Mississauga, Ontario, Canada L5G 2J6
Brzustowski, T.A., Univ. of Waterloo, Vice-President Academic, Waterloo, Ontario, Canada N2L 3G1

Chan, P., New Jersey Inst. of Technology, Environmental Eng., 323 High Street, Newark, New Jersey 07102, USA
Chandra, A.L., Bose Institute, Dept. of Microbiology, 93/1 Acharya P.C. Road, 35-2402 Calcutta 700 009, India
Chang, N.H., Esso Resources Canada Ltd., 339-50th Avenue S.E., Calgary, Alberta Canada, T2G 2B3
Chin, K.K., Univ. of Singapore, Dept. of Civil Eng., Singapore 0511
Constable, T.W., Canada Centre for Inland Waters, Burlington, Ontario, Canada
Cornwall, G.M., Environmental Protection Services, Policy Planning Group, Place Vincent Massey, Hull, Quebec, Canada
Craig, G., Ontario Ministry of the Environment, Water Resources Laboratory, Toxicity Unit, 1 St. Clair Avenue W., Toronto, Ontario, Canada

Dalgity, C., Univ. of Toronto, Toronto, Ontario, Canada
Dance, K., Ecologists Ltd., Environmental Planning, 309 Lancaster Street W., Kitchener, Ontario, Canada N2H 4V4
Daniell, P.L., McDonnell Douglas Canada Ltd., Materials and Process Engineering, P.O. Box 6013, Toronto AMF, Ontario, Canada L5P 1B7
Darcel, F., Ministry of the Environment, Laboratory Services Branch, Rexdale, Ontario, Canada

Darmawan, A., A. Darmawan Company, Consultant for Water and Waste Treatment, Hegarmanah No. 65, Bandung, Indonesia
Douglas, P., Univ. of Waterloo, Dept. of Chem. Eng., Waterloo, Ontario, Canada, N2L 3G1

Ekama, G., Univ. of Cape Town, Dept. of Civil Eng., Rondebosch 7700, South Africa
Elliot, K., Univ. of Waterloo, Man Environment Dept., Waterloo, Ontario, Canada N2L 3G1
Ellis, H., Univ. of Waterloo, Dept. of Civil Eng., Waterloo, Ontario, Canada N2L 3G1
Esener, A.A., Delft Univ. of Technology, SBR, Jaffalaan 9, Delft, The Netherlands

Farquhar, G., Univ. of Waterloo, Dept. of Chem. Eng., Waterloo, Ontario, Canada N2L 3G1
Felker, B., Univ. of Waterloo, Environmental Studies, Waterloo, Ontario, Canada N2L 3G1
Flaherty, G., Univ. of Waterloo, Dept. of Chem. Eng., Waterloo, Ontario, Canada N2L 3G1
Forgie, D.J.L., Univ. of Saskatchewan, Dept. of Civil Eng., Saskatoon, Saskatchewan, Canada
Fortier, J., Univ. of Western Ontario, London, Ontario, Canada N6A 5B9
Fredette, M., ERCO Industries Ltd., 2 Gibbs Road, Islington, Ontario, Canada M9B 1R1

Ganczarczyk, J., Univ. of Toronto, Dept. of Civil Eng., Toronto, Ontario, Canada M5S 1A4
Ghoche, C., Atlas Chemical, Brantford, Ontario, Canada
Gokhale, S.B., Diamond Shamrock Corp., Divisional Technical Center, P.O. Box 191, Painesville, Ohio, USA 44077
Gonzalez, A., Univ. of Waterloo, Dept. of Chem. Eng., Waterloo, Ontario, Canada N2L 3G1
Greenfield, D.H., Philips Planning and Engineering Ltd., 4020 Derry Road W., R.R.#2, Milton, Ontario, Canada
Griffin, P.J., Univ. of Alberta, Dept. of Mineral Engineering, Edmonton, Alberta, Canada T6G 2G6

Hart, O.O., Water Research Commission, P.O. Box 824, Pretoria, Rep. of South Africa
Hermanowicz, S.W., Univ. of Toronto, Dept. of Civil Eng., Toronto, Ontario, Canada M5S 1A4
Hilton, J., James F. MacLaren Ltd., Consulting Engineers, 1220 Sheppard Ave. E., Toronto, Ontario, Canada M2K 2T8
Hoekstra, J.A., Univ. of Waterloo, Optometry, Waterloo, Ontario, Canada N2L 3G1
Huck, P.M., Univ. of Regina, Faculty of Eng., Regina, Saskatchewan, Canada S4S 0A2
Hudgins, R.R., Univ. of Waterloo, Dept. of Chem. Eng., Waterloo, Ontario, Canada N2L 3G1
Hung, Y.T., Univ. of North Dakota, Dept. of Civil Eng., Grand Forks, North Dakota, USA 58202

Ireland, D.R., Ministry of Environment, Environmental Engineering, 637 Niagara St. N., Welland, Ontario, Canada L3C 1L9

Janson, A., Univ. of Waterloo, Dept. of Chem. Eng., Waterloo, Ontario, Canada N2L 3G1
Jawa, S.C., Northland Engineering Limited, 101 Durham Street South, Sudbury, Ontario, Canada P3E 3M9

Kestle, G., J.M. Schneider Inc., Plant Engineering, 321 Courtland Ave., Kitchener, Ontario, Canada N2G 3X8
Khettry, R.K., Ministry of the Environment, 135 St. Clair Avenue W., Toronto, Ontario, Canada M4V 1P5
King, B., Drew University, Dept. of Chem., Madison, New Jersey, USA 07940

Kobayashi, T., Nagoya Univ., Dept. of Food Sci. and Technol., Chikusa-ku, Nagoya 464, Japan

Komow, J., L.S. Love & Associates Canada Limited, 158 Kennedy Road S., Brampton, Ontario, Canada L6W 3G7

Kumar, A., Univ. of Toledo, Dept. of Civil Eng., Toledo, Ohio, USA 43606

Lamptey, J., Univ. of Waterloo, Dept. of Chem. Eng., Waterloo, Ontario, Canada N2L 3G1

Landine, R.C., ADI Limited, 1115 Regent St., Fredericton, New Brunswick, Canada E3B 4Y2

Laughlin, R.G.W., Ontario Research Foundation, Canadian Waste Materials Exchange, Sheridan Park Research Community, Mississauga, Ontario, Canada L5K 1B3

Lebel, G., Tioxide Canada Inc., C.P. 580, Sorel, Quebec, Canada J3P 5P8

Lever, N.A., AECI Limited, Environmental Services, P.O. Box 1122, Johannesburg 2000, Republic of South Africa

Litchfield, R.J., Surrey University, Dept. of Chem. Eng., Guildford, Surrey, England

Lyness, R., Ministry of Environment, 133 Dalton Street, Kingston, Ontario, Canada K7L 4X6

Marais, G.v.K., University of Cape Town, Dept. of Civil Eng., Rondebosch 7700, Republic of South Africa

Marchandise, P., Labo Central Des Ponts Et Chausses, B.P. 19, 44340 Bougenais, France

Margaritis, A., Univ. of Western Ontario, Chem. and Biochem. Eng., London, Ontario Canada N6A 5B9

Martin, H., Environment Canada, Atmospheric Environment Service, 4905 Dufferin St., Downsview, Ontario, Canada M3H 5T4

Mayfield, C.I., Univ. of Waterloo, Dept. of Biology, Waterloo, Ontario, Canada N2L 3G1

McBean, E.L., Univ. of Waterloo, Dept. of Civil Eng., Waterloo, Ontario, Canada N2L 3G1

McLaughlin, W.A., Univ. of Waterloo, Dean of Engineering, Waterloo, Ontario, Canada N2L 3G1

Melnichuk, P., Univ. of Waterloo, Waterloo Centre for Process Development, Waterloo, Ontario, Canada N2L 3G1

Mellary, A., Ministry of the Environment, Groundwater Evaluation, 150 Ferrand Drive, Don Mills, Ontario, Canada M3C 3C3

Metcalfe, D., Univ. of Waterloo, Dept. of Civil Eng., Waterloo, Ontario, Canada N2L 3G1

Mitwally, H., High Institute of Public Health, Environmental Health Dept., 165 Horria Ave., Alexandria, Egypt

Moo-Young, M., Univ. of Waterloo, Dept. of Chem. Eng., Waterloo, Ontario, Canada N2L 3G1

Moser, A., Inst. für Biotechnology, Technical Univ., Schlögelgasse 9, Graz, Austria

Mosoiu, M.A., Univ. of Waterloo, Dept. of Civil Eng., Waterloo, Ontario, Canada N2L 3G1

Mueller, J.A., Manhattan College, Bronx, New York, USA 10471

Myhre, M., Dunlop Research Centre, Sheridan Park, Mississauga, Ontario, Canada N2L 3G1

Nagendran, J., Water Quality Engineering, 5th Floor, 9820-106 Street, Edmonton, Alberta, Canada T5K 2J6

Nanda, M., Conestoga College, Kitchener, Ontario, Canada

Nazar, M.A., Univ. of Toronto, Dept. of Chem. Eng., Toronto, Ontario, Canada M5S 1A4

Neale, R.J., Environment Canada, Canadian Forestry Service, Ottawa, Ontario, Canada K1A 0E7

Ng, M., Walter, Fedy, McCargar, Hachborn Consulting Engineers, 391 Hazel Street, Waterloo, Ontario, Canada N2L 3P7

Nirdosh, I., McMaster Univ., Dept. of Chem. Eng., Hamilton, Ontario, Canada L8S 4L7

O'Neil, D., Metropolitan Service District, Solid Waste Coordinator, 527 S.W. Hall St., Portland, Oregon, USA 97201

Orchard, D., Eco/Log Week, 1450 Don Mills Rd., Don Mills, Ontario M3B 2X7 Canada

Paar, H., Ministry of the Environment, 150 Ferrand Drive, Don Mills, Ontario, Canada
Pearson, F., Science and Technology Research Office, 59 State Office Building, St. Paul, Minnesota, USA 55155
Pinder, K.L., Univ. of British Columbia, Dept. of Chem. Eng., Vancouver, B.C., Canad V6J 1W5
Pos, J., Univ. of Guelph, School of Eng., Guelph, Ontario

Rhodes, E., Univ. of Waterloo, Dept. of Chem. Eng., Waterloo, Ontario, Canada N2L 3G
Ribeiro, C.C., Centro de Tecnologia Promon, Praia do Flamengo 154 12, 20000 Rio de Janeiro RJ, Caixa Postal 1798, Brasil
Robinson, C.W., Univ. of Waterloo, Dept. of Chem. Eng., Waterloo, Ontario, Canada N2L 3G1
Roffel, B., Twente Univ. of Technology, P.O. Box 217, 7500 AE Enschede, The Netherlands
Rousseau, M., CDN Titanium Pigments Ltd., P.O. Box 5800, Varennes, Quebec, Canada JOL 2P0

Sadiq, M., Research Institute, Dhahran International Airport, UMP Box 417, Dhahran, Saudi Arabia
Sakagami, Y., Tokyo Inst. of Technology, 3-16-19-603, Jiugaoka, Meguroku, Tokyo, Japan
Scharer, J.M., Univ. of Waterloo, Dept. of Chem. Eng., Waterloo, Ontario, Canada, N2L 3G1
Schmidtke, K., Univ. of Waterloo, Dept. of Civil Eng., Waterloo, Ontario, Canada, N2L 3G1
Scott, D.S., Univ. of Waterloo, Dept. of Chem. Eng., Waterloo, Ontario, Canada N2L 3G1
Silver, M., CANMET - Research, 555 Booth Street, Ottawa, Ontario, Canada K1A 0G1
Singh, B.A., Ministry of the Environment, Technical Support Section, 150 Ferrand Dr. Don Mills, Ontario, Canada M3C 3C3
Sloan, D., Ministry of the Environment, 469 Bouchard Street, Sudbury, Ontario, Canada P3E 2K8
Smith, E.D., U.S. Army Construction Engineering, Sanitary/Environmental Research Laboratory, P.O. Box 4005, Champaign, Illinois, USA 61820
Spink, D.R., Univ. of Waterloo, Dept. of Chem. Eng., Waterloo, Ontario, Canada N2L G
Stanley, C., Pratt & Whitney Aircraft of Canada Ltd., Box 10, Longueuil, Quebec, Canada J4K 4X9
Stefan, S., Pratt & Whitney Aircraft of Canada Ltd., Box 10, Longueuil, Quebec, Canada J4K 4X9
Stephenson, S.R., Petrosar Limited, Sarnia, Ontario, Canada NON 1G0
Subba Narasiah, K., Univ. of Sherbrooke, Dept. of Civil Eng., Sherbrooke, Quebec, Canada
Surdia, N.M., Inst. of Technology Bandung, Dept. of Chem., Dago Timur 2, Bandung, Indonesia

Terlecky, P.M., Frontier Technology Association, 8675 Sheridan Drive, Williamsville, New York, USA 14221

Van Norman, A., Conestoga-Rovers & Associates, 651 Colby Drive, Waterloo, Ontario, Canada N2V 1B4
Varjavandi, J., Crooks Michell Peacock Stewart Pty. Limited, 2 Thomas Street, Chatswood, N.S.W. 2067, Australia

Walker, W., Catalytic Enterprises Limited, P.O. Box 3008, Sarnia, Ontario, Canada N7T 7M7
Watkiss, M., Ministry of the Environment, 445 Albert Street E., Sault Ste. Marie, Ontario, Canada P6A 2J9

Wilcox, L., Ministry of the Environment, 135 St. Clair Ave. West, 9th Floor, Toronto, Ontario, Canada M4V 1P5
Wilson, M., Alberta Environment Centre, Box 4000, Vegreville, Alberta, Canada TOB 4L0
Woods, G.S., Dow Chemical of Canada, Limited, Ontario District Sales Office, 170 Attwell Drive, Rexdale, Ontario, Canada M9W 5Z5

Zelver, N., Montana State Univ., Dept. of Civil Eng., Bozeman, Montana, USA 59717
Zoller, U., Haifa Univ., Division of Chemical Studies, Oranium, P.O. Kiryat Tivon, Israel

REVIEWERS

The editors are grateful for the assistance of the following who reviewed one or more of the technical papers.

Abou-El-Hassan, M.E., Cairo University, Egypt
André, G., University of Waterloo, Canada

Bewtra, J.K., University of Windsor, Canada
Bhattacharyya, A.K., Jawaharlal Nehru University, India
Brown, D.E., UMIST, UK

Chandra, A.L., Bose Institute, India
Characklis, W.F., Montana State University, USA
Constable, T.W., Canada Centre for Inland Waters, Burlington, Canada
Cornwall, G., Environmental Protection Service, Canada

Douglas, P.L., Queen's University, Canada

Esener, A.A., Technische Hogeschool Delft, The Netherlands

Farquhar, G., University of Waterloo, Canada
Fohr, P.G., Wageningen Agricultural University, The Netherlands

Ganczarczyk, J., University of Toronto, Canada
Gillham, R., University of Waterloo, Canada
Griffin, P.J., University of Alberta, Canada
Griffin, R.A., Illinois Institute of Natural Resources, USA

Henry, J.G., University of Toronto, Canada
Hung, Y.-T., University of North Dakota, USA
Huck, P.M., University of Regina, Canada

Jank, B.E., Wastewater Technology Centre, Canada
Jervis, R.E., University of Toronto, Canada

Kumar, A., Syncrude Canada Ltd., Canada

Laughlin, R.G.W., Ontario Research Foundation, Canada

Marais, G.v.R., University of Cape Town, Republic of South Africa
Martin, H.C., Environment Canada, Canada
Mayfield, C.I., University of Waterloo, Canada
McAdie, H.G., Ontario Research Foundation, Canada
Meyers, S.P., Louisiana State University, USA
Moo-Young, M., University of Waterloo, Canada
Moser, A., Technical University Graz, Austria
Mueller, J.A., Manhattan College, USA

Nestmann, E.R., Health and Welfare, Canada

Oldham, W., University of British Columbia, Canada

Pinder, K.L., University of British Columbia, Canada
Pos, J., University of Guelph, Canada

Rempel, G.L., University of Waterloo, Canada
Robinson, C.W., University of Waterloo, Canada
Roels, J.A., Technische Hogeschool Delft, The Netherlands

Scharer, J.M., University of Waterloo, Canada
Schmidt, J.W., Environment Canada, Canada
Scott, D.S., University of Waterloo, Canada
Schmidtke, N., Canada Center for Inland Waters, Canada
Spink, D.R., University of Waterloo, Canada
Stephenson, S.R., Petrosar Ltd., Canada

Tarlton, E.J., Domtar Inc., Canada
Telliard, W.A., U.S. Environmental Protection Agency, USA

Vermeulen, T., University of California, Berkeley, USA
Vincent, C.L., Domtar Inc., Canada

Webber, M.D., Wastewater Technology Centre, Burlington, Canada
Wynnyckyj, J.R., University of Waterloo, Canada

Young, J.C., University of Waterloo, Canada

AUTHOR INDEX

Numbers refer to Paper Numbers identified in the Table of Contents

SUBJECT INDEX

Numbers refer to Paper Numbers identified in the Table of Contents